Sustainable Water Treatment

Scrivener Publishing
100 Cummings Center, Suite 541J
Beverly, MA 01915-6106

Publishers at Scrivener
Martin Scrivener (martin@scrivenerpublishing.com)
Phillip Carmical (pcarmical@scrivenerpublishing.com)

Sustainable Water Treatment

Advances and Technological Interventions

Edited by
Siddhartha Moulik
Aditi Mullick
and
Anirban Roy

Scrivener Publishing

WILEY

This edition first published 2022 by John Wiley & Sons, Inc., 111 River Street, Hoboken, NJ 07030, USA and Scrivener Publishing LLC, 100 Cummings Center, Suite 541J, Beverly, MA 01915, USA

For more information about Scrivener publications please visit www.scrivenerpublishing.com.

Wiley Global Headquarters
111 River Street, Hoboken, NJ 07030, USA

For details of our global editorial offices, customer services, and more information about Wiley products visit us at www.wiley.com.

Library of Congress Cataloging-in-Publication Data

ISBN 9781119479987

Cover image: Pixabay.Com
Cover design by Kris Hackerott

Set in size of 11pt and Minion Pro by Manila Typesetting Company, Makati, Philippines

Printed in the USA

10 9 8 7 6 5 4 3 2 1

Contents

Introduction to Water Treatment: A Historical Perspective on Technological Development and Future Landscape

Anirban Roy[1*], Anupam Mukherjee[1], Aditi Mullick[2] and Siddhartha Moulik[2†]

[1] *Water-Energy Nexus Lab, Department of Chemical Engineering Goa, BITS Pilani, Goa, India*
[2] *Cavitation and Dynamics Lab, Process Engineering and Technology Transfer Division, Indian Institute of Chemical Technology, Hyderabad, India*

Abstract

Efficient water production, recovery, recycling and management underlines the degree of sustainability of the resource. In the last decade, water has been considered as the "new oil" and as the popularity of non-renewable energy sources are slowly starting to fade, there is no alternative to water. In this regard, this book focuses on technological advances for water treatment in four categories: (i) advanced oxidation technologies, (ii) nanoparticles for water treatment, (iii) membrane separations, and (iv) other emerging technologies or processes. A wide variety of pollutants have also been discussed like acidic wastewater treatment, metallurgical wastewater, and textile wastewater as well as groundwater. More often than not, a stand-alone technology is not the solution for sustainability and one requires a process to be developed and technologies to be integrated. In this regard, the book will give valuable insights into both understanding of technologies as well as possible integration of technologies for a given waste stream to recycle and purify water.

***Keywords*:** Advanced oxidation, membrane, nanoparticle, metallurgical wastewater, acidic wastewater, groundwater

**Corresponding author*: anirbanr@goa.bits-pilani.ac.in
†*Corresponding author*: smoulik@iict.res.in

Anirban Roy, Aditi Mullick and Siddartha Moulik (eds.) Sustainable Water Treatment: Advances and Technological Interventions, (xix–xxiv) © 2022 Scrivener Publishing LLC

Introduction

The current water crisis looming over a hapless human civilization can be attributed to (i) abuse of available sources of fresh water, (ii) wastage of water due to rapid urbanization and an improved quality of life, (iii) increased agricultural activities to feed an ever-growing population, depleting groundwater reserves, (iv) rapid industrialization, and (v) lack of coordinated policy implementations [1–3]. Adding to the woes, the current pandemic has challenged humanity like never before [4, 5]. This has posed a challenge of an entirely different nature, where water requirement for sanitization has increased manifold. The irreplaceable requirement of water in every sphere of life has made it imperative that to support an ever-increasing population, it is important to preserve, conserve, reuse and recycle the resource [6–9].

It is important to understand the technologies being employed to treat water in the current scenario. Water quality depends on the process it undergoes and therefore, depending on the constitutes of effluent stream, proper strategy can be framed and suitable process with minimum energy penalty can be designed [10]. The maturity of individual technology with its application potential and energetics involved must be considered before making a choice for a desired purpose [7, 11].

In this regard, the current volume is a collection of chapters belonging to mainly four categories of technologies: (i) Advanced Oxidation Processes (AOP), (ii) Nanoparticle-Based Treatment (NPT), (iii) Membrane-Based Treatment (MBT), and (iv) Emerging Technologies and Processes (EMT).

In Section I (comprising AOP technologies), a general overview is presented in Chapter 1, where fundamentals related to AOP are discussed along with global trends. Popular AOP technologies like Fenton process, photocatalysis, ozone, ultraviolet (UV) and H_2O_2 are discussed in detail. Chapter 1 ends with challenges posed in AOP technologies. Chapter 2 delves into a different genre of AOP, Hydrodynamic Cavitation (HC). This chapter discusses the mechanisms of ultrasound and HC processes and delves deep into integration of cavitation technologies with existing wastewater treatment technologies. The chapter also deals with integration of HC with adsorption and chemical oxidation processes and ends with future perspective of HC. Chapter 3 deals with a different aspect of HC, where, bubble dynamics, bubble collapse and thermodynamic effects of HC have been discussed. This gives an in-depth understanding of cavitation phenomena. The chapter also discusses various cavitating reactors and the toxicity effects of treatment process; it ends with future avenues. Chapter 4 (the last chapter in AOP) discusses fundamentals of ozonation technology. This chapter deals with various mass transfer models and mass transfer mechanism of ozone in water. The chapter handles the topic of ozone as a gas liquid contacting system and discusses the related mechanism as well as energetics associated. In the last part, the chapter deals with various sectors of application of ozone.

Section II of the book deals with NPT and in this regard, chapter 1 discusses application of Nanoparticles (NP) in fixed bed column filter system. This chapter discusses various target contaminants and the type of NP used to remove those contaminants. Then the chapter deals with design of fixed bed columns using the NP for removal of pollutants. Chapter 2 discusses the potential of NP-based water treatment processes. This chapter deals with various types of NPs employed to remove specific pollutant streams and the barriers posed in large-scale implementations.

Section III deals with membrane-based technology. In this regard, chapter 1 discusses microbial fuel cell (MFC) technology for wastewater treatment. This chapter discusses fundamentals related to working of an MFC and recent developments in anode and cathode material development. It also discusses development of membranes for MFC and challenges in MFC, and ends with case studies. Chapter 2 discusses the potential applications and challenges of ceramic membrane filtration. This chapter discusses the application of various ceramic membranes in industrial wastewater treatment as well as emerging contaminants. It also discusses the various configuration variants of ceramic membranes ultimately ending with pilot scale studies and challenges of ceramic membrane filtration units. Chapter 3 discusses an emerging area in the field of membrane separation—membrane distillation (MD). This chapter discusses the possibility of acidic wastewater treatment using MD. This chapter deals with the basics of the origin of acidic wastewater as well as various MD technologies employed in practice. This chapter also discusses various hybrid MD processes ultimately concluding with implications of MD in acidic wastewater treatment. The last chapter in this section (Chapter 4) discusses long-term assessment of vacuum membrane distillation (VMD) for textile wastewater treatment. This chapter covers the basics of transport mechanism and mass transfer involved in VMD and then discusses the influence of various operational parameters in textile wastewater treatment process in VMD.

The last section of the book is dedicated to few emerging processes in water treatment and recovery systems. Chapter 1 in this regard deals with application of zero valent iron to remove chromium and other heavy metals in metallurgical wastewater. This chapter deals with characterization of metallurgical wastewater in detail as well as preparation of zero valent iron. It then discusses in detail the batch experiments for treatment as well as detailed parametric investigations involved in treating the streams. The second chapter discusses removal of arsenic and fluoride using coagulation precipitation, ion exchange, and adsorption technologies. The chapter discusses detailed analysis of each genre of technology as well as its application and removal efficacies for arsenic and fluoride. The third chapter focuses on developing theoretical understanding of Zero Liquid Discharge (ZLD) process and its energy penalties. This chapter discusses in detail development of process flow sheet in Aspen and recovering maximum water from seawater feed using multistage flash coupled with evaporative crystallizer. This chapter discusses in detail development of the flow sheet and modeling of components along with energy analysis associated with a ZLD process for a feed whose salinity is equal to that of seawater. The fourth chapter focuses on a self-consistent approach in developing a water - energy - food nexus with an objective of highlighting the prospects and challenges in "conceptualization to technology transfer" for nutrient recovery from municipal wastewater for sustainable resource recovery and waste water management. Wastewater today is considered as a resource because of abundance of nutrients present in it. These nutrients like phosphorous and nitrogen if could be extracted economically, have the potential to solve multiple problems such as reutilization of natural resources and reduction in eutrophication in water bodies. In this regard the methods of extraction, based on the laws of thermodynamics shall be utilized for the same in order to maintain the crafted balance of water-energy-food nexus (WEFN). Throughout the chapter, nutrients, biomass, energy, and water are collectively termed as resources. The fifth and sixth chapter in this section highlights towards sustainable desalination. Desalination has largely been limited to affluent countries in the Middle East and has recently started making inroads in parts of the United States and Australia. In this regard, the initial efforts was to highlight the growing opportunities and scopes in the Indian subcontinent, while in the last chapter authors discussed on the least energy (work, heat, and fuel) required for desalination establishing a benchmark using the first and second laws of thermodynamics.

This book has been edited to bring forth a broad perspective in understanding the various state-of-the-art technologies being employed or developed in laboratory scale in treating a wide range of water stream and hopefully will benefit a broad section of readers.

Acknowledgement

Dr. Siddhartha Moulik would like to thank the Director, CSIR-IICT for the support [Ref. IICT/Pubs./2022/110].

References

1. X. Liu, H. Du, Z. Zhang, J.C. Crittenden, M.L. Lahr, J. Moreno-Cruz, D. Guan, Z. Mi, J. Zuo, Can virtual water trade save water resources?, *Water Res.* 163 (2019) 114848. https://doi.org/10.1016/j.watres.2019.07.015.
2. O.A. Aluko, E.E. Osei Opoku, M. Ibrahim, Investigating the environmental effect of globalization: Insights from selected industrialized countries, *J. Environ. Manage.* 281 (2021) 111892. https://doi.org/10.1016/j.jenvman.2020.111892.
3. S. Mondal, A. Roy, R. Mukherjee, M. Mondal, S. Karmakar, S. Chatterjee, M. Mukherjee, S. Bhattacharjee, S. De, A socio-economic study along with impact assessment for laterite based technology demonstration for arsenic mitigation, *Sci. Total Environ.* 583 (2017) 142–152. https://doi.org/10.1016/j.scitotenv.2017.01.042.
4. S. Lahrich, F. Laghrib, A. Farahi, M. Bakasse, S. Saqrane, M.A. El Mhammedi, Review on the contamination of wastewater by COVID-19 virus: Impact and treatment, *Sci. Total Environ.* 751 (2021) 142325. https://doi.org/10.1016/j.scitotenv.2020.142325.
5. C. Huo, A. Ahmed Dar, A. Nawaz, J. Hameed, G. albashar, B. Pan, C. Wang, Groundwater contamination with the threat of COVID-19: Insights into CSR theory of Carroll's pyramid, J. King Saud Univ. - *Sci.* 33 (2021) 101295. https://doi.org/10.1016/j.jksus.2020.101295.
6. S. Lanjewar, A. Mukherjee, P. Khandewal, A.K. Ghosh, A. Mullick, S. Moulik, A. Roy, Thermodynamics of synthesis and separation performance of Interfacially polymerized "loose" reverse osmosis membrane: Benchmarking for greywater treatment, *Chem. Eng. J.* (2020). https://doi.org/10.1016/j.cej.2020.127929.
7. A. Mukherjee, A. Mullick, R. Teja, P. Vadthya, A. Roy, S. Moulik, Performance and energetic analysis of hydrodynamic cavitation and potential integration with existing advanced oxidation processes: A case study for real life greywater treatment, *Ultrason. Sonochem.* 66 (2020). https://doi.org/10.1016/j.ultsonch.2020.105116.
8. Debarghya Halder, Subhajit Mondal, Anupam Mukherjee, Avijit Ghosh, An Overview on the Treatment of Waste Water Using Renewable Energy, *Int. J. Res. Eng. Technol.* 05 (2016) 78–83. https://doi.org/10.15623/ijret.2016.0513014.
9. A. Mukherjee, A. Mullick, P. Vadthya, S. Moulik, A. Roy, Surfactant degradation using hydrodynamic cavitation based hybrid advanced oxidation technology: A techno economic feasibility study, *Chem. Eng. J.* 398 (2020). https://doi.org/10.1016/j.cej.2020.125599.
10. L.M. Rehman, R. Dey, Z. Lai, A.K. Ghosh, A. Roy, Reliable and Novel Approach Based on Thermodynamic Property Estimation of Low to High Salinity Aqueous Sodium Chloride Solutions for Water-Energy Nexus

Applications, *Ind. Eng. Chem. Res.* (2020). https://doi.org/10.1021/acs.iecr.0c02575.

11. A. Mukherjee, S. Lanjewar, R. Kumar, A. Chakraborty, A. Abdelrasoul, A. Roy, Role of Thermodynamics and Membrane Separations in Water-Energy Nexus, in: *Model. Membr. Membr. Process.*, 2020: pp. 145–199. https://doi.org/10.1002/9781119536260.ch4.

Section I

ADVANCED OXIDATION PROCESSES

1

Advanced Oxidation Processes: Fundamental, Technologies, Applications and Recent Advances

Akshat Khandelwal and Saroj Sundar Baral*

Department of Chemical Engineering, BITS Pilani K K Birla Goa Campus, Goa, India

Abstract

Water waste globally triggers warnings, since a majority of the population has no access to clean potable water. Vast amounts of synthetic organic pollutants such as toxic chemicals, pesticides, pharmaceutical products and personal care products have been reported to be released into water supplies daily. Advance oxidation process (AOP) involves formation of reactive oxygen species which are very reactive and non-selective, and can degrade these hazardous compounds to CO_2 and inorganic ions. It can potentially reduce pollutant concentrations from hundreds of parts per million (ppm) to only a few parts per billion (ppb). As new developments in wastewater treatment, TiO_2, photocatalysis, Fenton and photo-Fenton approaches have appeared. This chapter discusses the fundamentals and role of the AOP in wastewater remediation. Further, background and global research trends of wastewater treatment employing AOP are discussed. Moreover, different AOP technologies along with challenges restricting their full-scale operation are addressed in detail.

***Keywords*:** Advanced oxidation processes (AOC), photocatalysis, fenton, photo-fenton, effluent treatment, reactive oxygen species

**Corresponding author*: ssbaral75@gmail.com

Siddhartha Moulik, Aditi Mullick and Anirban Roy (eds.) Sustainable Water Treatment: Advances and Technological Interventions, (3–26) © 2022 Scrivener Publishing LLC

1.1 Introduction

Water is a finite resource that is important for sustainable growth. While abounding on Earth as saltwater from the oceans and seas, only 1% is readily accessible for human consumption. Our society needs convenient access to more and more safe drinking water. Worldwide water pollution is raising alarms as 780 million people do not have access to safe drinking water [1]. Water contamination continues to be an omnipresent danger as vast amounts of synthetic organic contaminants, like industrial, agricultural, pharmaceutical chemicals and personal care products, are reported to be released on a daily basis into water streams [2, 3]. For this wastewater, treatments are necessary. After the treatment still, some pollutants are highly stable and are resistant to radiation, temperature and microbial decomposition [4–6]. These contaminants can be termed as pertinent organic pollutants that cannot be easily treated in a conventional wastewater treatment plant. Moreover, pertinent compounds possess unknown characteristics that raise a great threat to environmental and human health.

Wastewater management technology has seen a recent and conclusive awakening, and today's attention centers on designing quick, safe, reliable and cost-effective solutions to decompose POPs before disposal. A thorough overview of available eco-friendly approaches widely used for wastewater treatment techniques is provided in detail in many reviews [4–8]. In addition, in some literature these methods can be split into two groups: (i) physicochemical separation system, for example, membrane separation, and (ii) chemical and microbiological driven transformation systems. Transformation technologies are quite effective as they cleave the dangerous pollutants into less harmful compounds. In this chapter, we are going to deal with the advance oxidation process and its role in wastewater treatment.

Advanced oxidation processes (AOPs) can effectively degrade pollutants into least dangerous substances by oxidising toxic compounds. AOP are the oxidation processes that are connected to reactive oxygen species formation, such as hydroxyl radicals. Hydroxyl radical is widely explored in AOP as it is the second-strongest oxidising agent behind fluorine, with a standard reduction potential of 2.8 V/SHE. Moreover, hydroxyl radical is highly reactive and non-selective and can cause harmful materials to degrade to CO_2 and inorganic ions. Therefore, these radical hydroxyls can be used to degrade various wastewater effluents from industry and agriculture.

There are different approaches to bring in different effects when using multiple technologies. Specifically, ozone, hydrogen peroxide, and ultraviolet light are commonly used to create adequate levels of hydroxyl radical that can be used to degrade some organic and inorganic contaminants. Pollutant concentrations can be lowered by AOP, theoretically from hundreds of parts per million (ppm) to just a few parts per billion (ppb). As new developments in wastewater treatment, TiO2, photocatalysis, Fenton and photo-Fenton approaches have appeared. These AOPs are added to the purification and detoxification of many toxic waste materials containing pharmaceutical products. Photocatalysis and photo-Fenton are also suggested in urban effluent as tertiary remediation. Removing targeted micro-pollutants, such as naproxen, carbamazepine, diclofenac, gemfibrozil, ibuprofen, caffeine and mecoprop, was recorded with photolysis and ultraviolet/hydrogen gas processes. The average levels of oxidation for organic compounds are improved by the inclusion of hydrogen peroxide. They may be employed to eradicate the non-biodegradable contaminants after the biological procedure. Electrochemical AOPs, plasma, electron beam, ultrasound, and even micro-wave dependent AOPs have also been suggested and are commonly being recorded by researchers. However, a wide range of research, as well as a growing array of proposed innovations and process combinations, make it difficult to determine current AOPs cost, chemical inputs and sustainability. Also, the water matrix might impact radical generation processes, such as reducing UV defence or reactions with ozone.

Therefore, this chapter provides a critical summary of the numerous current and emerging AOPs based on data gathered from comprehensive literature survey. The purpose of this chapter is to provide readers with general understanding of the fundamentals of AOP. We explored the background and global trends of AOPs. Various AOP innovations are addressed in the next section, followed by a thorough comparison between them. Moreover, challenges to these AOP technologies are discussed side by side.

1.2 Background and Global Trend of Advanced Oxidation Process

The advanced oxidation process (AOP) employs the formation of highly reactive oxidizing radicals for example hydroxyl radical ($OH\cdot$), sulphate radical (SO_4^{2-}), hydroperoxyl radical ($HO_2\cdot$), etc., at ambient temperature and pressure in the solution for the degradation of pertinent organic pollutants (POPs) in the waterbodies. AOP set its footmark in the wastewater treatment process in 1987 when Glaze *et al.* defined AOP as "the process of forming suitable amount of OH radical for effective treatment of wastewater" [9, 10]. Initially, AOP studies were limited to the generation of •OH radical but later on degradation employing sulphate radicals was also explored [11, 12].

Various techniques have been studied to destroy compound aromatics such as phenols, chlorophenols, and nitrophenols by AOPs. The findings reveal that under optimal conditions, high removal efficiency can be achieved. Studies have been carried out evaluating the impact of certain operating parameters on AOP. The degradation mechanisms, along with the influence of the variation in the molecular structure of aromatics on degradation performance, have been deduced and demonstrated in experiments. Moreover, textile colouring may have adverse effects on microorganisms, marine life and human beings. These colouring dyes limit the penetration of light into polluted water. The decolorization and oxidation of dyes have gained rising research interest. The research was conducted for analyzing the effect of operational characteristics, decolourisation, and degradation processes on the overall efficiency of AOP. However, the mechanism by which the AOP dyes are degraded is unclear, but some stable intermediates do form in the solution.

Over the years, the concentration of antibiotics has risen in wastewater. A variety of medical prescriptions drugs are consumed worldwide each year. These drugs may have harmful effects on the aquatic environment and even on terrestrial beings. Various forms of concerns have arisen following the elimination of these compounds and AOP has arisen as a promising technology for treatment of these pharmaceutical wastewaters. Moreover, research has been focussed on the effects of operational characteristics, the function of radical promoters and scavengers, as well as the effect of these parameters on the chemical composition of the waters.

In addition, AOP has shown potential for the treatment of wastewater containing pharmaceutical effluents. The primary use of pesticides in agriculture is to increase crop production. Surface water and groundwater may contain residues from the runoff of agricultural sites. A number of experiments have evaluated the effects of operational requirements in order to obtain optimum working conditions to remove pesticides from the wastewater using the AOPs. In these studies degradation mechanisms are postulated by determining the role of scavengers and identifying intermediate reactions. Moreover, AOP may also be used to detoxify other hazardous wastes, such as carboxylic acid, heavy metals, bacteria, and actual industrial wastewater. The operational conditions have been optimized using different interventions, and important effects on the reaction kinetics and degradation processes have been investigated.

Recently research was published highlighting the worldwide research trend of AOP in wastewater treatment [13]. The worldwide trend based on the number of research studies published in a year in the period 1990-2018 is shown in Figure 1.1a. Findings indicate that there has been a slow increase in annual publications for the period 1990-2002 while the growth for the period 2002-2018 has been significantly higher with a publication rate exceeding 25 articles per year, reaching a total of 450 articles published in 2018. Further country-wise comparison of several publications was done shown in Figure 1.1b. Eighty-six countries have published scientific papers in this area of research. However, the 13 most relevant countries correspond to about 85% of the publications in the region. Moreover, it was also seen in Figure 1.1c how China and Spain publish the largest amount of scientific papers on AOP wastewater treatment, and how five Chinese and

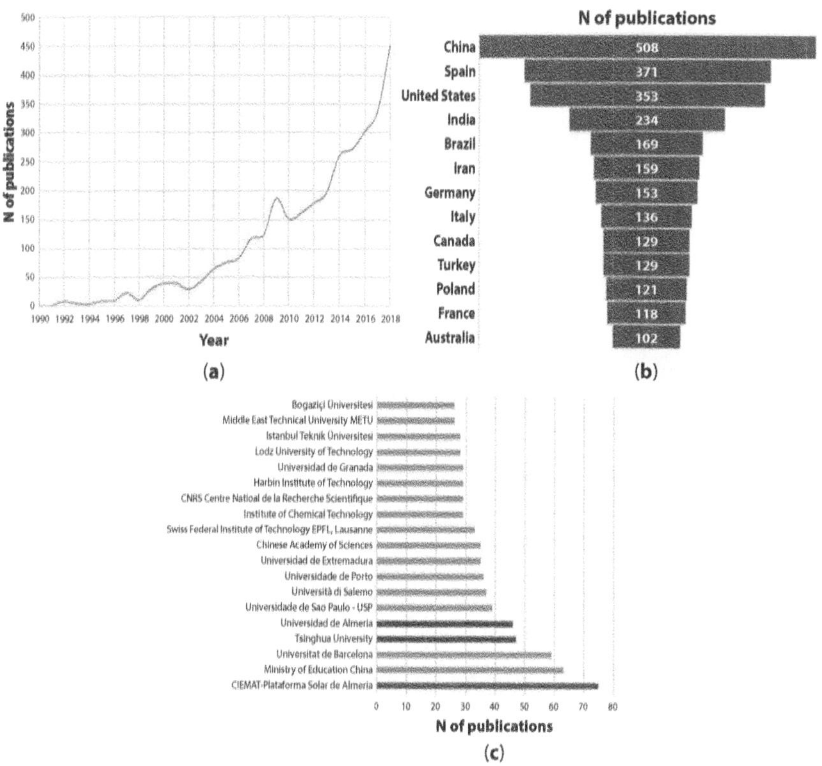

Figure 1.1 (a) The annual trend in the number of advanced oxidation process publications from 1990-2018. (b) Representation of the countries with the largest number of advanced oxidation process publications. (c) Major organizations relevant to scientific development in the advanced oxidation process of wastewater. Reproduced from ref [13].

three Spanish institutions lead the scientific output in this field, with the CIEMAT-Solar Platform of Almeria leading the ranking.

1.3 Advanced Oxidation Systems

Heterogeneous TiO_2-based photocatalyst and homogeneous photo-Fenton-based catalyst, under on UV or solar, visible irradiation, electrochemical

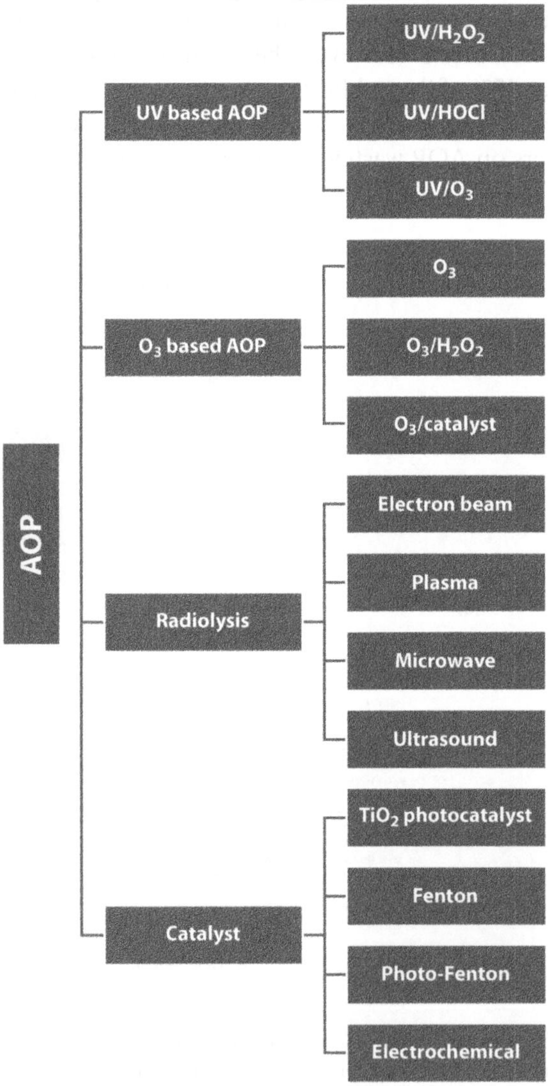

Figure 1.2 Classification of AOP systems.

AOPs, ozonation, Fenton, sonolysis, ionizing beam and μ-radiolysis radiation, microwaves, pulsed plasma is widely used for wastewater treatment. However, it is worth noting that the system of grouping shown in Figure 1.2 cannot be regarded as rigid as some systems require multiple methods and may thus be delegated to separate categories.

1.3.1 Ozone-Based AOP

There are ozone-based AOPs that are activated in combination with ultraviolet light, hydrogen peroxide, catalysts and ultrasound [14]. These types of systems are excellent for oxidising hazardous organic water pollutants that are not targeted by chlorine and ozone. O_3/UV is an effective technology for the removal of refractory contaminants. The procedure is initiated by ozone photolysis and accompanied by the creation of hydroxyl radical. The effects of pH on the decolorization of reactive red 2 by O_3, O_3/Fe(II), UV/O_3 and UV/O_3/Fe(II) systems were examined [15]. The result indicates that at all pH, O_3/UV system promotes reactive red 2 degradations. Besides, only at pH=4 can O_3/Fe(II) and UV/O_3/Fe(II) structures exert synergistic effects. An experiment has been carried out to study the ozonation of urban wastewaters [16]. Results show that the addition of H_2O_2 promotes the degradation of Para-chloro-benzoic acid. Moreover, a threshold dosage value of H_2O_2 is specified below which there is no increase in the rate of degradation. Also, catalytic ozonation is attracting considerable interest as an important method for extracting organics from water [17]. Much is still unclear about the dynamics of catalytic processes. In order to have this methodology for industrial-scale consumption, one must provide an understanding of the mechanisms. As an example in the ozonation of oxalic acid in the water, two iron catalysts were used at pH 2.5 [18]. The process will probably evolve by the creation of iron-oxalate complexes that react further with ozone without hydroxyl radicals.

In the heterogeneous process, some liquidation of metal was detected and quantified due to the stringent acidic conditions. The authors assume that the method leads to full mineralization in any event. In addition heterogeneous catalytic ozonation also showed potential for the degradation of refractory contaminants. For example, magnesium Iron Tetroxide (MgFe2O4) was successfully synthesized and used in ozonation for degradation of Acid Orange II (AOII) [19]. Over a broad pH spectrum of 4.6-9.6, the degradation performance of AOII approaches 90%, marginally influenced by the solution pH. Mg role in MgF2O3 was to maintain the high electron density on lattice oxygen. Carbon-based materials were also utilized for catalytic ozonation for the degradation of organic pollutants. The research assessed the impact of nitrogen doping on carbon nanofibers' catalytic activity [20]. Findings indicate that the inclusion of N-containing functionalities on the surface of the carbon nanofibers improves their catalytic properties.

1.3.2 UV/H_2O_2

The implementation of UV-based processes is likely to address the concern about the effect of micro-pollutants on the quality of drinking water [21]. The efficacy of UV processing approaches for handling pharmaceuticals is heavily dependent on their UV absorbance [22–24]. Also important parameters in the degradation of organic contaminants by UV light are the rate constants, quantum yield, and molar extinction coefficient [25]. UV along with H_2O_2 increases the effectiveness of separation from contaminants with low UV absorption. UV/H_2O_2 processes are regulated by the H_2O_2 concentration, the rate of HO radical formation, the UV light strength, the water constituents and the chemical composition of the pharmaceutical [25]. UV/H_2O_2 process was reported to moderately yield reduction in the concentration of volatile organic compounds (VOC) with degradation efficacy higher than 80% [26]. The superior efficiency of UV/H_2O_2 is attributed to the formation of hydroxyl radicals. UV/H_2O_2 can be used for the treatment of VOCs with effective and environmentally safe technologies. UV photolysis of hydrogen peroxide is a traditional advanced oxidation technique. Substitution of peroxide by HOCl promises cost-effectiveness [27]. In an experiment, the effects of pH and the molar ratio of a pollutant to an active chlorine ingredient on the UV dependent method were tested using 1,4-dioxane as a model pollutant [28]. According to the UV/HOCl/photolytic optimum pH in range of 3–6 was found. The reaction efficiency at a high molar ratio of initial chlorine to initial HOCl exceeded 100% at a pH of 3 to 6, implying feasibility of photolysis-UV/HOCl AOPs. UV along with photocatalytic method was employed to treat wastewater taken from anaerobic sedimentation tank [29]. The results indicate that the optimal UV power of 20 W could be used for the process's time needed for the optimal TiO_2dose with a contact time of 110 min.

1.3.3 Radiation

The use of electron beam irradiation has been tested since the 1980s for water treatment. Accelerated electrons enter the water surface producing electro-excited species, including ions and free radicals [25]. Gamma irradiation, unlike UV processes, is more penetrating and promotes the increased development of hydroxy radicals and hydrated electrons to support successful contaminant breakdown. In addition, the energy of the electrons concerned is directly proportional to the degree of penetration. Therefore, water is either radiated as a thin film or as an aerosol. This method due to high oxidising power and low interaction with the water matrix was found to be very successful in degrading many contaminants [25].

The effect of single gamma radiation in peroxy-monosulfate activation for carbamazepine degradation was explored. The results showed that the addition of peroxy-monosulfate caused carbamazepine oxidation, which was due to the formation of hydroxyl, per hydroxyl radicals and superoxide-radical anions. Further research has been conducted on the feasibility of the integration of electron beam radiation and hydrogen peroxide. The result indicates a 90% deterioration in the absence of H_2O_2 at a dosage of 1 kGy. Oxidation has increased in this situation by adding H_2O_2 (10 mM concentration). Moreover, a rise in H_2O_2 concentrations from 50 to 200 mM resulted in a reduction in the deterioration rates suggesting the importance of concentration of H_2O_2. To show the efficacy of fluoxetine degradation, a minimal irradiation dosage of 0.5 kGy proved to be highly efficient. In a wastewater treatment plant electron beam therapy using a linear electron accelerator successfully improved bacterial resistance to antibiotics. This has been shown by Szabo *et al.* in the elimination of piperacillin's antimicrobial action [30]. Owing to the high capital expense, the safety processes of an electron accelerator and the risks involved with rays, gamma rays and the electron beam process are not yet profitable.

Extremely energetic microwaves were also studied for the elimination of water pollutants. In order to promote the degradation of organic pollutants, the microwaves were combined with oxidants such as hydrogen peroxide and catalysts such as TiO_2 [31–33]. Microwaves can not only boost degradation rates of contaminants but also aid in the heating of particular contaminants. Unsurprisingly, microwaves offer low electrical efficiency as most energy gets converted to heat. Often, cooling equipment has to be utilized to ensure that the filtered water does not overheat.

Another component of AOP is ultrasound irradiation, also known as sonolysis [34, 35]. This is based on the radical hydrolysis of water due to the high strength of acoustic cavity bubbles, thus creating hydroxyl radicals from water. Sonication of water by ultrasound contributes to the creation and collapse of micro-bubbles leading to mediated acoustical wave compression and rarefaction. Upon passing a certain point, these bubbles implode violently and erupt with tremendous intensity. Water toxins are destroyed by decomposition and various reactive chemical reactions. Ultrasound irradiation experiences minimal interference from the water matrix and less heat transfer compared to UV irradiation. A degrading process of a known water pollutant, coomassie brilliant blue (CBB), has been performed in pure water and river water via ultrasound AOP [36]. The oxidation was carried out in the presence of inorganic ions such as chloride, sulphate, nitrate, bicarbonate, and carbonate ions. A potential degradation mechanism is suggested based on the product profile calculated by liquid chromatography-quadrupole-time-of-flight mass spectrometry. The poor productivity of oxidation of river water relative to that in pure water is mostly due to high levels of dissolved oxygen scavenging of hydroxyl radical by organic material and erosion of bubbles by dissolution.

1.3.4 Fenton Reaction

The potential use of a combination of H_2O_2 and Fe^{2+} to destroy tartaric acid was pointed out by Fenton in 1876. A radical mechanism for the catalytic decomposition of H_2O_2 by iron salts was introduced in the 1930s by Haber and Weiss [37, 38]. Mechanistic experiments about Fenton reaction since then have been extensively documented in reaction chemistry [39, 40]. The Fenton reaction takes advantage of iron salts and hydrogen peroxide to yield hydroxyl radicals.

In photo-Fenton technology photo-assisted reduction of Fe(III) takes place (at pH 2.8–3.5) under UV and visible light irradiation [40]. Quantum efficiency of photo-Fenton process is quite low, therefore to increase the quantum efficiency organic ligands are been employed [41]. For example, ferrioxalate which can absorb radiation up to 550 nm displays a good UV light absorption resulting in improved quantum efficiency [30].

Homogeneous photo-Fenton process

$$Fe(OH)^{2+} + hv \rightarrow Fe^{2+} + OH$$

Both Fenton and photo Fenton processes are successful for pharmaceuticals degradation. It has been shown that low concentrations of Fe^{2+} effectively degrade 100 mg/L ofloxacin and trimethoprim [42]. Moreover, the photoFenton performance was influenced by the water composition specifically during the elimination of 5-Fluorouracil from industrial effluent [43]. A combination of ultrasonic irradiation to the Fenton and photo-Fenton process can be considered as a good choice for degrading phenol derivatives obtained from petrochemical plants wastewater [44]. The electro-Fenton procedure is an evolving treatment technology. Fortunately, this approach is extremely effective in handling recalcitrant compounds that are not readily degradable. The electro-Fenton method for degradation of benzene, toluene, and p-xylene using the sacrificial precious metal, iron, and steel anodes attained 96.4%, 88.7%, and 87.6% removal efficiencies respectively at an initial concentration of 150 mg/L [45]. When the analysis was completed, treatment for one cubic meter of polluted water cost as much as $1.40 with the electro-fenton procedure. Moreover, degradation of reactive Black 5 dye and phenanthrene was tested in an electro-Fenton reactor, using the electricity provided by the microbial fuel cell. The findings allow for the conclusion that the integrated system is adequate in wastewater remediation to achieve high treatment performance with low electrical usage [46].

1.3.5 Photocatalytic

In recent decades the application of photocatalytic materials in water oxidation processes has been explored widely [47, 48]. Even though there are several alternative catalysts, studies centered on TiO_2 based photocatalyst.

Heterogeneous photocatalytic reaction

$$TiO_2 + hv \rightarrow e^-_{CB} + h^+_{VB}$$

$$h^+_{VB} + OH^-_{Ads} \rightarrow OH\cdot_{Ads}$$

In heterogeneous TiO_2-based photocatalysis, when the light of sufficient wavelength (greater than the wavelength of photocatalyst) strike semiconductor, electrons and holes are generated in conduction and valence band, respectively. These generated charge species then promote surface oxidation and reduction reaction of water constituents. Photocatalyst technology faces the challenge of poor light absorption and faster recombination of charge species which result in low quantum yield [49]. The addition of an electron donor will restrict the migration of holes and will reduce the recombination rate. For example, addition of citric acid to TiO_2 based photocatalyst restricts the charge recombination by filling the holes [50].

Moreover, researchers studied various forms of AOPs for handling refining wastes during bitumen production. Photocatalysis combined with peroxide was the most cost-effective and most efficient process employed. TiO_2 is used in the AOPs process that has the additional advantage of lowering the chemicals used [51]. TiO_2 is more efficient than other photocatalysts (Fe_2O_3, ZrO_2, CdS, WO_3 and SnO_2) and is chemically stable in an aqueous medium. Modification of TiO_2 photocatalysts to increase visible light absorbance is currently being studied by researchers. Recently, several experiments have concentrated on doping metals, metal ions, semiconductor oxides, and non-metal atoms into TiO_2 to expand the absorption spectrum to visible light, limit the recombination rate of the excitons, and improve the interfacial charge transferability. A novel method is reported for the synthesis of TiO_2nano-photocatalysts engineered by acoustic cavitation [52]. The effect of doping with Fe and Ce on the decolorization of crystal violet dye was also investigated. 0.8% of Fe-doped Titanium oxide shows optimum photochemical activity due to the presence of Fe in TiO_2 and serves as a Fenton catalyst. The research was based on X-ray diffraction and transmission electron microscopy to determine the structure and morphology of the sample. In another research Ce doping of TiO_2 and visible light activation of it were addressed for degradation of organic dyes [53]. The synthesis of various morphology of Ce-doped TiO_2 is also very well presented in the literature. This reaction process provides the groundwork for developing improved materials for visible-light conversion. Moreover, hybrid UV/TiO_2 photocatalysis offersthe advantage of low heating temperature, low-pressure requirements, low operating cost, and low energy usage [54]. For example, a combination of O_3 UV and TiO_2 was employed for the degradation of phenol [55]. It took 120 minutes for a successful decomposition to take place at 50 mg/dm^3 phenol in UV-TiO_2process. The UV-TiO_2 photocatalyzed method is inexpensive, sustainable and effective. The addition of hydrogen peroxides to generate reactive radicals has been shown to increase the degradation efficiency, as UV photocatalytic degradation is highly dependent on the strength of irradiation, reaction of pollutants and quantum performance. Ultraviolet H_2O_2 is a powerful photo-oxidant that also helps to reduce the UV energy needed to degrade organic pollutants.

1.3.6 Electrochemical Oxidation

As an appealing alternative to wastewater treatment, electrochemically based AOPs have been proposed. They produce reactive species by electricity and without the need forchemicals [56–58]. The use of Boron-doped diamond (BDD) anodes is now widespread in the electrochemical oxidation of pharmaceuticals. They are common due to their stability to corrosion, high oxygen-over capacity and inert surface [59–61]. Various experiments have shown that BDD anodes are efficient at removing pharmaceuticals. In the degradation of real drug wastewater sampled from the pharmaceutical facility, electrochemical oxidation using BDD was stated to be more successful than Fenton oxidation [62]. This research compared the degradation of four organic compounds in actual hospital wastewater and synthetic wastewater using electrochemical oxidation. Findings reveal that degradation was shown to be positively increased by a rise in applied current from 0.9 to 3.1 A. Additives may improve the degradation efficiency by enhancing the yield of radicals [63]. Electrode materials like BDD are costly and PbO_2 was identified as providing an optimized performance, although its fabrication on a wide scale is still a work in progress. Finding cheap substitutes would make for cost-effective implementation of the procedure. A closer look at the cost is required. This may be replaced by cheaper electrode materials, more solar power, etc. An inquiry into the development of chlorinated by-products at Ti_4O_7 anodes was carried out using resorcinol as a model organic contaminant [64]. Chlorinated by-products were closely researched during advanced oxidation procedures, and it was proposed that a multi-barrier treatment technique is important to ensure the contaminants are effectively eliminated during treatment.

Degradation efficiency is also affected by the existence of sulphate and nitrate ions. For example, Diuron is a poisonous herbicide for underground and surface water pollution [65]. It can be treated using sophisticated oxidation methods, i.e., via EAOPs. Diuron in concentrations of 10 ppm could degrade 91% in 100 s of residence time in the microreactor. Nitrate and sulphates slow the degradation of diurons by EAOP by scavenging hydroxyl radicals but do not modify the pathway of diuron degradation. The intermediate concentrations rise when the existence of nitrate or sulphate makes the EAOP less effective, but the intermediate is less toxic than the diuron. Moreover, the researchers tried to find out the utility of EAOPs for fixing furfural in aqueous solutions [66]. Taguchi procedure was used to minimise the number of tests and speed up the determination of optimal conditions. Results reveal that EAOP procedure can be applied for degrading and mineralizing furfural containing wastewater and is more efficient than the conventional treatment of furfural in industrial wastewaters.

1.4 Comparison and Challenges of AOP Technologies

AOPs are employed when a conventional oxidation system is not reliable. In conjunction with other conventional wastewater treatment methods, the biggest challenge is the amount of maintenance costs combined with AOPs. In this situation green energy is important and the use of waste materials as a catalyst will minimise costs. Advanced oxidation systems have gained intensively from advancements in materials science. AOP research must be multidisciplinary focussing on variety of technologies and their scaled up applications. In this segment we have made made a comparison of various AOP in Table 1.1. The advantages and disadvantages of AOPs technologies over other oxidation processes is addressed.

Table 1.1 Comparison of AOP systems.

AOP technologies	Advantages	Disadvantages
Ozone based AOP	• O_3 have strong oxidative power hence can be employed for degrading a range of pollutants. • Preferable when wastewater streams are poor in UV transmission. • Hydroxyl radicals formed arecapable of breaking down conjugated double bonds. • No generation of sludge. • Scaled-up application exists.	• Rate of degradation is low compared to other AOP technologies. • Generation of hazardous by-products possesses a risk. • Power intensive system having a high cost of operation. • Possible microbial growth may take place due to the presence of excess H_2O_2.
UV/H_2O_2	• For the same amount of energy used to supply oxidants with the source water, H_2O_2/UV processes can produce higher quantities of hydroxyl radicals than O_3/UV processes. • No possible bromated compound formation. • A scaled-up system exists.	• The high turbidity and high concentrations of nitrate can adversely affect UV light penetration. • Possible formation of bromated compounds. • Residual hydrogen peroxide may need to be separated from the water. • Theoretical yield is greatly affected by the presence of intervening compounds in water.

(Continued)

Table 1.1 Comparison of AOP systems. (*Continued*)

AOP technologies	Advantages	Disadvantages
UV/HOCl	• Free chlorine is cheap • Strong UV absorption of free chlorine	• Special reactors are required for UV illumination • Cannot oxidize perfluorinated compounds
UV/O_3	• Compared to O_3 or UV employed alone, UV/O_3 is more efficient in generating hydroxyl radicals. • Can be utilised as a disinfectant.	• Power intensive system compares to UV/H_2O_2 and O_3/H_2O_2. • Possible formation of bromated by-products. • The high turbidity and high concentrations of nitrate can adversely affect UV light penetration. • Power intensive system having a high cost of operation. • Mass transfer limitation due to gaseous O_3.
TiO_2 catalysed UV oxidation	• The catalyst has low costs and large commercial availability in diverse crystalline shapes and particle characteristics. • Environmental friendly photostable catalyst. • No possible bromated compound formation. • The photocatalyst can be recycled and utilised for next operation. • With doping with certain metal or non-metal the absorption region can be expanded to visible spectra.	• TiO_2 has a low quantum yield • Rapid recombination of excitons. • No scale-up application exists. • If the catalyst is applied as a slurry, it needs a separation step • Comprehensive studies are needed for determining optimal catalyst concentration.

(*Continued*)

Table 1.1 Comparison of AOP systems. (*Continued*)

AOP technologies	Advantages	Disadvantages
Fenton based Technology	• Ability to remove soluble and insoluble dyes in industrial wastewater. • No possible bromated compound formation.	• Production of iron sludge by mixed reagent flocculation and organic compounds • Low pH requirements. • Also altering pH raises the cost of operation.
Sonication	• Versatile technology. • Preferable for treating small volumes of wastewater. • Low heat transfer compared to UV technology. • No possible bromated compound formation if O_3 is not added.	• Power intensive system having a high cost of operation. • No scaled-up system exists. • Oxidant could be required to increase wastewater remediation efficacy which ultimately results in increased costs.
Radiation	• No additional chemical required • Can be operated at varying temperatures	• Use of gamma and X- rays
EAOP	• Robust • Easy to operate • Flexible in case of fluctuating wastewater streams • No additional chemicals • High energy efficiency	• Service time is short • Efficacy highly depends upon electrode surface

Moreover, the existence of additives in the solution matrix has an effect on pollutant oxidation by AOPs. In wastewater as ions, several additives are usually Fe^{2+}, Zn^{2+}, Ag^{+} and Na^{+}. Solvent sulphate anions can react with ·OH radicals, which contribute to a decrease inthe degradation percentage. Fe^{2+} ions most likely undergo hydroxyl radical reactions, reducing ·OH radicals concentration and efficiency of degradation. The presence of CO_3^{2+} ions normally applied to change the pH of the solution often reduces the effective degradation due to ·OH radical reactions. For scavenging hydroxyl radicals formed in solution, alcohols, such as methanol, have been used. Besides, nitrate and nitrite ions are also strong hydroxy-radical reagent which can induce pollutant photo-degradation in simulated sunlight. Moreover, the presence of NO^{3+} and NO^{2+} ions contribute to the formation of ·OH. In addition, excess 2-propanol is readily oxidised as a ·OH scavenger by hydroxyl radicals. Moreover, with superoxide O_2· these alcohols are less reactive than with ·H. The effect of different scavengers on the efficiency of degradation has generally been calculated to explore the potential pollutant degradation mechanism and pathway.

1.5 Conclusion and Perspective

It is very important to take into account the implications of various operating parameters of AOPs in order to maximise degradation efficiency and to reduce the overall treatment costs. A contaminant may be entirely extracted or refined into less toxic compounds that are ideal for biodegradation under optimal operating conditions. The major problem with AOPs, which restrict their industrial use, is the high cost of electricity supplies. The creation of new catalytic materials can contribute to the improvement and expansion of the use of different types of AOPs.

A restriction at the full-scale by the photocatalysis is primarily attributable to two factors: (1) isolation of colloidal catalyst from suspension during treatment, and (2) mass transfer restrictions on the immobilised catalyst on the substrate. After many research attempts, photocatalysis in industrial or municipal water treatment systems is not commonly used because of low OH radical performance. Currently, several experiments in photo-catalytical approaches have concentrated on photocatalysts that use solar radiation to minimise costs. In addition, the special emphasis on degradation is paid to the enhancement of optical and optoelectronic materials. Additionally, the impact on degradation efficiency of pollutant concentration, pH, catalyst loading, dissolved oxygen concentration and scavengers is examined.

There has been extensive research on the use of iron catalysts for hydroxyl radicals in Fenton, but it is limited in its use in real-life wastewater treatment as ideal pH is 2.8. Researchers are currently researching ways to overcome the low acid criteria of the Fenton and photo-Fenton reactions. By reworking on the "conventional" criterion, intuition on the performance can be gained and a challenge of slow degradation kinetics could be resolved. However, there is still a lot of analysis and discoveries to be made.

Although electrochemical oxidation has great advantages, the efficacy of the oxidation reaction is highly influenced by the electrode surface which decreases treatment effectiveness as a result of the formation of materials. There is a broad literature available on potential uses of electrochemical oxidation for wastewater treatment. Future studies should aim to gain mechanistic details via spectroscopic techniques, and also use in situ approaches. The study should concentrate on the implementation of innovative evidence-based approaches a less costly electrode and one that is reliable and is stable with the operation. Research is required on various novel electrodes in order to develop technology that avoids the above limitation on a pilot scale.

Wastewater treatment via radiation has been recorded as a clean process since no additional solvent is used to induce the reaction and the process can be conducted at various temperatures. Studies must concentrate on radiolytic degradation materials, mineralization and toxicity in addition to the degradation kinetics. It is recommended that the possibilities of ionising radiation such as electrons and gamma rays should be studied.

References

1. Global WASH Fast Facts | Global Water, Sanitation and Hygiene | Healthy Water | CDC. https://www.cdc.gov/healthywater/global/wash_statistics.html. Accessed 11 Feb 2021.
2. Daughton CG, Jones-Lepp TL (2001) *Pharmaceuticals and Personal Care Products in the Environment. Scientific and Regulatory Issues.*
3. Kümmerer K (2001) *Pharmaceuticals in the Environment. Sources, Fate and Risks.*
4. Alawi M, Masaad M, Al-Hussaini M (2018) Comparative study of persistent organic pollutant (POP) (chlorinated pesticides, PCBs, and dioxins/furans) concentrations in cancer-affected human organs with those of healthy organs. *Environ Monit Assess* 190. https://doi.org/10.1007/s10661-018-6856-3.
5. Kudom Donyinah S (2019) Introductory Chapter: Persistent Organic Pollutants (POPs). In: *Persistent Organic Pollutants.* IntechOpen.
6. T. Zacharia J (2019) Degradation Pathways of Persistent Organic Pollutants (POPs) in the Environment. In: *Persistent Organic Pollutants.* IntechOpen.
7. Ashraf MA (2017) Persistent organic pollutants (POPs): a global issue, a global challenge. *Environ. Sci. Pollut. Res.* 24:4223–4227.

8. Nguyen VH, Smith SM, Wantala K, Kajitvichyanukul P (2020) Photocatalytic remediation of persistent organic pollutants (POPs): A review. *Arab. J. Chem.* 13:8309–8337.
9. Glaze WH, Kang JW, Chapin DH (1987) No Title. *Ozone Sci Eng* 9:335.
10. Glaze WH (1994) No Title. *Chem Oxid* 2:44.
11. A. House D (2002) Kinetics and Mechanism of Oxidations by Peroxydisulfate. *Chem Rev* 62:185–203. https://doi.org/10.1021/cr60217a001.
12. Kolthoff IM, Miller IK (1951) The Chemistry of Persulfate. II. The Reaction of Persulfate with Mercaptans Solubilized in Solutions of Saturated Fatty Acid Soaps1. J Am Chem Soc 73:5118–5122. https://doi.org/10.1021/ja01155a030.
13. Garrido-cardenas JA, Agüera A (2020) Wastewater Treatment by Advanced Oxidation Process and Their Worldwide Research Trends.
14. Rekhate C V., Srivastava JK (2020) Recent advances in ozone-based advanced oxidation processes for treatment of wastewater- A review. *Chem Eng J Adv* 3:100031. https://doi.org/10.1016/j.ceja.2020.100031.
15. Wu CH (2008) Decolorization of C.I. Reactive Red 2 by ozonation catalyzed by Fe(II) and UV. *React Kinet Catal Lett* 93:35–42. https://doi.org/10.1007/s11144-008-5200-2.
16. Barry MC, Hristovski K, Westerhoff P (2014) Promoting Hydroxyl Radical Production during Ozonation of Municipal Wastewater. *Ozone Sci Eng* 36:229–237. https://doi.org/10.1080/01919512.2014.886938.
17. Nawrocki J, Kasprzyk-Hordern B (2010) The efficiency and mechanisms of catalytic ozonation. *Appl. Catal. B Environ.* 99:27–42.
18. Beltrán FJ, Rivas FJ, Montero-De-Espinosa R (2005) Iron type catalysts for the ozonation of oxalic acid in water. *Water Res* 39:3553–3564. https://doi.org/10.1016/j.watres.2005.06.018.
19. Lu J, Wei X, Chang Y, Tian S, Xiong Y (2016) Role of Mg in mesoporous $MgFe_2O_4$ for efficient catalytic ozonation of Acid Orange II. *J Chem Technol Biotechnol* 91:985–993. https://doi.org/10.1002/jctb.4667.
20. Restivo J, Garcia-Bordejé E, Órfão JJM, Pereira MFR (2016) Carbon nanofibers doped with nitrogen for the continuous catalytic ozonation of organic pollutants. *Chem Eng J* 293:102–111. https://doi.org/10.1016/j.cej.2016.02.055.
21. Yang W, Zhou H, Cicek N (2014) Treatment of Organic Micropollutants in Water and Wastewater by UV-Based Processes: A Literature Review. *Crit Rev Environ Sci Technol* 44:1443–1476. https://doi.org/10.1080/10643389.2013.790745.
22. Kim I, Yamashita N, Tanaka H (2009) Photodegradation of pharmaceuticals and personal care products during UV and UV/H2O2 treatments. *Chemosphere* 77:518–525. https://doi.org/10.1016/j.chemosphere.2009.07.041.
23. Kim IH, Yamashita N, Kato Y, Tanaka H (2009) Discussion on the application of UV/H2O2, O 3 and O3/UV processes as technologies for sewage reuse considering the removal of pharmaceuticals and personal care products. *Water Sci Technol* 59:945–955. https://doi.org/10.2166/wst.2009.076.

24. Kim I, Tanaka H (2009) Photodegradation characteristics of PPCPs in water with UV treatment. *Environ Int* 35:793–802. https://doi.org/10.1016/j.envint.2009.01.003.
25. Luo S, Wei Z, Spinney R, Zhang Z, Dionysiou DD, Gao L, Chai L, Wang D, Xiao R (2018) UV direct photolysis of sulfamethoxazole and ibuprofen: An experimental and modelling study. *J Hazard Mater* 343:132–139. https://doi.org/10.1016/j.jhazmat.2017.09.019.
26. Liu G, Ji J, Huang H, Xie R, Feng Q, Shu Y, Zhan Y, Fang R, He M, Liu S, Ye X, Leung DYC (2017) UV/H2O2: An efficient aqueous advanced oxidation process for VOCs removal. *Chem Eng J* 324:44–50. https://doi.org/10.1016/j.cej.2017.04.105.
27. Roback SL, Ishida KP, Chuang Y-H, Zhang Z, Mitch WA, Plumlee MH (2020) Pilot UV-AOP Comparison of UV/Hydrogen Peroxide, UV/Free Chlorine, and UV/Monochloramine for the Removal of N -Nitrosodimethylamine (NDMA) and NDMA Precursors . *ACS ES&T Water* 1:396–406. https://doi.org/10.1021/acsestwater.0c00155.
28. Kishimoto N, Nishimura H (2015) Effect of pH and molar ratio of pollutant to oxidant on a photochemical advanced oxidation process using hypochlorite. *Environ Technol* (United Kingdom) 36:2436–2442. https://doi.org/10.1080/09593330.2015.1034187.
29. Khaing ET, Htun MMH (2019) Distillery Waste Water Treatment using Photo-catalytic UV-TiO2 System. *Int J Adv Sci Res Eng* 05:210–217. https://doi.org/10.31695/ijasre.2019.33622.
30. Hislop KA, Bolton JR (1999) The photochemical generation of hydroxyl radicals in the UV- vis/ferrioxalate/H2O2 system. *Environ Sci Technol* 33:3119–3126. https://doi.org/10.1021/es9810134.
31. Han DH, Cha SY, Yang HY (2004) Improvement of oxidative decomposition of aqueous phenol by microwave irradiation in UV/H2O2 process and kinetic study. *Water Res* 38:2782–2790. https://doi.org/10.1016/j.watres.2004.03.025.
32. Zhihui A, Peng Y, Xiaohua L (2005) Degradation of 4-Chlorophenol by microwave irradiation enhanced advanced oxidation processes. In: *Chemosphere*. Elsevier Ltd, pp 824–827.
33. Bo L, Quan X, Chen S, Zhao H, Zhao Y (2006) Degradation of p-nitrophenol in aqueous solution by microwave assisted oxidation process through a granular activated carbon fixed bed. *Water Res* 40:3061–3068. https://doi.org/10.1016/j.watres.2006.06.030.
34. Mahamuni NN, Adewuyi YG (2010) Advanced oxidation processes (AOPs) involving ultrasound for waste water treatment: A review with emphasis on cost estimation. *Ultrason Sonochem* 17:990–1003. https://doi.org/10.1016/j.ultsonch.2009.09.005.
35. Babu SG, Ashokkumar M, Neppolian B (2016) The role of ultrasound on advanced oxidation processes. *Top. Curr. Chem.* 374:1–32.

36. Rayaroth MP, Aravind UK, Aravindakumar CT (2017) Ultrasound based AOP for emerging pollutants: from degradation to mechanism. *Environ Sci Pollut Res* 24:6261–6269. https://doi.org/10.1007/s11356-016-6606-4.
37. Haber F, Weiss J, Seph JO, Eiss W (1934) The catalytic decomposition of hydrogen peroxide by iron salts. Proc R Soc London Ser A - *Math Phys Sci* 147:332–351. https://doi.org/10.1098/rspa.1934.0221.
38. Haber F, Weiss J (1932) Über die Katalyse des Hydroperoxydes. Naturwissenschaften 20:948–950. https://doi.org/10.1007/BF01504715.
39. Brown RF, Jamison SE, Pandit UK, Pinkus J, White GR, Braendlin HP (1964) The Reaction of Fenton's Reagent with Phenoxyacetic Acid and Some Halogen-Substituted Phenoxyacetic Acids. *J Org Chem* 29:146–153. https://doi.org/10.1021/jo01024a034.
40. Brillas E, Sirés I, Oturan MA (2009) Electro-Fenton Process and Related Electrochemical Technologies Based on Fenton's Reaction Chemistry. *Chem Rev* 109:6570–6631. https://doi.org/10.1021/cr900136g.
41. Parsons S (2015) Advanced Oxidation Processes for Water and Wastewater Treatment. Water Intell Online 4:9781780403076–9781780403076. https://doi.org/10.2166/9781780403076.
42. Michael I, Hapeshi E, Aceña J, Perez S, Petrović M, Zapata A, Barceló D, Malato S, Fatta-Kassinos D (2013) Light-induced catalytic transformation of ofloxacin by solar Fenton in various water matrices at a pilot plant: Mineralization and characterization of major intermediate products. *Sci Total Environ* 461–462:39–48. https://doi.org/10.1016/j.scitotenv.2013.04.054.
43. Koltsakidou, Antonopoulou M, Sykiotou M, Evgenidou, Konstantinou I, Lambropoulou DA (2017) Photo-Fenton and Fenton-like processes for the treatment of the antineoplastic drug 5-fluorouracil under simulated solar radiation. *Environ Sci Pollut Res* 24:4791–4800. https://doi.org/10.1007/s11356-016-8138-3.
44. Shokri A (2018) Application of Sono–photo-Fenton process for degradation of phenol derivatives in petrochemical wastewater using full factorial design of experiment. *Int J Ind Chem* 9:295–303. https://doi.org/10.1007/s40090-018-0159-y.
45. Radwan M, Gar Alalm M, El-Etriby HK (2019) Application of electro-Fenton process for treatment of water contaminated with benzene, toluene, and p-xylene (BTX) using affordable electrodes. *J Water Process Eng* 31:100837. https://doi.org/10.1016/j.jwpe.2019.100837.
46. Fernández De Dios MÁ, Iglesias O, Pazos M, Sanromán MÁ (2014) Application of electro-fenton technology to remediation of polluted effluents by self-sustaining process. *Sci World J* 2014. https://doi.org/10.1155/2014/801870.
47. Pignatello JJ, Oliveros E, MacKay A (2006) Advanced oxidation processes for organic contaminant destruction based on the fenton reaction and related chemistry. *Crit. Rev. Environ. Sci. Technol.* 36:1–84.

48. Rahim Pouran S, Abdul Aziz AR, Wan Daud WMA (2015) Review on the main advances in photo-Fenton oxidation system for recalcitrant wastewaters. *J. Ind. Eng. Chem.* 21:53–69.
49. Sun L, Bolton JR (1996) Determination of the quantum yield for the photochemical generation of hydroxyl radicals in TiO2 suspensions. *J Phys Chem* 100:4127–4134. https://doi.org/10.1021/jp9505800.
50. Vohra MS, Davis AP (2000) TiO2-Assisted photocatalysis of lead-EDTA. *Water Res* 34:952–964. https://doi.org/10.1016/S0043-1354(99)00223-7.
51. Fernandes A, Makoś P, Wang Z, Boczkaj G (2020) Synergistic effect of TiO2 photocatalytic advanced oxidation processes in the treatment of refinery effluents. *Chem Eng J* 391:123488. https://doi.org/10.1016/j.cej.2019.123488.
52. Bethi B, Sonawane SH, Rohit GS, Holkar CR, Pinjari D V., Bhanvase BA, Pandit AB (2016) Investigation of TiO2 photocatalyst performance for decolorization in the presence of hydrodynamic cavitation as hybrid AOP. *Ultrason Sonochem* 28:150–160. https://doi.org/10.1016/j.ultsonch.2015.07.008.
53. Pawar M, Sendoğdular ST, Gouma P (2018) A brief overview of TiO2 photocatalyst for organic dye remediation: Case study of reaction mechanisms involved in Ce-TiO2 photocatalysts system. *J. Nanomater.* 2018.
54. Thiruvenkatachari R, Vigneswaran S, Moon IS (2008) A review on UV/TiO2 photocatalytic oxidation process. *Korean J Chem Eng* 25:64–72. https://doi.org/10.1007/s11814-008-0011-8.
55. Suzuki H, Araki S, Yamamoto H (2015) Evaluation of advanced oxidation processes (AOP) using O3, UV, and TiO2 for the degradation of phenol in water. *J Water Process* Eng 7:54–60. https://doi.org/10.1016/j.jwpe.2015.04.011.
56. Moreira FC, Boaventura RAR, Brillas E, Vilar VJP (2017) Electrochemical advanced oxidation processes: A review on their application to synthetic and real wastewaters. *Appl. Catal. B Environ.* 202:217–261.
57. Sirés I, Brillas E, Oturan MA, Rodrigo MA, Panizza M (2014) Electrochemical advanced oxidation processes: Today and tomorrow. A review. *Environ Sci Pollut Res* 21:8336–8367. https://doi.org/10.1007/s11356-014-2783-1.
58. Chaplin BP (2014) Critical review of electrochemical advanced oxidation processes for water treatment applications. *Environ. Sci. Process. Impacts* 16:1182–1203.
59. Švorc Ľ, Borovská K, Cinková K, Stanković DM, Planková A (2017) Advanced electrochemical platform for determination of cytostatic drug flutamide in various matrices using a boron-doped diamond electrode. *Electrochim Acta* 251:621–630. https://doi.org/10.1016/j.electacta.2017.08.077.
60. Kanakaraju D, Glass BD, Oelgemöller M (2018) Advanced oxidation process-mediated removal of pharmaceuticals from water: A review. *J. Environ. Manage.* 219:189–207.
61. Garcia-Segura S, Ocon JD, Chong MN (2018) Electrochemical oxidation remediation of real wastewater effluents — A review. *Process Saf. Environ. Prot.* 113:48–67.

62. Loos G, Scheers T, Van Eyck K, Van Schepdael A, Adams E, Van der Bruggen B, Cabooter D, Dewil R (2018) Electrochemical oxidation of key pharmaceuticals using a boron doped diamond electrode. *Sep Purif Technol* 195:184–191. https://doi.org/10.1016/j.seppur.2017.12.009.
63. Fang C, Megharaj M, Naidu R (2017) Electrochemical Advanced Oxidation Processes (EAOP) to degrade per- and polyfluoroalkyl substances (PFASs). *J Adv Oxid Technol* 20. https://doi.org/10.1515/jaots-2017-0014.
64. Lin MH, Bulman DM, Remucal CK, Chaplin BP (2020) Chlorinated Byproduct Formation during the Electrochemical Advanced Oxidation Process at Magnéli Phase Ti4O7Electrodes. *Environ Sci Technol* 54:12673–12683. https://doi.org/10.1021/acs.est.0c03916.
65. Bumroongsakulsawat P, Khongthon W, Pavarajarn V (2020) Degradation of diuron in water by electrochemical advanced oxidation in a microreactor: Effects of anion contamination on degradation and toxicity. *J Environ Chem Eng* 8:103824. https://doi.org/10.1016/j.jece.2020.103824.
66. Azarian G, Leili M, Godini K, Faradmal J, Bajalan S (2018) Furfural degradation using an electrochemical advanced oxidation process (EAOP): Optimization of operating parameters using taguchi approach. *Desalin Water Treat* 126:287–295. https://doi.org/10.5004/dwt.2018.22895.

2

A Historical Approach for Integration of Cavitation Technology with Conventional Wastewater Treatment Processes

Bhaskar Bethi[1]*, G. B. Radhika[1], Shirish H. Sonawane[2], Shrikant Barkade[3] and Ravindra Gaikwad[4]

[1]*Department of Chemical Engineering, B. V. Raju Institute of Technology, Narsapur, Medak (Dist.), Telangana State, India*
[2]*Department of Chemical Engineering, National institute of Technology, Warangal, Telangana State, India*
[3]*Department of Chemical Engineering, Sinhgad College of Engineering, Pune, M.S, India*
[4]*Department of Chemical Engieering, Pravara Rural Engineering College, Pravara Nagar, M.S, India*

Abstract

For conventional wastewater treatment plants, pollution discharge from manufacturing systems comprising complex and toxic substances is a major problem. The discharge of this hazardous wastewater into the surrounding environment is not only harmful to aquatic species but also raises major environmental problems. Traditional methods for treating wastewater, such as activated carbon adsorption, filtration and chemical oxidation have a few constraints such as partial usefulness and less effectiveness. A historical approach for designing advanced wastewater treatment technologies based on integration of cavitation processes and convention treatment techniques is explored in this section of the book. It involves the significance of combining modern methods with specialized treatment technologies in handling wastewater. Because of the nature of incompatibility and the unsynergetic effect, the technology integration in wastewater treatment has been introduced to limited studies. Fewer innovations, though, revealed an outstanding synergetic effect in terms of overall operation quality of the process. This chapter discusses in detail the integration of advanced ultrasound and hydrodynamic

**Corresponding author*: bhaskar.b@bvrit.ac.in

Siddhartha Moulik, Aditi Mullick and Anirban Roy (eds.) Sustainable Water Treatment: Advances and Technological Interventions, (27–56) © 2022 Scrivener Publishing LLC

cavitation technology with conventional processes (adsorption, filtration, chemical oxidation, etc.). Reviewing the exploratory work in this field and analyzing its feasibility would be useful in choosing whether the combined methods are efficient in meeting the lower-investment and treatment capacity. Therefore, each system should still be optimized based on industrial scale, considering various factors such as pollutant concentration, operational costs and energy usage. It was also discussed the scale-up issues specifically with ultrasound cavitation for the development of a hybrid wastewater treatment system.

Keywords: Ultrasound cavitation, hydrodynamic cavitation, conventional treatment, wastewater treatment, integrated technology

2.1 Introduction to Cavitation for Wastewater Treatment

2.1.1 Mechanistic Aspects of Ultrasound Cavitation

One type of cavitation process, in this case, high-frequency ultrasound waves (usually 20 kHz–100 MHz) will be used for the generation of cavities. The cyclic compression and expansion cycles creates the cavities in the liquid medium. Generally an ultrasonic probe is used as a cavitating device for production of cavities in the liquid. Subsequent growth and collapse of cavities generates the hydroxyl radicals that are highly oxidative in nature [1].

Ultrasound cavitation has two important effects in chemical transformations, in the case of water treatment; the generation of radicals is the chemical effect and the higher shearing due to micro-jets' formation is the physical effect. Due to these effects a number of chemical reactions and processes can be intensified in the chemical process industries.

Recently, ultrasonic cavitation is considered as an emerging advanced oxidation process since it generates the hydroxyl radicals as a result of ultrasonic vibrations [2]. In an aqueous solution ultrasonic cavitation can be created by the transmission of ultrasonic waves in the range of 20–100 kHz [3]. Ultrasonic cavitation is defined as the formation, growth, and subsequent collapse of vapor or gas-filled cavities within microseconds of time, releasing enormous magnitudes of energy [4, 5]. During the cavities collapse process, the hydrophobic volatile nature of organic pollutants such as phenol or any other organic solvents in wastewater will undergo thermal pyrolysis in the gas phase inside the cavity, and the non-volatile and hydrophilic organic polluted species can be degraded by hydroxyl radicals ($^{\bullet}OH$) at the interface of the bubble surface and bulk solution [6].

When ultrasonic waves are irradiated in a liquid medium, the molecules or particles swing at their mean positions. The molecules or particles come in contact with each other in the compression cycle and are drawn apart in the rarefaction cycle due to the variation of the acoustic pressure which involves a sinusoidal pattern [7]. If the applied acoustic pressure amplitudes are very large to break the molecular interaction, then cavities will form. This phenomenon is called as acoustic cavitation. Even though when the tensile strength of water is very large, then cavitation can occur at fairly low pressure conditions (< 20 atm) due to the presence of the dissolved gas molecules and very fine suspended particles acting as the extra nuclei for cavitation [7]. When a cavitation bubble is formed, due to the diffusion or bubbles coalescence the size of the bubble increases until it is dissolved in the liquid medium. Under the extreme conditions of cavitation such as at high temperatures (several thousand Kelvin) and high pressures (several hundred atmospheres) water molecules are thermally dissociated to hydrogen and hydroxyl radicals ($H^{\bullet}$ and $^{\bullet}OH$) [8, 9].

2.1.2 Mechanistic Aspects of Hydrodynamic Cavitation

In this case, cavities are produced due to the pressure variations in flow line, which is obtained by changing the flow geometry in the pipe line. When a fluid flows through a narrow constriction device which is provided in the flow line, the pressure at the narrow constriction equals or falls below the vapor pressure of the liquid generates the cavities; subsequent growth and collapse of bubbles produces the hydroxyl radicals [10]. Usually orifice, venturi are used as cavitating devices to generate the cavities leading to hydroxyl radicals. Researchers have also been developing some new kind of cavitating devices in the past decade such as vertex diode, rotating disc, etc. [10].

Application of hydrodynamic cavitation is limited on intensification of physical and chemical processes, even though this technique has several advantages, which makes it useful for application in wastewater treatment. Hydrodynamic cavitation can be easily scaled-up due to its capability in generating high density of hydroxyl radicals at ambient condition, and less material cost makes this technique more economical to use in the scale-up processes as compared to ultrasonic cavitation technique. In hydrodynamic cavitation, mainly cavities are formed due to the pressure variation in the flowing liquid in a pipe caused by a change in flow area such as venturi, and orifice plate [11, 12]. According to the Bernoulli principle, when the liquid flows through the narrow constriction such as venturi or orifice, the kinetic head increases at the expenses of pressure head. If the throttling of the constriction is sufficient, when the local pressure at the throat of the venturi or orifice equals or falls below the vapor pressure of that liquid; then the vaporous cavities are formed [13]. Further, lowering the pressure, the cavities continue to grow and reach their maximum size at the minimum pressure, and whenever the liquid jet expands by reducing the flow velocity as function of length of the pipe, the expanded pressure recovers at a particular point and this results in the collapse of earlier formed cavities [14]. The micro-bubbles/cavities collapse ability to generate localized "hot spots"; this phenomenon is known as "hot spot theory". Those hot spots consist of transient temperature of the order of 10,000 °K, and pressures of about 1000 atm over a microsecond interval [11]. Under this extreme conditions water molecules are able to form or dissociate into $H^{\bullet}$ and $^{\bullet}OH$ radicals. These $^{\bullet}OH$ radicals are then diffused into the bulk liquid medium which reacts with organic pollutants [11]. Even though the chances of thermal decomposition of organic pollutants also occur in the hydrodynamic cavitation due to the release of high temperature and pressure during the cavity collapse. Hence, under these extreme conditions larger molecules can break down into smaller molecules [13]. The ability of developing cavitational condition by hydrodynamic cavitation is equivalent to that developed by acoustic cavitation, but the hydrodynamic cavitation technique is a capable and useful process intensifying tool which can be explored for the degradation of various organic pollutants at an industrial scale. Hydrodynamic cavitation can be described by using Bernoulli's theorem.

$$P_{static} + P_{dynamic} = \text{Constant} = P_{static} + \frac{1}{2}\rho v^2 \quad (2.1)$$

Where P_{static} and $P_{dynamic}$ are static and dynamic pressures, 'ρ' and 'v' are the density and velocity of the liquid at the throat of the venturi or orifice. The sum of dynamic and static pressure is constant and increase in dynamic pressure subsequently reduces the static pressure. When the static pressure falls below threshold value of the liquid (i.e., the vapour pressure) then the cavities are generated.

2.2 Importance of Integrating Water Treatment Technology in Present Scenario

Wastewater generated from industries like textile, pharmacy, pesticide, and petrochemical processing contains large amounts of organic compounds such as textile dyes, aromatic compounds, chlorinated hydrocarbons, and phenolic compounds. These compounds are bio-refractory or toxic to the microorganisms and thus, conventional biochemical processes are not able to completely degrade them [15]. If these wastewaters are discharged to the environment without any treatment, these toxic pollutants can remain in the environment for an extended period of time due to their high stability to light and temperature. A number of methods have been investigated for degradation of bio-recalcitrant pollutants in wastewater such as chemical oxidation, chemical precipitation and flocculation, photolysis, adsorption, ion pair extraction, electrochemical treatment, advanced oxidation, and cavitation techniques, etc. [16, 17]. Advanced oxidation processes (AOPs) have been studied extensively both for recalcitrant wastewater in general; their commercialization has not yet been realized because of prevailing barriers.

However, biochemical method is the general economical method for degradation of various organic pollutants in wastewater. Though some organic wastewater effluents such as poly(ethylene terephthalate) (PET), poly (vinyl alcohol) (PVA), and some dispersed dyes are bio-refractory or toxic in nature, those pollutants cannot be treated with the help of biochemical method alone. Hence, the hydrodynamic cavitation can be retrofitted with minor modifications into existing biological-based wastewater treatment plants to enhance the biological degradation of bio-refractory pollutants. Retrofitting of hydrodynamic cavitation to any other existing effluent treatment plant can reduce the pretreatment time and costs, as well as the reduction in the usage of chemicals in an environmentally friendly way.

2.3 Integration Ultrasound Cavitation (UC) with Conventional Treatment Techniques

Ultrasound cavitation is an effective chemical-free and eco-friendly process used for water treatment which is contaminated by various toxic chemicals, pathogenic microorganisms, etc. Most of the research reports from literature have indicated that adsorption in combination with ultrasound cavitation yielded better results for water treatment rather than adsorption alone. However, low-frequency ultrasound enhances pore diffusion and initial rate of adsorption. The cavitation bubbles, and the associated micro-disturbances near the surface of solid, modify the boundary layer and increase the intra particle diffusion. The ultrasonic effects increase with increasing acoustic intensity.

2.3.1 Sonosorption (UC+ Adsorption)

Palm oil effluent was subjected to hybrid treatment combining ultrasound cavitation, coagulation, and adsorption. Both natural and synthetic coagulants were used, chitosan being the natural coagulant and ferric chloride the synthetic coagulant. Activated carbon was used as an adsorbent. The ultrasound cavitation alone was carried out in basic phase with a pH greater than 11; this helped in removal of COD due to the presence of Hydroxyl radical. At pH 11, the percentage reduction of COD, color and TSS in this case was 26.3%, 52.7% and 58.2%, respectively. Further coagulation studies were carried out in the acidic medium (pH maintained between 4.5 and 5), was observed that the performance of Ferric chloride was better than Chitosan. Studies were also carried out using activated carbon alone, though the removal of contaminants was reasonably good, dosage of activated carbon was higher. However, the combined hybrid treatment, ultrasound cavitation, coagulation using synthetic coagulant and activated carbon as adsorbent, the reduction in BOD5, COD, Color and TSS were in the acceptable range set by the Department of Environment of Malaysia. Further, it was reported that the efficiency of the treatment process will be improved by adopting the hybrid technologies [18].

Pharmaceutical wastewater of high strength was treated using combined sono-Fenton-sorption process. Studies were carried out individually and also in combination. Better performance was observed by this combination as compared to individual processes. The authors observed that the maximum reduction of chemical oxygen demand was after treating for 3hrs in the case of all the three processes. Maximum reduction was around 71.9% with sono-Fenton-sorption process and the reduction with sono-Fenton and sono-sorption were 48.4, and 43.75%, respectively. Parameter considered for the study were initial concentration, dosage of Fenton's reagent, sorbent dosage, and pattern of addition of iron. The effect of these parameters on the performance of the sono-Fenton-sorption process was reported. Langmuir–Hinshelwood isotherm was found to best fit the sonosorption and sono-Fenton-sorption data. Authors have concluded that the toxicity of the effluent was minimized to a considerable extent and the electrical energy consumption was found to reduce drastically by adopting the combination of sonolysis with sono-Fenton-sorption for the effluent treatment [19].

Activated carbon adsorption and ultrasound cavitation were employed individually in series and in combination to process the palm oil mill effluent. Percentage removal of chemical oxygen demand and total suspended solids evaluated the process efficiency. The optimization studies carried out by applying individual techniques in series resulted in 100% removal of chemical oxygen demand and 83.33% of total suspended solids for the conditions, ultrasound power amplitude of 50%, cavitation time 15min, activated carbon dosage 50 g/L and contact time 30 min. However, the combined operation by simultaneously applying the two techniques have shown improved result in the case of total suspended solids (98.33% removal) and 73.08% removal of chemical oxygen demand for the same conditions as individual in series. It was oncluded that this hybrid process would serve as an alternative technique for treating palm mill effluent with the availability of continuous and feasible sono chemical reactors [20].

Mesoporous alumina was used as adsorbent in combination with ultrasonic irradiation for the removal of tetracycline (TC) from the aqueous solution. The effect of parameters like pH, adsorbent dosage, ultrasound power and initial concentration of tetracycline on the performance of the combined process was studied. Studies indicated that the removal rate of tetracycline increased by adopting the combined process. Rate of removal increased with increasing pH, adsorbent dosage, ultrasound power and decreased with the tetracycline initial concentration. The Freundlich model was observed to best fit the data and the process followed the pseudo second order kinetic model [21].

Acid orange 52 is widely used in food, textile, paper, printing industries and used as a pH indicator. The effluent water from these industries which is discharged to the water streams, contains acid orange 52. This must be removed as it was found to be toxic for aquatic life. Hence an attempt has been made for the removal of acid orange 52 in aqueous solution by adsorption using Lapindo Volcanic mud as an adsorbent in the presence of ultrasound irradiation. Adsorption using Lapindo Volcanic mud was found to be effective in removal of acid orange 52. The combined method has increased the removal of the contaminant from 95.54 mg/g to 129.5 mg/g. This may be due to increased mass transfer rate in the presence of ultrasound waves. Langmuir isotherm was observed to best fit the data with an adsorption capacity of 833.33 mg/g. The adsorption data follows pseudo second order kinetic model [22].

The presence of phenol and its derivatives in water not only deteriorates its quality but also is toxic to human health. The present work focuses of the ultrasound aided adsorption of phenol on to the granular activated carbon and also conventional adsorption studies. Investigation was carried out at three different ultrasound frequencies. The influence of 35 kHz, 58 kHz, and 192 kHz frequencies, adsorbent dosage, initial concentration of phenol and RPM on the performance of the process was studied. Calorimetric studies were carried out to determine the power dissipated. The increase in temperature with respect to time over 10 min was recorded at regular intervals. The temperature of the bath and the system was monitored for every 30s using k-type thermocouples. From the studies of the cavitation mapping from the ultrasound baths they have concluded that the position of the adsorption vessel does not affect the cavitation on adsorption. Subsequently, for further studies the glass vessel was placed at the center of the bath. It was found by BET analysis that there is only a 5% increase in surface under the influence of ultrasound, so it was concluded that the influence of these has no significant impact on the equilibrium value of adsorption. Statistical analysis indicated that influence of adsorption dosage on the percentage removal was much higher than other factors [23].

The aqueous solution loaded with crystal violet pigment was treated for the removal of the pigment. Studies include sorption of the pigment on to activated charcoal in the presence and absence of ultrasound waves. The combined effect of sorption and sonolysis has resulted in higher performance than the individual processes. This could be due to the synergistic process achieved in the combined system. Influence of pigment concentration, ranging from 5 to 100 mg/L and the temperature (25, 65 and 80°C) on the percentage removal of the pigment was studied. Concentration of the samples before and after treatment was measured using UV- vis spectrophotometer at 590 nm. Experimental data was fitted to different isotherms and the Langmuir isotherm was found to best fit the sorption data. The maximum sorption capacity was reported as 50.1 mg/g. Thermodynamic properties were determined. The feasibility and spontaneity of the sorption of crystal violet pigment using activated carbon under the influence of ultrasound waves of high frequency was obvious from the negative values of the Gibbs free energy. The negative value of enthalpy indicated that the process is exothermic. The authors have concluded that the sorption or ultrasound cavitation alone could not remove the dye pigment to the desired levels whereas combinations of both were very efficient in removal of the dye pigment [24].

Groundwater used for municipal purposes was treated using granular activated carbon under the influence of ultrasound waves. Both laboratory and pilot plant scale were carried out. The effect of total organic carbon (TOC), permanganate index, UV absorbance, turbidity and color on the performance of the process was studied. It was inferred from the results that the influence of ultrasound was quite significant on the performance of the sorption process. The combined effect of sonication and sorption using activated carbon increased the removal efficiency of organic matter by 6–37% for TOC, and 18.6–27.9% for permanganate index. However, not much effect was felt in pilot scale [25].

Cadmium (Cd(II)) levels in groundwater were increased because of the effluent disposal from electroplating, smelting, plastics, pigments, ceramics, battery, fertilizers and mining industries. Due its toxic effects, many attempts were made for the removal of this metal ion from water. One such technique is adsorption. The present study focused on the removal of Cd(II) by adsorption using magnetic clay composite under the influence of ultrasound irradiation. Batch mode of operation was adopted for adsorption. Ultrasound device and a rotary shaker were used to carry out the experiments. The performance of the process was evaluated by varying the pH of the solution, contact time, initial metal ions concentration and temperature. The combined effect has shown a 35% increase in removal of the metal ion as compared to the conventional method. The effect of temperature on the performance of the process was reported to be negligible. The maximum adsorption capacity at different temperatures was calculated by using the linearized Langmuir model. The capacity was found to be 96, 99 and 100 mg Cd^{2+} g^{-1} at 298 K, 323 K and 343 K, respectively [26].

Adsorption of methylene blue onto different cellulosic materials like waste newspaper in the presence and absence of ultrasound was studied and compared and the results obtained. Effluents from the dye, textile industries, etc., releases methylene blue to the water streams. Due to its toxic effects, it needs to be treated. This attempt has been made towards the journey of developing low-cost processes. Data best Sono-sorption had resulted in 98% of the dye removal with respect to the conventional adsorption. The effect of the parameters like amount of sorbent, type of cellulosic sorbents, initial dye concentration, temperature, and contact time on the performance of the process was studied. It was concluded that the results obtained by using waste newspaper as a sorbent were most promising as compared to other cellulosic materials. Removal of the dye was faster and more efficient in the presence of ultrasound as the cavitation process facilitated the mass transfer which in turn led to increased surface area of the sorbent [27].

A well-known fact is that the granular activated carbon is commonly used as adsorbent from ages in treating water for domestic purposes and also industrial wastewaters. The present study focused on adsorption and desorption studies of p-chlorophenol onto and from granular activated carbon in the presence and absence of ultrasound of 21 kHz. Adsorption equilibrium and kinetic studies inferred that the adsorption was favorable in the absence of ultrasound waves. However, the desorption of p-chlorophenol was appreciably good with the combined effect. Increase in the ultrasound intensity increased the desorption rate. Desorption was further favored by increase in temperature and by addition of chemicals like sodium hydroxide and ethanol in the presence of ultrasound [28].

Color is one of the major contaminants in the textile effluent streams. The presence of dark colored pigments like reactive blue in such effluent streams hinders the process of photosynthesis by stopping the sunlight, thereby affecting the aquatic plant life. These coloring pigments also cause toxic effects in human beings on consumption of water contaminated by these pigments. Hence, degradation or removal of these contaminants is essential. The present study focuses on adsorption of reactive blue 19 onto activated carbon in combination with ultrasound. The equilibrium data resulted from the adsorption studies was evaluated using Langmuir and Freundlich isotherms. The equilibrium dimensionless parameter (R_L = 1/ (1+bC_0; b is the Langmuir constant and C_0 is the initial concentration of the dye solution)) with respect to Langmuir isotherm was found to be between o to 1 indicating favorable adsorption of reactive blue 19 on to activated carbon. The value of the Freundlich constants indicates the performance of the adsorption. The higher the value the greater is the adsorption. The Freundlich constant, K_f was found to be 10 and 35.5 in case of adsorption and sono-sorption respectively. This indicates higher adsorption rates for the combined effect. Subsequently, an attempt was made to determine the thermodynamic parameters. The decrease in the change in Gibbs free energy proves that the adsorption process is spontaneous. However, the Gibbs free energy decreases with increase in temperature. Decrease in the change of Gibbs free energy was more in the case of ultrasound than on to the adsorption of activated carbon alone. The results of the evaluated thermodynamic parameters implied that the adsorption was most effective in the presence of ultrasound [29].

Adsorption studies on removal of p-nitro phenol from aqueous solution were carried out under the influence and absence of ultrasonic irradiation using activated carbon (AC) and magnetized activated carbon (MAC) prepared from carrot dross. MAC was prepared by treating activated carbon with Fe_3O_4 in two different proportions. The surface area of AC was observed to be larger than the surface areas of MACs. The lower surface area of the MAC may be attributed due to the presence of magnetite nano particles on the surface of the activated carbon. However, due to the magnetic property of MAC, separation of solid phase from aqueous phase has become much easier as compared to AC. The effect of pH, adsorbent dosage, contact time and initial concentration on the adsorption performance was studied in the presence and absence of ultrasound irradiation. The adsorption data was fit to Langmuir and Freundlich isotherms. Data by using AC in the presence of ultrasound irradiation best fit well with the Langmuir isotherm. Similarly, the results of adsorption using AC and MAC in the absence of ultrasound irradiation and the results using MAC in the presence of ultrasound irradiation was observed to fit well with the Freundlich model [30].

Im *et al.* [31] investigated the application of US/absorbent system for the degradation of pharmaceuticals (PhACs) compounds such as Acetaminophen (AAP) and Naproxen (NPX) in the aqueous solution. Batch experiments show that biochar adsorbent was more efficient for removal of PhACs than activated carbon in the presence of US irradiation. US/absorbent system shows a potential method for degrading PhACs. Milenkovic *et al.* [32] studied the ultrasound-assisted adsorption removal of 4-dodecylbenzene sulfonate (DBS) ions from aqueous solutions using carbonized corn cob (AC). It was found that maximum adsorption capacities of the adsorbent for DBSwere 29.41 mg/g and 27.78 mg/g in the presence of US and its absence, respectively. Hamdaoui *et al.* [33] investigated the application of *Pinus halepensis* sorbent for the sorption of malachite green (MG) dye from aqueous media in the presence of ultrasound and in the absence of ultrasound. It was observed that sorption assisted with ultrasound shows the removal of dye can be improved bychanging the concentration of sorbate and temperature. The combined approach of ultrasound, stirring and sorption shows significant enhancement in the removal of dye.

In addition to the above studies, a few researchers also explored the combination of ultrasound cavitation and polymer hydrogels adsorption for the removal of dye pollutants from aqueous solution. Shirsath *et al.* [34] have investigated the synthesis and application of poly-acrylamide-kaolin (PAAm-K) nanocomposite hydrogel for the crystal violet dye removal from aqueous solution. Ultrasound polymerization method was used for the synthesis of nanocomposite hydrogel. The following operating parameters such as dye concentration, kaolin clay content, pH, temperature, and cavitation conditions was investigated on the removal of CV dye. It was observed that the following conditions such as higher pH, temperature, lower dye concentration, and higher quantity of hydrogel loading enhanced the dye removal efficiency. It was found that 80% removal of CV dye was obtained with in 5 h, whereas the combined approach of ultrasound and hydrogel adsorption showed 89% removal in only 0.5 h. The individual and combined performance of the hydrogel and UC+hydrogel adsorption is shown in Figure 2.1.

Shirsath *et al.* [35] have studied the synthesis and application of Bentonite-FeCo incorporated poly-acrylic acid nanocomposite hydrogel for the removal ofcrystal violet (CV) dye. Poly-acrylic acid nanocomposite hydrogel was synthesized using ultrasound emulsion polymerization method. Incorporation of clay and Fe–Co into the hydrogel matrix can enhance the strength of the polymer hydrogel and increase the adsorption

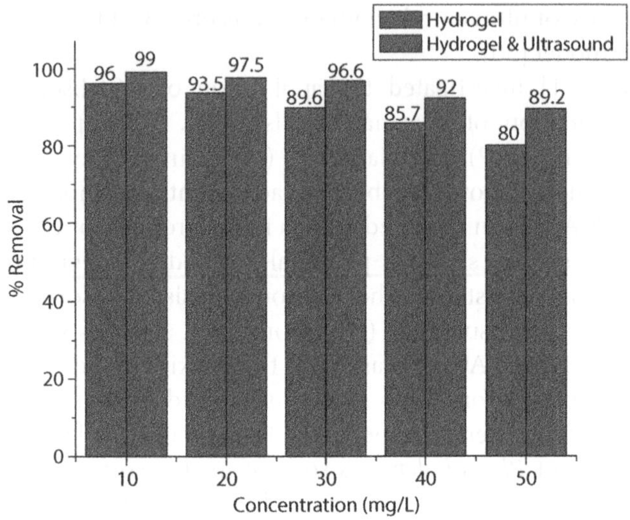

Figure 2.1 Comparison of hydrogel & hydrogel with ultrasound on adsorption of CV at pH 10 with temperature 35 °C [34].

efficiency of the hydrogel. It was found that 87% decolourization of CV dye was achieved within 15 h by stand-alone PAA-B-FeCo composite hydrogel, whereas the ultrasound coupled with hydrogel adsorption showed 97% removal of CV dye in 5 h. Sonawane *et al.* [36] studied the application of ultrasound assisted synthesis of poly acrylic acid (PAA) nanocomposite hydrogel for removal of malachite green (MG) dye from aqueous media. Their investigation reported that combined approach of ultrasound cavitation and (PAA) nanocomposite shows 73% of removal of MG dye.

2.3.2 Son-Chemical Oxidation (UC + Chemical Oxidation)

The degradation of various organic pollutants in the presence of ultrasound cavitation can be accelerated with the addition of chemical oxidants such as H_2O_2, Ozone, chlorine and salt ions, etc. In the past, many studies have reported the application of ultrasound cavitation and chemical oxidizing additives together for degradation of different organic pollutants in aqueous solution. Here, some of the works on the combination of ultrasound and chemical oxidation for the degradation of organic pollutants are summarized along with the details of operating parameters, optimum conditions and important results.

Voncina *et al.* [37] studied the decolourization of six reactive dyes such as C.I. Reactive Yellow 15, C.I. Reactive Red 22, C.I. Reactive Blue

28, Remazol Dark Black N 150%, C.I. Reactive Blue 220 and C.I. Reactive Black 5 in aqueous solution using stand-alone ultrasound (US) cavitation and with addition of H_2O_2. It was reported that decolourization of all the dyes in combined process shows two times more than the decolourization obtained by the stand-alone US (ultrasound) cavitation. Gogate *et al.* [38] investigated the degradation of orange acid II (OA-II) dye in the presence of US cavitation alone and its combination with addition of H_2O_2. It was found that increasing the addition of H_2O_2 increases the removal OA-II from aqueous solution till an optimum loading of H_2O_2 is 95.2 mg/L. US cavitation alone shows 49.5% degradation of OA-II dye and US + H_2O_2 process shows 75% of degradation in 120 min. Ghodbane and Hamdaoui [39] demonstrated the degradation of Acid Blue 25 (AB25) in aqueous solution in the presence of high-frequency ultrasound waves (1700 kHz). It was observed that the degradation rate of AB25 increases by 2.4 times when the addition of H_2O_2 is 386 mg/L and the degradation rate decreases with further loading of H_2O_2 above this value.

2.3.3 UC+Filtration

Lim and Bai [40] reported the vibrations which are generated during the dissipation of acoustic energy by sonication can separate the clogged material from the surface of the membrane filter and provide the free surface for solvent percolation in the case of wastewater treatment using a combined approach of ultrasound cavitation and membrane separation. Rocha *et al.* [41] investigated the effect of ultrasonic waves on the performance of the filtration of the oilfield wastewater through a porous ceramic filter. It was found that combined operation of ultrasonic waves and filtration process shows 99.5% of oil and grease (O&G) removal for the initial concentration of TSS 250 mg/L and 20 mg/L of O&G. The obtained results indicated that coupling of ultrasound with filter medium can enhance the removal efficiency of filtration process and also regenerate the filter medium surface. Li *et al.* [42] studied the treatment of oil emulsion wastewater using a combined approach of ultrasound and ZrO_2(Zicronium dioxide) ceramic membrane. In this work, authors introduced the ultrasound probe (length=7cm) into the tubular channel of the membrane. It was observed that 99.9% oil was rejected using membrane under the irradiation of ultrasonic waves in the flow direction of wastewater and optimal ultrasonic conditions. Kertesz *et al.* [43] investigated the feasibility of model dairy wastewater treatment using combination of membrane filtration and ultrasonication. In this work authors used the polyethersulfone based pressure-driven ultrafiltration membrane module in conjunction with ultrasonication and operated them in simultaneous mode to enhance the removal of organic pollutants from aqueous solution. The obtained results explained that stand-alone ultrafiltration experiments show 53.1% COD removal, the combined approach of ultrafiltration with half intermittent operation of ultrasound irradiation increases the removal of COD from 53.1 to 65.4%, whereas the ultrafiltration with continuous mode of ultrasound irradiation further increases the removal of COD from 53.1 to 74.4%.

2.4 Integration of Hydrodynamic Cavitation (HC) with Conventional Treatment Techniques

Cavitation technology can be effectively used to treat industrial effluent for removal of COD, ammoniacal nitrogen and/or color. The geometry of cavitating device and operating parameters such as pressure drop and initial concentration play an important role in process performance and optimization and in terms of these is crucial. Further, cavitation can be employed alone or in combination (process integration) with other processes such as biological treatment process, oxidation, adsorption, ion exchange or coagulation. Cavitation coupled with other methods, such as coagulation or adsorption, can be effective in water treatment and pollutant removal. Thus, process intensification can work wonders if cavitation and suitable other methods are integrated, especially in treating wastewaters containing refractory pollutants and/or having unusually high COD.

2.4.1 Hydrodynamic Cavitation + Adsorption

Although the combination of US cavitation and adsorption in wastewater treatment shows promising results, the cost and operation of the US and adsorption are neither inexpensive nor easy to commercialize industrially. The disadvantages of the above-proposed treatment method necessitates the finding of an alternative method for the efficient treatment of wastewater.

To date, research investigations on HC combined with adsorption is limited and the lack of lab-scale studies. Only a few lab-scale studies have been reported so far on combination of HC and adsorption for removal of organic pollutants.

Bethi *et al.* [44] reported a novel hybrid technique (hydrodynamic cavitation + hydrogels packed bed adsorption) for dye degradation and subsequent adsorption of dye molecules. Initially, hydrodynamic cavitation (HC) was employed for the degradation of crystal violet dye molecules and subsequent adsorption of dye molecules has been carried out by adopting synthesized polymer hydrogels. Poly-acrylic acid (PAA)/nano composite hydrogels have been synthesized via ultrasound assisted emulsion polymerization technique. The bentonite clay loaded PAA hydrogel showed better performance than the bare PAA hydrogel towards the removal of CV dye in the hybrid system. The percentage removal of CV dye is higher for lower clay content of hydrogel, and lower pH and flow rate. It was observed that 20% decolourization of CV dye and 17% of TOC removal was attained in HC only. The extent of decolourization of CV dye and TOC removal was achieved from 20 to 96% and 17 to 70%, respectively, when the hydrogels were combined with HC at pH 2.3. Figure 2.2 illustrates the hybrid system involving hydrodynamic cavitation and adsorption column.

Aman *et al.* [45] studied a novel hybrid treatment technique involving hydrodynamic cavitation (HC), and hydrogel adsorption. Hydrodynamic cavitation (HC) combined with poly-acrylamide (PAM) hydrogel nano-

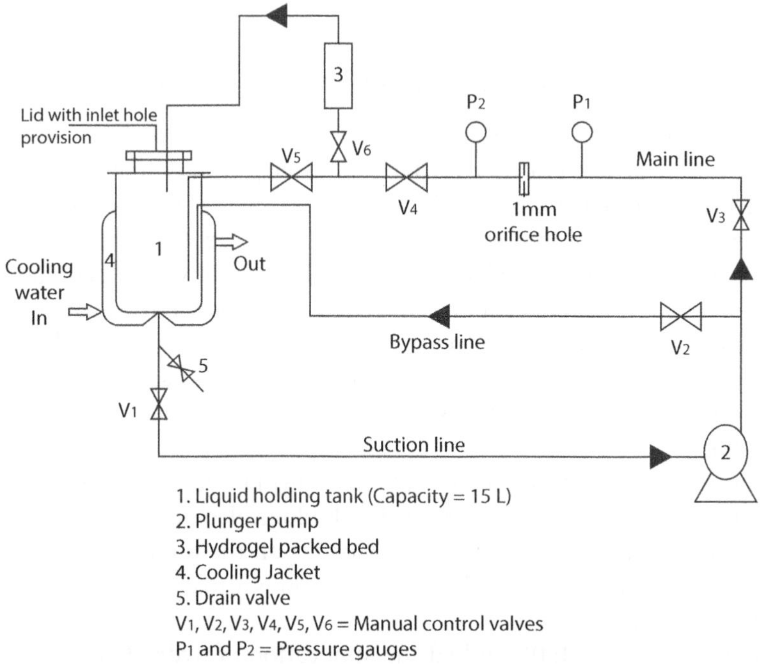

Figure 2.2 In-house fabricated experimental setup (hybrid treatment system) [44].

composite was studied for CV dye removal. TiO_2 nano-particles embedded poly-acrylamide (PAM) nanocomposite hydrogel was prepared by using the ultrasound polymerization technique. The individual effect of PAM-TiO_2 nano-composite shows 26% of TOC removal within 90 min for optimum TiO_2 loading, i.e., 0.5 wt. % in polymer hydrogel. Whereas, hydrodynamic cavitation alone showed 19% of TOC removal during the same period for an optimum inlet pressure to orifice at 3 bar. The combined effect of both these system shows a total of 73.60% of TOC removal in 90 min.

Bethi *et al.* [46] studied the decolorization of rhodamine-B (Rh-B) dye by using a hybrid treatment technique involving hydrodynamic cavitation (HC) and hydrogel adsorption. Halloysite nanoclay embedded polyacrylic acid (PAA) nanocomposite hydrogel was prepared by using the ultrasound polymerization technique. It was found that the HC assisted hydrogel nanocomposite adsorption shows the decolorization of 65% at the optimized operating conditions such as, pH 7.62, 0.5 g clay loaded nanocomposite hydrogel, and 25 g of hydrogels loading in adsorption column. Addition of H_2O_2 into the hybrid system further enhances the decolourization of Rh-B dye to 72.85% with 45% of total organic carbon removal at the optimum conditions.

2.4.2 Hydrodynamic Cavitation + Biological Oxidation

Due to stringent wastewater and water quality law regulations, the expanded investigation has been focused on developing the processes that are viable and can efficiently deal with toxic and refractory organic pollutants in wastewater and on improving the existing wastewater treatment technologies. Numerous studies have been conducted to enhance biodegradability and treat municipal wastewater [47]. The conventional and commonly known treatment methods of wastewater are not capable of reaching complete transformation of the pollutants to water and carbon dioxide, so the development in wastewater treatment methods and technologies is focused to meet the requirements of environmental law [48].

Sometimes the intermediates obtained during treatments based on HC may be more effectively treated using biological oxidation which helps in reducing the energy requirement along with the total treatment costs. Combination of HC with biological oxidation is also considered as an efficient hybrid process [49].

Magdalena *et al.* [50] studied the application of hydrodynamic cavitation to improve the biodegradability index of mechanically pre-treated municipal wastewater. The experiments were carried out at the pressure of 3.5, 5, and 7 bar. The use of an inlet pressure of 3.5 bar increased the BI index from 0.42 to 0.48 in the 5th minute of the process; a similar effect was obtained for the pressure of 7 bar (in the 5th minute of cavitation, a 13% increase in the BI index was obtained). The highest value of the BI index increase was observed for a pressure of 5 bar in 30 min of process. The biodegradability ratio increased by 32%, from 0.53 to 0.7. Similar results were obtained by Padoley *et al.* [51] for the enhancement of distillery wastewater biodegradability. They observed that higher inlet pressure (13 bar) provided the maximum value of the BOD5/COD rate. Dhanke *et al.* [52] used the hydrodynamic cavitation process to treatment fish processing industry wastewater. They observed a significant increase in the BI index, up to 0.93 in the 120-minute duration of the experiment at an inlet pressure of 8 bar. The same researchers applied the hydrodynamic cavitation to treatment vegetable oil refinery wastewater, the biodegradability index in this study reaches the maximum value of 0.69, and the reduction of COD was up to 71.8%. These parameters were obtained for an inlet pressure of 8 bar and in 120 min reaction time [53].

Thanekar *et al.* [54] used the HC as the pretreatment for the biological oxidation of dichlorvos present in the aqueous solution using slit venturi as a cavitating device. Under the optimized set of operating parameters of HC reactor as inlet pressure of 5 bar and pH of 4, the effect of HC alone on the removal of dichlorvos was investigated. It was reported that the HC alone gave 41.4% of degradation and 27.9% of TOC reduction. Subsequently, the effluents obtained after treatment of HC alone were subjected to biological oxidation. The maximum TOC reduction using HC combined with biological oxidation resulted in 50.6%. Only 14.4% of TOC reduction was obtained using direct biological oxidation (without any pretreatment).

2.4.3 Hydrodynamic Cavitation + Chemical Treatment

The degradation of various organic pollutants in the presence of ultrasound cavitation can be accelerated with the addition of chemical oxidants such as H_2O_2, Ozone, chlorine and salt ions, etc. In the past, many studies have reported the application of ultrasound cavitation and chemical oxidizing additives together for degradation of different organic pollutants in aqueous solution. However, a few studies which are specifically relevant to the degradation of dyes are discussed here. Table 2.1 summarizes some of the earlier works on the application of combination of hydrodynamic

Table 2.1 Overview of some previous works on combination of HC and chemical additives for the degradation of various organic pollutants.

Model pollutant/Operating parameters	Optimum parameters	Important results	Ref.
Model pollutant: Reactive orange 4 (RO4) dye **Operating Parameters:** Dyeconcentration: 40 mg/L Inlet pressure: 3–8 bar pH: 2–10 Oxidant dosage: RO4:H_2O_2 (1:5 to 1:50) Ozone: 1-8 g/h Volume: 5 L Reaction time: 120 min (HC+H_2O_2); 60 min (HC+Ozonation)	Dye concentration:40 mg/L Inlet pressure = 5 bar pH =2 Oxidant dosage: RO4:H_2O_2:1:30 Ozone: 3 g/h	**HC:** 37.23% decolourization; 22.22% mineralization in 120 min **H_2O_2 alone:** 4.65% decolourization and 2.72% mineralization in 120 min **Ozone alone:** 20.75% mineralization in 60 min **HC+H_2O_2:** 99.56% decolourization; 50.73% mineralization in 120 min **HC+Ozonation:** 76.25% mineralization in 60 min.	[55]
Model pollutant: Rhodamine B **Operating Parameters:** Inlet pressure: 0.2-0.6MPa Temperature:30-60 °C H_2O_2 dosage: 0-150 mg/L Dye concentration:10-50 mg/L pH: 2-10 Volume: 25 L Reaction time: 180 min	Dye concentration:10 mg/L Temperature: 50 °C Inlet pressure: 0.6 MPa H_2O_2: 100 mg/L pH:3	**HC alone:** 65% decolourization **H_2O_2 alone:** 13% decolourization **HC+H_2O_2:** 98.9% decolourization	[56]

(Continued)

Table 2.1 Overview of some previous works on combination of HC and chemical additives for the degradation of various organic pollutants. (*Continued*)

Model pollutant/Operating parameters	Optimum parameters	Important results	Ref.
Model pollutant: Reactive Red 120 (RR120) dye **Operating Parameters:** Concentration: 34 μM Inlet pressure: 3-10 bar. pH: 2–11. H_2O_2: 0-3400 μM volume: 6 L Reaction time: 120 min	Inlet Pressure: 5 bar Concentration: 34 μM, pH: 2 H_2O_2: 2040 μM	**HC:** 28% TOC reduction; 60% decolourization **HC+H_2O_2:** 60% TOC removal; 100% decolourization.	[57]
Model pollutant: orange acid-II (OA-II) **Operating Parameters:** Inlet pressure: 3 – 7 kg/cm^2 Temperature: 20 – 40 °C pH: 2-8 Concentration: 20 mg/L H_2O_2: 95.2 -571.2 mg/L Reaction time: 120 min	Inlet pressure: 5 kg/cm^2 Concentration: 20 mg/L Temperature: 20 °C pH: 2 H_2O_2: 244.8 mg/L	**HC:** 34.2% decolourization **HC+H_2O_2:** 86%decolourization; 96% degradation	[58]

(*Continued*)

Table 2.1 Overview of some previous works on combination of HC and chemical additives for the degradation of various organic pollutants. (*Continued*)

Model pollutant/Operating parameters	Optimum parameters	Important results	Ref.
Model pollutant: Imidacloprid **Operating Parameters:** Inlet pressure: 5- 20 bar Volume: 5 L pH:2-7.5 Imidacloprid: H_2O_2: 1:1 to 1:60 Reaction time: 180 min	Initial concentration of imidacloprid- 25 mg/L, Inlet pressure: 15 bar pH: 2.7 Imidacloprid: H_2O_2 : 1:40	**HC:** 26.50% degradation **HC+H_2O_2:** 100% degradation of imidacloprid; 37% TOC removal	[59]
Model pollutant: Methyl orange (MO) dye **Operating Parameters:** Inlet pressure: 0.2-0.6 MPa Temperature: 20 -40 °C Chlorine dioxide (ClO_2) dosage: 2 -10 mg/L pH=3.2 - 9.2 Volume: 5 L Reaction time: 90 min	Inlet pressure: 0.4 MPa Temperature: 35 °C ClO_2dosage: 8 mg/L pH= 7.2	**HC:** 18% degradation **ClO_2:** 48% degradation **HC+ ClO_2:** 91% degradation	[60]

(*Continued*)

Table 2.1 Overview of some previous works on combination of HC and chemical additives for the degradation of various organic pollutants. (*Continued*)

Model pollutant/Operating parameters	Optimum parameters	Important results	Ref.
Model pollutant: Reactive brilliant red K-2BP **Operating Parameters:** Concentration: 10-50 mg/L Inlet pressure: 0.2-0.6 MPa Volume: 25 L H_2O_2 doage: 0-500 mg/L pH: 2-10 Temperature: 303-333 K Reaction time: 120 min	Concentration: 20 mg/L Inlet pressure: 0.6 MPa H_2O_2 dosage: 300 mg/L Temperature: 323 K pH: 5.5	**H_2O_2:** 2.5% degradation **Swirling jet cavitation:** 10% degradation **Swirling jet cavitation+ H_2O_2:** 97.2% degradation at pH 2.	[61]

cavitation and chemical additives for the degradation of organic pollutants. It summarizes the details of operating parameters, optimum conditions and important results.

The combination of hydrodynamic cavitation and chemical oxidizing agents has been studied by very few researchers for the degradation of different organic pollutants. Gore *et al.* [55] studied the combination of HC with other advanced oxidation processes for the degradation of reactive orange 4 (RO4) dye. In that work the authors used the circular venturi as cavitation device in HC for the degradation of RO4 dye. Inlet pressure, pH, and cavitation number parameters were optimized for the degradation of RO4 dye. It was found that the combined approach of HC + H_2O_2 and HC + Ozone has shown the synergetic coefficient more than one and efficient treatment method for degradation of RO4. Wang *et al.* [56] studied the application of swirling jet-induced cavitation along with addition of H_2O_2 for the mineralization of rhodamine B dye. Dosage of H_2O_2, pH of the solution temperature of the solution, inlet pressure to the cavitating device and the initial dye concentration was optimized for the removal of rhodamine B dye. It was observed that increasing the dosage of H_2O_2 and inlet fluid pressure to the cavitation device increases the rhodamine B dye degradation, whereas the degradation of rhodamine B decreased when increasing the pH of the dye solution and initial concentration of dye. Saharan *et al.* [57] investigated the application of HC with the addition of H_2O_2 as an oxidizing agent for the Reactive Red 120 (RR120) dye degradation. Transparent venturi made up of acrylic material was used as the cavitating device in the HC for the RR120 degradation. Behavior of cavity in the fluid flow across the venturi at different inlet fluid pressure was analyzed through the flow visualization study. It was found that RR120 degradation is highly dependent on the pH of the dye solution, acidic medium of dye solution favors the higher degradation of RR120 and the dosage of H_2O_2 extended the degradation rate of RR120 dye. The percentage removal of TOC using HC alone and addition of H_2O_2 is shown in Figure 2.3.

Gogate and Bhosale [58] studied the degradation of different dyes such as orange acid-II (OA-II) and brilliant green (BG) in the presence of HC and various oxidizing agents, namely H_2O_2, NaOCl and $Na_2S_2O_8$. In this study, a 2 mm centered hole orifice was used as a caviation device in HC. The effect of various operating parameters such as temperature of the dye solution, pH of the dye solution and inlet pressure to the cavitation device on the degradation of dyes was studied. It was found that the extent of decolourization of dyes can be enhanced by the addition

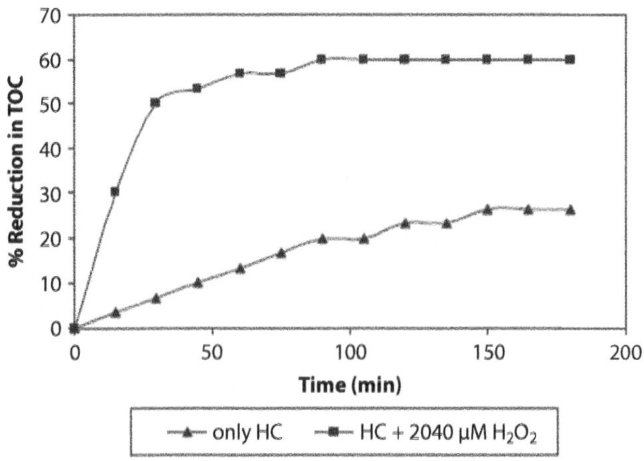

Figure 2.3 Percentage reduction in TOC as a function of time at different optimized conditions [57].

of the above-mentioned chemical oxidizing agents in hydrodynamic cavitation. It was observed that decolourization of dyes are more at lower solution temperature and in acidic pH condition of dye solution. Jadhav *et al.* [59] studied the application of HC along with addition of chemical oxidizing agent (H_2O_2) for the removal of imidacloprid from aqueous media. Circular venturi having 2 mm throat was used in the HC for the degradation studies of imidacloprid. It was found that complete removal of imidacloprid was obtained within 45 min of reaction time using HC + H_2O_2.Yang *et al.* [60] studied the removal of methyl orange (MO) dye from aqueous solution using orifice cavitator type hydrodynamic caviation reactor with chlorine dioxide (ClO_2) injection. In this study, different operating parameters such as inlet fluid pressure, temperature of the solution, pH of the solution, dosage of ClO_2were optimized with the aim of achieving maximum degradation of MO dye. It was found that increasing the dye solution temperature increases the degradation of MO dye. With continuous increase of dosage of ClO_2 the degradation rate of MO dye was also increased linearly. Wang *et al.* [61] investigated the application of swirling jet cavitation along with the addition of H_2O_2 for the reactive briliant red K-2BP (K-2BP) degradation in aqueous media. It was found that high temperature, pressure of the fluid and higher H_2O_2 concentration shows higher degradation of K-2BP. Similarly, the lower pH of the dye solution and less concentrated dye solution (10 mg/L) also shows maximum degradation of K-2BP.

2.5 Scale-Up Issues with Ultrasound Cavitation Process

As we discussed in previous sections about the wastewater treatment by cavitation techniques such as acoustic and hydrodynamic cavitation, from them one can observe that acoustic cavitations is unable to scale up to industrial scale due to production of less cavitational yield, difficult design and material of construction. Due to these drawbacks no one has attempted to make a pilot-scale wastewater treatment using ultrasonic cavitation. Presently research is going in the way of making pilot-scale continuous wastewater treatment using ultrasonic cavitation; the imaginary design follows a plug flow reactor with the incorporation of ultrasound probe tips or transducers with the equal distances to cover the total length of the reactor. The distributed probe tips or transducers in the plug flow reactor make the process continuous and treat the large volumes of wastewater by the uniform distribution of cavities throughout the length of the reactor. But this is not a possible way to treat the large volumes of wastewater due to high energy consumption and high capital investment.

2.6 Conclusion and Future Perspectives: Hydrodynamic Cavitation as a Future Technology

Researchers have reported that hydrodynamic cavitation is an energy-efficient and more suitable technology than the acoustic cavitation for treatment of bulk wastewater. Scale-up of hydrodynamic cavitation equipment is relatively easy as compared to acoustic cavitation equipment due to simplicity in design, inexpensive material of construction and higher cavitation yield, which induce this technique to scale up the process. It has been concluded that hydrodynamic cavitation could be easy to couple with the other advanced oxidation processes as well as conventional treatment techniques. The operating parameters, such as pH, Inlet pressure, addition of H_2O_2 changes the rate of degradation of organic pollutants. The combination of the cavitation technique with other processes such as photocatalysis, adsorption, filtration and H_2O_2 has shown the significant synergetic effects on the degradation of various organic pollutants in the aqueous solution. The magnitude of various operating parameters such as pressure, temperature, pH and the number of free radicals could affect the degradation rate of different organic pollutants. Hence, it is necessary to optimize those parameters through the theoretical and experimental studies before putting into practice any new technology related to wastewater treatment based on hydrodynamic cavitation. To our knowledge, limited literature has been found on the combination of hydrodynamic cavitation with adsorption, membrane filtration, photocatalytic membrane reactor, and thin film photocatalytic reactor in wastewater treatment. This could give future directions to scholars and scientists to work on the above hybrid systems.

A hybrid treatment scheme based on hydrodynamic cavitation and its retrofitting into existing methods of wastewater treatment will be the great option in future. One of the future prospectives can be seen from the above case studies, i.e., combining the membrane separation as a pretreatment and posttreatment technique. With this future scope, it can be possible to make the process of wastewater treatment based on hydrodynamic cavitation continuous as well as reaching the wastewater dischargeable limits set up by the government.

Acknowledgements

Dr. Bhaskar Bethi acknowledges the B V Raju Institute of Technology, Narsapur, for providing necessary support for contributing a chapter in this book.

References

1. Y.L. Song, J.T. Li, H. Chen, Degradation of C.I. Acid Red 88 aqueous solution by combination of Fenton's reagent and ultrasound irradiation, *Journal of Chemical Technology and Biotechnology* 84 (2009) 578–583.
2. Y.G. Adewuyi, Sonochemistry: environmental science and engineering applications, *Industrial & Engineering Chemistry Research* 40 (2001) 4681–4715.
3. T.J. Mason, J.P. Lorimer, *Applied Sonochemistry – The Uses of Power Ultrasound in Chemistry and Processing*, Wiley-VCH, Verlag GmbH, Weinheim, 2002.
4. V.K. Saharan, A.B. Pandit, P.S.S. Kumar, S. Anandan, Hydrodynamic cavitation as an advanced oxidation technique for the degradation of Acid Red 88 dye, *Industrial & Engineering Chemistry Research* 51 (2011) 1981–1989.
5. M.V. Bagal, P.R. Gogate, Degradation of 2,4-dinitrophenol using a combination of hydrodynamic cavitation, chemical and advanced oxidation processes, *Ultrasonics Sonochemistry* 20 (2013) 1226–1235.
6. Y.L. Song, J.T. Li, H. Chen, Degradation of C.I. Acid Red 88 aqueous solution by combination of Fenton's reagent and ultrasound irradiation, *Journal of Chemical Technology and Biotechnology* 84 (2009) 578–583.

7. G. Cravotto, P. Cintas, Harnessing mechano-chemical effects with ultrasound induced reactions, *Chemical Science* 3 (2012) 295–307.
8. B. Yang, J.N. Zuo, X.H. Tang, F. Liu, X. Yu, X.Y. Tang, H. Jiang, L.L. Gan, Effective ultrasound electrochemical degradation of methylene blue wastewater using a nanocoated electrode, *Ultrasonics Sonochemistry* 21 (2014) 1310–1317.
9. L. Zhao, J. Ma, X.D. Zhai, Enhanced mechanism of catalytic ozonation by ultrasound with orthogonal dual frequencies for the degradation of nitrobenzene in aqueous solution, *Ultrasonics Sonochemistry* 17 (2010) 84 91.
10. V.K. Saharan, A.B. Pandit, P.S.S. Kumar, S. Anandan, Hydrodynamic cavitation as an advanced oxidation technique for the degradation of Acid Red 88 dye, *Industrial & Engineering Chemistry Research* 51 (2011) 1981–1989.
11. A.G. Chakinala, P.R. Gogate, A.E. Burgess, D.H. Bremner, Industrial wastewater treatment using hydrodynamic cavitation and heterogeneous advanced Fenton processing, *Chemical Engineering Journal* 152 (2009) 498–502.
12. P.R. Gogate, A.B. Pandit, Engineering design methods for cavitation reactors II: hydrodynamic cavitation, *American Institute of Chemical Engineers Journal* 46 (2000) 1641–1649.
13. V.K. Saharan, A.B. Pandit, P.S.S. Kumar, S. Anandan, Hydrodynamic cavitation as an advanced oxidation technique for the degradation of Acid Red 88 dye, *Industrial & Engineering Chemistry Research* 51 (2011) 1981–1989.
14. P.R. Gogate, Treatment of wastewater streams containing phenolic compounds using hybrid techniques based on cavitation: A review of the current status and the way forward, *Ultrasonics Sonochemistry* 15 (2008) 1–15.
15. Yuequn Tao, Jun Cai, Xiulan Huai, Bin Liu, Zhixiong Guo, Application of Hydrodynamic Cavitation to Wastewater Treatment, *Chem. Eng. Technol.* 2016, 39, No. 8, 1363–1376.
16. K. Li *et al., Appl. Catal., B* 2015, 164, 82–91.
17. S. S. Garcia *et al., J. Hazard. Mater.* 2015, 283, 551–557.
18. Wong Ging King, Lee man Djun, Augustine Chioma Affam, Wong Chee Chung, Ir. Wong Chee Swee, Johnson Olufemi Adebayo, Application of Hybrid Ultrasonic Cavitation/Adsorption and Coagulation for Treatment of Palm Oil Mill Effluent, *AIP Conference Proceedings* 2124, 020008 (2019).
19. Senthilnathan Nachiappan, Karuppan Muthukumar, Treatment of Pharmaceutical Effluent by Sono-Fenton-Sorption, *CLEAN – Soil, Air, Water* 2014, 42 (11), 1526–1533.
20. Shridharan Parthasarathy, Rafie Rushdy Mohammed, Chong Mei Fong, Rachel L. Gomes, Sivakumar Manickam, A novel hybrid approach of activated carbon and ultrasound cavitation for the intensification of palm oil mill effluent (POME) polishing, *Journal of Cleaner Production* 112 (2016) 1218-1226.

21. Huiyuan Li, Xin Zhong, Hui Zhang, Luojing Xiang, Sebastien Royer, Sabine Valange and Joel Barrault, Ultrasound-assisted removal of tetracycline from aqueous solution by mesoporous alumina, *Water Science & Technology*, 2014, 819-824.
22. Nur Hidayatul Nazirah Kamarudin1, Herma Dina Setiabudi, Aishah Abdul Jalil, Siti Hazirah Adam, Nur Fatien Muhamad Salleh, Utilization of Lapindo Volcanic Mud for Enhanced Sono-sorption Removal of Acid Orange 52, *Bulletin of Chemical Reaction Engineering & Catalysis*, 14 (1) 2019, 189-195.
23. P. R. Rahul, A. Kannan, A DOE Study of Ultrasound Intensified Removal of Phenol, *International Scholarly and Scientific Research & Innovation* 9(7) 2015.
24. Sobhy M. Yakout, Mohamed R. Hassan, Ahmed A. Abdeltawab, Mohamed I. Aly, Sono-sorption efficiencies and equilibrium removal of triphenylmethane (crystal violet) dye from aqueous solution by activated charcoal, *Journal of Cleaner Production* 234 (2019) 124–131.
25. Andżelika Pietrzyk, Dorota Papciak, Removal of organic compounds from natural underground water in sorption and sonosorption processes on selected activated carbons, *E3S Web of Conferences* 17, 00074 (2017).
26. Zuzana Danková, Annamária Mockovčiaková, Mária Orolínová, Cd(II) Adsorption by Magnetic Clay Composite under the Ultrasound Irradiation, *Energy and Environmental Engineering* 1(2):74–80, 2013.
27. M.H. Entezari, Z. Sharif Al-Hoseini, Sono-sorption as a new method for the removal of methylene blue from aqueous solution, *Ultrasonics Sonochemistry* 14 (2007) 599–604.
28. O. Hamdaoui, E. Naffrechoux, L. Tifouti, C. Petrier, Effects of ultrasound on adsorption–desorption of p-chlorophenol on granular activated carbon, *Ultrasonics Sonochemistry* 10 (2003) 109–114.
29. Enes Şayan, Osman Nuri Ata, Meryem Esra Edecan, Investigation of Adsorption Thermodynamics on Removal of Reactive Blue 19 onto Activated Carbon under Ultrasonic Irradiation, *J. Chem. Soc. Pak.*, Vol. 36, No. 2, 2014.
30. Tahereh Rohani Bastami, Mohammad H. Entezari, Activated carbon from carrot dross combined with magnetite nanoparticles for the efficient removal of p-nitrophenol from aqueous solution, *Chemical Engineering Journal* 210 (2012) 510–519.
31. J.K. Im, L.K. Boateng, J.R.V. Flora, N. Her, K. DukZoh, A. Son, Y. Yoon, Enhanced ultrasonic degradation of acetaminophen and naproxen in the presence of powdered activated carbon and biochar adsorbents, *Sep. Purif. Technol.* 123 (2014) 96–105.
32. D.D. Milenkovic, A.L. Bojic, V.B. Veljkovic, Ultrasound-assisted adsorption of 4-dodecylbenzene sulfonate from aqueous solutions by corn cob activated carbon, *Ultrasonics Sonochemistry* 20 (2013) 955–962.

33. O. Hamdaoui, M. Chiha, E. Naffrechoux, Ultrasound-assisted removal of malachite green from aqueous solution by dead pine needles, *Ultrasonics Sonochemistry* 15 (2008) 799–807.
34. S.R. Shirsath, A.P. Patil, B.A. Bhanvase, S.H. Sonawane, Ultrasonically prepared poly(acrylamide)-kaolin composite hydrogel for removal of Crystal Violet dye from waste water, *Journal of Environmental Chemical Engineering* 3 (2015) 1152–1162.
35. S.R. Shirsath, A.P. Hage, M. Zhou, S.H. Sonawane, M. Ashokkumar, Ultrasound assisted preparation of nanoclay Bentonite-FeCo nanocomposite hybrid hydrogel: A potential responsive sorbent for removal of organic pollutant from water, *Desalination* 281 (2011) 429–437.
36. S.H. Sonawane, P.L. Chaudhari, S.A. Ghodke, M.G. Parande, V.M. Bhandari, S. Mishra, R.D. Kulkarni, Ultrasound assisted synthesis of polyacrylic acid–nanoclay nanocomposite and its application in sonosorption studies of Malachite Green dye, *Ultrasonics Sonochemistry* 16 (2009) 351–355.
37. D.B. Voncina, A.M.L Marechal, Reactive dye decolourization using combined ultrasound /H_2O_2, *Dyes and Pigments* 59 (2003) 173–179.
38. P.R. Gogate, G.S. Bhosale, Comparison of effectiveness of acoustic and hydrodynamic cavitation in combined treatment schemes for degradation of dye wastewaters, *Chemical Engineering and Processing* 71 (2013) 59–69.
39. H. Ghodbane, O. Hamdaoui, Degradation of Acid Blue 25 in aqueous media using 1700 kHz ultrasonic irradiation: ultrasound/Fe(II) and ultrasound/H_2O_2 combinations, *Ultrasonics Sonochemistry* 16 (2009) 593–598.
40. A. L. Lim, R. Bai, Membrane fouling and cleaning in microfiltration of activated sludge wastewater, *Journal of Membrane Science* 216 (2003) 279–290.
41. I. C. C. Rocha, J. J. Marques, A. S. Silva, Effects of ultrasound on the performance improvement of wastewater microfiltration through a porous ceramic filter, *Brazilian Journal of Chemical Engineering* 26 (2009) 641–648.
42. S. Li, X. Weihong, X. Nanping, Effect of ultrasound on the treatment of emulsification wastewater by ceramic membranes, *Chinese Journal of Chemical Engineering* 15(6) 855–860 (2007).
43. S. Kertesz, I. Kovacs, C. Hodur, G.K. Szabo, G. Vereb, Z. Laszlo, Ultrasound membrane hybrid processes for dairy wastewater treatment: pilot-scale analysis, *Desalination and Water Treatment* (2016) 1–8.
44. Bhaskar Bethi, S. H. Sonawane, Irina Potoroko, Bharat A. Bhanvase, Shriram S. Sonawane, Novel hybrid system based on hydrodynamic cavitation for treatment of dye waste water: A First report on bench scale study, *Journal of Environmental Chemical Engineering* 5 (2017) 1874–1884.
45. Aman Raj, Bhaskar Bethi, Shirish H. Sonawane, Investigation of removal of crystal violet dye using novel hybrid technique involving hydrodynamic cavitation and hydrogel, *Journal of Environmental Chemical Engineering* 6 (2018) 5311–5319.

46. Bhaskar Bethi, V. Manasa, K. Srinija, Shirish H.Sonawane, Intensification of Rhodamine-B dye removal by combination of Hydrodynamic cavitation and hydrogel adsorption, *Journal of Chemical Engineering and Processing - Process Intensification* 134 (2018) 51–57.
47. Korpe S., Bethi B., Sonawane S.H., Jayakumar K.V. Tannery wastewater treatment by cavitation combined with advanced oxidation process (AOP), *Ultrasonics Sonochemistry* 59, 2019, 104723.
48. Thanekar P., Gogate P. Application of Hydrodynamic Cavitation Reactors for Treatment of Wastewater Containing Organic Pollutants: Intensification Using Hybrid Approaches, *Fluids* 3, 2018, 98.
49. Oller, I.; Malato, S.; Sanchez-Perez, J.A. Combination of Advanced Oxidation Processes and biological treatments for wastewater decontamination - A review. *Sci. Total Environ.* 2011, 409, 4141–4166.
50. Magdalena Lebiocka, Application of Hydrodynamic Cavitation to Improve the Biodegradability of Municipal Wastewater, *Journal of Ecological Engineering* 21(6), 2020.
51. Padoley K.V.,Saharan V.K., Mudliara S.N., Pandey R.A., Pandit A.B. 2012. Cavitationally induced biodegradability enhancement of a distillery wastewater. *Journal of Hazardous Materials* 219–220, 69–74.
52. Dhanke P., Wagh S., Patil A. 2020. Treatment of fish processing industry wastewater using hydrodynamic cavitational reactor with biodegradability improvement. *Water Science and Technology* doi: https://doi.org/10.2166/wst.2020.049.
53. Dhanke P., Wagh S. 2019. Treatment of vegetable oil refinery wastewater with biodegradability index improvement. *Materials Today: Proceedings* doi: https://doi.org/10.1016/j.matpr.2019.10.004
54. Thanekar, P.; Murugesan, P.; Gogate, R.P. Improvement in biological oxidation process for the removal of dichlorvos from aqueous solutions using pretreatment based on Hydrodynamic Cavitation. *J. Water Process Eng.* 2018, 23, 20–26.
55. M.M. Gore, V.K. Saharan, D.V. Pinjari, P.K. Chavan, A.B. Pandit, Degradation of reactive orange 4 dye using hydrodynamic cavitation based hybrid techiniques, *Ultrasonics Sonochemistry* 21 (2014) 1075–1082.
56. X. Wang, J. Wang, P. Guo, W. Guo, C. Wang, Degradation of rhodamine B in aqueous solution by using swirling jet-induced cavitation combined with H_2O_2, *Journal of Hazardous* Materials 169 (2009) 486-491.
57. V.K. Saharan, M.P. Badve, A.B. Pandit, Degradation of Reactive red 120 dye using hydrodynamic caviatation, *Chemical Engineering Journal* 178 (2011) 100-107.
58. P.R. Gogate, G.S. Bhosale, Comparison of effectiveness of acoustic and hydrodynamic cavitation in combined treatment schemes for degradation of dye wastewaters, *Chemical Engineering and Processing: Process Intensification* 71 (2013) 59-69.

59. S.R. Jadhav, V.K. Saharan, D.V. Pinjari, S.H. Sonawane, D.R. Saini, A.B. Pandit, Synergetic effect of combination of AOP's (hydrodynamic cavitation and H_2O_2) on the degradation of neonicotinoid class of insecticide, *Journal of Hazardous Materials* 261 (2013) 139–147.
60. S. Yang, R. Jin, Z. He, Y. Qiao, S. Shi, W. Kong, Y. Wang, X. Liu, An Experimental Study on the Degradation of Methyl Orange by Combining Hydrodynamic Cavitation and Chlorine Dioxide Treatments, *Chemical Engineering Transactions* 59 (2017) 289–294.
61. J. Wang, X. Wang, P. Guo, J. Yu, Degradation of reactive brilliant red K-2BP in aqueous solution using swirling jet-induced cavitation combined with H_2O_2, *Ultrasonics Sonochemistry* 18 (2011) 494–500.

3

Hydrodynamic Cavitation: Route to Greener Technology for Wastewater Treatment

Anupam Mukherjee[1*], Ravi Teja[1], Aditi Mullick[2], Subhankar Roy[3], Siddhartha Moulik[2†] and Anirban Roy[1‡]

[1] *Water-Energy Nexus Lab, Department of Chemical Engineering Goa, BITS Pilani, Goa, India*

[2] *Cavitation and Dynamics Lab, Process Engineering and Technology Transfer Division, Indian Institute of Chemical Technology, Hyderabad, India*

[3] *Department of Chemical Engineering, Pandit Deendayal Energy University, Gandhinagar, India*

Abstract

Global water supply and demand trends point towards developing sustainable management of water sources. In this regard, high standards of environmental regulations are also being set, hence implementing new technologies. The present work highlights using one such promising technology, Hydrodynamic cavitation (HC), for wastewater treatment applications. HC has emerged as a state-of-the-art technology due to higher energy efficiency, low cost, no added chemicals, and large-scale implementations. The present chapter is divided into three major areas: (i) the first part provides a general overview of the cavitation phenomenon, the science behind the growth and collapse of cavitating bubbles, and their physico-chemical effects; (ii) the second part deals with different HC reactors presently been employed in the laboratory and large-scale applications and the influence of various operating parameters on HC performance; (iii) the third part illustrates the techno-ecological and techno-economic feasibility of the HC process, HC coupled with other Advanced Oxidation Processes (AOPs) and their recent applications. Based on the detailed review analysis carried out, it can be conclusively

**Corresponding author*: p20170408@goa.bits-pilani.ac.in

†*Corresponding author*: smoulik@iict.res.in

‡*Corresponding author*: anirbanr@goa.bits-pilani.ac.in

Siddhartha Moulik, Aditi Mullick and Anirban Roy (eds.) Sustainable Water Treatment: Advances and Technological Interventions, (57–116) © 2022 Scrivener Publishing LLC

proven that HC serves as the potential route to treat and recycle wastewater to meet the production-demand cycle.

***Keywords*:** Hydrodynamic cavitation, advanced oxidation process, wastewater treatment, bubble dynamics, water-energy-food-health nexus

3.1 Introduction

The history of cavitation dates back to the mid-18th century when a renowned scientist and professor, Euler, came up with his idea of a possible phenomenon for a particular water wheel design [1]. Marine technology forms the basis for the fundamental knowledge of this science [2]. At the end of the 18th century, there was a need for technological developments based on the design aspect in manufacturing ship propellers to control the speed and improve propeller performance. The English Navy successfully created a high-speed boat that had the mechanism to drive the marine propeller at high speed, which gave it an edge over the enemy [3]. Posthaste, engineers faced a challenge where they noticed that when the propeller started to work, there was an occurrence of high-pressure gradients. The liquid around the propeller was torn apart, giving rise to bubbles that consumed large amounts of energy. This also damaged the propeller blades as a consequence of high temperature and pressure.

Following this, in 1895, Thorneycroft and Barnaby [4] described this harmful phenomenon as cavitation, and the term was coined. It comes from the Latin word *cavus,* which means *hollow* in English [1]. In the mid-19th century, a series of papers related to marine engineering concerning engine racing in screw-propelled steamers were written by Reynolds, which made the subject of cavitation familiar [2]. To investigate these phenomena, Lord Rayleigh was appointed, and he published his recognized report on cavitation in 1915 [5]. From the context of fluid mechanics, cavitation is a general phenomenon that can happen whenever a liquid is in any machine where there is a fluctuation in fluid velocity and pressure. Later on, Parsons performed a series of model experiments to understand the nature of cavitation [4]. Since then, colossal works were performed by researchers in this domain to avoid cavitation in hydraulic equipment.

Recently, a few research groups worldwide have started converting cavitation from curse to boon to strength. Even nature uses cavitation in its own way. The exciting phenomena were notably reported by Lohse and his group, where they studied how snapping shrimp hunts its prey using the phenomenon of cavitation [6]. From their high-speed photographs, it was observed that the shrimp snapped its jaws frequently and speedily, which led the water to get thrown away at very high velocities capable of creating cavitation bubbles which traveled along with water and got collapsed near the prey, generating a shock wave which gave shrimp the edge over the prey. After that, the last few decades witnessed extensive works on cavitation [7, 8]. They ensured that cavitation phenomena played an imperative role in different science and technology areas, ranging from marine engineering and industrial processes to medical sciences [9, 10]. Cavitation is defined as the nucleation, growth, and subsequent collapse of gas vapor bubbles or cavities occurring in a small interval of time (milliseconds), releasing large energy magnitudes [11].

For a process to be feasible the physical or chemical changes involved can be identified by the energy input to the system. Apart from the quantity of energy input to the system, several other critical factors dominate the process's overall energy, like the method of introduction of energy and its availability in the required form at the particular location [12]. In this regard, cavitation can be fruitfully implemented to improve the energy efficiency of the process. In an application where chemical changes occur and are restricted by either mass transfer resistance or the activation energy, the availability of energy is in the form of fluid turbulence and pressure and heat, respectively [13]. Applications involving physical changes like the synthesis of nanoparticles and cell disruption the availability of energy are in the form of turbulence which imparts stress [10]. Hence it is evident that there is a methodological way of inputting energy to the system to minimize the wastage of energy, distinguishing this technology as better than other traditional ones.

This chapter focuses on hydrodynamic cavitation (HC), and the authors believe that HC can provide sustainable solutions to the challenges posed by existing technologies. Thus, this chapter is based on a viewpoint with a futuristic perspective. This chapter firstly explains the fundamentals of cavitation and the types of cavitation involved. Then, the authors provide an in-depth analysis of HC and discuss the bubble dynamics involved in the process, followed by the physio-chemical effects. The authors also discuss various HC-based reactors presently used in lab-scale and industrial scale and the impact of operating parameters on HC performance. The techno-ecological and techno-economic feasibility has also been discussed to highlight the technology's importance for scale-up applications. Finally, the authors have provided a summary of the application of hydrodynamic cavitation in the last few years (Table 3.1).

Table 3.1 Summary of current applications of HC and other AOPs for treatment of different pollutants in waste water.

Mode of pollutant	Technology used	Experimental conditions	Results	Highlights and remarks	References
Greywater	(i)HC (ii)AOP's - O_3, H_2O_2	(i)HC device- orifice plate (ii)O_3 = 3 g/h (iii)H_2O_2 = 1:5 molar ratio	**% COD and TOC removal:** (i)HC alone= 25% of COD and 15% of TOC (ii)HC+ H_2O_2= 87.5% of COD and 57% TOC (iii)HC+ O_3= 57.5% of COD, 35% of TOC (iv) HC+ H_2O_2+ O_3= 98.25% of COD, 76% of TOC.	(i) Cavitation induced with H_2O_2 (AOP) and Ozone as chemical oxidant was used for treatment of greywater. An orifice plate of 1.6 mm diameter was used as the cavitating device (ii) The % reduction in TOC and COD values at optimum conditions of 4 bar pressure and pH of 3 with 5 g/hr dosage of H_2O_2 were 98.25% and 76% respectively. (iii) The total cost was reported as 1.16 $/m^3$	[122]

(*Continued*)

Table 3.1 Summary of current applications of HC and other AOPs for treatment of different pollutants in waste water. (*Continued*)

Mode of pollutant	Technology used	Experimental conditions	Results	Highlights and remarks	References
Sodium dodecyl Sulphate (SDS)	(i)HC (ii)AOP's - H_2O_2 (iii)Fenton	(i)HC device- orifice plate (ii)Fenton = 1:3 molar ratio (iii)H_2O_2 = 1:5 molar ratio	**% Degradation** (i)HC alone= 64.93% (ii)HC+ Fenton= 995% (iii) HC+ H_2O_2=98.37%	(i) The degradation process of SDS with cavitation induced with AOP and Fenton was conducted using orifice plate of 1.6mm. (ii) Optimum inlet pressure of 5bar and a pH of 2 were observed to be giving best results. (iii) Most efficient process costed up to 2.47$/m^3.	[101]

(*Continued*)

Table 3.1 Summary of current applications of HC and other AOPs for treatment of different pollutants in waste water. (*Continued*)

Mode of pollutant	Technology used	Experimental conditions	Results	Highlights and remarks	References
Crystal violet	(i) HC (ii) Fenton Reagent (iii) UV lamps	(i) HC device-Slit Venturi (ii) US horn - rated power of 125W (iii) Fe-TiO_2 = 600ppm	**% Decolourization:** (i)HC alone = 44.5% (ii)HC+-Fe-TiO_2= 98%	(i) Hydrodynamic cavitation resulted in better decolourization compared to any other process used for crystal violet. (ii) The synergistic value calculated for the combined process is 3.52. (iii) The kinetic study of the experiment followed pseudo-first order kinetics.	[9]

(*Continued*)

Table 3.1 Summary of current applications of HC and other AOPs for treatment of different pollutants in waste water. (*Continued*)

Mode of pollutant	Technology used	Experimental conditions	Results	Highlights and remarks	References
Reactive red 180	(i)HC (ii) Photocatalytic- TiO_2, ZnO (iii) UV lamps	(i) HC device- circular Venturi (ii) US horn - rated power of 125W (iii) ZnO-TiO_2 = 1 g/l	**% COD and TOC removal:** (i)HC+ UV+ TiO_2= 45% COD, 38.7% TOC (ii)HC+ UV+ ZnO= 77% COD, 66% TOC	(i) Highest degradation was obtained in the HC+UV combined ZnO compared to photocatalytic reactor (ii) The synergistic Coeffecient of the process is 1.8. (iii) The mineralization of the organic compounds took much longer time compared to degradation.	[123]

(*Continued*)

Table 3.1 Summary of current applications of HC and other AOPs for treatment of different pollutants in waste water. (*Continued*)

Mode of pollutant	Technology used	Experimental conditions	Results	Highlights and remarks	References
Orange 4 dye	(i)HC (ii)AOP's - O_3, H_2O_2	(i) HC device-Slit Venturi (ii)O_3 = 3 g/h (iii)H_2O_2 = 1:5 molar ratio	**% Mineralisation:** (i)HC alone= 14.26% (ii) HC+ H_2O_2+ O_3=76.25%	(i) Inducing AOPs into any technology increase the % reduction or mineralization. (ii) The combined process proved to more energy efficient than individual process resulting in only 12.36×10^{-6} mg/J energy efficiency. (iii)The synergistic coefficient was 3.03 suggesting that combined process resulted in better degradation.	[124]

(*Continued*)

Table 3.1 Summary of current applications of HC and other AOPs for treatment of different pollutants in waste water. (*Continued*)

Mode of pollutant	Technology used	Experimental conditions	Results	Highlights and remarks	References
Alachor	(i)HC	(i)HC device: Swirling jet	**% Degradation rate constant K** (i)HC (pH = 2) = 4.45×10^{-2} min^{-1} (ii)HC (pH = 12) = 5.77×10^{-2} min^{-2}	(i) The Kinetic study of degradation of alachor observed to have pseudo-first order kinetics. (ii) It was observed that degradation increased with increase in pressure and decreased with increasing initial concentrations. (iii) Possible pathway for degradation of alachor was proposed and by products has also formed during the degradation process.	[71]

(*Continued*)

Table 3.1 Summary of current applications of HC and other AOPs for treatment of different pollutants in waste water. (*Continued*)

Mode of pollutant	Technology used	Experimental conditions	Results	Highlights and remarks	References
Dichlorvos	(i)HC (ii) Fenton	(i)HC device: circular venturi (i)$FeSO_4$-H_2O_2: 1:3 molar ratio	**% Degradation** (i)HC alone= 12.5% (ii)HC+ Fenton= 91.5%	(i) Optimum pressure for degradation of dichlorvos was calculated to be 5 bar. (ii) Lower temperature favoured degradation of dichlorvos. (iii) The cavitational yield of the combined process was calculated to be 0.074 KJ/m^3.	[125]

(*Continued*)

Table 3.1 Summary of current applications of HC and other AOPs for treatment of different pollutants in waste water. (*Continued*)

Mode of pollutant	Technology used	Experimental conditions	Results	Highlights and remarks	References
Methyl parathion	(i)HC (ii) AOP's-, H_2O_2 (iii)CCl_4	(i)HC device: Orifice plate (ii)H_2O_2: 1:4 molar ratio, (iii) CCl_4: 6 gm/l,	**% Degradation** (i)HC alone= 44.4% (ii)HC+ H_2O_2 = 100%	(i) The kinetic study of the combined process for degradation was observed to have pseudo first order kinetics with a rate constant of 5.4×10^3 min^{-1}. (ii) The cavitational yield is calculated to be 4.44×10^{-6} mg/J. (iii) The total power dissipation per unit of volume was calculated to 275 W/L	[126]

(*Continued*)

Table 3.1 Summary of current applications of HC and other AOPs for treatment of different pollutants in waste water. (*Continued*)

Mode of pollutant	Technology used	Experimental conditions	Results	Highlights and remarks	References
Carbamazepine (CBZ)	(i)HC (ii)AOP"s: O_3,H_2O_2 (iii)UV	(i)**HC device:** slit venturi (ii)**O_3:** 0.4 g/hr, **H_2O_2:** 1:5 molar ratio, (iii) **UV:** 16W	% **Degradation** (i)HC alone= 38.7% (ii)HC+ H_2O_2 = 52.9% (iii)HC+ O_3 + H_2O_2 = 100%	(i) HC and the combined process resulted in better degradation with optimum pressure of 4 bar in acidic medium. (ii) The cavitational yield was calculated to be 15.7×10^{-5}mg/J. (iii) The operating cost for combined process to achieve 100% degradation 100 min is 0.16 Rs/L.	[127]

(*Continued*)

Table 3.1 Summary of current applications of HC and other AOPs for treatment of different pollutants in waste water. (*Continued*)

Mode of pollutant	Technology used	Experimental conditions	Results	Highlights and remarks	References
Sulfadiazine	(i)HC (ii)Fenton (iii)persulfate	(i)**HC device:** circular venturi (ii)**$N_2S_2O_5$:** 348.5 mg/l, (iii) **H_2O_2:** 0.65 ml, α- Fe_2O_3: 181.8 ppm	**% Degradation** (i)HC+ Fenton + persulfate = 81%	(i) The combined process has resulted in better degradation compared to individual process at an optimum pressure of 10 bar in acidic medium. (ii) The kinetic study of the experiment was observed to follow pseudo first order kinetics with rate constant 41.50×10^{-3} min^{-1}. (iii) The cavitation number achieved for optimum pressure was 0.25	[128]

(*Continued*)

Table 3.1 Summary of current applications of HC and other AOPs for treatment of different pollutants in waste water. (*Continued*)

Mode of pollutant	Technology used	Experimental conditions	Results	Highlights and remarks	References
Diclofenac	(i)HC (ii)AOP's:H_2O_2 (iii)Photocatalytic (iv)UV	(i)**HC device:** slit venturi (ii)**H_2O_2:** 0.2 g/l, (iii) **TiO_2:** 0.2 g/l (iv)**UV:** 250W	% **Degradation** (i)HC alone= 26.85% (ii)HC+UV+H_2O_2+ TiO_2 = 94.78%	(i) Induced process of HC, AOP, UV and photo catalyst has resulted almost complete degradation of diclofenac. (ii) At optimum pressure of 4 bar in acidic medium at a pH of 4 highest % of degradation was observed. (iii) The synergistic value for combined process was calculated to be 2.5.	[129]

(*Continued*)

Table 3.1 Summary of current applications of HC and other AOPs for treatment of different pollutants in waste water. (*Continued*)

Mode of pollutant	Technology used	Experimental conditions	Results	Highlights and remarks	References
Waste Orange peel	Extraction	(i)Extraction of bio –active compounds like Polyphenols, Pectin, Terpenes	(i)HC device- venturi (ii)T=20^0C (iii)RPM = 200	(i) Two peels were used for extraction secured from different places, the yield % were 60% and 45%. (ii) The experiment resulted in excellent valorization of waste orange peel. (iii) The methane produced to the energy consumed resulted higher ratio of energy contained in methane	[130]
Micro Algae	Extraction	(i)Extraction of Nanochloropids or lipids (wet)	(i)HC device- orifice plate with 13 holes of 0.5 mm (ii)pressure = 4 bar (iii)Reaction time= 30 minutes	(i) Highest % yield of lipids was 49.7% for a reaction time of 30 minutes. (ii)The cavitation number as calculated to 1 for the yield obtained. (iii) The process was proved to be energy efficient for extraction of lipids from micro algae	[131]

3.2 Cavitation: General Perspective

3.2.1 Phase Transition

It is a well-known fact that liquid water evaporates when the temperature is increased above the boiling point, i.e., 100 °C at a constant pressure of 1 bar. But evaporation of water is not only possible when the temperature is increased at a constant pressure. The other possibility is to keep the temperature constant and reduce the pressure below the vapor pressure. An instance of such a phenomenon is the boiling of a kettle of water at a low temperature when it is taken at higher altitudes [14]. The reason behind this is that as we go up in altitude, air pressure decreases and molecules of water require a smaller amount of energy to become free from the water's bulk that permits water to boil at lower temperature. Similarly, when we cook in a pressure cooker the internal pressure rises, which promotes high cooking temperatures. To comprehend this phenomenon, it is essential to understand the typical phase equilibrium curve. The typical phase diagram is shown in Figure 3.1(a) and Figure 3.1(b)

Phase diagrams demonstrate the variations among the states of matter of compounds or elements with respect to temperature and pressure [15]. The lines represent the region where two phases can exist in equilibrium, concerning both temperature and pressure. In other words, these lines explain phase change points and are referred to as co-existence curve. The red lines (sublimation curve) separate the solid and gas phase regions, signifies sublimation (solid to gas) and deposition (gas to solid) route. The green line (melting curve) splits the solid and liquid phase regions and represents melting (solid to liquid) and freezing (liquid to solid) route. The blue (boiling curve) divides the liquid and gas phase regions, characterizes vaporization (liquid to gas) and condensation (gas to liquid) process.

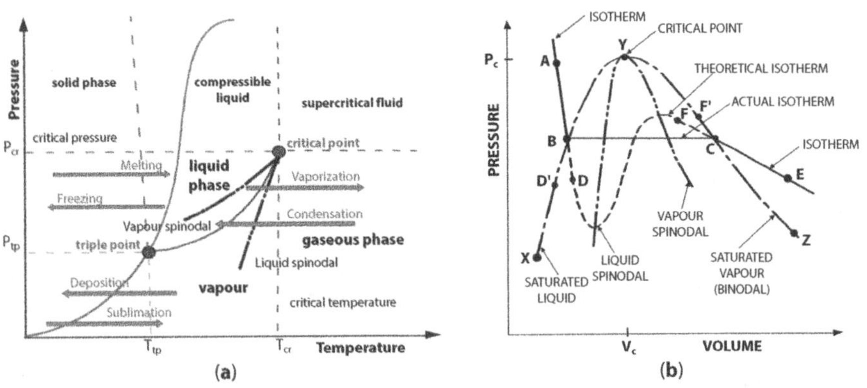

Figure 3.1 Phase diagram of water.

The diagram also consists of two important points, termed as the triple point and the critical point. The triple point facilitates co-existence of all three phases of matter at equilibrium. The critical point represents co-existence of two phases of matter, i.e., liquid and gas and relays to the critical pressure, the pressure which is above supercritical fluid form [16].

Generally, for most of the substances, the temperature and pressure associated to the triple point lie below standard temperature and pressure and the pressure connected to the critical point lies above standard pressure. Therefore, at standard pressure as temperature rises, substances start converting from solid to liquid and ultimately to gaseous form, and at standard temperature as pressure increases, substances start transforming from gas to liquid and finally to solid form [17]. Moving along the temperature line tells relative densities of the phases. When moving from the foot of the diagram to the head, the relative density upsurges. Moving through the pressure line exposes relative energies of the phases. Going from the left to the right of the diagram, the relative energies increase [18].

3.2.2 Types of Cavitation

Cavitation can be characterized among different kinds, on the basis of cavity production [19]. The four basic categories of cavitation and their sources can be briefly summarized as follows [20]:

(a) *Hydrodynamic cavitation (HC):* HC is caused due to rapid alterations of pressure in a liquid by manipulating system design leading to the velocity variation.
(b) *Acoustic cavitation (AC):* AC is driven by high-intensity ultrasound irradiation usually ranging from 16 KHz- 100 MHz's
(c) *Optic Cavitation (OC):* In optic cavitation (OC), only a single cavitation bubble is produced in transparent fluid by applying high-intensity light from a low-power pulsed laser.
(d) *Particle cavitation (PC):* Particle cavitation (PC) is caused by any other type of the beam of the simple particles such as proton in a liquid tank emits ionization trail within a fraction of time G.

According to W. Lauterborn and H. Bolle [21], tensions prevalent in a liquid motivate HC and AC generation, whereas consequence of local deposition of energy promote OC and PC phenomenon. Further cavity development is attained by lowering the ambient pressure using dynamic or static means. The bubble may enclose only vapor or a mixture of gas and vapor or only gas. On classifying HC from a physical vantage point, three kinds of cavitation exist [22–24]. The three classes of HC cavitation are briefly explained below:

(a) *Vaporous cavitation:* Vaporous cavitation is an extremely rapid process, which occurs in homogenous materials when a series of phase transition processes take place from liquid to vapor and from vapor to liquid. In this condition, the liquid remains attached with its wet saturated vapor inside the cavitation zone. The pressure level rapidly lowers from the vapor pressure of the liquid and prevents diffusion of any gases dissolved in that liquid.

(b) *Gas-vapor cavitation:* Gas-vapor cavitation occurs when liquid undergoes phase transition changes from liquid to wet saturated vapor to again liquid. In this condition, evaporation of water and diffusion of gas both the process take place through the phase transition boundary layer which results in compact and filled cavity formation. Conversely, gases produced from a liquid phase diffuse at wider cavitation zone due to variances in the partial pressures outside the cavity in that region. The rate of diffusion process differs from the rate of vaporization process, which elongates the gas diffusion duration and releases similar amounts of gas and vapor within the cavity.

(c) *Gaseous cavitation:* Gaseous cavitation, which can also be termed as diffusion process, occurs when non-condensable gas dissolved in the liquid is diffused at cavitation zone. It depends upon the degree of fluid circulation (convection) present in it which slow the process. Such a situation happens when the pressure of the gas becomes lower than its saturation pressure.

3.2.3 Hydrodynamic Cavitation

HC is formed by sudden fluctuation of the pressure and the flow, which is created by either specific construction such as nozzle, orifice plate and venture tube or by mechanical rotation of the rotating-type devices such as rotor and stator [25–28]. Here the authors have tried to explain the physical mechanism of the working principle of HC using an orifice as a cavitating device as shown in Figure 3.2.

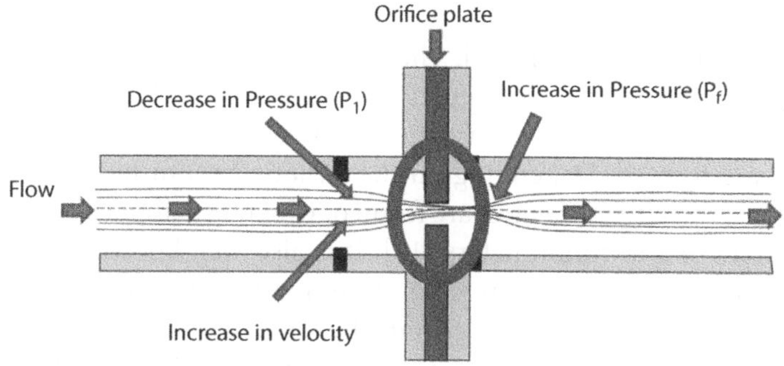

Figure 3.2 Working principle of cavitating device (Orifice plate).

An orifice is a device that can artificially manipulate the pressure when a fluid flows through it. The differential pressure created due to an orifice is a function of the beta ratio (b) which is mathematically defined as the ratio of orifice bore diameter to the inside pipe diameter [28]. The pressure distribution and pressure recovery profile on the downstream of the orifice is greatly affected by the geometric conditions which ultimately affect the active cavitation volume in the holistic overview of the hydrodynamic cavitation reactor setup [29]. These orifice-based constrictions basically work as a throttle valve. When the fluid enters at the throttle of the reactor, the cross-sectional area reduces, the fluid experiences some increase in pressure, which upsurges the flow rate or kinetic energy [30]. This is governed by the continuity equation as stated in Eq. 3.1

$$A * v = contant \tag{3.1}$$

Where A and v represent the area and velocity respectively around the regions of orifice. The intensity of the liquid turbulence depends on flowing parameters and geometrical design of the constructions. Turbulence and boundary layer separation of the liquid occurs at the constriction passage of the orifice, leading to substantial amount of energy loss because of permanent pressure drop. The interchange of pressure and velocity near the edge of the orifice is governed by the Bernoulli's Principle stated in Eq. 3.2.

$$P + \frac{1}{2}\rho v^2 + \rho gh = constant \tag{3.2}$$

Where, P = pressure of fluid, v = velocity of fluid, ρ = density of fluid and h = elevation.

Upstream of a restriction, the static pressure is well above the vapor pressure, thus water remains liquid. The static pressure in the flow decreases as it reaches the narrow pipe section. The pressure at the exposed edge is below the vapor pressure. Consequently, water evaporates and cavitation bubbles are formed. Depending on the various circumstances of flowing conditions, the sizes of cavities are formed, which usually varies from a few nanometers to a few millimeters [13]. It is also well known that liquid is a thousand times denser than vapor, thus at a pressure of 2300 Pa (room temperature being 20°C), water from the vapor bubbles occupies 50,000 times less volume after having condensed in its liquid state [11]. Hence, as the cavity size goes on reducing, the liquid will start rushing from all directions to fill up the vacant space as a result of a reduction in the volume of the cavity. As a result, a huge amount of momentum is associated with these on-rushing liquids. This momentum will keep on compressing the cavity and the cavity compresses at a velocity higher than the velocity of sound. When the cavities collapse, the compressibility and inertia of the gas-vapor bubbles releases an enormous implosion force, which creates some local hotspots and produces large magnitudes of energy. The temperature and the pressure can range from 500 to 15000 K and 100 - 5000 atmospheres, respectively [31].

The destruction/oxidation of organic pollutants using cavitation takes place through two mechanisms: (1) thermal decomposition/pyrolysis of the volatile pollutant molecule in and around the collapsing bubbles, and (2) oxidation of pollutant molecules by reactive free radicals (such as HO•, O•, and HOO• radicals) generated during the cavity collapse. Considering a single cavity and its inside-outside condition, there exist three locations around the bubble [32], along with the cavitational effects and free radical formation as shown in Figure 3.3. These locations are described below:

(i) *Hot region:* In this core location, the temperature and pressure attempt very high values. This adverse condition pyrolyzed the water molecules into •OH and •H radicals, which are strong oxidizing agents. The number of radicals generated in this location plays a major role in pollutant abasement.

(ii) *Interfacial surface area:* In the interfacial surface area of cavity and bulk liquid, the temperature reduces in comparison with the interior hot region of the vapor bubble. Depending on the amount of •OH radicals released in this location, organic pollutant gets removed from wastewater.

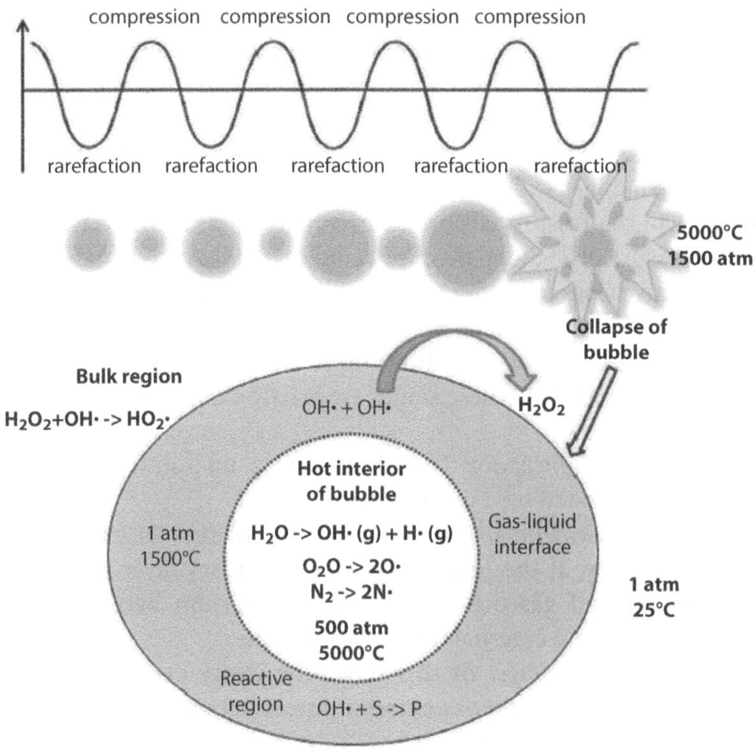

Figure 3.3 Mechanism of free radical development within a cavity [Reproduced with permission from A. Mullick and S.Neogi [32]].

(iii) *Region of bulk liquid medium:* At this outer location, recombination of •H and •OH radicals take place, resulting in H_2O_2 formation, which is a noble OH· scavenger.

In extreme cavitational circumstances, water molecules putrefied into a wide variety of species with high oxidation potential such as •OH, •OOH, and H_2O_2, that reacts with other organic compounds present in the wastewater. Along with the tremendous implosion force, shear forces of cavities also assist in breaking the molecular bonds of the pollutants and making them thermally decomposed and kill microorganisms, to attain the purpose of macromolecule organic substance removal, that exist in wastewater. Moreover, during vapor bubble collapse, the micro-jets create local turbulence and micro-circulation; by this means, the gas-liquid-mass transfer of reactants increases and enhances the exclusion of mass transfer resistances methods in the system. As the pressure recovers further downstream, the cavities will start collapsing. The typical pressure axis profile is shown in Figure 3.4. At the inlet of orifice, vapor pressure decreases with temperature but at the outlet, local pressure tries to recuperate beyond the liquid pressure. If the downstream pressure of the fluid could not recover above the vapor pressure, the process condition is called flashing and if the pressure recovers, the bubbles suddenly collapse or implode, resulting in cavitation. From the typical pressure axis profile graph below, it can be concluded that initially pressure drop along the pipe reduces rapidly with distance, having a steep negative slope, but after a certain distance, pressure drop as the pipe rises linearly with distance having a positive slope and regains its pressure.

During the process of cavitation, three characteristic and consecutive stages are observed [33]. These stages are briefly explained below:

(a) *Cavitation Inception Stage:* In this stage, micron size cavities are generated on a macro-scale.
(b) *Bubble Growth:* In this stage, cavity enlarges in size, consequently, at the phase transition boundary layer between liquid and gaseous phases, the equilibrium between the internal and external forces misbalance.
(c) *Dissolution Stage:* At this stage, the cavity experiences an implosive collapse, as external forces acting on them suddenly escalates.

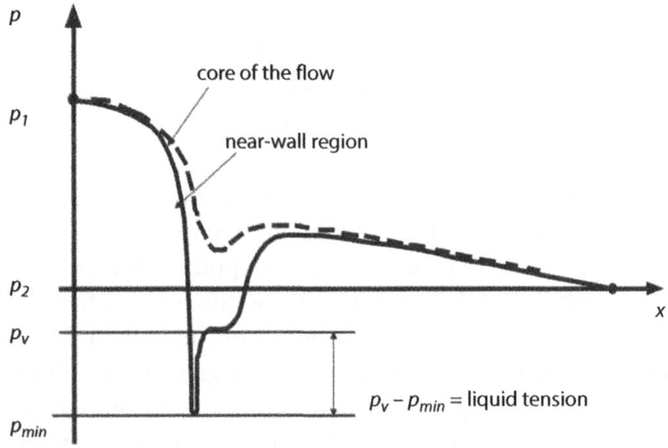

Figure 3.4 Typical pressure axis profile [Reproduced with permission from A.B. Pandit and P.R. Gogate [24]].

From the above sections it is inferred that cavitation takes place in the low-pressure region that forms at the orifice entry and in the zone around the section at which the jet flow area is at the minimum known as the vena contracta. The degree of cavitation development can be distinguishing by a non-dimensional parameter termed as the cavitation number [12], denoted by σ and is shown in Eq. 3.3

$$\sigma = \frac{P_{ref} - P_v}{\frac{1}{2}\rho v^2} \tag{3.3}$$

Here, p_{ref} = Pressure taken at a given point in the fluid flow, v = *Velocity*, ρ = *Density*, p_v = Vapor pressure of the fluid.

On plotting a graph between different length of cavitation regimes (L_{cav}) against cavitation number, A. Ferrari [33] studied four kind of cavitation regimes and are classified as shown in Figure 3.5.

At the orifice entrance, as soon as cavitation starts, incipient cavitation is observed. When the vapor tends to cover the separation region of axial extension length of separation zone, CN diminishes and sub-cavitation stage happens. Further length of cavitation region extends significantly downstream from the vena contracta and transitional cavitation occurs.

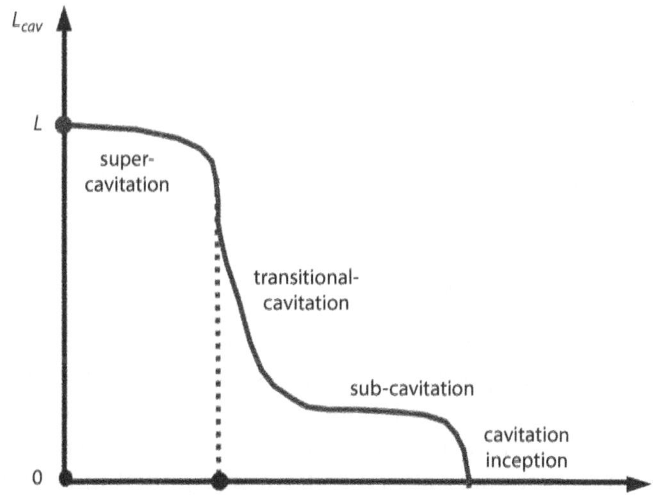

Figure 3.5 Different cavitation regimes [Reproduced with permission from A. Ferrari [33]].

When the cavity length develops up to or close to the orifice exit, supercavitation is observed.

3.2.4 Bubble Dynamics Model

3.2.4.1 Rayleigh-Plesset Equation

In this section, the authors shed light on the dynamics of bubble formation from the extension of the discussion carried out in the earlier section. Bubble dynamics plays a crucial role in geometric parameters along with operational conditions of the cavitation technique [11, 12, 23]. The overall effect of the hydrodynamic cavitating system depends on two important factors, viz., firstly due to each bubble's reaction and the overall quantity of vapor bubbles produced [11]. Bubble dynamics predict the time-dependent radius of a single spherical cavity R(t) and various other considerations like a cluster of bubbles, cavitational intensity, gravitation, viscosity, surface tension, and effects of turbulence, the compressibility of the medium, etc., which are applied in various numerical schemes [12, 34]. In fluid mechanics, the Rayleigh-Plesset equation derived from the Navier–Stokes equations under the theory of spherical symmetry governs the dynamics of a spherical bubble in an infinite body of incompressible Newtonian liquid [35, 36]. The derivation is explained somewhere else [37]. From the relation of fluid flow with bubbles, the Rayleigh-Plesset equation for bubble dynamics can be stated as follows (B):

$$\frac{P_B(t) - P_\infty(t)}{P_L} = R\frac{d^2R}{dt^2} + \frac{3}{2}\left(\frac{dR}{dt}\right)^2 + \frac{4\nu_L dR}{Rdt} + \frac{2S}{\rho_L R} \tag{3.4}$$

Where, P_B = Interior pressure, P = Reference pressure, R = Exterior bubble radius, S = Surface tension, ρ_L = Liquid density.

3.2.4.2 Bubble Contents

Apart from the Rayleigh-Plesset equation, discussed in the last section, bubble contents assumptions are also necessary to understand the bubble dynamics. The necessary assumptions considered in this case are constant liquid density (ρ_L), constant dynamic viscosity (μ_L), homogeneous bubbles and uniform temperature (TB (t)) and pressure pB (t)) in the interior of the bubble. The interior bubble pressure was determined using the van der Waals state equation [34].

$$PB = \frac{(N_{va} + N_{nc})R_g T}{\frac{4\pi}{3}(R^3 - h^3)} \tag{3.5}$$

Where, P_B = Interior pressure, h = Van der Waals hard core radius (h R_o/ 8.86).

R_g = Gas constant, N_{va} = Quantities of water vapor and N_{nc} = Quantities of non-condensable gas and T = Temperature of heat transfer mode between the bubble and the outside flow field.

The bubble interior is composed of both vapor and some quantity of contaminant gas in an unknown ratio [38]. Another factor, heat transfer between cavity and neighboring ambient liquid was determined by the model proposed by Qin *et al.* with Stefan-Boltzmann law [29, 34]. The temperature convection inside the cavity with the surrounding fluid could be described as [34]:

$$\Delta T = \frac{1}{C_{va}N_{va} + C_{nc}N_{nc}}\left[-p_b\Delta V - \frac{\kappa A(T - T_\infty)}{R}\Delta t - e\sigma A(T^4 - T_\infty{}^4)\Delta t\right] \tag{3.6}$$

Where, C_{va} = Specific heat of water vapor, C_{nc} = Specific heat of non-condensable gas, κ = Thermal conductivity ($\kappa = 0.6$ W/(m*K)), e = Emissivity (0.95) and σ = *Stefan-Boltzmannconstatnt* (5.67 x 10^{-8} W/(m^2 * K^4).

Alternatively, when the pressure inside the cavity reduces, and cavity condensed to liquid, the quantity of non-condensable gases changes and this unbalance concentration leads to mass transfer [34]. This mass transfer between cavity and neighboring ambient fluid can be calculated by diffusion for cavity, as proposed by Storey and Szeri as [34]:

$$\frac{dN}{dt} = 4\pi R^2 D\frac{(C_{wi} - C_i)}{l_d} \tag{3.7}$$

where N = *Quantityofgas*, D = *Diffusioncoefficient*, l_d = *Diffusion length*, C_{wi} = Gas concentration inside ambient fluid and C_i = Gas concentration inside cavity.

It is considered in the numerical scheme that there is a relative motion between the gas and liquid phase known as slip velocity, thus the liquid gas interfacial friction loss is negligible [38, 39]. As cavitation is significantly fast method, heat and mass transfer effects are considered to be negligible and hence Rayleigh-Plesset equation follows [11]:

$$P_B(t) = P_v(T_B) + P_{Go}\left(\frac{T_B}{T_\infty}\right)\left(\frac{R_o}{R}\right)^3 \tag{3.8}$$

Where, P_B = Interior pressure, P_v = Saturated vapor pressure, T_B = Bubble temperature, T_∞ = Liquid temperature, P_{GO} = Partial pressure of contaminant gas inside the bubble, R_o = Original nuclei size and R = Exterior bubble radius.

At different vapor pressure, (P_v (T_B), temperature difference, (T_B (t) – T_∞) value mitigate the rate of growth or violent collapse of cavitation bubbles. On substituting Equation (3.8) into (3.4) we get the below equation [11]:

$$\frac{P_v(T_\infty)-P_\infty(t)}{\rho_L}+\frac{p_{Go}}{\rho_L}\left(\frac{T_B}{T_\infty}\right)\left(\frac{R_o}{R}\right)^3+\frac{P_v(T_B)-P_v(T_\infty)}{\rho_L}=R\frac{d^2R}{dt^2}+\frac{3}{2}\left(\frac{dR}{dt}\right)^2+\frac{4v_L dR}{Rdt}+\frac{2S}{\rho_L R} \tag{3.9}$$

Where, P_B = Interior pressure, P_∞ = Reference pressure, T_∞ = Liquid temperature, ρ_L = Liquid density, P_{GO} = Partial pressure of contaminant gas inside the bubble, T_B = Bubble temperature, R_o = Original nuclei size, R = Exterior bubble radius, P_v = Saturated vapor pressure, S = Surface tension

During bubble growth, $\frac{P_v(T_B)-P_v(T_\infty)}{\rho_L}$ part of the Equation (3.9) in the Rayleigh-Plesset Equation becomes [11]:

$$(T_B - T_\infty)\frac{P_v L}{P_L T_\infty} = -\Sigma(T_\infty)C(n)R*t^{n-\frac{1}{2}} \tag{3.10}$$

Where, T_B = Bubble temperature, T_∞ = Liquid temperature, P_v = Saturated vapor pressure, T_∞ = Liquid temperature, R^* = Constant, C = Constant, n = Constant ($0 < n < 1$ *for of bubble growth condition*).

From Eqn. (3.10), it can be observed that, Σ, plays a crucial parameter in evaluating the bubble dynamic behavior.

From a case study of zero gas content in bubble interior, the bubble will reach smallest size with an immeasurable internal velocity [11, 38]. On integrating Rayleigh Equation, in the absence of both gas content and surface tension, we will get the time, t_{TC}, the needed time for complete collapse of bubble from R_o to null [40]:

$$t_{TC} = 0.915\left(\frac{P_v R_o^{\ 2}}{P_\infty - P_v}\right)^2 \tag{3.11}$$

Where, t_{TC} = Time for complete collapse, P_v = Saturated vapor pressure, R_o = Original nuclei size, P_∞ = Reference pressure,

Albeit, the above situation is completely a hypothetical event.

The pressure variation outside the cavity was analyzed by [34]

$$P(t) = P_{amb} - P_a \exp\left[\frac{a(t-t_0)^2}{\Delta t_{span}}\right] \tag{3.12}$$

Where, $P(t)$ = Time dependent pressure, P_{amb} = Ambient pressure in the bottom hole, P_a = Amplitude of pressure variation, and Δt_{span} = Duration of the pressure drop

Equilibrium Conditions

Along with the characteristic bubble contents, it is also important to recognize that the stability phenomenon of gas/ vapor bubbles [11].

$$p_v - p_\infty + p_{GE} - \frac{2S}{\mathrm{R}} = 0 \tag{3.13}$$

Where, P_∞ = Reference pressure, P_v = Saturated vapor pressure, S = Surface tension, R = Exterior bubble radius P_{GE} = Partial pressure of gas.

This equilibrium effects have vital significances in numerous cavitating flows. Two probable cases are likely to be taken care for any tiny perturbation in bubble size: firstly, the partial pressure of the gas should remain equal at P_{GE}; secondly, the temperature of gas in the bubble T_B should be equal with the mass of the gas. For isothermal situation, the cavity will go through a sequence of constant equilibrium conditions till the particular critical pressure of the cavity reached to its maximum value. Any slight decrease in P_∞ will alter the explosive cavitation growth. By observing cavitation nuclei of different sizes at lower pressure region of the reactor. Due to lower in pressure level, critical size will shrink and more nuclei get activated; as a result bubble number increases in the cavitation flow. Indeed, the critical tension for a fluid should have range between $4S/3R$ and $2S/R$ [11, 40].

$$\frac{4S}{3R} < p_v - p_\infty < \frac{2S}{R} \tag{3.14}$$

When fluid have subcritical tension value, dual equilibrium states occur alternatively, one a lesser stable state another a more unstable state [11]. Cavity growth phenomenon is a thermally controlled growth process. Because at super-heat or super-cool, critical time duration becomes longer or quicker than cavity formation duration respectively [11, 12, 23, 40–42]. Water at 20°C has $1\,m/s^{\frac{3}{2}}\sum$ value, and having critical time very much longer than the bubble growth time [11]. Hence the bubble growth occurring smoothly without any obstacles by thermal effects; so it is "inertially controlled" bubble growth [11, 38, 40, 41, 43, 44].

3.2.4.3 *Nonequilibrium Effects*

It is considered that vapor in the cavity is in thermal equilibrium with the fluid at the interface [11, 40]. But according to T. Theofanous's [45] suggestion, it is not true for each case. Because of the high evaporation rate of fluid, temperature of the liquid at the interface, T_{LS} can differ from saturated vapor in the bubble, T_B [44–46]. This evolved the below equation [11]:

$$G_v = \Lambda\left[\left(\frac{1}{2\pi K_v}\right)^{\frac{1}{2}}\left(\frac{P_v(T_{LS})}{T_{LS}^{\frac{1}{2}}} - \frac{P_v(T_B)}{T_B^{\frac{1}{2}}}\right)\right] \tag{3.15}$$

Where, Λ = Accommodation coefficient, P_v = Saturated vapor pressure, T_B = Bubble temperature, G_v = Evaporative mass flux, and K_v = Gas constant of the vapor, T_{LS} = Temperature of the saturated vapor in the bubble.

3.2.5 Physio-Chemical Effects

As the cavity implodes, its potential energy reduces which feeds into kinetic energy of the liquid body [47]. The implosions of cavities can release a huge amount of energy densities on the order of $1 - 10^{18}$ kW/m^3 into surrounding liquids [47]. The released energy can cause the following effects:

3.2.5.1 *Thermodynamic Effects*

(a) *Generation of temperature (*1000 to 5000 *K):* Cavitation phenomenon results in energy dissipation which initiates the temperature rise at the location of the implosion. The maximum temperature (T_{max}) emission is stimulated when bubble having tiniest nuclei size radius, (R_o). The temperature released by the vapor - cavity at radius (R_o) can be given as:

$$T_{max} = T_0\left[\frac{(k-1)\left(P^*{}_{\infty} - P_v - P_{GO} + \dfrac{3S}{R_0}\right)}{P_{GO}}\right] \tag{3.16}$$

Where, P_v = Saturated vapor pressure, S = Surface tension, $P^*{}_{\infty}$ = Surface e constant reference pressure, P_{GO} = Partial pressure of contaminant gas inside the bubble. k = *Constatnt.* Maximum temperature release would be nearly $4 * 10^4$ times the ambient temperature [48].

(b) *Emergence of pressure (*100 *to* 5000 *bar):* From nuclei size radius (R_o), when a cavitation bubble enlarges and reached to its maximum radius, (R_{max}), the collapsing process of the bubble starts with a small amount of partial pressure of gas, (P_{GO}). The pressure discharges by the vapor-cavity with respect to a constant reference pressure, ($P^*{}_{\infty}$) of a uniform, upstream flow can be predicted by the following equation:

$$P_{max} = P_{GO}\left[\frac{(k-1)\left(P^*{}_{\infty} - P_v - P_{GO} + \dfrac{3S}{R_0}\right)}{P_{GO}}\right] k/(K-1) \tag{3.17}$$

P_{max} = *Maximum pressure,* K = *Constant,* P_v = Saturated vapor pressure, S = Surface tension

Approximately, 10^{10} bar pressure is generated during cavity implosion [48].

3.2.5.2 Mechanical Effects

(a) *Increase in the surface area of the phase transition boundary:* The shockwaves during cavity implosion lead to a significant extension of the phase transition boundary surface, that alters the chemical composition, accelerates chemical reactions and mass transfer.

(b) *Production of tension and interfacial turbulence:* Based on reflection principle, a detonation water shock wave goes down the orifice-tube and get reflected at the water interface, leading to a strong compression wave generation. By the support of diaphragm in the orifice these waves stretch to the bottom of the column, which get reflected and travelled back upwards as a wave of tension. This interfacial tension is very sensitive to the concentration of solutes present in the water. These compressional waves grounds irregular flow of the liquid in the column, resulting in interfacial turbulence [49].

(c) *Creation of self-excited oscillations:* Due to irradiation of intensive periodic pressure wave, cavitation bubbles are forced to oscillate. The self-excited oscillations vapor-bubbles radiate pressure waves consist of steep, short and high pressure pulses into the surrounding liquid [50].

(d) *Occurrence of cumulative liquid micro-jets and formation of intensive shock waves:* Few cavities implosions take place near the surfaces of orifice and drives a shock-induced collapse These collapse at different locations can affect in production of re-entrant liquid micro-jets with a high water-hammer pressure of about 450 MP and velocity around 120m/s along with microjects generation; it also manifests shock waves production with velocity propagation of 2000m/s. in the direction of microjets occurrence. Basically, two types of shockwaves are produced; primary shockwaves, that are produced when the molecular speed is up to 1700 m/s and the pressure amplitude is about 240 Mpa, whereas secondary shockwaves are produced when the molecular speed is around 1800 m/s and the pressure is up to 70 GPa [51].

(e) *Develop noises:* Cavitation noise is exhibited under the assumption that bubble oscillations control some parameters and hence radiate pressure waves, which are termed as noise. Cavitation noise can be used to evaluate some information on physical processes that are occurring during cavitation. Cavitation noise can be created over a frequency ranging from 100 Hz to 100 kH. The mathematical model of cavitation noise $p(t)$ can be written as:

$$p(t) = \Sigma P_k \ \exp(-(t - kT - \varphi_k)/\theta_k) \tag{3.18}$$

Where, P_k = Peak pressure of the pulse in the Kth period,
φ_k = random distance of the kth pulse from a reference point in the kth period,
θ_k = Random time constant of the Kth pulse [50].

(f) *Cavitation corrosion:* Cavitation erosions and cavitation corrosions are other consequences that generally occur when high-speed fluid strikes the metal surface of orifice during the violent implosions of cavities. The occurrence of cavitation erosion is more favors at lower frequencies. These two damages can disturb the flow rate by pits, cracks, leading-edge tear or bends formation on orifice surface [51].

(g) *Sonoluminiscence and sonochemical processes:* Sonoluminiscence and sonochemical conditions occurred during cavitation bubble implosion. This process can dramatically increase liquid-solid reaction rates of reactive metals of mixed-phase reactions to create nanostructured inorganic materials which can remediate unclean water and light emission is observed depending on the cavity implosion intensity [52].

3.2.5.3 Chemical Effects

There are mainly three locations inside a cavity where chemical reactions can be seen [32]. First at cavity interior site, where temperature and pressure is around 500 K and 500 atm respectively. Second effects can be seen at interfacial region, where temperature is around 2000 K and third, at bulk phase, where temperature is around 300K [48].

(a) *Break H_2O and generation of •H and •OH radicals:* Due to powerful collapse of cavitational bubbles, some hot spots are created in the orifice where water vapor gets dissociated [53]. This dissociation can result in breakdown of van der Waals bonding and ruptures the water molecules by heterolytic cleavage and leads to generation of chemical products such as •OH, •O, •H, as well as H_2O_2 formation by •OH recombination reaction with a standard oxidation potential of 2.8 V [53].

(b) *Chemical reactions:* Violent collapse of vapor bubbles causes hot spots formation where water molecules split by heterolytic cleavage and emerge •OH radicals. These •OH radicals are oxidizing agent and acts as an excellent initiators of chain reactions react with pollutant molecules and convert into simple inorganic or organic compound with lower toxicity in a short time. The substrate inside the cavitation bubble either reacts with the •OH or go through pyrolysis [53].

3.2.5.4 *Biological Effects*

(a) *Microbial cell death:* High shear manifested during the cavity collapse can physically rupture the microbial membrane and can break the outer shell of microbes. This disruption of outer membrane in microbes leads to release of cytoplasmic and periplasmic matter from the cell and therefore cause disinfection [54, 55].

3.3 Hydrodynamic Cavitation Reactors

Hydrodynamic cavitation reactors are less complex to plan and furthermore for the scale-up when contrasted with their acoustic partners. The different lab scale/pilot plant scale reactors which have been utilized for the experiments for studying uses of cavitation applications have been examined with both merits and demerits. A few points of interest of hydrodynamic cavitation reactors can be given as: a) uniformity during the cavitation process is because of the efficient mixing; consequently there is no issue of directional sensitivity, b) cavitation is at the shear layer of the mass fluid and consequently metal erosion issues are less serious, c) quantification and command over the design boundaries is not very complicated, prompting development of the required cavitating conditions [56].

3.3.1 Liquid Whistle Reactors

One of the first reactors to depict principle of hydrodynamic cavitation was the Liquid Whistle Reactor [24]. The use of this reactor is mixing/homogenization. It works on the principle of vibration; liquid is passed over vibrating metal plates with high velocity. The liquid flow couples itself with the vibration producing cavitation bubbles; this method also results in efficient mixing. The major merits of the reactor are: (a) the flow systems used in the reactor for operation makes it flexible to be used as a mixing device; (b) the maintenance cost of the device is reduced as there are no moving parts [57]. Though the device is suited for efficient homogenization or mixing material, these suffered from few shortcomings like low vibration power, erosion of blade in the presence of particular substances, high pumping costs [58]. Not many reports have been published on liquid whistle reactors as the intensity of the cavitation produced in these types of reactors is very low.

3.3.2 High-Speed Homogenizers

High-speed homogenizers showed that cavitation can be generated using rotating equipments. In a rotating device the local pressure at the bottom

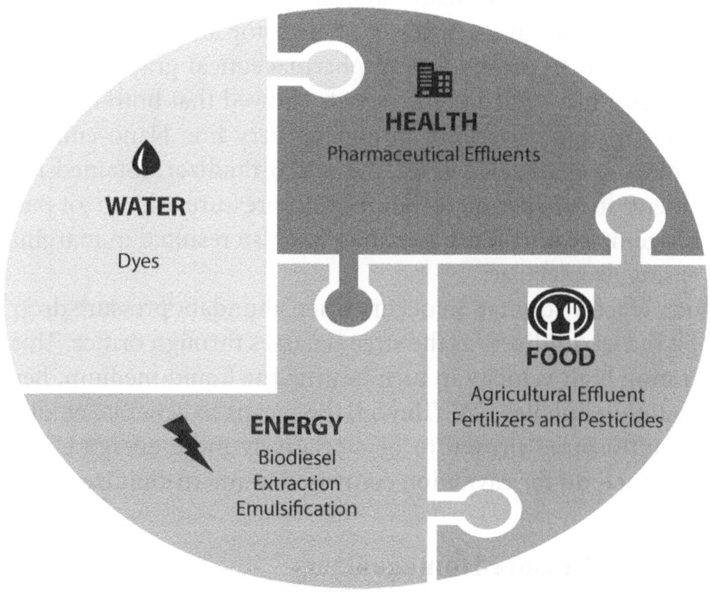

Figure 3.6 HC-based technologies in Water-Energy-Food-Health Nexus.

edge of impeller falls down closely till vapor pressure when the rotating device reaches critical speed producing cavitation bubbles. This principle can be attributed effectively by the bubbles occurring on the propellers of ship and erosion of the blades of the propeller or shaft. In a closed system, the liquid moves towards tank boundaries away from impeller recovering the local pressure with expense of velocity head, this resulting in implosion of cavity bubbles [59]. A typical high-speed homogenizer consists of an impeller and a stator made of stainless steel and the impeller is driven by a voltage motor. The speed of rotation of the impeller can be changed changing the applied voltage on the power supplier [60]. Though the intensity of the cavitation is high in a high-speed homogenizer, it is noted that the energy consumption of a high-speed homogenizer reactor is very high and also has a major problem of erosion of impeller. Such reactors have less flexibility over design parameter when compared to orifice plate's setup [61].

3.3.3 Micro-Fluidizers

Micro-fluidizers have led to numerous applications in physical process of emulsification as high jet fluidizers are capable of generating pressure drops as high as 2kbar and jet velocities upto 200 m/sec almost [59]. Sian Ying Tang and Parathasarthy Shridharan (2013) [62] have worked on emulsification of Aspirin in to nano-emulsions using ultrasound and micro-fluidizer. The operation efficiency of a bench-top air driven micro-fluidizer has been studied in production of pharmaceutical grade nano-emulsion of Aspirin as a model drug. The reports showed that both the emulsification methods were capable of producing very fine Nano-emulsion with minimum droplet size. However, the micro-fluidizer obtained result was independent of applied micro-fluidization pressure and no. of passes, the pre-homogenization and drug loading has also resulted in marginal effect of increasing droplet size.

In such reactors cavitation occurs when abundant pressure drop is generated by the jet fluidizers as the stream passes through orifice. This results in generating high velocity in turn heating the liquid medium, hence creating cavitation bubbles [24]. Though the means to generate emulsions are different, ultrasound proved to be a relatively more energy efficient and promising method for cavitation compared to micro-fluidizers.

3.3.4 High-Pressure Homogenizers

A high-pressure homogenizer works on the principle of high-pressure relief technique. A high-pressure homogenizer reactor can be termed as a high-pressure positive displacement pump with a throttling device. A basic high-pressure homogenizer consists of a feed tank and double throttling valves each designed for two different stages, respectively. The pump installed in the set-up is used to drive the feed from feed tank to the first throttling device and pressure up to 70atm can be achieved by the throttling device. The stream is circulated back to the feed tank through a second throttling device; this leads to increase in pressure values and increase in temperature. A critical discharge pressure is developed at which cavitation events occur and cavitational yield is obtained beyond this pressure [59]. The cavitational effect led due to discharge pressure is dependent completely on geometry of the throttling valve and feed used for the application. Though the high-pressure homogenizer reactor produces cavitation, the required cavitational intensities are not achieved as there is not enough control over the feed tank volume and the pressure generated at the end of cavitational occurrence [63]. This reactor can be used for emulsification process in pharmaceutical and biochemical industries [64, 65] No report has been published on this type of reactor carrying out chemical processes except for modeling studies.

3.3.5 Orifice Plates Setup

The orifice plate set-up reactors are studied extensively and many reports have been reported previously. In such reactors a constraint or an orifice is fitted to the flow pipe. When the stream passes through the constraint the velocity of the flow is suddenly raised to a point where the pressure is reduced to below vapor pressure resulting into cavity bubbles [66]. The constraint or an orifice consists of a single hole or multiple holes. A basic set-up of the orifice plates reactor consists of a closed loop with feed tank; a centrifugal pump is used to pump the feed to reactor, control valves in either side of the cavitating device, flanges to fit the orifice. The feed discharged from feed tank using the centrifugal pump is branched into two streams, these streams control the inlet pressure of the stream flowing towards the orifice. The branched streams are brought to the feed tank and recirculate again. The volume of the feed tank should be maintained in order to avoid any induction of air into pump and the flow pipes. The inlet pressure and the discharge pressure can be measured using pressure gauges fitted on either side of the flanges which is used to accommodate the orifice plate. The setup described above has tremendous flexibility over operating conditions, i.e., control over inlet pressure, temperature, flow rate and geometric condition, i.e., arrangement of orifice and number of holes in orifice plate [59]. A multiple holes orifice plate having different arrangement of holes and diameter of holes helps in attainting numerous cavitational intensities of a single orifice plate. It also helps in increasing the number of cavitational events generated [67]. The cavitational effect produced by the orifice plates is mainly studied from the following parameters, i.e., (i) effect of the ratio of the total perimeter to the total flow rate (α); (ii) effect of the ratio of the total flow area to the cross-sectional area of the pipe.

3.3.5.1 Effect of the Ratio of Total Perimeter to Total Flow Area

The effect of total perimeter and total flow area is an important parameter as these parameters quantify the effect of the orifice plate geometry. The total perimeter of the hole defines the area of shear layer or the layer of the fluid flow. The turbulence caused by the shear layer and the area of the layer of fluid flow are important factors in deciding the cavitational yield. The increase in the total perimeter of the orifice is directly proportional to cavitational yield. For a given operating time, the number of passes through the fluid is determined by the flow area. The flow area is the flow rate of the fluid in the given operating time. The ratio of the total perimeter of the orifice and the total flow area of the hole is denoted by "α". The ratio of the total perimeter to the flow area defines the effect of the orifice geometry on the cavitational events resulting in a better analysis of cavitational yield. The ratio α depends upon size, shape and number of holes of the orifice plate [67].

3.3.5.2 Effect of Flow Area to the Cross-Sectional Area of the Pipe

The ratio of the total flow area to the cross-sectional area of the pipe is denoted by "β". The value of β depends upon the number of holes on the orifice plate and the diameter of each hole. Since the diameter of the orifice holes is fixed β depends defines the area of the flow of the orifice. So the change in β value changes the number of rotations and the orifice flux. The bubble life is affected by β value as β is related to turbulence flow of the fluid [68]. The researchers studied the design parameters of any orifice plate by focusing on α and β parameters. These parameters depend upon number of holes and the flow area of the orifice plate.

3.3.6 Venture Device Setup

A venture type HC reactor mainly consists of a diverging section, a converging section and a throat. The converging section and diverging section of a venture generates higher velocity at the throat, when compared to an orifice plate HC reactor, for a given pressure drop and also results in a low cavitational number [26]. The bubbles or cavities are given sufficient time to reach their maximum size before collapsing in the diverging section. Similar to orifice when the fluid enters the converging section the velocity of the fluid is increased, resulting in generating low pressure. The pressure is reduced below the vapor pressure of the fluid resulting in cavitational activity. The cavities produced gradually come out of the throat in divergent section recovering the pressure and collapsing. The collapse of the cavity results in pressure drop which influences the cavitational activity [59]. The basic set-up of a venture-type HC reactor is similar to that of orifice-type HC reactor. It is a closed circuit with a feed tank, a centrifugal pump, two valves, two pressure gauges, a Rotameter, and cavitating device. The feed tank is covered with a coolant jacket to maintain the ambient temperature of the feed. The flow leaving the feed tank is brought back to the feed tank using the valve before the cavitating device a bypass flow is created to adjust the inlet pressure of the fluid. The pressure gauge before the cavitating device measures the inlet pressure and the pressure gauge after the cavitating device measures the discharge pressure. The geometry of the venture is classified into three different types: (i) slit venture; (ii) circular venture; (iii) elliptical venture. The study of the geometry is essential as it affects the cavitational activity. The effect of optimizing the venture geometry depends on two parameters: (i) effect of divergence angle; (ii) effect of ratio of the diameter of the throat section to throat length; (iii) effect of the ratio of perimeter of the throat to its open area (α).

3.3.6.1 Effect of Divergence Angle

For a venture the number of cavities produced depends on the convergent section. The size and shape of the convergent section decides the number of cavities formed and their dynamic behavior [30]. The cavitation effect depends on the number of cavities generated and the intensity of the collapse of the cavities; the collapse of the cavities depends on the divergent section of the venture. The shape of the divergent section decides the life of the bubble or cavity. The divergence angle of the section decides the pressure recovery rate. A greater divergence angle results in faster pressure recovery rate and rapid collapse of the bubble. The pressure recovery rate increases with the increase in the shape of the divergent section. For a smaller or an optimal divergent section the pressure recovery is slow and the bubble tends to reach its maximum shapes before collapsing resulting in generating greater cavitational effect [27].

3.3.6.2 *Effect of the Ratio of Throat Diameter/Height to Length*

The effect of the ratio of the throat diameter/heights to length decides the maximum size of the cavity before the collapse. The length of the throat section decides the maximum time the cavity stays in low pressure zone. A higher number of cavities will be produced when the throat length of the venture is bigger [56].

3.3.7 Vortex-Based HC Reactor

A vortex-based HC reactor basically generates vortex and produces a cavitational effect. Swirling jet cavitating device and vortex diode are two types of vortex-based HC reactor. The vortex-based reactors do not have any moving parts thus the maintenance cost of the vortex reactor is very low. The vortex HC reactor mainly consists of three components; inlet tangential point, vortex diode chamber, and an axial port. The fluid is passed into the vortex of the chamber, which produced a cavitational effect, from tangential point and low pressure is generated in the vortex chamber where the cavities are produced [31].

3.4 Effect of Operating Parameters of HC

3.4.1 Effect of Inlet Pressure

The most important operating parameter which determines the bubble formation or cavitational event is inlet pressure. The bubble collapse or the implosion of cavity is mainly dependent on the change rate of pressure generated in downstream flow of the cavitating device. The pressure changes increase when there is increase in upstream pressure of the cavitating device leading to increase in downstream pressure [69]. The increase in upstream pressure also results in increase of liquid flow rate, high power input per unit mass of liquid and high-pressure drop is generated across the cavitating device due to high energy dissipation rate. The number of cavities produced reduces with increase in the velocity and the liquid flow rate. The increase in inlet pressure increases the pressure generated after the final collapse of cavity [20]. The increase in collapse pressure is assigned to turbulence pressure and energy dissipation rate for high inlet pressure; this caused the cavities to collapse with greater intensity. Although the C_v values decrease with the increase in inlet pressure, increase in number of cavities formed and collapsed for given time also decreases the C_v values. The change in cavitational yield due to inlet pressure depends upon the number of cavities formed and intensity of cavity collapse. Also, the number of liquid passes at the same given time increases with increase in inlet pressure resulting in increase in residence time of pollutant in cavitation zone [28]. However, it is very important to maintain optimal inlet pressure to avoid supercavitation, which is undesirable. The optimal pressure attained results in greater extent of degradation of pollutants, once the optimal pressure is surpassed the effect of hydrodynamic cavitation is reduced gradually resulting in less degradation of pollutants [70]. The optimal inlet pressure results in a greater degree of dissociation in turn increasing the number of hydroxyl molecules. Once the optimal inlet pressure is crossed the cavity bubbles combined to form an indiscriminant sized cavity which tends to vaporize the flow downstream the cavitating device [56]. Many reports have been published confirming the influence of optimal inlet pressure for better hydrodynamic cavitation.

3.4.2 Effect of Temperature

The cavitation yield of the hydrodynamic cavitation (HC) is affected by temperature. The effect of temperature plays a crucial role in determining the removal efficiency or percentage degradation. The effect of temperature on the pollutants mainly depends on the boiling point and vapor pressure of pollutants and the media. The formation of the cavities increases with increase in temperature as at higher temperature the formation of vaporous cavities increases [71]. Also, higher temperatures dissolve gas into the solution, reducing the formation of cavity nucleus for formation of cavitation. Meanwhile, at very low temperature the transformation rate into vaporous cavities would reduce, which results in less degradation efficiency [68]. However, at very high temperatures (above optimal temperature) the constrain or the cavitating device is filled with water vapors cushioning the implosion of bubbles. In addition, the effect of temperature also influences the density, surface tension and viscosity which may hinder cavitational events and cavitational yields. At high temperature the apparatus consumes higher thermal energy and increases the operational costs [20]. Many reports have been published, reporting the effect of optimum temperature for better degradation of pollutants. However, a few reports suggest that temperature has no effect on cavitation process. Therefore, the effect of temperature on HC depends completely on pollutants and also should be studied depending on the boiling point of water and the pollutants [72].

3.4.3 Effect of Initial Concentration of Pollutant

The percentage degradation of pollutant is varied by varying the initial concentration of pollutants. The effect of initial concentration of the pollutants is inversely proportional to the degradation efficiency. The hydroxyl ions produced during HC sustain under specific condition, with high initial concentration the number hydroxyl ion (•OH) may not be sufficient to degrade the complete concentration of pollutant resulting in less removal [73]. The industrial effluent contains more pollutants or has higher COD value, thus, to obtain higher efficiency dilution is needed. However, by diluting the liquid total quantity of effluent to be treated will increase, also the apparatus will consume more energy as the total volume increases so it is necessary to estimate optimum initial concentration with perfect dilution ratio to achieve maximum removal efficiency [74]. Many reports have been published reporting that the lower the initial concentration, higher is the degradation percentage. However, a few reports suggest that maintaining optimum initial concentration results in higher degradation of pollutants. At the end the effect of initial concentration depends upon the properties of the pollutants and process requirement [75].

3.4.4 Effect of Treatment Time

The liquid flow rate is determined by the treatment time through the cavitation device when the flow rate is constant. As the number of passes increases, the more pollutant molecules are revealed to the cavitational zone. Therefore, the degree of pollutant removal is directly proportional to the treatment time as reported by many reports [73]. These studies have shown that by regulating the treatment time, the residence time of the liquid inside the cavitating device can be altered [7, 76], although different results may be observed if there is a change in liquid flow rate as the outcome is influenced by both the pass number and the liquid flow rate. It has also been studied that the incomplete or early collapse of the cavities is caused due to the higher static pressure which again is caused by the decrease in flow rate as the complete cavitating device is filled with water [77]. The extent of collapse can be reduced if the outlet valve is completely open as the liquid is perhaps exposed to extreme cavitation circumstances. This is close to supercavitation. It should be kept in mind that more energy will be exhausted if the water treatment time is increased to achieve higher efficiency of pollutant removal. Hence, the optimization of treatment time should be dependent upon the amalgamation of removal rate and energy efficiency to achieve maximum utilization of resources [75].

3.4.5 Effect of pH

The effect of pH has a great influence on the degradation of pollutants. The pH of solution determines whether the pollutant is hydrophobic or hydrophilic in nature since it affects the pollutants [78]. Mostly the effluent exists in molecular state or ionic state; if the effluent exists in molecular state, then it is mostly concentrated at the cavity and water interface and is attacked by •OH radical. The concentration of •OH radicals is often more at cavity and water interface. The effluent when in molecular state is often hydrophobic in nature. And if the effluent is in ionic state then the effluent is native to be hydrophilic in nature. When the effluent is in ionic state it is mostly substituted into bulk solution, where the % of •OH radicals is very low because only 10-15% radicals generated are diffused into bulk solution, the rest recombine into H_2O_2 due to short half-life [79]. Many reports observed that at acidic conditions the formation of •OH radicals is greater and also the recombination of the generated •OH radicals reduces and also offers high oxidation potential. To get efficient degradation acidic medium is very conducive; however, the effect of pH, in acidic medium, on degradation of pollutants is not the same always [80]. A few reports indicated that the effect of pH is dependent on the characteristics of the pollutant. If the pollutant exists in molecular state and is hydrophobic in nature alkaline conditions are highly favorable for the degradation process. The % degradation reduces in acidic medium for hydrophobic pollutants as the generated •OH radicals combine with hydroxide ions in the solution, also protons and the •OH radicals recombine when the pH of the solution decreases. Thus •OH radicals, which are string oxidizing agents for cavitation process reduce, resulting in less degradation of the pollutants [71]. Overall, the effect of pH on HC totally depends on the pollutant and its characteristics. Whether acidic medium or basic medium is favorable for the process has to be determined by the pollutant and the degradation pathways.

3.5 Toxicity Assessment

This section discusses toxicity assessment in the ecosystem by evaluating toxic properties of a substance. Organic compounds from commercial processing and industries play a substantial source of toxicants [81]. These organic micro-pollutants can often include a variety of raw materials and their by-products such as antibiotics, dyes, pesticides, halogenated organic pollutants, drugs, chemicals, polycyclic aromatic hydrocarbons, etc. Monitoring the usable water and its sources is a crucial step to prevent pollution and to supply safe water. Wastewater is a highly complex matrix, consisting of lots of compounds, which need to be identified and detected in matrices [60]. Even discharge from industries needs to be detected before release in the environment. To detect toxicity on aquatic organisms, various effective methods are applied to measure wastewater's effects on the survival, growing and reproducing ability of living aquatic organisms. By toxicity assessment, we can get information about how much of an element can cause what kind of harm in the environment [82]. Apart from direct ingestion of contaminated drinking water, humans can also get affected from these toxicants through the food chain. This also includes a study of how the toxicity is responding to another subgroup population. Its effect can last for a short term (acute effects) or over a long term (chronic effects). Acute toxic effects are observed after a solo exposure, such as redness, swelling, sweating, nausea, itching, paralysis, and even death [83, 84], whereas, chronic toxic effects are observed after frequent longtime exposure and can be categorised into two carcinogenic effects, such as cancer and non-carcinogenic effects such as organ damage, nervous breakdown, and reproductive difficulties [82, 85]. Toxicity assessment is based on two steps [86]:

(a) *Hazard identification:* In this step, process hazard analysis (PHA) techniques like fault tree analysis, failure mode analysis, and safety audits are applied to determine poisonous substances present in the environment.
(b) *Dose-response assessment:* The judgment of dose and its response is based on the equations [86]:

$$Response = 1 - \exp(-k * dose) \tag{3.19}$$

Where, k = identical probability.

For identification of hazards that affect the environment, data are collected from animal studies, epidemiologic studies, and by experimental work [87, 88], while for dose-response assessment, detailed effects of each toxic compounds on the quantitative relationship is determined flanked by toxic response and dose. While performing toxicity assessments, there are two widely used models as proposed by various researchers from this domain: non-threshold and threshold model. Substituting the animal studies in these model gives a pre-knowledge about the effects of low doses on the human condition. For toxicity assessment, a non-threshold model is used for single exposure and a threshold model is used to repeated exposures to a chemical. For toxicity evolution, it is important to understand the quantitative relationship between chemical dose and its extent of toxic response by proper designing of chemical exposure pathway and duration. Toxicity experiment of a chemical is performed on animals before epidemiologic experiment to become more assured in characterization of the toxicity of an element. Animals and animal tissues are used by pathologists and toxicologists to determine the full range of toxic effects of a chemical. LD_{50}, Lethal Dose for 50% of the animals, N_{50} is a popular toxicity assessment test carried out by scientists on animals to assess acute toxicity in them [89]. Toxicity studies are based on a particular chemical's dose-response on different species of animals following the equation below [90]:

$$N_{50} = \beta * \left[2^{\frac{1}{\alpha}} - 1 \right] \tag{3.20}$$

Where, β and α are pathogen survival probabilities

Certainly, in many countries, laws and acts are implemented and enforced for practicing toxicity test on animals before marketing such as the Cruel Treatment of Cattle Act 1822 in the UK [91], the Animal Drug Availability Act 1996 1996 (ADAA) in the United States [92], the Cruelty to Animals Act 1960 and Animal Welfare Act, 2011 in India [93]. Toxicity can be categorized in specific forms based on their cause and evolutional conditions such as neurotoxicity, carcinogenicity, etc. Scientific experimentation helps to determine the various nature of exposure such as "no-observed-effect level" (NOEL), "no-observed-adverse-effect level" (NOAEL), "lowest observable effect level" (LOEL), "lowest-observed-adverse-effect level" (LOAEL) [86]. Also, different living beings having different rates and degrees of absorption, metabolism, and excretion, so toxicology assessment provides information about type and degree of harm caused by different quantity of elements on different species [81]. Discussed below are a few works highlighting the importance of toxicity assessment using hydrodynamic cavitation [86].

Y.N. Liu *et al.* [90] represented a plasmonic nanomechanical sensing (PNMS) system for detection of a variety of toxic substances in water by sensing the amplitude of nano-mechanical movement of bacterial cells. The physical and chemical characteristics are also observed during the toxicity assessment. The result of X.Yi *et al.* [94] showed that, temperature, pH, and hardness of water has no major correlation with luminescence inhibition rate at the 43 sampling sites. Even animals are also tested to assume human toxicity assessment. G.A. Van Norman [87] stated that animal testing is insufficient to predict human safety and toxicity for therapeutics, whereas H. Olson *et al.* [89] reported that animal studies have 71% potential to identify human toxicity. Perhaps the most challenging consideration is the presence of toxicants even after raw wastewater treatment. HC provides high temperature and pressure which promote suitable conditions for detox process [95]. J. Carpenter *et al.* [13] stated that HC combined with biological processes can remove the toxicity of water. To predict that toxicity of a water source, fishes, invertebrates, algae, and natural bacteria are used as an indicator to detect the toxic effects of compounds [94]. M.Farré *et al.* [96] highlights different bioassays and biosensors based on acute toxicity measurements. P Samantha *et al.* [84] demonstrated that different levels of toxic sensitivity of *Euphlyctiscyanophlyctis* was observed, under different concentration of two heavy metals, arsenic (As^{3+}) and chromium (Cr^{6+})] and an herbicide (Almix). Presently scientists are involved in drug designing by applying hERG cardiac potassium channel to avoid over or minor exposures [97]. During drug development, a pharmacological bases toxicity assessment is performed for better understanding the mechanisms underlying safety and toxicity [83]. G.C. Kuriakose *et al.* [98] has performed a sub-acute toxicity assessment from Fusarium solani, an endophytic Fungus of Taxus in wistar rats and analyzed its apoptotic potential and cytotoxicity in lung cancer cells. Various innovative toxicity tests are inventing. X. Miao *et al.* [88] shows that the ethanol extract *of Peganum harmala L.* seeds exerts toxic effects on *C. elegans*, and provide innovative concepts for the biological evaluation of the toxicity of Chinese medicinal materials. Even marine plants are tested for toxicity and standard test protocols have been developed, identifying species with low salinity and toxicant interaction [99].

3.6 Techno-Economic Feasibility

Techno-economic feasibility is another crucial study to understand the implementation of any technology at large scale. It refers to the estimation of potential of any project, choice of optimum technology, and analysis of existing market [100]. It analyses different aspects and criteria and sets the stage for detailed design development. In this regard, this section will highlight the different techno-economic feasibility approach carried out previously for potential implementation of HC technology at industrial scale [101]. X. Sun *et al.* [10] reported that HC is biased towards industrial scale more than lab-scale experimental setup as it can fuse into a continuous flow process. M.G. De Giorgi *et al.* [48] performed an experiment by using low-cost cryogenic fluid as working fluid for the nitrogen cavitating flow in an orifice, so that less temperature was applied from outside during the process. K. Li *et al.* [51] HC improved rheological effect in the process, which significantly eliminates the energy cost operating cost during the test [51, 102]. M.B. Mane *et al.* [55] stated that exploring use of natural oils during HC for disinfection had positively impacted in reducing the cost of operation consequently to reduce the time of operation. The work I. Gonclaves *et al.* [53] developed a novel and environmental-friendly technology using HC reactor for peroxide cotton bleaching with reduced processing costs. The experiment reduced 40°C in terms of energy consumption with reduction of 90 min operation time. Due to its low usage of electricity, bigger bubble size and high cloud density, the effective reaction volume of HC is high [54]. According to M Gagol *et al.* [54], the cost of disinfection of 1m^3 of water by HC is 3.7 USD, which is approximately ten times lower than AC. Self-resonating cavitating jet (SRCJ) technology, promotes strong cavitation inception in situ and emits enormous pressure pulse and temperature peak, which having better energy, cost efficiency with easy scale up of HC reactors [34, 37, 103–105]. Most often, HC aided by AOPs is utilized to improve the effectiveness and reduce the cost of the process. A. Mukherjee *et al.* [101] has studied that the energy of degradation was decreased noticeably from 8.81 kwh/m^3 with respect to HC alone process to 1.76 kwh/m^3 with respect to HC with 5 g/L H_2O_2. Maximum of 18.5*10^4 mg/J of cavitation yield was obtained in comparison with other hybridized process united with HC. 0.59 \$/m^3 was obtained for 95% SDS degradation by HC alone method whereas 6.96 \$/m^3 was spent by HC combined with fenton technology. A. Mukherjee *et al.* [101] concluded that HC process with H_2O_2 (5 g/L) was one of the most economically feasible combined techniques. R.A. Torres *et al.* [106] performed an experiment where electric energy cost was calculated as 1033 KWh/m^3, whereas P. Thanekar *et al.* [107] performed a test where energy required for degradation of NAP was 0.056 KWh, also had operational cost of 0.21 Rs/L. Similarly, combining HC with other AOP techniques, disinfection cost of water was USD per 1m^3 of water. M.V. Bagal *et al.* [108] suggested that combined HC-Fenton is more cost effective as compared to individual treatment cost of water.

3.7 Applications

Conventional technologies find certain limitations such as replacement and disposal issues of reject-stream in adsorption [109, 110] and membrane-based processes [111–114], high maintenance cost in chemical-based process [115, 116] and energy intensive, time consuming in biological-based processes [117–120]. The minute presence of micropollutants such as endocrine disrupting compounds even after conventional treatment is another potential challenge that is nowadays acknowledged as an emerging environmental problem [121]. Thus, this section describes a review of recent applications where HC-based treatments have been applied. Figure 3.6 shows a schematic of state-of-the-art application with HC-based technologies.

3.8 Conclusions and Thoughts About the Future

With the recent developments in technological advancement, HC has been widely proven to be a sustainable and effective technology for the treatment of wastewater streams. The current chapter majorly spotlights the current HC scenario both from the techno-ecological and techno-economic point of view and its feasibility for industrial-scale applications. Herein, the authors have primarily discussed the general perspectives of cavitation, followed by the types of cavitation and the science behind the growth and collapse of the cavitating bubbles. Furthermore, the authors have attempted to provide an insight into the different HC-based reactors and their influencing parameters. The detailed review indicated that optimization of process parameters and selection of proper cavitating device based on the application is an important step for effective HC performance. The study also indicated that the future of hydrodynamic cavitation in wastewater treatment depends on the design of energy-efficient and economic reactors. However, for transferring the fundamental knowledge of HC into practice efforts are needed in terms of theoretical studies where realistic models (and not the assumption of single spherical bubble) pertaining to the bubble-liquid interface need to be carried out for real-life effluent streams, simulation of the degradation process on the basis of chemical reaction kinetics. Moreover, to have an effective degradation and control over the HC process, a generalized correlation relating hydroxyl radical generation with respect to the operating parameters must be developed, and CFD simulations of various flow fields, turbulence characteristics, and geometrical parameters should be carried out for designing efficient HC reactors. Although numerous studies on HC and efficient design of HC reactors have been carried out in laboratory scale to prove the establishment of the technology, still, it remains more of an art than a science. In a nutshell, it can be concluded that positive efforts towards transformation of HC technology from lab scale to commercialization can definitely provide a green route towards a sustainable source of water management.

3.9 Acknowledgement

The authors would like to thank the Director, BITS Pilani Goa Campus and HOD, Department of Chemical Engineering, BITS Pilani Goa Campus, for the continuous support and motivation. The authors would also like to express sincere gratitude towards Director CSIR-IICT [Ref. IICT/Pubs./2022/109] for the support, institutional infrastructure and encouragement for writing this chapter. The authors would further like to thank the industrial startups Epione Swajal Solutions LLP and Prasinos Tech Innovations India LLP for taking an initiative in the development of this technology.

3.10 Disclosure

Mr. Anupam Mukherjee would like to thank the BRIC Idea Exposition Grant [Ref. BIRAC Innovation Centre/IKP Knowledge Park/GIM/Goa Cluster dt. 23.02.2020] for carrying out the work. Dr. Siddhartha Moulik would also like to thank DST-WTI Grant [Ref.DST/TMD/EWO/WTI/2K19/EWFH/2019/143] for carrying out the work. Dr. Anirban Roy would further thank the OPERA grant and Additional Competitive Grant [Ref. GOA/ACG/2018–2019/Oct/08 dt. 30.10.2018] by BITS Pilani Goa for carrying out the work.

Nomenclature

1. Roman Symbols

A = Area around the region of orifice (m^2)
C_{wi} = Gas concentration inside ambient fluid(m^3/mol)
C_l = Gas concentration inside cavity(m^3/mol)
C_{va} = Specific heat of water vapor (Nm)
C_{nc} = Specific heat of non-condensable gas (Nm)
D = *Diffusioncoefficient* (m^2/s)
e = Emissivity (0.95)
G_v = Evaporative mass flux (kg/s)
h = elevation (m)

K_v = Gas constent of the vapor$\left(\frac{J}{k*mol}\right)$
K = *Constant*
l_d = Diffusion length (m)
L_{cav} = Length of cavitation regimes (m)
N = *Quantityofgas*
N_{va} = Quantities of water vapor$\left(\frac{kg}{m^3}\right)$

N_{nc} = Quantities of non-condensable gas$\left(\frac{kg}{m^3}\right)$
$pB(t)$ = homogeneous bubbles and uniform pressure (Pa)
P_{GO} = Partial pressure of contaminant gas inside the bubble (Pa)
p_{ref} = Pressure taken at a given point in the fluid flow(Pa)
p_v = *Standard* vapor pressure of the fluid(Pa)
P = pressure of fluid(Pa)
P_B = Interior pressure(Pa)
P = Reference pressure(Pa)
P^*_∞ = Surface constant reference pressure (Pa)
P_{GO} = Partial pressure of contaminant gas inside the bubble(Pa)
P_{max} = *Maximum pressure*(Pa)
P_v = Saturated vapor pressure(Pa)
R = Exterior bubble radius (m)
R_g = Gas constant$\left(\frac{J}{k*mol}\right)$
R_o = Original nuclei size (m)
S = Surface tension$\frac{J}{m^3}$
T_{LS} = Temperature of the saturated vapor in the bubble (K)
$TB(t)$ = homogeneous bubbles and uniform temperature (K)
T_∞ = Liquid temperature (K)
T = Temperature of heat transfer mode between the bubble and the outside flow field (K)
T_B = Bubble temperature (K)
v = velocity of fluid $\left(\frac{m}{s}\right)$
v = Velocity around the region of orifice$\left(\frac{m}{s}\right)$

2. *Greek Symbols*

α = pathogen survival probabilities
β = pathogen survival probabilities

$\kappa = \text{Thermal conductivity}\left(\kappa = 0.6\frac{\text{W}}{\text{m} * \text{K}}\right)$

Λ	= Accommodation coefficient
k	= identical probability
μ_L	= dynamic viscosity $\left(\frac{N\,s}{m^2}\right)$
ρ_L	= Liquid density $\left(\frac{kg}{m^3}\right)$
ρ	= density of fluid $\frac{kg}{m^3}\Big)$
σ	= *Stefan-Boltzmannconstatnt* (5.67 x 10^{-8} W/(m^2 * K^4)

3. *Subscripts*

B	= bubble
C_{av}	= cavitation regimes
D	= diffusion
G	= gas
i	= inside
L	= liquid
L_s	= saturated vapor in the bubble
Max	= maximum
N_c	= non-condensable
O	= original
out	= outlet
V	= vapor

References

1. E. Weitendorf, On the History of Propeller Cavitation and Cavitation Tunnels, *Proceedings of the Fourth International Symposium on Cavitation*. (2001).
2. A. Alhelfi, B. Sunden, The cavitation phenomenon: A literature survey, in: *WIT Trans. Eng. Sci.*, 2014. https://doi.org/10.2495/HT140311.

3. S. Li, Tiny bubbles challenge giant turbines: Three gorges puzzle, *Interface Focus.* (2015). https://doi.org/10.1098/rsfs.2015.0020.
4. S. Li, C.E. Brennen, Y. Matsumoto, Introduction for amazing (cavitation) bubbles, *Interface Focus.* (2015). https://doi.org/10.1098/rsfs.2015.0059.
5. R.B. LINDSAY, On the Pressure Developed in a Liquid During the Collapse of a Spherical Cavity (1917), in: *Men of Physics Lord Rayleigh–The Man His Work*, 1970. https://doi.org/10.1016/b978-0-08-006821-3.50028-3.
6. M. Versluis, B. Schmitz, A. Von der Heydt, D. Lohse, How snapping shrimp snap: Through cavitating bubbles, *Science* (80-.). (2000). https://doi.org/10.1126/science.289.5487.2114.
7. M. Badve, P. Gogate, A. Pandit, L. Csoka, Hydrodynamic cavitation as a novel approach for wastewater treatment in wood finishing industry, *Sep. Purif. Technol.* (2013). https://doi.org/10.1016/j.seppur.2012.12.029.
8. P. Jain, V.M. Bhandari, K. Balapure, J. Jena, V. V. Ranade, D.J. Killedar, Hydrodynamic cavitation using vortex diode: An efficient approach for elimination of pathogenic bacteria from water, *J. Environ. Manage.* (2019). https://doi.org/10.1016/j.jenvman.2019.04.057.
9. B. Bethi, S.H. Sonawane, G.S. Rohit, C.R. Holkar, D. V. Pinjari, B.A. Bhanvase, A.B. Pandit, Investigation of TiO2 photocatalyst performance for decolorization in the presence of hydrodynamic cavitation as hybrid AOP, *Ultrason. Sonochem.* (2016). https://doi.org/10.1016/j.ultsonch.2015.07.008.
10. X. Sun, X. Xuan, L. Ji, S. Chen, J. Liu, S. Zhao, S. Park, J.Y. Yoon, A.S. Om, A novel continuous hydrodynamic cavitation technology for the inactivation of pathogens in milk, *Ultrason. Sonochem.* (2021). https://doi.org/10.1016/j.ultsonch.2020.105382.
11. C.E. Brennen, *Cavitation and bubble dynamics*, 2013. https://doi.org/10.1017/CBO9781107338760.
12. E. Webster, *Cavitation, Ultrasonics.* (1963). https://doi.org/10.1016/0041-624X(63)90116-1.
13. J. Carpenter, M. Badve, S. Rajoriya, S. George, V.K. Saharan, A.B. Pandit, Hydrodynamic cavitation: An emerging technology for the intensification of various chemical and physical processes in a chemical process industry, *Rev. Chem. Eng.* (2017). https://doi.org/10.1515/revce-2016-0032.
14. E. Herbert, S. Balibar, F. Caupin, Cavitation pressure in water, *Phys. Rev. E - Stat. Nonlinear, Soft Matter Phys.* (2006). https://doi.org/10.1103/PhysRevE.74.041603.
15. F. Caupin, Liquid-vapor interface, cavitation, and the phase diagram of water, *Phys. Rev. E - Stat. Nonlinear, Soft Matter Phys.* (2005). https://doi.org/10.1103/PhysRevE.71.051605.
16. F. Caupin, E. Herbert, Cavitation in water: a review, *Comptes Rendus Phys.* (2006). https://doi.org/10.1016/j.crhy.2006.10.015.
17. M. Dular, O. Coutier-Delgosha, Thermodynamic effects during growth and collapse of a single cavitation bubble, *J. Fluid Mech.* (2013). https://doi.org/10.1017/jfm.2013.525.

18. M.J. Moran, Engineering thermodynamics, in: *Mechatronics Handb.,* 2002. https://doi.org/10.4324/9781315119717-1.
19. A. Karimi, F. Avellan, Comparison of erosion mechanisms in different types of cavitation, *Wear.* (1986). https://doi.org/10.1016/0043-1648(86)90031-1.
20. B. Randhavane Shrikant, A.K. Khambete, Hydrodynamic Cavitation: A Novel Treatment Approach, in: *Mater. Today Proc.,* 2017. https://doi.org/10.1016/j.matpr.2017.06.246.
21. W. Lauterborn, H. Bolle, Experimental investigations of cavitation-bubble collapse in the neighbourhood of a solid boundary, *J. Fluid Mech.* (1975). https://doi.org/10.1017/S0022112075003448.
22. T.M. Gondal, Z. Hameed, M.U. Shah, H. Khan, Cavitation phenomenon and its effects in francis turbines and amassed adeptness of hydel power plant, in: *2019 2nd Int. Conf. Comput. Math. Eng. Technol. ICoMET 2019,* 2019. https://doi.org/10.1109/ICOMET.2019.8673454.
23. J. Ozonek, Application of hydrodynamic cavitation in environmental engineering, 2012. https://doi.org/10.1201/b11825.
24. P.R. Gogate, A.B. Pandit, Hydrodynamic cavitation reactors: A state of the art review, *Rev. Chem. Eng.* (2001). https://doi.org/10.1515/REVCE.2001.17.1.1.
25. P.B. Patil, V.M. Bhandari, V. V. Ranade, Improving efficiency for removal of ammoniacal nitrogen from wastewaters using hydrodynamic cavitation, *Ultrason. Sonochem.* (2021). https://doi.org/10.1016/j.ultsonch.2020.105306.
26. V.K. Saharan, M.A. Rizwani, A.A. Malani, A.B. Pandit, Effect of geometry of hydrodynamically cavitating device on degradation of orange-G, *Ultrason. Sonochem.* (2013). https://doi.org/10.1016/j.ultsonch.2012.08.011.
27. E. Abbasi, S. Saadat, A. Karimi Jashni, M.H. Shafaei, A novel method for optimization of slit Venturi dimensions through CFD simulation and RSM design, *Ultrason. Sonochem.* (2020). https://doi.org/10.1016/j.ultsonch.2020.105088.
28. P.R. Gogate, A.B. Pandit, Engineering design methods for cavitation reactors II: Hydrodynamic cavitation, *AIChE J.* (2000). https://doi.org/10.1002/aic.690460815.
29. Z. Qin, K. Bremhorst, H. Alehossien, T. Meyer, Simulation of cavitation bubbles in a convergent - Divergent nozzle water jet, *J. Fluid Mech.* (2007). https://doi.org/10.1017/S002211200600351X.
30. Kuldeep, V.K. Saharan, Computational study of different venturi and orifice type hydrodynamic cavitating devices, *J. Hydrodyn.* (2016). https://doi.org/10.1016/S1001-6058(16)60631-5.
31. A.A. Kulkarni, V. V. Ranade, R. Rajeev, S.B. Koganti, Pressure drop across vortex diodes: Experiments and design guidelines, *Chem. Eng. Sci.* (2009). https://doi.org/10.1016/j.ces.2008.10.060.
32. A. Mullick, S. Neogi, A review on acoustic methods of algal growth control by ultrasonication through existing and novel emerging technologies, *Rev. Chem. Eng.* (2017). https://doi.org/10.1515/revce-2016-0010.

33. A. Ferrari, Fluid dynamics of acoustic and hydrodynamic cavitation in hydraulic power systems, *Proc. R. Soc. A Math. Phys. Eng. Sci.* (2017). https://doi.org/10.1098/rspa.2016.0345.
34. K. PENG, S. TIAN, G. LI, Z. HUANG, R. YANG, Z. GUO, Bubble dynamics characteristics and influencing factors on the cavitation collapse intensity for self-resonating cavitating jets, *Pet. Explor. Dev.* (2018). https://doi.org/10.1016/S1876-3804(18)30038-7.
35. T.G. Leighton, Derivation of the Rayleigh-Plesset Equation in Terms of Volume, 2007.
36. A. Prosperetti, A generalization of the Rayleigh-Plesset equation of bubble dynamics, *Phys. Fluids.* (1982). https://doi.org/10.1063/1.863775.
37. Z. Li, Criteria for jet cavitation and cavitation jet drilling, *Int. J. Rock Mech. Min. Sci.* (2014). https://doi.org/10.1016/j.ijrmms.2014.03.021.
38. A.A. Kulkarni, J.B. Joshi, Bubble formation and bubble rise velocity in gas-liquid systems: A review, *Ind. Eng. Chem. Res.* (2005). https://doi.org/10.1021/ie049131p.
39. Muchammad, M. Tauviqirrahman, A.W. Pratomo, J. Jamari, D.J. Schipper, The effect of boundary slip and cavitation on hydrodynamic pressure generation in pocket bearings, in: *AIP Conf. Proc.*, 2016. https://doi.org/10.1063/1.4945502.
40. C.E. Brennen, Bubble Growth and Collapse, in: *Fundam. Multiph. Flow,* 2014. https://doi.org/10.1017/cbo9780511807169.005.
41. L. Ming, N. Zhi, S. Chunhua, Numerical simulation of cavitation bubble collapse within a droplet, *Comput. Fluids.* (2017). https://doi.org/10.1016/j.compfluid.2017.04.019.
42. A.R. Klotz, Bubble dynamics in N dimensions, *Phys. Fluids.* (2013). https://doi.org/10.1063/1.4817803.
43. A. Tezel, S. Mitragotri, Interactions of Inertial Cavitation Bubbles with Stratum Corneum Lipid Bilayers during Low-Frequency Sonophoresis, *Biophys. J.* (2003). https://doi.org/10.1016/S0006-3495(03)74770-5.
44. J.J. Kwan, S. Graham, C.C. Coussios, Inertial cavitation at the nanoscale, in: *Proc. Meet. Acoust.*, 2013. https://doi.org/10.1121/1.4800019.
45. T. Theofanous, L. Biasi, H.S. Isbin, H. Fauske, A theoretical study on bubble growth in constant and time-dependent pressure fields, *Chem. Eng. Sci.* (1969). https://doi.org/10.1016/0009-2509(69)85008-6.
46. A. Prosperetti, M.S. Plesset, Vapour-bubble growth in a superheated liquid, *J. Fluid Mech.* (1978). https://doi.org/10.1017/S0022112078000671.
47. S. Schenke, T. Melissaris, T.J.C. Van Terwisga, On the relevance of kinematics for cavitation implosion loads, *Phys. Fluids.* (2019). https://doi.org/10.1063/1.5092711.
48. M.G. De Giorgi, D. Bello, A. Ficarella, Analysis of thermal effects in a cavitating orifice using rayleigh equation and experiments, *J. Eng. Gas Turbines Power.* (2010). https://doi.org/10.1115/1.4000367.
49. D.H. Trevena, Cavitation and the generation of tension in liquids, *J. Phys. D. Appl.* Phys. (1984). https://doi.org/10.1088/0022-3727/17/11/003.

50. K. Vokurka, Cavitation noise modeling and analyzing, *Proc. Forum Acoust.* 2002 (Sociedad Esp. Acustica). (2002).
51. K. Li, M.W. Woo, H. Patel, L. Metzger, C. Selomulya, Improvement of rheological and functional properties of milk protein concentrate by hydrodynamic cavitation, *J. Food Eng.* (2018). https://doi.org/10.1016/j.jfoodeng.2017.10.005.
52. K.S. Suslick, D.J. Flannigan, Inside a collapsing bubble: Sonoluminescence and the conditions during cavitation, *Annu. Rev. Phys. Chem.* (2008). https://doi.org/10.1146/annurev.physchem.59.032607.093739.
53. I. Gonçalves, M. Martins, A. Loureiro, A. Gomes, A. Cavaco-Paulo, C. Silva, Sonochemical and hydrodynamic cavitation reactors for laccase/hydrogen peroxide cotton bleaching, *Ultrason. Sonochem.* (2014). https://doi.org/10.1016/j.ultsonch.2013.08.006.
54. M. Gągol, A. Przyjazny, G. Boczkaj, Wastewater treatment by means of advanced oxidation processes based on cavitation – A review, *Chem. Eng. J.* (2018). https://doi.org/10.1016/j.cej.2018.01.049.
55. M.B. Mane, V.M. Bhandari, K. Balapure, V. V. Ranade, A novel hybrid cavitation process for enhancing and altering rate of disinfection by use of natural oils derived from plants, *Ultrason. Sonochem.* (2020). https://doi.org/10.1016/j.ultsonch.2019.104820.
56. P.S. Kumar, A.B. Pandit, Modeling hydrodynamic cavitation, *Chem. Eng. Technol.* (1999). https://doi.org/10.1002/(SICI)1521-4125(199912)22:12<1017::AID-CEAT1017>3.0.CO;2-L.
57. R. Chand, D.H. Bremner, K.C. Namkung, P.J. Collier, P.R. Gogate, Water disinfection using the novel approach of ozone and a liquid whistle reactor, *Biochem. Eng. J.* (2007). https://doi.org/10.1016/j.bej.2007.01.032.
58. A.G. Chakinala, P.R. Gogate, R. Chand, D.H. Bremner, R. Molina, A.E. Burgess, Intensification of oxidation capacity using chloroalkanes as additives in hydrodynamic and acoustic cavitation reactors, *Ultrason. Sonochem.* (2008). https://doi.org/10.1016/j.ultsonch.2007.02.008.
59. P.R. Gogate, I.Z. Shirgaonkar, M. Sivakumar, P. Senthilkumar, N.P. Vichare, A.B. Pandit, Cavitation reactors: Efficiency assessment using a model reaction, *AIChE J.* (2001). https://doi.org/10.1002/aic.690471115.
60. K.K. Jyoti, A.B. Pandit, Water disinfection by acoustic and hydrodynamic cavitation, *Biochem. Eng. J.* (2001). https://doi.org/10.1016/S1369-703X(00)00128-5.
61. L. Patil, P.R. Gogate, Large scale emulsification of turmeric oil in skimmed milk using different cavitational reactors: A comparative analysis, *Chem. Eng. Process. Process Intensif.* (2018). https://doi.org/10.1016/j.cep.2018.02.019.
62. S.Y. Tang, P. Shridharan, M. Sivakumar, Impact of process parameters in the generation of novel aspirin nanoemulsions - Comparative studies between ultrasound cavitation and microfluidizer, *Ultrason. Sonochem.* (2013). https://doi.org/10.1016/j.ultsonch.2012.04.005.
63. P.R. Gogate, Hydrodynamic Cavitation for Food and Water Processing, *Food Bioprocess Technol.* (2011). https://doi.org/10.1007/s11947-010-0418-1.

64. Y. Li, X. Zhao, Y. Zu, Y. Zhang, Preparation and characterization of paclitaxel nanosuspension using novel emulsification method by combining high speed homogenizer and high pressure homogenization, *Int. J. Pharm.* (2015). https://doi.org/10.1016/j.ijpharm.2015.05.070.
65. D. Quintanar-Guerrero, H. Fessi, E. Allémann, E. Doelker, Influence of stabilizing agents and preparative variables on the formation of poly(D,L-lactic acid) nanoparticles by an emulsification-diffusion technique, *Int. J. Pharm.* (1996). https://doi.org/10.1016/S0378-5173(96)04697-2.
66. S. Rajoriya, S. Bargole, V.K. Saharan, Degradation of reactive blue 13 using hydrodynamic cavitation: Effect of geometrical parameters and different oxidizing additives, *Ultrason. Sonochem.* (2017). https://doi.org/10.1016/j.ultsonch.2017.01.005.
67. N.P. Vichare, P.R. Gogate, A.B. Pandit, Optimization of hydrodynamic cavitation using a model reaction, *Chem. Eng. Technol.* (2000). https://doi.org/10.1002/1521-4125(200008)23:8<683::AID-CEAT683>3.0.CO;2-9.
68. G. Lu, L. Zhao, M. Zhu, C. Deng, H. Liu, J. Ma, Y. Li, Effect of Cavitation Hydrodynamic Parameters on Bisphenol A Removal, *Environ. Eng. Sci.* (2019). https://doi.org/10.1089/ees.2018.0371.
69. A. Sharma, P.R. Gogate, A. Mahulkar, A.B. Pandit, Modeling of hydrodynamic cavitation reactors based on orifice plates considering hydrodynamics and chemical reactions occurring in bubble, *Chem. Eng. J.* (2008). https://doi.org/10.1016/j.cej.2008.04.005.
70. A. Raj, B. Bethi, S.H. Sonawane, Investigation of removal of crystal violet dye using novel hybrid technique involving hydrodynamic cavitation and hydrogel, *J. Environ. Chem. Eng.* (2018). https://doi.org/10.1016/j.jece.2018.08.016.
71. X. Wang, Y. Zhang, Degradation of alachlor in aqueous solution by using hydrodynamic cavitation, *J. Hazard. Mater.* (2009). https://doi.org/10.1016/j.jhazmat.2008.03.073.
72. P.N. Patil, S.D. Bote, P.R. Gogate, Degradation of imidacloprid using combined advanced oxidation processes based on hydrodynamic cavitation, *Ultrason. Sonochem.* (2014). https://doi.org/10.1016/j.ultsonch.2014.02.024.
73. S. Rajoriya, S. Bargole, V.K. Saharan, Degradation of a cationic dye (Rhodamine 6G) using hydrodynamic cavitation coupled with other oxidative agents: Reaction mechanism and pathway, *Ultrason. Sonochem.* (2017). https://doi.org/10.1016/j.ultsonch.2016.05.028.
74. J. Wang, H. Chen, R. Yuan, F. Wang, F. Ma, B. Zhou, Intensified degradation of textile wastewater using a novel treatment of hydrodynamic cavitation with the combination of ozone, *J. Environ. Chem. Eng.* (2020). https://doi.org/10.1016/j.jece.2020.103959.
75. L. V. Malade, U.B. Deshannavar, Decolorisation of Reactive Red 120 by hydrodynamic cavitation, in: *Mater. Today Proc.*, 2018. https://doi.org/10.1016/j.matpr.2018.06.180.

76. A.A. Pradhan, P.R. Gogate, Removal of p-nitrophenol using hydrodynamic cavitation and Fenton chemistry at pilot scale operation, *Chem. Eng. J.* (2010). https://doi.org/10.1016/j.cej.2009.09.042.
77. K.P. Mishra, P.R. Gogate, Intensification of degradation of Rhodamine B using hydrodynamic cavitation in the presence of additives, *Sep. Purif. Technol.* (2010). https://doi.org/10.1016/j.seppur.2010.09.008.
78. D. Musmarra, M. Prisciandaro, M. Capocelli, D. Karatza, P. Iovino, S. Canzano, A. Lancia, Degradation of ibuprofen by hydrodynamic cavitation: Reaction pathways and effect of operational parameters, *Ultrason. Sonochem.* (2016). https://doi.org/10.1016/j.ultsonch.2015.09.002.
79. V. Innocenzi, M. Prisciandaro, F. Tortora, F. Vegliò, Optimization of hydrodynamic cavitation process of azo dye reduction in the presence of metal ions, *J. Environ. Chem. Eng.* (2018). https://doi.org/10.1016/j.jece.2018.10.046.
80. R.H. Jawale, P.R. Gogate, Novel approaches based on hydrodynamic cavitation for treatment of wastewater containing potassium thiocyanate, *Ultrason. Sonochem.* (2019). https://doi.org/10.1016/j.ultsonch.2018.11.019.
81. L.S. McCarty, C.J. Borgert, L.D. Burgoon, Evaluation of the Inherent Toxicity Concept in Environmental Toxicology and Risk Assessment, *Environ. Toxicol. Chem.* (2020). https://doi.org/10.1002/etc.4881.
82. Y. Yang, Z. Qin, W. Zeng, T. Yang, Y. Cao, C. Mei, Y. Kuang, Toxicity assessment of nanoparticles in various systems and organs, *Nanotechnol. Rev.* (2017). https://doi.org/10.1515/ntrev-2016-0047.
83. T. Sahota, M. Danhof, O. Della Pasqua, Pharmacology-based toxicity assessment: Towards quantitative risk prediction in humans, *Mutagenesis.* (2016). https://doi.org/10.1093/mutage/gev081.
84. P. Samanta, S. Pal, A.K. Mukherjee, A.R. Ghosh, Acute toxicity assessment of arsenic, chromium and almix 20WP in Euphlyctis cyanophlyctis tadpoles, *Ecotoxicol. Environ. Saf.* (2020). https://doi.org/10.1016/j.ecoenv.2020.110209.
85. S.S. Amarasiri, A.P. Attanayake, L.D.A.M. Arawwawala, K.A.P.W. Jayatilaka, L.K.B. Mudduwa, Acute and 28-Day Repeated-Dose Oral Toxicity Assessment of Abelmoschus moschatus Medik. In Healthy Wistar Rats, *Evidence-Based Complement. Altern. Med.* (2020). https://doi.org/10.1155/2020/1359050.
86. G.S. King, Handbook of Toxicologic Pathology, *Arch. Pathol. Lab. Med.* (2002). https://doi.org/10.5858/2002-126-1138b-hotp.
87. G.A. Van Norman, Limitations of Animal Studies for Predicting Toxicity in Clinical Trials: Is it Time to Rethink Our Current Approach?, *JACC Basic to Transl. Sci.* (2019). https://doi.org/10.1016/j.jacbts.2019.10.008.
88. X. Miao, X. Zhang, Y. Yuan, Y. Zhang, J. Gao, N. Kang, X. Liu, J. Wu, Y. Liu, P. Tan, The toxicity assessment of extract of Peganum harmala L. seeds in Caenorhabditis elegans, *BMC Complement. Med. Ther.* (2020). https://doi.org/10.1186/s12906-020-03051-x.
89. H. Olson, G. Betton, D. Robinson, K. Thomas, A. Monro, G. Kolaja, P. Lilly, J. Sanders, G. Sipes, W. Bracken, M. Dorato, K. Van Deun, P. Smith, B. Berger, A. Heller, Concordance of the toxicity of pharmaceuticals in humans

and in animals, *Regul. Toxicol. Pharmacol.* (2000). https://doi.org/10.1006/rtph.2000.1399.
90. Y.N. Liu, H.B. Chen, X.W. Liu, Rapid Assessment of Water Toxicity by Plasmonic Nanomechanical Sensing, *Anal. Chem.* (2020). https://doi.org/10.1021/acs.analchem.9b04476.
91. S. Conway, A Short History of American Revolutionary War (2020) 347–390. https://doi.org/10.5040/9780755603817.0010.
92. E.J. Allera, L. Journal, FDA's Combination Animal Drug Policy - Is It Feasible? Or Should Elsie Be the Only One FDA's Combination Animal Drug Policy — Is It Feasible? Or Should Elsie be the Only One Getting Milked?, 33 (1978).
93. C. Ii, A. Welfare, C. Iii, S. Animal, W. Boards, THE ANIMAL WELFARE ACT , 2011 Arrangement of Sections, Anim. Welf. (2011).
94. X. Yi, Z. Gao, L. Liu, Q. Zhu, G. Hu, X. Zhou, Acute toxicity assessment of drinking water source with luminescent bacteria: Impact of environmental conditions and a case study in Luoma Lake, East China, *Front. Environ. Sci. Eng.* (2020). https://doi.org/10.1007/s11783-020-1288-z.
95. E.R. Bandala, O.M. Rodriguez-Narvaez, On the Nature of Hydrodynamic Cavitation Process and Its Application for the Removal of Water Pollutants, *Air, Soil Water Res.* (2019). https://doi.org/10.1177/1178622119880488.
96. M. Farré, D. Barceló, Toxicity testing of wastewater and sewage sludge by biosensors, bioassays and chemical analysis, *TrAC - Trends Anal. Chem.* (2003). https://doi.org/10.1016/S0165-9936(03)00504-1.
97. A. Garrido, A. Lepailleur, S.M. Mignani, P. Dallemagne, C. Rochais, hERG toxicity assessment: Useful guidelines for drug design, *Eur. J. Med. Chem.* (2020). https://doi.org/10.1016/j.ejmech.2020.112290.
98. G.C. Kuriakose, B.P. Arathi, M. Divya Lakshmanan, M. V. Jiby, R.S. Gudde, C. Jayabhaskaran, Sub-acute Toxicity Assessment of Taxol Isolated From Fusarium Solani, an Endophytic Fungus of Taxus Brevifolia, in Wistar Rats and Analyzing Its Cytotoxicity and Apoptotic Potential in Lung Cancer Cells, *Front. Oncol.* (2020). https://doi.org/10.3389/fonc.2020.538865.
99. V.K. Saharan, A.B. Pandit, P.S. Satish Kumar, S. Anandan, Hydrodynamic cavitation as an advanced oxidation technique for the degradation of Acid Red 88 dye, *Ind. Eng. Chem. Res.* (2012). https://doi.org/10.1021/ie200249k.
100. N.B. Suryawanshi, V.M. Bhandari, L.G. Sorokhaibam, V. V. Ranade, Developing techno-economically sustainable methodologies for deep desulfurization using hydrodynamic cavitation, *Fuel.* (2017). https://doi.org/10.1016/j.fuel.2017.08.106.
101. A. Mukherjee, A. Mullick, P. Vadthya, S. Moulik, A. Roy, Surfactant degradation using hydrodynamic cavitation based hybrid advanced oxidation technology: A techno economic feasibility study, *Chem. Eng. J.* (2020). https://doi.org/10.1016/j.cej.2020.125599.

102. R. Dolas, C. Saravanan, B.P. Kaur, Emergence and era of ultrasonic's in fruit juice preservation: A review, *Ultrason. Sonochem.* (2019). https://doi.org/10.1016/j.ultsonch.2019.05.026.
103. L. Gensheng, S. Zhonghou, Z. Changshan, Z. Debin, C. Hongbing, Investigation and application of self-resonating cavitating water jet in petroleum engineering, *Pet. Sci. Technol.* (2005). https://doi.org/10.1081/LFT-200027961.
104. K. Peng, S. Tian, G. Li, Z. Huang, R. Yang, Z. Guo, Bubble dynamics characteristics and influencing factors on the cavitation collapse intensity for self-resonating cavitating jets, Shiyou Kantan Yu Kaifa/Petroleum Explor. Dev. (2018). https://doi.org/10.11698/PED.2018.02.16.
105. L. Gensheng, S. Zhonghou, Z. Changshan, Z. Debin, C. Hongbing, Investigation and Application of Self-Resonating Cavitating Water Jet in Petroleum Engineering, *Pet. Sci. Technol.* (2005). https://doi.org/10.1081/lft-20009686218.
106. R.A. Torres, C. Pétrier, E. Combet, F. Moulet, C. Pulgarin, Bisphenol A mineralization by integrated ultrasound-UV-iron (II) treatment, *Environ. Sci. Technol.* (2007). https://doi.org/10.1021/es061440e.
107. P. Thanekar, S. Garg, P.R. Gogate, Hybrid Treatment Strategies Based on Hydrodynamic Cavitation, Advanced Oxidation Processes, and Aerobic Oxidation for Efficient Removal of Naproxen, *Ind. Eng. Chem. Res.* (2020). https://doi.org/10.1021/acs.iecr.9b01395.
108. M. V. Bagal, P.R. Gogate, Wastewater treatment using hybrid treatment schemes based on cavitation and Fenton chemistry: A review, *Ultrason. Sonochem.* (2014). https://doi.org/10.1016/j.ultsonch.2013.07.009.
109. G. Zhao, H. Zhang, Q. Fan, X. Ren, J. Li, Y. Chen, X. Wang, Sorption of copper(II) onto super-adsorbent of bentonite-polyacrylamide composites, *J. Hazard. Mater.* (2010). https://doi.org/10.1016/j.jhazmat.2009.08.135.
110. M.G.A. Vieira, A.F.A. Neto, M.L. Gimenes, M.G.C. da Silva, Sorption kinetics and equilibrium for the removal of nickel ions from aqueous phase on calcined Bofe bentonite clay, *J. Hazard. Mater.* (2010). https://doi.org/10.1016/j.jhazmat.2009.12.040.
111. S. Chakraborty, A. Mukherjee, S. Das, N.R. Maddela, S. Iram, P. Das, Study on isotherm, kinetics, and thermodynamics of adsorption of crystal violet dye by calcium oxide modified fly ash, *Environ. Eng. Res.* (2021). https://doi.org/10.4491/eer.2019.372.
112. S. Lanjewar, A. Mukherjee, P. Khandewal, A.K. Ghosh, A. Mullick, S. Moulik, A. Roy, Thermodynamics of synthesis and separation performance of Interfacially polymerized "loose" reverse osmosis membrane: Benchmarking for greywater treatment, *Chem. Eng. J.* (2020). https://doi.org/10.1016/j.cej.2020.127929.
113. K. Roy, A. Mukherjee, N.R. Maddela, S. Chakraborty, B. Shen, M. Li, D. Du, Y. Peng, F. Lu, L.C. Garciá Cruzatty, Outlook on the bottleneck of carbon nanotube in desalination and membrane-based water

treatment-A review, *J. Environ. Chem. Eng.* (2020). https://doi.org/10.1016/j.jece.2019.103572.

114. R. Das, A. Mukherjee, I. Sinha, K. Roy, B.K. Dutta, Synthesis of potential bio-adsorbent from Indian Neem leaves (Azadirachta indica) and its optimization for malachite green dye removal from industrial wastes using response surface methodology: kinetics, isotherms and thermodynamic studies, *Appl. Water Sci.* (2020). https://doi.org/10.1007/s13201-020-01184-5.
115. G. Vijayaraghavan, T. Sivakumar, A. Vimal Kumar Application of Plant Based Coagulants for Waste Water Treatment, *Int. J. Adv. Eng. Res. Stud.* (2011).
116. H.R. Aerni, B. Kobler, B. V. Rutishauser, F.E. Wettstein, R. Fischer, W. Giger, A. Hungerbühler, M.D. Marazuela, A. Peter, R. Schönenberger, A.C. Vögeli, M.J.F. Suter, R.I.L. Eggen, Combined biological and chemical assessment of estrogenic activities in wastewater treatment plant effluents, *Anal. Bioanal. Chem.* (2004). https://doi.org/10.1007/s00216-003-2276-4.
117. M.B. Ahmed, J.L. Zhou, H.H. Ngo, W. Guo, N.S. Thomaidis, J. Xu, Progress in the biological and chemical treatment technologies for emerging contaminant removal from wastewater: A critical review, *J. Hazard. Mater.* (2017). https://doi.org/10.1016/j.jhazmat.2016.04.045.
118. N.R. Maddela, L.C. García Cruzatty, D.A. Leal-Alvarado, J.C. Olaya, S. Chakraborty, A. Mukherjee, Quorum Quenching for Sustainable Environment: Biology, Mechanisms, and Applications, in: 2020. https://doi.org/10.1007/978-981-15-2679-4_4.
119. T. Kamiya, J. Hirotsuji, New combined system of biological process and intermittent ozonation for advanced wastewater treatment, in: *Water Sci. Technol.*, 1998. https://doi.org/10.1016/S0273-1223(98)00688-X.
120. D.Halder, S.Mondal, A.Mukherjee, A.Ghosh, An Overview on the Treatment of Waste water using Renewable energy, *Int. J. Res. Eng. Technol.* vol. 5, no. 1, pp. 78-83,(2016).
121. M. Dular, T. Griessler-Bulc, I. Gutierrez-Aguirre, E. Heath, T. Kosjek, A. Krivograd Klemenčič, M. Oder, M. Petkovšek, N. Rački, M. Ravnikar, A. Šarc, B. Širok, M. Zupanc, M. Žitnik, B. Kompare, Use of hydrodynamic cavitation in (waste)water treatment, *Ultrason. Sonochem.* (2016). https://doi.org/10.1016/j.ultsonch.2015.10.010.
122. A. Mukherjee, A. Mullick, R. Teja, P. Vadthya, A. Roy, S. Moulik, Performance and energetic analysis of hydrodynamic cavitation and potential integration with existing advanced oxidation processes: A case study for real life greywater treatment, *Ultrason. Sonochem.* (2020). https://doi.org/10.1016/j.ultsonch.2020.105116.
123. Y. Çalışkan, H.C. Yatmaz, N. Bektaş, Photocatalytic oxidation of high concentrated dye solutions enhanced by hydrodynamic cavitation in a pilot reactor, *Process Saf. Environ. Prot.* (2017). https://doi.org/10.1016/j.psep.2017.08.003.
124. M.M. Gore, V.K. Saharan, D. V. Pinjari, P. V. Chavan, A.B. Pandit, Degradation of reactive orange 4 dye using hydrodynamic cavitation based hybrid techniques, *Ultrason. Sonochem.* (2014). https://doi.org/10.1016/j.ultsonch.2013.11.015.

125. R.K. Joshi, P.R. Gogate, Degradation of dichlorvos using hydrodynamic cavitation based treatment strategies, *Ultrason. Sonochem.* (2012). https://doi.org/10.1016/j.ultsonch.2011.11.005.
126. P.N. Patil, P.R. Gogate, Degradation of methyl parathion using hydrodynamic cavitation: Effect of operating parameters and intensification using additives, *Sep. Purif. Technol.* (2012). https://doi.org/10.1016/j.seppur.2012.04.019.
127. P. Thanekar, M. Panda, P.R. Gogate, Degradation of carbamazepine using hydrodynamic cavitation combined with advanced oxidation processes, *Ultrason. Sonochem.* (2018). https://doi.org/10.1016/j.ultsonch.2017.08.001.
128. K. Roy, V.S. Moholkar, Sulfadiazine degradation using hybrid AOP of heterogeneous Fenton/persulfate system coupled with hydrodynamic cavitation, *Chem. Eng. J.* (2020). https://doi.org/10.1016/j.cej.2019.03.170.
129. M. V. Bagal, P.R. Gogate, Degradation of diclofenac sodium using combined processes based on hydrodynamic cavitation and heterogeneous photocatalysis, *Ultrason. Sonochem.* (2014). https://doi.org/10.1016/j.ultsonch.2013.10.020.
130. F. Meneguzzo, C. Brunetti, A. Fidalgo, R. Ciriminna, R. Delisi, L. Albanese, F. Zabini, A. Gori, L.B. dos S. Nascimento, A. De Carlo, F. Ferrini, L.M. Ilharco, M. Pagliaro, Real-scale integral valorization of waste orange peel via hydrodynamic cavitation, *Processes.* (2019). https://doi.org/10.3390/pr7090581.
131. I. Lee, J.I. Han, Simultaneous treatment (cell disruption and lipid extraction) of wet microalgae using hydrodynamic cavitation for enhancing the lipid yield, *Bioresour. Technol.* (2015). https://doi.org/10.1016/j.biortech.2015.03.045.

4

Recent Trends in Ozonation Technology: Theory and Application

Anupam Mukherjee[1], Dror Avisar[2] and Anirban Roy[1]*

[1]Water-Energy Nexus Lab, Department of Chemical Engineering Goa, BITS Pilani, Goa, India
[2]School of Earth Sciences, The Water Research Center, Hydrochemistry, Tel Aviv University, Tel Aviv, Israel

Abstract
The treatment of wastewater by ozonation is one of the growing and reliable technologies from the last few years. One of the critical considerable aspects in an ozonation system is the gas-liquid contacting system which is used in a variety of industrial applications and has been studied from many points of view. Therefore, considerations of mass transfer fundamentals are very crucial by considering the reaction dynamics. The mass transfer taking place in these systems is affected by a number of individual factors like volumetric mass transfer coefficient, Henry's coefficient, and solubility of ozone, which play a critical role in effective performance of the ozonation process. In this work a comprehensive assessment has been illuminated on gas-liquid interfacial mass transfer, in particular ozone-water environment. An insight into the design parameters (gas holdup, bubble characteristics) affecting the mass transfer mechanism in a typical bubble column gas-liquid contactor is discussed along with a mathematical model on a single bubble for estimation of mass transfer efficiencies. The chapter ends with case studies on the application of ozonation in greywater treatment.

***Keywords*:** Ozone, advance oxidation process, mass transfer, water treatment, bubble dynamics

**Corresponding author*: anirbanr@goa.bits-pilani.ac.in

Siddhartha Moulik, Aditi Mullick and Anirban Roy (eds.) Sustainable Water Treatment: Advances and Technological Interventions, (117–170) © 2022 Scrivener Publishing LLC

4.1 Introduction

Utilization of ozone for treatment of wastewater and potable water production has been expanded extensively through the last decade [1–3]. In water and wastewater treatment, ozonation is usually performed by dissolving gaseous ozone into water to oxidize target contaminants and microorganism typically with passage of oxygen or dry air stream through two electrodes, which generates electric discharge [4]. Treatment with ozone alone in some cases is sufficient enough to degrade targeted contaminants to meet water quality standards [5]. The high oxidizing and disinfectant character of ozone makes it very attractive to diminish and check the organohalogen formation potential of the water and is an attractive alternative to traditional technologies such as chlorination [6], adsorption [7, 8], membranes [9, 10] etc., and has been of potential use for the water-energy-food-health nexus. Furthermore, ozone has the potential to combine with other agents like hydrogen peroxide or UV radiation, to facilitate the formation of hydroxyl radical. This process is termed as advanced oxidation process. High operational cost and formation of by-products have been the primary concern for limitation in the current treatment systems using ozone. To compensate for this limitation the systems are designed accordingly which results in considerable cost of maintenance as well as operating along with initial capital investment. Hence design of the contacting system plays a vital role for significant performance of the ozonation process.

The knowledge of ozone mass transfer is derived from chemical engineering principles. An examination of the literature tells us that the destruction of pollutants in water treatment is the limiting step but there is lack of data available on the mass transfer coefficient of ozone-water in different contact systems. The use of appropriate mathematical models using different parameters like time of contact, mass transfer, and rate of reaction is very useful in designing of the system. Besides, the overall rate of ozone mass transfer is significantly affected not only by different operating conditions but also the water quality and setup facility. These preludes to understanding led us to the conclusion that the gas liquid contact remains the critical aspect of the ozonation process. Since the interfacial area regulates the size of the bubble, mass transfer process involved, degree and degree of coalescence play a vital role in determining the mass transfer rate and the potential of ozone to oxidize. Bubble size and degree of coalescence are in fact dependent on parameters like gas hold up, bubble characteristics, column diameter. Utilizing knowledge gathered from above parameters, and applying appropriate mathematical models, an efficient system can be designed [11, 12]. If the mass transfer rates can be increased by proper designing of the system it will reduce the cost associated with the water treatment. In an attempt to predict the performance of ozonation process various mass transfer correlation along with the kinetics of ozone decay are presented in the subsequent section. Hence in this chapter the authors have presented a comprehensive review of the ozonation technology from a fundamental understanding of mass transfer to applications in water treatment.

4.2 Fundamentals of Mass Transfer

Diffusive mass transfer usually results from the tendency of a component in a mixture to migrate from a region of high concentration to one of low concentration [13]. The diffusive flux of a substance across a unit area of interface per unit time is given in one dimension by Fick's first law:

$$J_A = -c_m D_{AB} \frac{\partial_{\chi A}}{\partial_z} \tag{4.1}$$

Where,

J_A = *Diffusive flux of component A, (moles/m².s)*
c_m = *molar diffusivuty of fluid, (moles/m³)*
D_{AB} = *Free liquid diffusivity of component A in B, (m²/s)*
χA = *mole fraction of component A,*
z = *distance in the z direction, (m)*

The constant of proportionality (D) is described as the diffusion coefficient. The diffusion coefficient for gases are easier to calculate than to determine the diffusion coefficient for liquids because the molecules are much closer together and so there are many more intermolecular forces. Fick's first law has therefore served as a general basis for many models formulated to describe mass transfer at phase boundaries [14]. Mass transfer between liquid and gas phase can be well described by theoretical models like two film model [15], penetration models [16], "The rate of absorption of a pure gas into a still liquid during short periods of exposure)" [17], combined models [18, 19], eddy cell models [20–22], and eddy diffusivity model [23, 24]. The essential features of these models are listed in Table 4.1.

Table 4.1 Different models of mass transfer in the liquid phase.

Mass transfer model	**Fundamental expression for k_L**	**Remarks**
Two-film Model	$k_L = \frac{D_L}{\delta_L}$ [24]	D_L - liquid molecular diffusivity δ_L - liquid film thickness
Penetration Models	$\overline{k_L} = 2\sqrt{\frac{D_L}{\pi\theta}}$ [16]	θ - Exposure time of surface elements
	$\overline{k_L} = \sqrt{D_L s}$ [17]	s- rate of surface renewals
Combined Models	$\overline{k_L} = \sqrt{D_L s}\cot h\sqrt{\frac{s\delta_L}{D_L}}$ [18]	
	$\overline{k_L} = \frac{D_L}{\delta_L}\lfloor 1+2\sum_{n=1}^{\infty}(1+n^2\pi^2 D_L/(sD_L))^{-1}\rfloor$ [19]	$V=\sqrt{u'^2}$ - velocity fluctuation L – eddy length scale
Eddy Cell Models	$\overline{k_L} = 1.46\left(\frac{D_L V}{L}\right)^{1/2}$ [20]	
	$\overline{k_L} = 0.4(\varepsilon v_L)^{1/4} Sc^{-1/2}$ [21]	ε - rate of energy dissipation per unit mass of fluid v_L – kinematic viscosity of liquid

(Continued)

Table 4.1 Different models of mass transfer in the liquid phase. (*Continued*)

Mass transfer model	Fundamental expression for k_L	Remarks
	$\overline{k_L} = C_1 D_L^{1/2}\left(\frac{\varepsilon}{v_L}\right)^{1/4}$ for Ret>>1 $k_L = C_2 F(\alpha)\left(\frac{D_L V}{L}\right)^{1/2}$ for Re_t that is not >>1 [22]	$Re_t = \frac{V_L}{v_L}$ V and L – velocity length scales of turbulence $\alpha = \frac{\sqrt{2}\theta_{\exp col} V}{L}$ $F(\alpha)$ Given in a graphical form
Eddy Diffusivity Models	$k_L = \frac{n}{\pi}\sin\left(\frac{\pi}{n}\right)\alpha^{1/n} D_L^{1-1/n}$ [23] For n=2 $k_L = \frac{2}{\pi}\alpha^{1/2} D_L^{1/2}$	Long contact time asymptote $D_t = ay^n$ a depends on hydrodynamic conditions.
	$k_L = 0.32\, D_L^{1/2} \rho_L^{1/2} \upsilon_o^{3/2} \sigma_r^{-1/2}$ [25]	$D_t = \upsilon_o y^2/\lambda$ (Levich) υ_o - characteristic turbulent velocity λ - thickness of a zone at the interface where turbulent fluctuations of velocity and concentration are damped

The simplest and most commonly used in practice is the two-film model which assumes that two hypothetical laminar films exist in each of the two fluids as shown in Figure 4.1. This model can be used to explain the rate of mass transfer of a material from one phase to another. The main basis of this theory is that the layers at the phase boundary are in a state of laminar flow, rather than in a turbulent state [16]. Outside this layer, in the

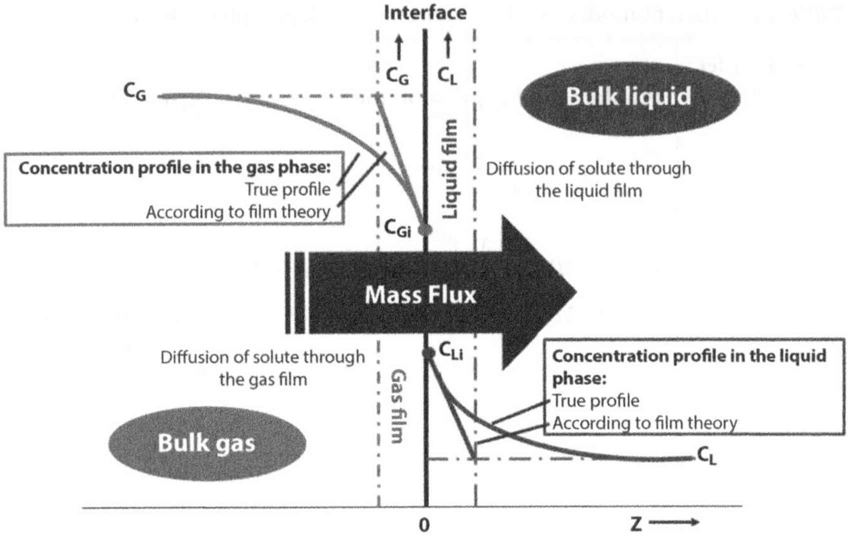

Figure 4.1 Concentration profile of the gas phase and liquid phase on the two side of the interface in a dilute solution.

bulk of that particular phase, the natural random motion of the particles is enhanced or supplemented by turbulent eddy-like motion. This eddy-like motion decreases the resistance to mass transfer. As can be seen, the model comprises two thin layers either side of the phase interface, from which the mass transfer occurs. From the figure it is evident that the concentration gradient is near-linear at the phase interface, gradually decreasing until the bulk phase concentration is reached [26].

The transport equation can be written as:

(Amount transferred per unit time) = (Mass Transfer Coefficient) *(Area)*(Driving Force)

Hence the physical mass transfer across the boundaries without a chemical reaction can be expressed by the equations mentioned below

$$N_A = k_G A(C_G - C_{Gi} = k_L A(C_{Li} - C_L) \quad (4.2)$$

$$N_A = K_G A(C_G - C_G^* = K_L A(C_L^* - C_L) \quad (4.3)$$

Where,

N_A = *Mass transfer per unit time, (moles/s)*

$k_G = Gas\ film\ co-efficient, \left(\frac{mol}{m^2}\right) s.\ Pa$

k_L = *Liquid film co-efficient*, (m/s)

$K_g = Overall\ mass\ transfer\ co-efficient\ based\ on\ gas\ phase\ concentration,$

$\left(\frac{mol}{m^2}\right) s.\ Pa$

K_L = *Overall mass transfer co-efficient based on liquid phase concentration*, (m/s)

A = *Mass Transfer area*, (m^2)

$C_G = Bulk\ Concentration\ of\ solute\ in\ gas\ phase, \left(\frac{mol}{lit}\right)$

$C_{Gi} = Interphase\ gas\ concentration, \left(\frac{mol}{lit}\right)$

$C_{Li} = Interphase\ Liquid\ concentration, \left(\frac{mol}{lit}\right)$

$C_L = Bulk\ concentration\ of\ solute\ in\ liquid\ phase, \left(\frac{mol}{lit}\right)$

$C_G^* = Concentration\ of\ solute\ gas\ in\ liquid\ which\ will\ be\ in\ equilibrium$

$with\ gas \left(\frac{mol}{lit}\right)$

$C_L^* = Concentration\ of\ solute\ gas\ in\ gaseous\ phase\ which\ will\ be\ in$

$equilibrium\ with\ liquid \left(\frac{mol}{lit}\right)$

Reciprocal to mass transfer coefficient are the mass transfer resistance which includes the local phase mass transfer resistance as well as overall mass transfer resistance. Physical significance can be given to restricted interfacial mass transfer coefficient by using the kinetic theory of gases where it is possible to compute the rate at which soluble molecules will strike and leave the surface [27]. The net rate difference can be calculated by the given below equation

$$k_s' = \frac{\propto H}{\sqrt{2\pi RTM}} \tag{4.4}$$

Where ∝ is the condensation coefficient and is the fraction of the molecules striking the surface which actually condense. At the interface between the two phases a thermodynamic equilibrium exists which is well described by the Henry's Law [28].

$$C_{Gi} = HC_{Li} \tag{4.5}$$

Where H is the Henry's Law constant. The solubility ratio, $S = 1/H$. It is noted that the values of C_G^* *and* C_L^* are estimated from Equation (4.5) using the corresponding bulk concentration C_G & C_L respectively. Equations (4.2) and (4.3) is further solved by simple summation of the mass transfer resistances connected in series to yield a general relationship existing between the overall mass transfer coefficients and the films.

$$\frac{1}{K_G} = \frac{1}{k_G} + \frac{H}{k_L} \tag{4.6}$$

$$\frac{1}{K_L} = \frac{1}{k_L} + \frac{1}{Hk_G} \tag{4.7}$$

If the component A is slightly soluble in water, the mass transfer is mainly controlled by the liquid phase, as goes in the case of ozone. Hence,

$$\frac{1}{K_L} = \frac{1}{k_L} \& C_{Li} = C_L^* \tag{4.8}$$

In a gas liquid contacting system, it is a bit arduous for determination of the mass transfer area. Therefore, it is more reasonable to operate with the liquid phase volume and to combine the interfacial area with the mass transfer coefficient to obtain the so-called volumetric mass transfer coefficient, k_La.

Hence Equation (4.3) can now be written as

$$N_A = k_La(C_L^* - C_L)V_L \tag{4.9}$$

Hydrodynamic condition of the liquid phase is utilized to calculate the value of mass transfer coefficient, k_La [2]. Determination of value of $k_L.a$ involves consideration of two conditions:-Steady state & *Unsteady state.*

For unsteady state it is estimated by studying the rate at which ozone concentration is enhanced after having all the ozone removed. In case of steady state it is calculated by amount of ozone transferred to water which is being fed and collected from gas liquid contactor. The concentration profile C_L in the liquid phase in a gas liquid contactor depends on the mixing condition in that phase and equilibrium concentration C_L^* depends on the gas phase concentration, temperature, pressure and the liquid phase matrix [2].

4.3 Mass Transfer of Ozone in Water

Several studies involving the mass transfer phenomena of ozone in a water have been investigated earlier [29–31]. To carry out treatment of water and wastewater involving ozonation, it is necessary to introduce ozone into liquid phase from gaseous one. Hence the process of ozonation fits the absorption process. The water treatment behavior of ozone is characterized by three main stages; the first stage includes the movement of the ozone in the gas stream to the gas/liquid interface involving the diffusion of ozone molecule through the gas. This diffusion of ozone molecule through the gas is affected by many factors like temperature and size of the molecules. The second stage involves the ozone transfer across the liquid gas boundary. This stage is characterized by various resistances on either the liquid or the gas side of the boundary. The third stage involves movement of the ozone from the liquid side or the liquid boundary into the bulk of water to be treated. According to penetration theory [16] the average rate of ozone gas absorption into water is a function of the diffusivity of the dissolved gas in the liquid, the gas-liquid contact time and the solubility of the gas and is of the form:

$$N_{O3} = 2a(C_{03}^* - C_{03})\sqrt{D/\pi\theta} \tag{4.10}$$

Where,

N_{o3} = *Rate of ozone gas adsorption,* $(mol\ m^{-2}\ s^{-1})$
D = *Diffusivity of ozone gas,* (m^2/s)
θ = *Exposure time,* (s)

The absorption rate of ozone in water can be expressed in general as

$$N_{o3} = k_L a(C_{o3}^* - C_{o3}) \tag{4.11}$$

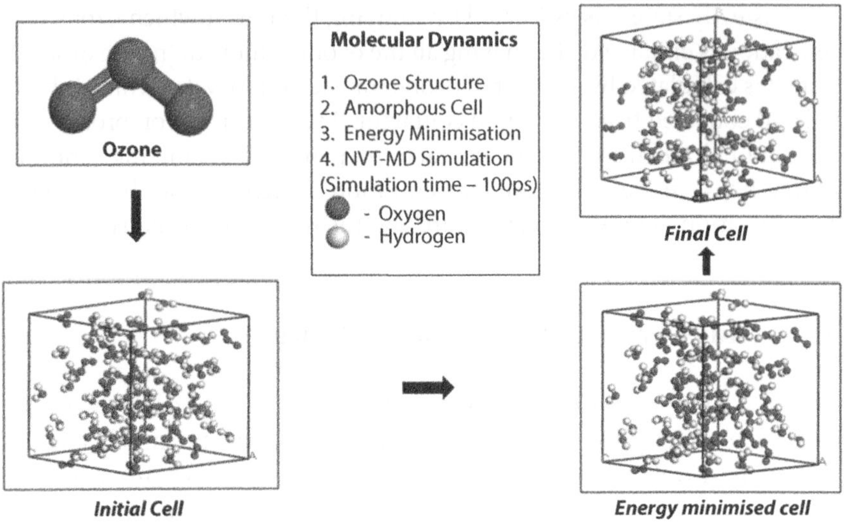

Figure 4.2 Schematic diagram of reaction of ozone molecule with water.

Where k_L is the ozone mass transfer coefficient and a the specific gas–liquid interfacial surface inside the ozonation reactor. Also, in an ozonation process, the most important variable is the volumetric mass transfer coefficient denoted by $k_L a$. A typical schematic diagram of ozone molecule in reaction with water is shown in Figure 4.2. The detailed modeling of mass transfer of ozone is discussed later in this chapter. The value of C_{o3} is determined by the ozone solubility in water [32].

4.3.1 Solubility of Ozone in Water

In a typical laboratory ozonation system, several phenomena play a significant role in determining the solubility characteristics of ozone which includes the mass transfer of the gaseous ozone into the aqueous phase and the rate of auto decomposition of the ozone absorbed into solution [33, 34]. For a non-dissociating solute, in a dilute isothermal solution, Henry's Law states that the fugacity of the solute is proportional to the mole fraction at equilibrium. A model to describe the true solubility is explained below.

4.3.1.1 Model for Determining the True Solubility Concentration

Considering the case where there is simultaneous interaction of gas-liquid undergoing a reaction. The mass balance around a completely mixed semi-batch reactor on the reacting species yields:

$$K_L a = (C_i - C)V = rV + V\frac{dC}{dt} \tag{4.12}$$

Where,

C_i = *Concentration of solute at the interface, (mol L^{-1}))*
C = *Concentration of solute, (mol L^{-1})*
r = *rate of reaction, (mol L^{-1} s^{-1})*

When the solute concentration reaches steady state, the accumulation term, $\frac{dC}{dt}$ goes to zero. Assuming nth order kinetics for Equation (4.12), it becomes

$$K_L a(C_i - C^*) = \psi C^{*n} \tag{4.13}$$

Where

ψ = *Specific rate Constant [s^{-1}) (molL^{-1})$^{1-n}$]*
C^* = *Steady State concentration of the solute, (mol L^{-1})*

Subtracting Equation (4.13) from (4.12),

$$\frac{dC}{dt} = K_L a(C^* - C) + \psi(C^{*n} - C^n) \tag{4.14}$$

The intrinsic rate expression for ozone is

$$-r = \psi C \tag{4.15}$$

Where ψ is the intrinsic kinetic rate constant and is a function of [OH^-] and temperature. When kinetics for ozone decomposition are applied it becomes

$$\frac{dC}{dt} = K_L a + \psi(C^* - C) \tag{4.16}$$

Equation (4.16) then provides a linear relationship through the origin which can be used to evaluate the slope by linear regression. Substituting equation (4.13) for first order kinetics and rearranging

$$C^* = \left(\frac{1}{1+Da}\right)C = \eta C \tag{4.17}$$

Where

Da = Damkohler Number for first order reaction and
η = *interphase effectiveness factor.*

Rewriting Equation (4.12), using the applicable assumption for ozone decomposition in an aqueous system,

$$\frac{dC}{dt} = K_L a(C^*) - (K_L a + \psi)C \tag{4.18}$$

Integrating Equation (4.18) and assuming the ozone concentration to be zero at time zero to yield

$$C = C^* \, [1 - \exp(-K_L a)/\eta)t] \tag{4.19}$$

4.3.2 Mass Transfer Model of Ozone in Water

For the mass transfer of ozone in water it follows the two-resistance theory model which is discussed in the previous section. Using an overall mass transfer coefficient, based on overall driving force, the ozone mass transfer rate is:

$$N_A = K_G a(P_G - P^*) \tag{4.20}$$

Where,

P_G = *Partial Pressure of Ozone in gas phase (Pa)*
P^* = *Partial Pressure of Ozone in equilibrium with bulk liquid concentration in the liquid phase (Pa)*
Or in terms of liquid phase concentration

$$N_A = K_L a(C^* - C_L) \tag{4.21}$$

Where,

C^* = Ozone concentration in equilibrium with bulk liquid concentration in the gas phase (mg L^{-1})
C_L = Ozone concentration in the liquid phase (mg L^{-1})

For a dilute ozone concentration, we can assume the validity of Henry's Law

$$Y = Hx^* \text{ or } P_G = HC^* \tag{4.22}$$

Considering the reactor to behave as an ideal stirred reactor. Without considering any reaction of ozone, the ozone mass balance in the liquid phase is:

$$(K_La)V(C^* - C_L) = QC_L \tag{4.23}$$

The mass balance in the gas phase is:

$$g_{in} - (K_La)V(C^* - C_L) = g_{out} \tag{4.24}$$

The overall mass balance of ozone is:

$$g_{out} = g_{in} - QC_L \tag{4.25}$$

Since the ozone concentration in the water is very low then,

$$x^* = \frac{C}{M_{03}} \bigg/ \left(\frac{C^*}{M_{o3}} + \frac{V_w * \delta H_2o}{M_{H_2o}} \right) \tag{4.26}$$

When the volume of water is 10^3 ml, molecular weights of water and ozone are approximately 18 and 48 gm, respectively, and neglecting $\frac{C^*}{M_{o3}}$ as being very small, the above equation (4.26) becomes

$$x^* = \frac{C^*}{\rho_{H_2o} 10^3} * \frac{M_{H_2o}}{M_{o3}} = \frac{18}{4.8 * 10^7} C^* \tag{4.27}$$

Thus, Henry's Law becomes:

$$Y = \frac{18}{4.8 * 10^7} * H * C^* \tag{4.28}$$

The relation between Y and g_{out} is:

$$Y=\frac{g_{out}}{G*\rho_{(oxygen)}10^{3}}*\frac{(M_{oxygen})}{(M_{o3})}=\frac{32*10^{-3}}{1.3*48}*\frac{g_{out}}{G} \tag{4.29}$$

The overall mass balance equation (4.25) becomes

$$Y=(g_{in}-QC_{L})*\frac{32*10^{-3}}{1.3*48}*\frac{1}{G} \tag{4.30}$$

From Equation (4.28) and (4.30)

$$C^{*}=(g_{in}-QC_{L})*\frac{1367}{HG} \tag{4.31}$$

And from the mass balance in the liquid phase:

$$(K_{L}a)V(g_{in}-QC_{L})*\frac{1367}{HG}-C_{L}=QC_{L} \tag{4.32}$$

Average densities of water and ozone at 20°C is considered. Ozone diffuses much deliberately in the gas phase in comparison to liquid phase, henceforth mass transfer in the liquid phase becomes the rate controlling step and the overall mass transfer coefficient is represented by the local liquid mass transfer coefficient. Ozone exhibits extremely oxidative properties, due to which once dissolved in water it gets decayed by means of auto-decomposition and oxidation of impurities [35]. The rate of decomposition depends on a multitude of system variables, including the water pH, temperature, the organic and inorganic constituents of waters, and the presence of organic and inorganic initiators and inhibitors. By assuming simple kinetics of nth order for the ozone the auto decomposition reaction might be considered.

The decomposition reaction rate expression for ozone is

$$\frac{dN}{dt}=KVC_{L}^{n} \tag{4.33}$$

$$\frac{dC_L}{dt} = KC_L^n \tag{4.34}$$

This rate of ozone disappearance Equation (4.33) should be incorporated into the mass balance equations,

$$(K_L a)V(C^* - C_L) - KVC_L = QC_L \tag{4.35}$$

$$g_{in} - (QC_L + KVC_L) = q_{out} \tag{4.36}$$

Combining the above Equation (4.35) and (4.36) into the final equation (4.32), resulting in:

$$(K_L a)V\left\{(g_{in} - QC_L - KVC_L) - \frac{1367}{HG} - C_L\right\} - KVC_L = QC_L \tag{4.37}$$

This equation can also be applied for distilled water, in the absence of oxidizable matter.

In the presence of any oxidation or disinfection reaction, Equation (4.33) is changed to

$$\frac{dN_L}{dt} = KVC_L^n + K_R VC_L^{n'} \tag{4.38-a}$$

Or

$$\frac{dC_L}{dt} = KC_L^n + K_R C_L^{n'} \tag{4.38-b}$$

Where, K_R = *Global Chemical reaction rate constant,* $(mg\ s^{-1}\ L^{-1})$

Apart from this there is a pseudo first order expression for a modified rate equation where the rate constant is replaced by a variable known as specific ozone-utilization rate and is represented as

$$\frac{dC_L}{dt} = K_w C_L \tag{4.39}$$

Where,

K_w = *Ozone decay rate constant,* (*sec*$^{-1}$)

For an irreversible first order reaction, there is a negligible effect of chemical reactions on the mass transfer coefficient [17] and is given by:

$$H_a^2 = \frac{DK_w}{2k_L} \ll 1 \tag{4.40}$$

Where H_a^2 is the dimensionless Hatta number and D = molecular diffusivity of ozone. The diffusivity of ozone can also be calculated by other co-relations proposed by researchers [35] with the following expression

$$D_{o3} = A\frac{T}{\mu_L} \tag{4.41}$$

Where D_{o3} is in m^{2}s^{-1}, T is in K and μ_L, the solution viscosity in Poise. The mass transfer coefficient K_L can be estimated from equations for stirred tanks in which μ_L and ρ_L, the viscosity and density of the aqueous solutions are stated in SI units

$$K_L = 0.42^3\sqrt{\frac{\mu_L g}{\rho_L}}Sc^{-0.5} \tag{4.42}$$

The variables like the gas phase retention h', the average size of bubbles "d_b" and the specific exchange surface A is related to each other by the below equation when ozone is sparged as bubbles in the column

$$A = \frac{6}{d_b}\frac{h'}{1-h'} \tag{4.43}$$

The bubble diameter can be estimated from operation parameters u_g (superficial gas velocity) and $\in_g$ (gas hold-up into the coloumn) and the liquid-phase properties ρ_L (density) and σ_L (surface tension):

$$\frac{6(1-\epsilon_g)}{d_b} = 2\left(\frac{\rho_L g}{\sigma_L}\right)^{0.5} \frac{\mu_g}{\left(\frac{\sigma_L g}{\rho}\right)^{0.25}} \tag{4.44}$$

If all bubbles are spheres with the same size, A can be calculated by

$$A = \frac{6\epsilon_g}{d_b} \tag{4.45}$$

In bubble columns and for bubble sizes $d_b < 2mm$, Equation (4.42) is used to estimate k_L and for bubbles with $d_b > 2mm$ [36]:

$$k_L = k_{L(d_b=2mm)} 500 d_b \tag{4.46}$$

Once the mean bubble diameter d_b and gaseous retention h is determined it becomes easier to obtain "A" and therefore k_L if $k_L a$ is known. This knowledge of k_L helps us to evaluate the importance of chemical reaction [37]. The application of a single bubble model in the estimation for ozone transfer efficiency in water is discussed later along with the effects in chemical reaction. The hydrodynamic behavior of fluids affecting the efficiency of ozone mass transfer in a bubble column is also discussed in a subsequent section. Although from Equation (4.41) to (4.46) the contribution of pollutants in wastewater has not been taken into account. The determination of Henry and Mass transfer coefficient in wastewater is explained in detail in the later section.

4.3.3 Henry and Volumetric Mass Transfer Coefficient Determination

As discussed in the previous section, parameters like $K_L a$ and H depends on the content of wastewater [38]. Practically for a thorough study both Henry and Volumetric Mass transfer coefficient should be determined from experimental data. There are numerous ways to determine the same which varies depending on the degree of observations and experiments. Mentioned below are three different and commonly used approaches.

4.3.3.1 Microscopic Ozone Balance in the Gas Phase

As stated earlier, the absorption of ozone in the wastewater containing dissolved matter follows irreversible chemical reaction [39]. Considering the plug flow behavior and perfect mixing in a cylindrical [40] bubble column reactor the absorption of ozone gives us

$$-\frac{m_T}{P_T} = dP_{o3} = N_{o3}\beta\sigma\, dh \tag{4.47}$$

When the Hatta Number (Ha) [41] is lower than 0.3 the rate of ozone absorption at the mass transfer interface is

$$N_{o3} = k_L a C^*_{o3}\frac{Ha}{\tanh(Ha)}\left(1-\frac{C_{o3}}{C^*_{03}\cosh(Ha)}\right) \tag{4.48}$$

Equation (4.48) simplifies to Equation (4.11), when it is rearranged taking Henry Equation as a characteristic variable of ozone partial pressure and is

$$\frac{-dP_{o3}}{P_{o3}-C_{03}H} = \frac{k_L a P_T}{Hm_T} A\, dh \tag{4.49}$$

Which on integrating taking the boundary condition as h = 0, $P_{o3} = P_{o3i}$ *and* $h = h_T$, $P_{o3} = P_{o3o}$ results in

$$P_{o3_o} = \frac{P_{o3_i}}{\exp\left[\frac{k_L a P_T A h_T}{Hm_T}\right]} + \left[H - \frac{H}{\exp\left[\frac{k_L a P_T S h_T}{Hm_T}\right]}\right] C_{o3} \tag{4.50}$$

From the above equation it can be concluded that the plot of ozone partial pressure versus the dissolved ozone concentration will yield a straight line whose slope and intercept will lead to *H and* $k_L a$ respectively.

4.3.3.2 Macroscopic Ozone Balance in the Gas Phase

Macroscopic ozone balance in gas phase [39] is given in equation 4.51 and is the second most frequent approach

$$U_g A C_{o3_i} = U_g A C_{o3_o} + k_L a V \beta (C_{o3}^* - C_{o3}) \tag{4.51}$$

Three approximations for C_{o3}^* can be assumed as expressed below by Equation (4.52) to (4.54)

$$C_{o3}^* = \gamma \frac{(C_{o3_i} - C_{o3_o})}{2} \tag{4.52}$$

$$C_{o3}^* = \gamma \frac{(C_{o3_i} - C_{o3_o})}{ln \dfrac{C_{o3_i}}{C_{o3_o}}} \tag{4.53}$$

$$C_{o3}^* - C_{o3} = \frac{\gamma (C_{o3_i} - C_{o3_o})}{ln \dfrac{\gamma (C_{o3_i} - C_{o3})}{\gamma (C_{o3_0} - C_{o3})}} \tag{4.54}$$

Where $\gamma = {RT}/{H}$, substituting these back in Equation (4.51) leads to expressions of the form, $y = b + cx$,

$$C_{o3_0} = C_{o3_i} \left[\frac{U_g A - k_L a V \beta \gamma / 2}{U_g A + k_L a V \beta \gamma / 2} \right] + \left[\frac{k_L a V \beta}{U_g A + k_L a V \beta \gamma / 2} \right] C_{o3} \tag{4.55}$$

$$\frac{C_{o3}}{(C_{o3_i} - C_{o3_0})} = -\frac{U_g A}{k_L a V \beta} + \frac{\gamma}{ln \dfrac{C_{o3_i}}{C_{o3_0}}} \tag{4.56}$$

$$C_{o3_0} = C_{o3_i} \exp\left[-\frac{k_L a V \beta \gamma}{U_g A} \right] + \left[1 - \exp\left(\frac{-k_L a V \beta \gamma}{U_g A} \right) \right] C_{o3} \tag{4.57}$$

Though Equations 4.55 to 4.57 are similar yet the use of Equation 4.50 is advantageous due to the assumed average value for C_{o3}^*.

4.3.3.3 Ozone Balance at Constant Ozone Concentrations

Additionally, from the experiments carried out on ozonation at a constant reaction rate H can be determined [4.42]. In such scenario, the chemical rate is given by

$$N_{03} = -\frac{1}{z}\frac{dC}{dt} = -r_{03};\, z\ \textit{being the stochiometric factor} \tag{4.58}$$

As discussed above, the ozone balance in water can be expressed by Equation (4.11), the ozone balance in the gas on the other hand is as follows:

$$\frac{m_{in} - m_{out}}{V} = k_L a(C_{o3}^{*} - C_{o3}) \tag{4.59}$$

Where m_i, m_o *and* V represents the molar flux rates of ozone at the reactor inlet and outlet and the reaction volume, respectively.

Comparing Equation (4.11) and (4.59) the derived equation obtained is given by:

$$C_{o3}^{*} = \frac{-r_{o3}}{k_L a} + C_{o3} \tag{4.60}$$

As assumed the water to be perfectly mixed, Henry's Equation can be related as [43]:

$$C_{o3}^{*} = \frac{P_{oSat}}{H} \tag{4.61}$$

Integrating Equation (4.49) between the bottom of column and a partial height h will yield:

$$P_{o3} = \frac{P_{o3} - HC_{03}}{\exp\left[\frac{k_L a P_T \sigma h}{H m_T}\right]} + HC_{o3} \tag{4.62}$$

Hence an average value for P_{o3} can be obtained as follows:

$$P_{o3_{avg}} = \frac{1}{h_T}\int_0^{h_T} P_{o3}\, dh \tag{4.63}$$

h represents the total height of the column.

Integrating Equation (4.63) after substitution of Equation (4.62) gives:

$$P_{o3_{avg}} = \frac{(P_{o3} - HC_{03})Hm_T}{k_L a P_T \sigma h}\left\{1 - \exp\left\{-\frac{K_L a P_T \sigma h}{Hm_T}\right\}\right\} + HC_{03} \tag{4.64}$$

On the other hand, from Equation (4.59)

$$C_{o3}^{*} = \frac{m_{in} - m_{out}}{k_L V} + C_{03} \tag{4.65}$$

Hence considering Equation (4.61) and Equation (4.65), for the whole column:

$$P_{o3_{avg}} = H\frac{m_{in} - m_{out}}{k_L V} + HC_{03} \tag{4.66}$$

Finally, from Equation (4.64) and (4.66) values of H can be calculated.

4.3.4 Single Bubble Model of Mass Transfer

A single bubble model of mass transfer in gas liquid system facilitates in the estimation of mass transfer efficiencies under variable process conditions [44, 45]. The effects of bubble size, value of the mass transfer coefficient, kinetics of reactions taking place in water and depth of the contact chamber can be simulated by this model. The results gathered utilizing this modelling in terms of mass transfer efficiency represent absorption of ozone in water at different hydrodynamic conditions. A single bubble model of mass transfer assumes that a solute is absorbed from a single bubble during its vertical rising through a liquid layer. The mass transfer rate for a component A is given as:

$$\frac{dn_A}{dt} = -k_L A_b (C_{Ai} - C_A) \tag{4.67}$$

The bubble loses a part of its mass due to absorption of a solute; also, the bubble volume changes because of hydrostatic effects. The equation of state describes these changes as follows:

$$P_A V_B = n_A RT \tag{4.68}$$

Similar equations are valid for the inert components in the bubble. The total pressure within the bubble is given as:

$$P_t = P_{atm} + Q_L gh + \frac{4\sigma}{d_b} = P_A + P_{inerts} \tag{4.69}$$

For bubbles not smaller than 0.1 mm in sizes the term $\frac{4\sigma}{d_b}$ can be neglected. We assume that Henry's law is obeyed by component A:

$$P_A = HC_{Ai} \tag{4.70}$$

If spherical shape of the bubble is postulated, Equations (4.67), (4.68) and (4.70) can be combined to yield:

$$\frac{dn_a}{dt} = -\frac{6k_L RT}{He_a d_b} n_A \left(1 - \frac{C_A}{C_{Ai}}\right) \tag{4.71}$$

Equation (4.71) can be solved analytically for the simple cases (k_L = const., d_b = const., $C_A = 0$). A substitution of $dh = v_b dt$ is convenient and Equation (4.71) takes the form:

$$\frac{dn_a}{dh} = -\frac{6k_L RT}{He_a d_b v_b} n_A \left(1 - \frac{C_A}{C_{Ai}}\right) \tag{4.72}$$

For k_L = constant, d_b = constant, $C_A = 0$, and v_b = *constant*, after integration one obtains

$$n_A = n_A^0 \exp\left(-\frac{6k_L RTh_0}{He_a d_b v_b}\right) \tag{4.73}$$

The absorption efficiency of the component A can be calculated as:

$$TE = 1 - \frac{n_A}{n_A^0} = 1 - \exp\left(-\frac{6k_L RT h_0}{H_A d_b v_b}\right) \tag{4.74}$$

In the more general situation, Equation (4.71) can be solved numerically, e.g., using the simple Euler method, taking a suitably small distance step Δh, and, if necessary, accounting for variations of the parameters, d_b, v_b, k_L, y_A at each step [39].

Inspection of Equation (4.74) shows that the component transfer efficiency is first of all dependent on the height of the liquid layer, h_0 on the mass transfer coefficient, k_L, and on the bubble size, d_b. A typical diagram of a bubble generated during ozonation is shown in Figure 4.3.

A prior calculation for k_L for the case of bubble column is possible to perform in which gravity plays a role in the rise of bubbles [16] and the equation for it is as follows:-

$$k_L = 1.13\sqrt{\frac{D}{t_c}} \tag{4.75}$$

Where t_c is denoted as the contact time and can be calculated by using the following relationship

$$t_c = \frac{d_b}{U_s} \tag{4.76}$$

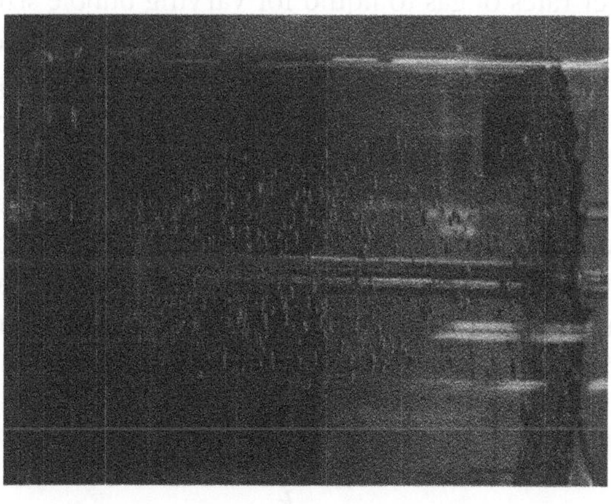

Figure 4.3 Bubble formation in a typical gas-liquid contactor.

For a bubble column U_s denotes the rise velocity of the bubble with respect to the liquid. For d_b close to 2mm the value of k_L is observed to be maximum and drop rapidly for smaller bubbles. As a result of a buildup of surface contaminations the bubble interface may become immobile. Since the presence of such contaminations is difficult to control, the experimental data collected for k_L varies widely. For bubbles greater than 2 mm in size, the interface is always mobile and the effect of the interface contamination on k_L should be irrelevant. For this range the experimental data can be approximated by the equation [46]:

$$sh = 1.128 Re_b^{1/2} Sc^{1/2} \tag{4.77}$$

A second suggestion [47] for calculation of k_L is the co-relation having the following form

$$Sh = 2 + a' \left\{ Re^{0.484} Sc^{0.339} \left(\frac{d_B g^{0.33}}{D^{0.66}} \right)^{0.072} \right\}^b \tag{4.78}$$

The exponent b has a value of 1.61 for bubble columns and proposed two values of a', for an individual bubble and swarm of bubbles, $a' = 0.061$ *and* 0.0187 respectively [47]. For the ozone-water system, reported values [46] of k_L (converted to 20°C) in the range of 2-3 x 10^{-4}m/s, collected in the bubble columns of 0.1 and 0.2 m in diameter, equipped with perforated gas distributors. The degree of coalescence and size of bubble determines the mass transfer rate and oxidative potential. There are numerous literatures on mass transfer rates of gas to liquid for varying bubble sizes both experimentally as well as using computer simulation and it has been showed the smaller the bubble size, the greater the mass transfer rate of gas. The larger surface area to volume ratio of very small bubbles provides an overall larger area for ozone mass transfer to occur. Other co-relations for the mass transfer coefficient k_L from single bubbles are given below in Table 4.2.

In the case of absorption with chemical reaction, the typical approach is based on the idea of the enhancement factor, E, which expresses the ratio of the mass transfer coefficient with a chemical reaction taking place in the liquid phase to the physical mass transfer coefficient:

$$E = \frac{k_L^*}{k_L} \tag{4.79}$$

Table 4.2 Correlations for the mass transfer coefficients, k_L, from single gas bubbles. (Reproduced with permission from [16]).

Investigator	Correlation	Remarks	Symbols
Boussinesq (1905)	$Sh = 1.128\ Re^{1/2}\ Sc^{1/2}$	$d_b > 2.5$mm	$Sh = k_L\ d_b / D_L$ $Re = d_b\ u_b \rho_L / \mu_L$ $Sc = \nu_L / D_L$ d_b – bubble diameter u_b – bubble rise velocity
Frössling (1938)	$Sh = 0.6\ Re^{1/2}\ Sc^{1/3}$	Modified	
Calderbank & Moo-Young (1962)	$Sh = 0.42\ Sc^{2/3}\ Gr^{1/3}$ $Sh = 0.31\ Ra^{1/3}$	$d_b < 2.5$mm $d_b > 2.5$mm	$Gr = d_b^3\ \rho_L\ \Delta\rho g / \mu_L^2$ $Ra = d_b^3\ \Delta\rho g / (D_L\ \mu_L)$
Calderbank & Lochiel (1964)	$Sh = 1.13\left(1 - \frac{2.90}{Re^{1/2}}\right)^{0.072} Pe^{1/2}$	100<Re<400	$Pe = R_b\ Sc$
Hughmark (1967)	$Sh = 2 + 0.061\left[Sc^{0.339} Re^{0.484}\left(\frac{d_b g^{1/3}}{D_L^{2/3}}\right)^{0.072}\right]^{1.61}$		g – acceleration due to gravity D_L – diffusivity in the liquid phase

(Continued)

Table 4.2 Correlations for the mass transfer coefficients, k_L, from single gas bubbles. (Reproduced with permission from [16]). (*Continued*)

Investigator	Correlation	Remarks	Symbols
Levins & Glastonbury (1972)	$Sh = 2 + 0.47\left(\frac{d_b^{4/3}\varepsilon_{diss}^{1/3}}{\nu_L}\right)^{0.072} Sc^{0.36}$		ε_{diss} – energy dissipation per unit volume $[W / m^3]$
Akita & Yoshida (1974)	$Sh = 0.5\,Sc^{1/3}\left(\frac{gd_b^3}{\nu_L^2}\right)^{1/4}\left(\frac{gd_b^2\rho_L}{\sigma}\right)^{3/8}$	For bubble columns	ν_L – liquid kinematic viscosity σ – surface tension
Cockx *et al.* (1995)	$k_L = (0.1 \pm 0.02)\,U_i^*\,Sc^{-0.5}$	For dispersed bubble systems	U_i^* - interfacial momentum transfer velocity
Wesselingh & Bollen (1999)	$k^{\%} = 4.8 \cdot 10^{-3} d_{\%}^{-0.132}$ @ $D_{\%} = 0.001$ $k^{\%} = 4.0 \cdot 10^{-2} d_{\%}^{-0.245}$ @ $D_{\%} = 0.001$ $k^{\%} \approx 0.4\,D_{\%}^{1/2}$	"dirty" system "clean" system Rigid sphere approx.	$k^{\%}$, $d_{\%}$, $D_{\%}$ - dimensionless parameters

This equation is used for calculating the reactions for known kinetics and for the case of a first order reaction [17] the evaluation will give the following results:

$$E = \sqrt{1+M} \tag{4.80}$$

Where $M = \frac{D_{o3}k}{k_L^2}$

The numerical values of this factor are dependent on the reaction order, the reaction rate constant, species diffusivity, and the physical mass transfer coefficient. For the first-order irreversible reaction (often assumed for ozone reactions with a number of pollutants), under the typical practical conditions (reactions of ozone with organic compounds) the values of the enhancement factor are on the order of magnitude of unity [48]. A brief discussion of the same for different reaction rates is presented below:

- In a case of very slow reaction where the consumption of ozone is much less, the concentration of ozone in the liquid phase can be determined by mass balance on the liquid with no chemical reaction in the mass transfer film, hence the transfer equation becomes

$$N_A = k_L a(C_L^* - C_L)V_L \tag{4.81}$$

Provided the following condition is satisfied

$$\frac{D_{o3}k}{k_L^2} \ll 1 \tag{4.82}$$

- From Equation (4.82), it is evident that though the chemical reaction is slow it is sufficiently enough to maintain the bulk concentration in the liquid phase and thus this forms a distinctive case of slow reactions. Hence Equation (4.81) becomes

$$N_A = k_L a C_L^* V_L \tag{4.83}$$

- In case of a fast reaction, some major fraction of ozone gets absorbed near the interface region. If all the ozone reacts in this film $C_L = 0$ & $E = M$

$$\therefore k_L a = \sqrt{D_{o3} k a}$$

Hence,

$$N_A = C_L \sqrt{D_{o3} k}\, a \tag{4.84}$$

4.3.5 Decomposition of Ozone in Water

Decomposition of ozone in aqueous media is carried out though a complex free radical mechanism instigated by the reaction between ozone and hydroxide ion though the formation of some oxidizing free radicals like HO, HO_2, HO_3 [34]. At the time of reaction, structural information about $O_{3 \text{ and } HOx}$ in the reaction medium (aqueous) remain unpredictable. In oxidation reaction between ozone and HO_x water plays an important role as solvent in this mechanism to modify their electronic and geometrical structures [32]. Figure 4.4 represents the schematic mechanism of ozone molecules with pollutants interaction.

Buhler *et al.* [49], Staehelin *et al.* [50], Tomiyasu *et al.* [51] and Hoigne [52] have studied widely on the principle mechanism and degradation kinetics of ozone through the chain process of reaction. The degradation initiates through the reaction between ozone and hydroxide ion in absence of light or catalysts to obtain hydroperoxide ($H\dot{O}_2$) and superoxide ion ($\dot{O}_2$) [2].

$$O_3 + HO^- \xrightarrow{k_{i,1}} H\dot{O}_2 + \dot{O}_2^- k_{OH^-} = 70\ Mol^{-1}s^{-1} \tag{4.85}$$

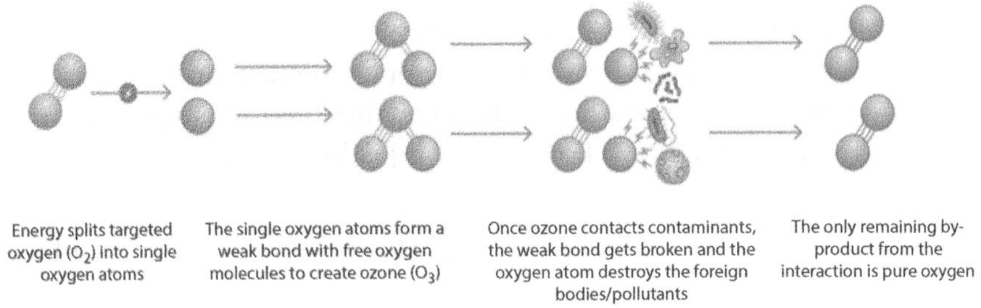

Figure 4.4 Schematic representation of reaction mechanism of ozone molecules with pollutants.

Then the reaction takes place between O_3 and $H\dot{O}_2$ i.e., conjugate base of hydrogen peroxide:

$$O_3 + HO_2^- \overset{k_{I,2}}{\rightarrow} H\dot{O}_2 + \dot{O}_2^- k_{OH^-} = 2.2 \times 10^6 M^{-1} s^{-1} \quad (4.86)$$

Propagation goes on by the formation of several radical ions as an intermediate product due to degradation reaction with hydrogen peroxide:

$$H\dot{O}_2 \rightleftarrows \dot{O}_2^- + H^+ \quad (4.87)$$

$$O_3 + \dot{O}_2^- \rightarrow \dot{O}_3^- + O_2 \quad (4.88)$$

$$HO_2^- + H^+ \leftrightarrows H_2O_2 \quad (4.89)$$

$$H\dot{O} + H_2O_2 \rightarrow H\dot{O}_2 + H_2O \quad (4.90)$$

$$H\dot{O} + HO_2^- \rightarrow H\dot{O}_2 + HO^- \quad (4.91)$$

A homogeneous termination of the reaction takes place by taking the radicals:

$$H\dot{O}_4 + H\dot{O}_4 \rightarrow H_2\dot{O}_2 + 2O_3 \quad (4.92)$$

$$H\dot{O}_4 + H\dot{O}_3 \rightarrow H_2\dot{O}_2 + O_2 + O_3 \quad (4.93)$$

Basically, the inhibiting compounds of free-radicals are the compounds efficient to consume hydroxyl ions without redeveloping the superoxide anion which include carbonate ions, bicarbonate ions, tert-alcohols and some humic materials [32].

In aqueous solution, metal oxides permit water molecules to coordinate with their surface due to the presence of Lewis acid sites [53]. Since the ozone molecules require the displacement of coordinated water molecules during adsorption and hence it is strongly dependent on thc availability of other bases. That time Lewis site becomes accessible to ozone for its adsorption/degradation reaction on catalytic surface [32]:

$$O_3 \rightarrow (O_3)_{ads} \tag{4.94}$$

$$O_{3ads} \rightarrow (O)_{ads} + O_2 \tag{4.95}$$

The interaction of ozone molecule with an oxidized site:

$$O_3 + (O)_{ads} \rightarrow O_2 + (O_2)_{ads} \rightarrow 2O_2 \tag{4.96}$$

The ozone molecule may interact with the basic delocalized Π-electron system or pyrrolic groups to form hydroxyl radicals, since the Lewis sites are located at Π-electron rich regions [54].

$$\cdots \cdot NH + O_3 \rightarrow \cdots N^+O^- + H\dot{O}_2 \tag{4.97}$$

$$H\dot{O}_2 \rightarrow H^+ + \dot{O}_2^- \tag{4.98}$$

$$\dot{O}_2^- + O_3 \rightarrow \dot{O}_3^- + O_2 \tag{4.99}$$

$$\dot{O}_2^- + H^+ \rightarrow H\dot{O} + O_2 \tag{4.100}$$

Though Sánchez-Polo *et al.* [55] showed the interaction between ozone and groups on the surface of activated carbon, activated carbon behaves as a conventional promote/initiator instead of catalyst for the transformation of ozone into radicals, since it decreased with ozone exposure.

4.3.6 Ozone Contactors and Energy Requirement

From recent reviews of the literature it is indicated by several researchers that there are several gas-liquid contactors for ozone treatment [46]. The type of treatment by ozone determines the requirement of the ozone contactor. Among all the known gas-liquid contactors for ozone treatment the bubble column contactors are the most frequently used. It is also popularly known as diffusion contactor, sparged tank or water column. The main advantages of this bubble column contactor over others include low cost, minimum energy consumption, minimum attention by operator (automatic operation), low maintenance, efficient mass transfer, simple design and flexibility to allow the variation of operating parameters. Treatment by

Table 4.3 Energetics database.

Contactor	Operating flexibility		Operating pressure (psi)		Operating energy (Kwh)/MG		Applications
	Water	Ozone	Water	Ozone	Required	Overall	
Bubble Column	0-100%	10-100%	None	6-12 psi	0-35	May-37	General Purpose
Positive Pressure Injector	80-100%	10-100%	6-12 psi	8-14 psi	0-130	12-175	Disinfection
Venturi Ejectors	85-100%	10-100%	5-12 psi	None	0-80	37-220	Low Pressure Ozone
Stationary Mixers	80-100%	10-100%	6-20 psi	8-22 psi	0-250	25-365	Small Flows
Stirred Reactors	0-100%	10-100%	None	2-5 psi	250-500	260-512	Small Flows

ozone is a relatively energy-intensive process and bubble column contactor is the most energy-efficient device of all. Energetics of ozone contactors depend mainly on the capital cost and operating cost [46].

For a bubble column contactor, the operating energy may be provided by the pressure, whereas for other gas-liquid contactors, namely positive pressure injector, venturi and stationary mixer, the operating energy may be provided by the water pressure and by the electrical power for the agitator of the stirred reactor. The overall energy requirement includes the operating energy and ozone loss of the off gas. Table 4.3 shows the energetics data based on operation of 3 mg/L ozone dosage, and 1% ozone generated from air a 24 hr/day basis along with the different gas-liquid contactors characteristics.

4.4 Factors Affecting Hydrodynamics and Mass Transfer in Bubble Column Reactor

In this section the authors present an insight into the factors affecting the hydrodynamics and mass transfer in a bubble column contactor [46]. A bubble column contactor is mostly a vessel with a cylindrical shaper having a gas distributor at the bottom. The ozone gas is sparged with a sparger in the form of bubbles into a liquid phase. These types of contactors are also known as slurry bubble column reactors when a solid phase exists.

Chemical, petrochemical, biochemical and metallurgical industries utilize bubble columns as multiphase contactors and reactors [56]. There are numerous factors affecting the mass transfer in gas-liquid dispersion which includes physical properties of gas and liquid, type of gas distributor, orifice diameter, spacing, dimension of column, gas flow rate and presence of chemical reactions [57]. For a given bubble system the dependent parameters include bubble size, gas holdup, bubble velocity rise and power input [58–60].

4.4.1 Fluid Dynamics and Regime Analysis

The operation and performance of bubble column is significantly affected by the fluid dynamic characterization of bubble column contactors. According to literature, the investigations of the parameters followed by experimental results led to the conclusion that analysis of fluid regime is very much essential [61]. The superficial gas velocity inside the bubble column plays an important role in categorizing the different flow regimes in it. In a bubble column typically, there are three types of flow regimes homogeneous (bubbly flow) regime; the heterogeneous (churn-turbulent) regime and slug flow regime. Apart from that there also exists a regime known as "foaming regime" but is not commonly termed in bubble column. In the homogeneous flow regime, the superficial gas velocity is much less (less than 5 cm/s in batch columns) and here the rise and size of bubbles are relatively uniform. Throughout the cross-sectional area of the column the mixing is gentle and there is uniform bubble distribution with no bubble coalescence which implies that this regime is controlled by the design of sparger and the system properties. In the heterogeneous flow regime, the superficial gas velocity is comparatively higher (greater than 5 cm/s in batch columns) [61]. Here there is a turbulent motion of gas bubbles and liquid recirculation which results in unsteady flow patterns, due to which there is a generation of large bubbles with short residence time [62]. This flow regime is thus sometimes referred to as coalesced bubble flow regime because of their difference in size of bubbles [63]. The average size of the bubble is governed by the energy dissipation rate which is a function of bubble coalescence. The slug flow regime is only observed at high gas flow rates and as the name goes it is due to the formation of bubble slags. As stated the fluid regime is an important aspect in bubble column but an interesting feature lies in detecting the transition from homogeneous to heterogeneous regime because in this particular phase only significant changes take place in the hydrodynamic behaviour of the system [64, 65].

4.4.2 Gas Holdup

The design and analysis of a bubble column reactor depends primarily upon the gas holdup which is defined as the volume fraction of gas phase occupied by the gas bubbles. The transport phenomena in a bubble column is characterized by this gas holdup [66]. In gas holdup there is a spatial variation which gives rise to variation in pressure as well as liquid recirculation. The heat and mass transfer in a bubble column is a strong function of liquid recirculation and extrapolations of the gas holdup profile will lead to better understanding in the design aspect of scaling up of bubble column. This gas holdup profile is further dependent on various factors which includes the physical properties of the system, the average velocity of the gas that is sparged into the bubble column popularly known as the superficial gas velocity and the column diameter [67]. Superficial gas velocity is simply expressed as the volumetric flow rate divided by the cross-sectional area of the column [68]. From the above studies it can be concluded with an increase in gas velocity and operating pressure and decrease in liquid viscosity and solid concentration, gas holdup increases. The addition of a surface-active reagent will increase the gas holdup. The aspect ratio of the bubble column larger than 5 will have no effect on the gas holdup. Gas holdup also depends on the pitch and diameter of the orifice holes, with insignificant effect with orifice diameter larger than 1 mm [46].

4.4.3 Bubble Characteristics

In a bubble column, clustering of bubbles along with their holdup contributions and rise velocities have considerable influence on amending the hydrodynamics, as well as heat and mass transfer coefficients. Researches have concluded that physical factors like bubble sizes such as gas density, liquid viscosity, surface tension and operating conditions like pressure and temperature profoundly affect bubble size distributions and bubble size and are widely reported in the following literatures [60, 69, 70]. For study of bubble holdup structures, rise velocities and bubble groups, dynamic gas disengagement (DGD) technique [71] is widely used. The principle includes tracking relative drop in dispersion height after the flow of gas is restricted. The disengagement profile can be utilized for separating the bubbles of different sizes contributing to the total gas holdup [72]. To be more specific, an accurate measurement of it is required along with the rate of gas-liquid dispersion. As per the published work till date it can be summarized that with an increase in a certain value of solid concentration, liquid viscosity, superficial gas velocity and surface tension, the bubble size increases. Along with increase in concentration of foaming agents, there is a decrease in average bubble size. In churn-turbulent regime, the contribution of small-sized bubble to total holdup is constant [73]. Large bubble holdup solely is dependent on the column diameter and not on solids concentration whereas vice versa holds valid for small bubble holdup.

4.4.4 Mass Transfer Coefficient

In a bubble column, the overall rate of mass transfer for unit volume of dispersion is solely regulated by the mass transfer coefficient for liquid side, i.e., $k_L a$ with an assumption of negligible gas side resistance. In a bubble column reactor with the variation of interfacial area a variation in $k_L a$ is also noticed. Assuming the bubbles to be spherical in shape, the Sauter mean bubble diameter and gas holdup is related to the specific gas-liquid interfacial area by Equation (4.45) [74]. In case of gas-liquid contacting reactors, efficient transfer of mass from gas to liquid phase plays the key role for operating the process. For both characterization and designing of industrial stirred and non-stirred gas-liquid contacting reactors, volumetric mass transfer coefficient holds significance as well [67]. With literature studies it is perceived that volumetric mass transfer coefficient, $k_L a$ enhances with increase in velocity, density of gas and pressure exerted whereas decreases with increase in solid concentration present and liquid viscosity. It can also be interpreted that with the presence of sufficient amount of surfactants an increase of $k_L a$ is noticed, due to formation of smaller bubbles. Henceforth, formation of larger bubbles in industrial column must be prevented to exalt the effectiveness of the mass transfer operation [75].

4.5 Application

Moving from the fundamental science to engineering application ozone is one of the most promising advanced oxidation processes and has a wide range of applications, either stand-alone or in combination with other advanced oxidation processes such as photolysis [76], fenton [77], cavitation [78, 79], etc. In regard to the water purification techniques, ozone can be used efficiently for water loaded with higher concentrations of industrial by-products. Due to degree of variation in contaminant loading, wastewater treatment equipment is simply specified after scaled down pilot experimentation has been performed to examine the average requirement of ozone dosages. From among the multiple benefits of ozone the most significant is the reduction in processing costs as well as the reduction of

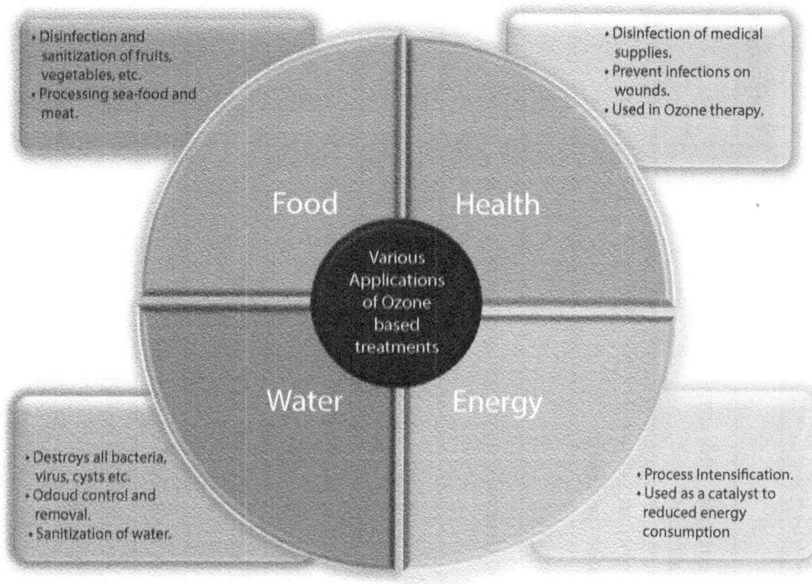

Figure 4.5 Different applications based on ozone treatment for water-food-health-energy nexus.

penalties due to the disposal of contaminated wastewater. Additionally, this technology is clean and environmental friendly compared to other technologies. Figure 4.5 highlights some major widespread applications of ozonation technology in daily life at small scale as well as industrial scale.

Ozonation is a widely used technique for disinfection in a number of industrial-scale water treatments due to its high reactive hydrolyse power [35, 80]. In the United States, this technology has been applied for more than hundred years to treat drinking water [80], because this technology is mainly expedient for potable reusability and water reclamation [81, 82]. Basically, ozone is a highly unstable and reactive gaseous chemical which can be generated by the conversion of oxygen by the application of high energetic power/electrical spark like electrical discharge and therefore, it is an even more powerful hydrolyze agent than H_2O_2. The main reason behind this highly reactive nature of ozone is the formation of free radicals and hence, this method can be enhanced to treat wastewater including hydrogen per-oxide, UV radiation and catalysts [83, 84]. Table 4.4 below provides a list of work done in the treatment of wastewater effluent from domestic, agricultural along with industrial sources by means of ozonation process. Figure 4.6 provides a graphical representation of the number of articles published on the same in the past decades.

Table 4.4 Different applications based on ozone treatment irrespective of pilot and industrial scale.

Type of water	Type of oxidation treatment	Key note	References
Domestic/ municipal wastewater	Ozone	Applied on pharmaceuticals, endocrine disruptors during drinking water treatment	[85]
	Aeration	Focused on the emission and control of N_2O from biological wastewater treatment	[86]
	Ozonation	Phenol degradation study	[87]
	Photo-fenton oxidation	polycyclic aromatic hydrocarbons degradation study from potable water	[88]
	Ozonation and photo-fenton process	Micropollutant removal and tertiary treatment of domestic wastewater	[89]
	Constructed wetland coupled with ozonation	Ibuprofen and Naproxen removal from domestic wastewater	[90]
	photolytic and photocatalytic ozonation	Enhanced micropollutants removal in municipal wastewaters	[91]
	UV/H_2O_2, ozonation	Removal of antibiotics from real wastewater	[92]
	Ozonation and advanced photocatalytic process	Removal of pharmaceuticals and personal care products from domestic wastewater	[93]

(Continued)

Table 4.4 Different applications based on ozone treatment irrespective of pilot and industrial scale. (*Continued*)

Type of water	Type of oxidation treatment	Key note	References
Industrial wastewater	Activated sludge with ozonation	High microbial inactivation with ozonation achieving microbial drinking water standards	[94]
	Fe-based catalytic ozonation	Biodegradation for the treatment of industrial wastewater	[95]
	Ozone/UV and ozone/H_2O_2	Treatment of mineral oil-contaminated wastewaters	[96]
	UV/H_2O_2	Biological wastewater treatment	[97]
	UV, O_3, H_2O_2	Review on AOPs involving hydrolysed hydrolyse of wastewater	[98]
	Ozonation and electrochemical process	Synergestic effect on industrial wastewater treatment has been evaluated	[99]
	Electrochemical AOP + electro-fenton + persulfate	Mixed industrial wastewater treatment to treat COD	[100]
	Hydrodynamic cavitation + AOPs using H_2O_2	Tannery wastewater treatment	[101]
	Photocatalytic ozonation	Treatment of highly polluted industrial wastewater	[102]

(*Continued*)

Table 4.4 Different applications based on ozone treatment irrespective of pilot and industrial scale. (*Continued*)

Type of water	Type of oxidation treatment	Key note	References
Textile industry wastewater	AOPs	Review on AOPs in industrial textile wastewater treatment	[103]
	Ozone	Textile wastewater treatment for brine reuse	[104]
	UV/H_2O_2, UV/TiO_2 and VIS/TiO_2	Photodegradation of the aminoazobenzene acid orange 52	[105]
	UV/H_2O_2	Decolorization of azo dye wastewater	[106]
	UV/H_2O_2	Photooxidative decolorization of four textile dye stuffs, Acid Orange 7 (AO7), Acid Orange 8 (AO8), Acid Orange 52 (AO52) and Acid Blue 74 (AB74)	[107]
	Fenton-oxidation and ozone	Acute toxicity removal in textile finishing wastewater	[108]
	Ozone	Removal of COD and color from hydrolysed textile azo dye	[109]
	UV, O_3, H_2O_2	Polishing of biologically treated textile wastewater	[110]
	Ozone integrated membrane bio-reactor	Real textile wastewater treatment and COD removal	[111]

(*Continued*)

Table 4.4 Different applications based on ozone treatment irrespective of pilot and industrial scale. (*Continued*)

Type of water	Type of oxidation treatment	Key note	References
Agricultural wastewater	Conmbined ozonation and photocatalysis	Degradation of nitrogen containing organic compounds	[112]
	Photo-fenton oxidation	Treatment of water containing pesticides	[113]
	Photo-fenton/TiO_2 + solar energy	Treatment of water soluble pesticides	[114]
	Fe^{2+}/H_2O_2, Fe^{3+}/H_2O_2, TiO_2 and $TiO_2/Na_2S_2O_8$	Degradation of pesticides in water	[115]
	Ozone/UV	Removal of pesticides from aqueous solutions	[116]
	Photo-fenton oxidation + nanofiltration	Treatment of malathion pesticide wastewater	[117]
	heterogeneous Fenton/-persulfate + hydrodynamic cavitation	Study on Sulfadiazine degradation	[118]
	Sunlight activated $Na_2S_2O_8$	Treatment of agro-wastewater polluted with pesticide residues	[119]
	Hydroxyl and sulfate radical-based AOPs	Treatment of phenolic compound present in olive mill wastewater	[120]
	UV/H_2O_2, UV/TiO_2	Urban wastewater disinfection for agricultural reuse	[121]

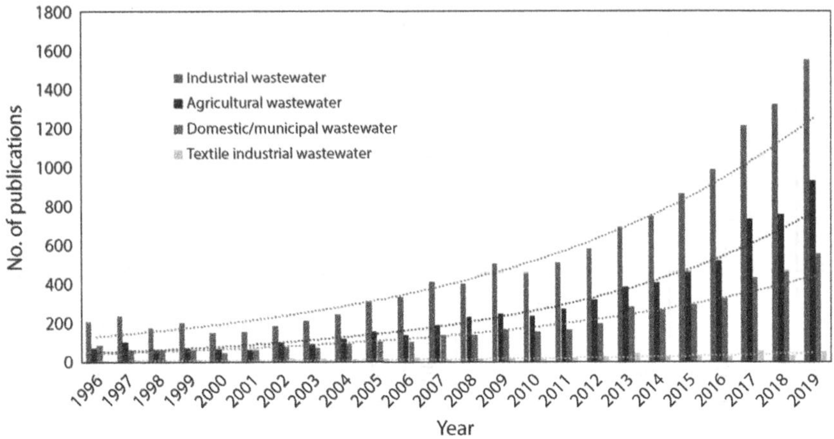

Figure 4.6 Comparison between the articles on treatment of effluents derived from various sources by ozonation published over the past 20 years.

There are plenty of research works available in the literature describing the broad application of ozonation in the field of wastewater treatment which are derived from domestic, agricultural and industrial sources. But the application of ozonation in grey water treatment has not been given that much importance compared to traditional systems. Since India is one of the growing economic and populated countries in the world as well as a developing country in the third world category, alternative as well as efficient technology is very important to keep the conventional systems and technology safe. India is also one of the countries suffering from water scarcity and hence, an alternative approach is necessary to eradicate the problem. With the development of technology for wastewater treatment from various effluents, a reusing system can also be very helpful for India like other highly populated countries. Therefore, in such point of objective, grey water can be the most prominent target source of water due to its high availability and less contaminated than industrial wastewater. Grey water basically contains household wastewater except fecal contaminants which include soap, detergents, hair and bacteria, etc., and hence this wastewater can be recycled by means of technology which consists of filters and purifiers. Once this process is completed, the water will be clean enough to be used for watering gardens including vegetable patches and other purposes. Among various processes for the treatment of grey water by disinfection, some widespread treatment processes are chemical, physico-chemical, biological and physical processes [122–125]. Usually, the above-mentioned methods are used for the treatment of different solids, organic and inorganic pollutants in effluent, whereas, to disinfect the contaminated water, mainly chlorination, UV, H_2O_2, Ozonation and advanced oxidation process (AOP) are applied [126, 127] to ensure the safety of the water microbiologically. The major advantages of application of ozonation in grey-water treatment over other technologies includes strong oxidizing power and short reaction time without addition of any chemicals, thereby having its effect over a varied pH range reacting rapidly with a variety of inorganic, organic and microbiological problems and taste and odor problems. Because of these several advantages of this technique over other oxidation processes, there are some case studies currently focusing on grey-water treatment for the implementation in industrial scale and the literature of the same has been reported below.

Hernandez Leal *et al.* [128] have studied the removal of micro-pollutants from aerobically treated grey water via ozone. They ozonated the grey water with a dosing rate of 1.22 to treat the micro-pollutants having the concentration of 100-1600 μgL^{-1} for 45 min and effectively reduced the concentration of parabens, bisphenol-A, hexylcinnamic aldehyde, 4-methylbenzylidene-camphor (4MBC), benzophenone-3 (BP3), triclosan, galaxolide and ethylhexyl methoxycinnamate up to 99%. After completion of 1 hr, they reduced the concentration of benzalkonium chloride at 98%, tonalide and nonylphenol at 95%, octocrylene at 92% and2-phenyl-5-benzimidazolesulfonic acid (PBSA) at 84%. After that, increasing the ozone dosage at 15 mgL^{-1} removed the presence of octocrylene, nonylphenol, triclosan, galaxolide, tonalide and 4 methylbenzylidene-camphor to below limits of quantification, with removal efficiencies of at least 79%.

Sevimli and Sarikaya [129] reported on the treatment of textile effluents and dyes by the application of ozonation to investigate the effect of ozone dosage, pH and dye concentration. From the investigation, results revealed that the apparent rate constants were raised with increase in ozone dose, pH and decrease in initial dye concentration. Colour removal efficiencies of wastewater for 15 min ozonation at high ozone dosage (~ 130 $mg.dm^{-3}.min^{-1}$) were 95 to 97% and at low dosage (~ 20 $mg.dm^{-3}.min^{-1}$) were 80 to 87%, respectively. Chemical oxygen demand (COD) and dissolved organic carbon (DOC) removal efficiencies were 15-46% and 10-20% for two different sample of wastewater for 15 min ozonation at different dosage of ozone. Specific ozone consumption for Acid Red 183 dye solution with a concentration of 50 $mg.dm^{-3}$ increased from 0.32 to 0.72 mg O_3 per mg of dye decomposed with the increase in dye concentration to 500 $mg.dm^{-3}$.

Ding *et al.* [130] have enlightened their investigation on the exploration of physiological and biochemical characteristics and some environmental influence like pH, temperature, and turbidity. Here the authors treated seven strains of chlorine-resistant bacteria isolated from drinking water.

Ozone was applied to investigate the inactivation effect of bacteria and spores present in the drinking water by the analysis of DNA concentration and cell structure variations of organisms. The ozone resistance of bacteria was lower than that of spores at an ozone concentration of 1.5 mg/L. The increase in ozone concentration and treatment time inactivated more than 99.9% of Bacillus cereus spores. Hence, from the results they concluded that ozone disinfection can be a promising method for inactivating chlorine-resistant bacteria and spores in drinking water.

4.6 Conclusion and Thoughts About the Future

This chapter is an effort to provide an in-depth understanding of the fundamental science of ozonation to engineering aspects and industrial applications. It can be concluded that the chemistry of ozone is explored pretty well; however, from the engineering point of view, the reactions involved during the typical gas-liquid contact involving mass transfer mechanism is still a thriving area of research. Additionally, the role of generated bubbles during the sparging of ozone which plays a major role in the mass transfer mechanism and kinetics is a thriving area of research. Considering the applications of ozone, energetics and thereby cost is a crucial factor, hence the future works could involve the integrated studies which can lead to process intensification. Hence, it is strongly believed that this work will attract special attention from researchers working in this field to develop an effective system for the sake of society which will improve the water treatment sector in developing countries like India.

Acknowledgement

Dr. Anirban Roy would like to thank the Additional Competitive Grant [Ref. GOA/ACG/2018-2019/Oct/08 dt. 30.10.2018] by BITS Pilani Goa for carrying out the work. Mr. Anupam Mukherjee would also like to thank the BRIC Idea Exposition Grant [Ref. BIRAC Innovation Centre/IKP Knowledge Park/GIM/Goa Cluster dt. 23.02.2020] for the partial support.

Nomenclature

1. Roman Symbols

J_A = Diffusive flux of component A, (moles/m^2.s)

c_m = molar diffusivuty of fluid, (moles/m^3)
D_{AB} = Free liquid diffusivity of component A in B, (m^2/s)
χA = mole fraction of component A,
z = distance in the z direction (m)
k'_s = restricted Interficial mass transfer coefficient (m/s)

T = Temperature, (°K)
R = Gas constant, (N – m / (k – mole) (K)
H = Henry' s Law Constant, [KPa(mg/L)$^{-1}$]
N_A = Mass transfer per unit time, (moles/s)
k_G = Gas film coefficient, $\left(\frac{\text{mol}}{\text{m}^2}\right)$s. Pa

k_L = Liquid film coefficient (m/s)
K_G = Overall mass transfer coefficient based on gas phase concentration,
$\left(\frac{\text{mol}}{\text{m}^2}\right)$s. Pa

K_L = Overall mass transfer coefficient based on liquid phase concentration, (m/s)
A = Mass Transfer area (m^2)

C_G = Bulk Concentration of solute in gas phase, $\left(\frac{\text{mol}}{\text{lit}}\right)$

C_{Gi} = Interphase gas concentration, $\left(\frac{\text{mol}}{\text{lit}}\right)$

C_{Li} = Interphase Liquid concentration, $\left(\frac{\text{mol}}{\text{lit}}\right)$

C_L = Bulk concentration of solute in liquid phase, $\left(\frac{\text{mol}}{\text{lit}}\right)$

C_G^* = Concentration of solute gas in liquid which will be in equilibrium with gas $\left(\frac{\text{mol}}{\text{lit}}\right)$

C_L^* = Concentration of solute gas in gaseous phase which will be in equilibrium with liquid $\left(\frac{\text{mol}}{\text{lit}}\right)$

M = Molecular Weight
V = Volume (m^3)
N_{o3} = Adsorption Rate or flux of ozone (mol m^{-2} s^{-1})

D = Diffusivity of ozone gas $m^2/(sec)$
C^*_{o3} = Equilibrium concentration of dissolved ozone in water (M^{-1})
C_{o3} = Concentration of dissolved ozone in water (M^{-1})
P_{o3} = Ozone Partial Pressure (Pa)
$k_L a$ = Volumetric Mass transfer coefficient (s^{-1})
a = Specific gas-liquid interficial area (m-1)
r = rate of reaction, (gmol L^{-1} s^{-1})
n = Order of reaction for power law rate expression
K_w = Ozone decay rate constant (sec^{-1})
$[OH^{-1}]$ = Hydroxide Ion Concentration (g-ml L^{-1})
Da = Damkohler Number
Ha = Hatta Number
Sc = Schmidt Number
Sh = Sherwood Number
Re = Reynold Number
P_G = Partial Pressure of Ozone in gas phase (Pa)
P^* = Partial Pressure of Ozone in equilibrium with bulk liquid concentration in the liquid phase (Pa)
g_{in} = Ozone Supply Rate (mg s^{-1})
g^{out} = Ozone output flow rate (mg s^{-1})
Y = Ozone mole fraction in the gas phase
G =. Gas flow rate (L sec^{-1})
Q = Water flow rate (L sec^{-1})
E = Enhancement Factor
h = height of the column (m)
d_b = average size of bubbles (mm)
u_g = superficial gas velocity (m/s)
ϵ_g = gas hold-up into the column
g = acceleration due to gravity (m^2/s)

2. Greek Letters

α = Condensation Coefficient
θ = Exposure Time
ψ = Specific rate Constant $[s^{-1} (gmolL^{-1})^{1-n}]$
η = interphase effectiveness factor
γ = Partition coefficient for ozone (dimensionless)
ρ = density, (kg/m3)
σ = Surface tention, (N m^{-1})
μ = Solution viscosity, (poise)

β = Liquid Holdup

3. Subscripts

L = Liquids
G = Gas
O_3 = Ozone Gas
i = interface
b = bubble
in = inlet
out = outlet
A = Solute
e = equivalent

References

1. K. Ikehata, Y. Li, Ozone-Based Processes, in: *Adv. Oxid. Process. Wastewater Treat. Emerg. Green Chem. Technol.*, 2018. https://doi.org/10.1016/B978-0-12-810499-6.00005-X.
2. H. Zhou, D.W. Smith, Ozone mass transfer in water and wastewater treatment: Experimental observations using a 2D laser particle dynamics analyzer, *Water Res.* (2000). https://doi.org/10.1016/S0043-1354(99)00196-7.
3. Debarghya Halder, Subhajit Mondal, Anupam Mukherjee, Abhijit Ghosh, An Overview on the Treatment of Waste Water Using Renewable Energy, *Int. J. Res. Eng. Technol.* 05 (2016) 78–83. https://doi.org/10.15623/ijret.2016.0513014.
4. H. El Abd, A. Hafez, E. El Zanati, G. El Diwani, Modelling of water-ozone reaction system, *Environ. Int.* (1985). https://doi.org/10.1016/0160-4120(85)90183-7.
5. M.M. Mitani, A.A. Keller, O.C. Sandall, R.G. Rinker, Mass transfer of ozone using a microporous diffuser reactor system, *Ozone Sci. Eng.* (2005). https://doi.org/10.1080/01919510590908995.
6. H. Gallard, U. Von Gunten, Chlorination of natural organic matter: Kinetics of chlorination and of THM formation, *Water Res.* (2002). https://doi.org/10.1016/S0043-1354(01)00187-7.
7. S. Chakraborty, A. Mukherjee, S. Das, N. Raju Maddela, S. Iram, P. Das, Study on isotherm, kinetics, and thermodynamics of adsorption of crystal violet dye by calcium oxide modified fly ash, *Environ. Eng. Res.* (2020). https://doi.org/10.4491/eer.2019.372.
8. R. Das, A. Mukherjee, I. Sinha, K. Roy, B.K. Dutta, Synthesis of potential bio-adsorbent from Indian Neem leaves (Azadirachta indica) and its optimization for malachite green dye removal from industrial wastes using response

surface methodology: kinetics, isotherms and thermodynamic studies, *Appl. Water Sci.* (2020). https://doi.org/10.1007/s13201-020-01184-5.

9. K. Roy, A. Mukherjee, N.R. Maddela, S. Chakraborty, B. Shen, M. Li, D. Du, Y. Peng, F. Lu, L.C. Garciá Cruzatty, Outlook on the bottleneck of carbon nanotube in desalination and membrane-based water treatment-A review, *J. Environ. Chem. Eng.* (2020). https://doi.org/10.1016/j.jece.2019.103572.
10. A. Mukherjee, S. Lanjewar, R. Kumar, A. Chakraborty, A. Abdelrasoul, A. Roy, Role of Thermodynamics and Membrane Separations in Water-Energy Nexus, in: *Model. Membr. Membr. Process.*, 2020. https://doi.org/10.1002/9781119536260.ch4.
11. R. Das, A. Mukherjee, I. Sinha, A. De, K. Roy, A. Sirkar, Studies on Boiling Point Elevation of Water and its Correlation with Salt Concentrations for Application in Evaporator Design, *SSRN Electron. J.* (2020). https://doi.org/10.2139/ssrn.3710394.
12. A. Mukherjee, K. Roy, D.K. Jana, S.A. Hossain, Qualitative model optimization of almond (Terminalia catappa) oil using soxhlet extraction in type-2 fuzzy environment, *Soft Comput.* (2020). https://doi.org/10.1007/s00500-019-04158-1.
13. D. Grasso, E. Fujikawa, W.J. Weber, Ozone mass transfer in a gas-sparged turbine reactor, *Res. J. Water Pollut. Control Fed.* (1990).
14. K.B. Bischoff, D.M. Himmelblau, Mass Transfer, *Ind. Eng. Chem.* (1968). https://doi.org/10.1021/ie50697a008.
15. W.K. Lewis, W.G. Whitman, Principles of Gas Absorption, *Ind. Eng. Chem.* (1924). https://doi.org/10.1021/ie50180a002.
16. A.K. Bin, M. Roustan, Basic Chemical Engineering Concepts for the Design of Ozone Gas-Liquid Reactors, *Int. Spec. Symp. IOA 2000 Fundamental Eng. Concepts Ozone React. Eng.* (2005) 99–131.
17. P. V. Danckwerts, Significance of Liquid-Film Coefficients in Gas Absorption, *Ind. Eng. Chem.* (1951). https://doi.org/10.1021/ie50498a055.
18. W.E. Dobbins, Mechanism of Gas Absorption by Turbulent Liquids, in: *Adv. Water Pollut. Res.*, 1964. https://doi.org/10.1016/b978-1-4832-8391-3.50027-0.
19. H.L. Toor, J.M. Marchello, Film-penetration model for mass and heat transfer, *AIChE J.* (1958). https://doi.org/10.1002/aic.690040118.
20. G.E. Fortescue, J.R.A. Pearson, On gas absorption into a turbulent liquid, *Chem. Eng. Sci.* (1967). https://doi.org/10.1016/0009-2509(67)80183-0.
21. J.C. Lamont, D.S. Scott, An eddy cell model of mass transfer into the surface of a turbulent liquid, *AIChE J.* (1970). https://doi.org/10.1002/aic.690160403.
22. L.K. Brumfield, T.G. Theofanous, Turbulent mass transfer in jet flow and bubble flow: A reappraisal of levich's theory, *AIChE J.* (1976). https://doi.org/10.1002/aic.690220332.
23. C.J. King, Turbulent liquid phase mass transfer at free gas-liquid interface, *Ind. Eng. Chem. Fundam.* (1966). https://doi.org/10.1021/i160017a001.

24. H.D. Spriggs, Turbulence phenomena, J. T. Davies, Academic Press, New York (1972). 412 pages.$19.50, *AIChE J.* (1973). https://doi.org/10.1002/aic.690190142.
25. J.T. Davies, Turbulence phenomena at free surfaces, *AIChE J.* (1972). https://doi.org/10.1002/aic.690180131.
26. M. Roustan, H. Roques, J. Mallevialle, J.P. Jones, Mass transfer of ozone to water: A fundamental study, *Ozone Sci. Eng.* (1980). https://doi.org/10.1080/01919518008550895.
27. P. Raimondi, H.L. Toor, Interfacial resistance in gas absorption, *AIChE J.* (1959). https://doi.org/10.1002/aic.690050119.
28. R.E. Treybal, *Mass Transfer Operations* 3rd edition, 1981.
29. A.K. Biń, B. Duczmal, P. Machniewski, Hydrodynamics and ozone mass transfer in a tall bubble column, *Chem. Eng. Sci.* (2001). https://doi.org/10.1016/S0009-2509(01)00213-5.
30. J.L. Sotelo, F.J. Beltrán, F.J. Benitez, J. Beltrán-Heredia, Henry's law constant for the ozone-water system, *Water Res.* (1989). https://doi.org/10.1016/0043-1354(89)90186-3.
31. G.A. Ratcliff, Unit operations of chemical engineering., *Chem. Eng. Sci.* (1957). https://doi.org/10.1016/0009-2509(57)85034-9.
32. A. Rodríguez, R. Rosal, J.A. Perdigón-Melón, M. Mezcua, A. Agüera, M.D. Hernando, P. Letón, A.R. Fernández-Alba, E. García-Calvo, Ozone-based technologies in water and wastewater treatment, *Handb. Environ. Chem. Vol. 5 Water Pollut.* (2008). https://doi.org/10.1007/698_5_103.
33. J.A. Roth, D.E. Sullivan, Solubility of ozone in water, *Ind. Eng. Chem. Fundam.* (1981). https://doi.org/10.1021/i100002a004.
34. M. Peleg, The chemistry of ozone in the treatment of water, *Water Res.* (1976). https://doi.org/10.1016/0043-1354(76)90052-X.
35. B.L. Loeb, Ozone: Science and engineering: Thirty-three years and growing, Ozone Sci. Eng. (2011). https://doi.org/10.1080/01919512.2011.584302.
36. P.H. Calderbank, J.M.O. Lewis, Ozone-decomposition catalysis, *Chem. Eng. Sci.* (1976). https://doi.org/10.1016/0009-2509(76)85035-X.
37. F.J. Beltrán, F.J. Rivas, R. Montero-De-Espinosa, Iron type catalysts for the ozonation of oxalic acid in water, *Water Res.* (2005). https://doi.org/10.1016/j.watres.2005.06.018.
38. M.D. Gurol, S. Nekouinaini, Effect of organic substances on mass transfer in bubble aeration, *J. Water Pollut. Control Fed.* (1985).
39. F.J. Beltrán, A. Aguinaco, J.F. García-Araya, Mechanism and kinetics of sulfamethoxazole photocatalytic ozonation in water, *Water Res.* (2009). https://doi.org/10.1016/j.watres.2008.12.015.
40. B.W. Brooks, Chemical reactor analysis and design, *Chem. Eng. J.* (1991). https://doi.org/10.1016/0300-9467(91)85016-o.
41. F.J. Beltrán, J.M. Encinar, J.F. García-Araya, M.G. Munñoz, Modelling Industrial Wastewater Ozonation in Bubble Contactors. 1. Rate

Coefficient Determination, *Ozone Sci. Eng.* (1995). https://doi.org/10.1080/01919519508547344.
42. A.G. Hill, H.T. Spencer, Mass Transfer in a Gas Sparged Ozone Reactor, in: 1975.
43. A. Laplanche, N.L. Le Sauze, G. Martin, B. Langlais, Simulation of Ozone Transfer in Water. Comparison with a Pilot Unit, *Ozone Sci. Eng.* (1991). https://doi.org/10.1080/01919512.1991.10555700.
44. M. Roustan, R.Y. Wang, D. Wolbert, Modeling hydrodynamics and mass transfer parameters in a continuous ozone bubble column, *Ozone Sci. Eng.* (1996). https://doi.org/10.1080/01919519608547331.
45. A.K. Biń, Application of A Single-Bubble Model in Estimation of Ozone Transfer Efficiency in Water, *Ozone Sci. Eng.* (1995). https://doi.org/10.1080/01919512.1995.10555761.
46. N. Kantarci, F. Borak, K.O. Ulgen, Bubble column reactors, *Process Biochem.* (2005). https://doi.org/10.1016/j.procbio.2004.10.004.
47. G.A. Hughmark, Holdup and mass transfer in bubble columns, *Ind. Eng. Chem. Process Des. Dev.* (1967). https://doi.org/10.1021/i260022a011.
48. W.J. Masschelein, Unit Processes in Drinking Water Treatment, 2020. https://doi.org/10.1201/9781003066880.
49. R.E. Bühler, J. Staehelin, J. Hoigné, Ozone decomposition in water studied by pulse radiolysis. 1. HO2/O2- and HO3/O3- as intermediates, *J. Phys. Chem.* (1984). https://doi.org/10.1021/j150656a026.
50. J. Staehelin, R.E. Bühler, J. Hoigné, Ozone decomposition in water studied by pulse radiolysis. 2. OH and HO4 as chain intermediates, *J. Phys. Chem.* (1984). https://doi.org/10.1021/j150668a051.
51. H. Tomiyasu, H. Fukutomi, G. Gordon, Kinetics and Mechanism of Ozone Decomposition in Basic Aqueous Solution, *Inorg. Chem.* (1985). https://doi.org/10.1021/ic00213a018.
52. J. Hoigné, Chemistry of Aqueous Ozone and Transformation of Pollutants by Ozonation and Advanced Oxidation Processes, in: 1998. https://doi.org/10.1007/978-3-540-68089-5_5.
53. B. Kasprzyk-Hordern, M. Ziółek, J. Nawrocki, Catalytic ozonation and methods of enhancing molecular ozone reactions in water treatment, *Appl. Catal. B Environ.* (2003). https://doi.org/10.1016/S0926-3373(03)00326-6.
54. U. Von Gunten, Ozonation of drinking water: Part I. Oxidation kinetics and product formation, *Water Res.* (2003). https://doi.org/10.1016/S0043-1354(02)00457-8.
55. M. Sánchez-Polo, U. Von Gunten, J. Rivera-Utrilla, Efficiency of activated carbon to transform ozone into ·OH radicals: Influence of operational parameters, *Water Res.* (2005). https://doi.org/10.1016/j.watres.2005.05.026.
56. S. Degaleesan, M. Dudukovic, Y. Pan, Experimental study of gas-induced liquid-flow structures in bubble columns, *AIChE J.* (2001). https://doi.org/10.1002/aic.690470904.

57. A. Prakash, A. Margaritis, H. Li, M.A. Bergougnou, Hydrodynamics and local heat transfer measurements in a bubble column with suspension of yeast, *Biochem. Eng. J.* (2001). https://doi.org/10.1016/S1369-703X(01)00137-1.
58. M. Bouaifi, G. Hebrard, D. Bastoul, M. Roustan, A comparative study of gas hold-up, bubble size, interfacial area and mass transfer coefficients in stirred gas-liquid reactors and bubble columns, *Chem. Eng. Process.* (2001). https://doi.org/10.1016/S0255-2701(00)00129-X.
59. K. Shimizu, S. Takada, K. Minekawa, Y. Kawase, Phenomenological model for bubble column reactors: Prediction of gas hold-ups and volumetric mass transfer coefficients, *Chem. Eng. J.* (2000). https://doi.org/10.1016/S1385-8947(99)00165-5.
60. R. Schäfer, C. Merten, G. Eigenberger, Bubble size distributions in a bubble column reactor under industrial conditions, *Exp. Therm. Fluid Sci.* (2002). https://doi.org/10.1016/S0894-1777(02)00189-9.
61. C.L. Hyndman, F. Larachi, C. Guy, Understanding gas-phase hydrodynamics in bubble columns: A convective model based on kinetic theory, *Chem. Eng. Sci.* (1997). https://doi.org/10.1016/S0009-2509(96)00387-9.
62. J.H. Hills, Radial non-uniformity of velocity and voidage in a bubble column., *Trans Inst Chem Eng.* (1974).
63. L.-S. Fan, Classification and Significance, in: *Gas–Liquid–Solid Fluid. Eng.*, 1989. https://doi.org/10.1016/b978-0-409-95179-0.50007-6.
64. A. Schumpe, G. Grund, The gas disengagement technique for studying gas holdup structure in bubble columns, *Can. J. Chem. Eng.* (1986). https://doi.org/10.1002/cjce.5450640602.
65. A. Matsuura, L. -S Fan, Distribution of bubble properties in a gas-liquid-solid fluidized bed, *AIChE J.* (1984). https://doi.org/10.1002/aic.690300604.
66. X. Luo, D.J. Lee, R. Lau, G. Yang, L.S. Fan, Maximum stable bubble size and gas holdup in high-pressure slurry bubble columns, *AIChE J.* (1999). https://doi.org/10.1002/aic.690450402.
67. Y. Kawase, S. Umeno, T. Kumagai, The prediction of gas hold-up in bubble column reactors: Newtonian and non-newtonian fluids, *Chem. Eng. J.* (1992). https://doi.org/10.1016/0300-9467(92)80001-Q.
68. Y.T. Shah, B.G. Kelkar, S.P. Godbole, W. -D Deckwer, Design parameters estimations for bubble column reactors, *AIChE J.* (1982). https://doi.org/10.1002/aic.690280302.
69. D.N. Miller, Scale-up of agitated vessels gas-liquid mass transfer, *AIChE J.* (1974). https://doi.org/10.1002/aic.690200303.
70. R. Kumar, N.K. Kuloor, The formation of bubbles and drops, *Adv. Chem. Eng.* (1970). https://doi.org/10.1016/S0065-2377(08)60186-6.
71. J.G. Daly, S.A. Patel, D.B. Bukur, Measurement of gas holdups and sauter mean bubble diameters in bubble column reactors by dynamics gas disengagement method, *Chem. Eng. Sci.* (1992). https://doi.org/10.1016/0009-2509(92)85081-L.

72. K. Sriram, R. Mann, Dynamic gas disengagement: A new technique for assessing the behaviour of bubble columns, *Chem. Eng. Sci.* (1977). https://doi.org/10.1016/0009-2509(77)80222-4.
73. W.D. Deckwer, Bubble column reactors. (1985). https://doi.org/10.1201/9780203736739-9.
74. J.B. Joshi, M.M. Sharma, Circulation Cell Model for Bubble Columns, *Trans. Inst. Chem. Eng.* (1979).
75. P. Zehner, Momentum, Mass and Heat Transfer in Bubble Columns. Part 2. Axial Blending and Heat Transfer, *Int. Chem. Eng.* (1986).
76. Z. Jing, S. Cao, Combined application of UV photolysis and ozonation with biological aerating filter in tertiary wastewater treatment, *Int. J. Photoenergy.* (2012). https://doi.org/10.1155/2012/140605.
77. C. Wu, W. Chen, Z. Gu, Q. Li, A review of the characteristics of Fenton and ozonation systems in landfill leachate treatment, *Sci. Total Environ.* (2021). https://doi.org/10.1016/j.scitotenv.2020.143131.
78. A. Mukherjee, A. Mullick, P. Vadthya, S. Moulik, A. Roy, Surfactant degradation using hydrodynamic cavitation based hybrid advanced oxidation technology: A techno economic feasibility study, *Chem. Eng. J.* (2020). https://doi.org/10.1016/j.cej.2020.125599.
79. A. Mukherjee, A. Mullick, R. Teja, P. Vadthya, A. Roy, S. Moulik, Performance and energetic analysis of hydrodynamic cavitation and potential integration with existing advanced oxidation processes: A case study for real life greywater treatment, *Ultrason. Sonochem.* (2020). https://doi.org/10.1016/j.ultsonch.2020.105116.
80. B.L. Loeb, C.M. Thompson, J. Drago, H. Takahara, S. Baig, Worldwide Ozone Capacity for Treatment of Drinking Water and Wastewater: A Review, *Ozone Sci. Eng.* (2012). https://doi.org/10.1080/01919512.2012.640251.
81. D. Gerrity, E. Owens-Bennett, T. Venezia, B.D. Stanford, M.H. Plumlee, J. Debroux, R.S. Trussell, Applicability of Ozone and Biological Activated Carbon for Potable Reuse, *Ozone Sci. Eng.* (2014). https://doi.org/10.1080/01919512.2013.866886.
82. D. Gerrity, S. Snyder, Review of ozone for water reuse applications: Toxicity, regulations, and trace organic contaminant oxidation, *Ozone Sci. Eng.* (2011). https://doi.org/10.1080/01919512.2011.578038.
83. M.A. Oturan, J.J. Aaron, Advanced oxidation processes in water/wastewater treatment: Principles and applications. A review, *Crit. Rev. Environ. Sci. Technol.* (2014). https://doi.org/10.1080/10643389.2013.829765.
84. F.J. Beltran, Ozone Reaction Kinetics for Water and Wastewater Systems, 2003. https://doi.org/10.1201/9780203509173.
85. R. Broséus, S. Vincent, K. Aboulfadl, A. Daneshvar, S. Sauvé, B. Barbeau, M. Prévost, Ozone oxidation of pharmaceuticals, endocrine disruptors and pesticides during drinking water treatment, *Water Res.* (2009). https://doi.org/10.1016/j.watres.2009.07.031.

86. K.Y. Park, Y. Inamori, M. Mizuochi, K.H. Ahn, Emission and control of nitrous oxide from a biological wastewater treatment system with intermittent aeration, *J. Biosci. Bioeng.* (2000). https://doi.org/10.1016/S1389-1723(00)80077-8.
87. C.D. Adams, R.A. Cozzens, B.J. Kim, Effects of ozonation on the biodegradability of substituted phenols, *Water Res.* (1997). https://doi.org/10.1016/S0043-1354(97)00114-0.
88. T.S.B. Abd Manan, T. Khan, S. Sivapalan, H. Jusoh, N. Sapari, A. Sarwono, R.M. Ramli, S. Harimurti, S. Beddu, S.N. Sadon, N.L.M. Kamal, A. Malakahmad, Application of response surface methodology for the optimization of polycyclic aromatic hydrocarbons degradation from potable water using photo-Fenton oxidation process, *Sci. Total Environ.* (2019). https://doi.org/10.1016/j.scitotenv.2019.02.060.
89. S. Arzate, S. Pfister, C. Oberschelp, J.A. Sánchez-Pérez, Environmental impacts of an advanced oxidation process as tertiary treatment in a wastewater treatment plant, *Sci. Total Environ.* (2019). https://doi.org/10.1016/j.scitotenv.2019.07.378.
90. J.C. Lancheros, C.A. Madera-Parra, A. Caselles-Osorio, W.A. Torres-López, X.M. Vargas-Ramírez, Ibuprofen and Naproxen removal from domestic wastewater using a Horizontal Subsurface Flow Constructed Wetland coupled to Ozonation, *Ecol. Eng.* (2019). https://doi.org/10.1016/j.ecoleng.2019.05.007.
91. D. Bertagna Silva, A. Cruz-Alcalde, C. Sans, J. Giménez, S. Esplugas, Performance and kinetic modelling of photolytic and photocatalytic ozonation for enhanced micropollutants removal in municipal wastewaters, *Appl. Catal. B Environ.* (2019). https://doi.org/10.1016/j.apcatb.2019.02.072.
92. R. Anjali, S. Shanthakumar, Insights on the current status of occurrence and removal of antibiotics in wastewater by advanced oxidation processes, *J. Environ. Manage.* (2019). https://doi.org/10.1016/j.jenvman.2019.05.090.
93. H. Kaur, G. Hippargi, G.R. Pophali, A.K. Bansiwal, Treatment methods for removal of pharmaceuticals and personal care products from domestic wastewater, in: *Pharm. Pers. Care Prod. Waste Manag. Treat. Technol. Emerg. Contam. Micro Pollut.*, 2019. https://doi.org/10.1016/B978-0-12-816189-0.00006-8.
94. P. Alfonso-Muniozguren, J. Lee, M. Bussemaker, R. Chadeesingh, C. Jones, D. Oakley, D. Saroj, A combined activated sludge-filtration-ozonation process for abattoir wastewater treatment, *J. Water Process Eng.* (2018). https://doi.org/10.1016/j.jwpe.2018.07.009.
95. L. Yang, M. Sheng, Y. Li, W. Xue, K. Li, G. Cao, A hybrid process of Fe-based catalytic ozonation and biodegradation for the treatment of industrial wastewater reverse osmosis concentrate, *Chemosphere.* (2020). https://doi.org/10.1016/j.chemosphere.2019.124639.
96. R. Andreozzi, V. Caprio, A. Insola, R. Marotta, R. Sanchirico, Advanced oxidation processes for the treatment of mineral oil-contaminated wastewaters, *Water Res.* (2000). https://doi.org/10.1016/S0043-1354(99)00169-4.

97. C.B. Chidambara Raj, L.H. Quen, Advanced oxidation processes for wastewater treatment: Optimization of UV/H2O2 process through a statistical technique, *Chem. Eng. Sci.* (2005). https://doi.org/10.1016/j.ces.2005.03.065.
98. N.N. Mahamuni, Y.G. Adewuyi, Advanced oxidation processes (AOPs) involving ultrasound for waste water treatment: A review with emphasis on cost estimation, *Ultrason. Sonochem.* (2010). https://doi.org/10.1016/j.ultsonch.2009.09.005.
99. L.A. Bernal-Martínez, C. Barrera-Díaz, C. Solís-Morelos, R. Natividad, Synergy of electrochemical and ozonation processes in industrial wastewater treatment, *Chem. Eng. J.* (2010). https://doi.org/10.1016/j.cej.2010.08.062.
100. A. Popat, P. V. Nidheesh, T.S. Anantha Singh, M. Suresh Kumar, Mixed industrial wastewater treatment by combined electrochemical advanced oxidation and biological processes, *Chemosphere.* (2019). https://doi.org/10.1016/j.chemosphere.2019.124419.
101. S. Korpe, B. Bethi, S.H. Sonawane, K. V. Jayakumar, Tannery wastewater treatment by cavitation combined with advanced oxidation process (AOP), *Ultrason. Sonochem.* (2019). https://doi.org/10.1016/j.ultsonch.2019.104723.
102. A.M. Chávez, O. Gimeno, A. Rey, G. Pliego, A.L. Oropesa, P.M. Álvarez, F.J. Beltrán, Treatment of highly polluted industrial wastewater by means of sequential aerobic biological oxidation-ozone based AOPs, *Chem. Eng. J.* (2019). https://doi.org/10.1016/j.cej.2018.12.064.
103. K. Paździor, L. Bilińska, S. Ledakowicz, A review of the existing and emerging technologies in the combination of AOPs and biological processes in industrial textile wastewater treatment, *Chem. Eng. J.* (2019). https://doi.org/10.1016/j.cej.2018.12.057.
104. L. Bilińska, M. Gmurek, S. Ledakowicz, Textile wastewater treatment by AOPs for brine reuse, *Process Saf. Environ. Prot.* (2017). https://doi.org/10.1016/j.psep.2017.04.019.
105. C. Galindo, P. Jacques, A. Kalt, Photodegradation of the aminoazobenzene acid orange 52 by three advanced oxidation processes: UV/H2O2, UV/TiO2 and VIS/TiO2. Comparative mechanistic and kinetic investigations, *J. Photochem. Photobiol. A Chem.* (2000). https://doi.org/10.1016/S1010-6030(99)00199-9.
106. H.Y. Shu, M.C. Chang, Pilot scale annular plug flow photoreactor by UV/H2O2 for the decolorization of azo dye wastewater, *J. Hazard. Mater.* (2005). https://doi.org/10.1016/j.jhazmat.2005.05.038.
107. N. Daneshvar, A. Aleboyeh, A.R. Khataee, The evaluation of electrical energy per order (EEo) for photooxidative decolorization of four textile dye solutions by the kinetic model, *Chemosphere.* (2005). https://doi.org/10.1016/j.chemosphere.2004.11.012.
108. S. Meriç, H. Selçuk, V. Belgiorno, Acute toxicity removal in textile finishing wastewater by Fenton's oxidation, ozone and coagulation-flocculation processes, *Water Res.* (2005). https://doi.org/10.1016/j.watres.2004.12.021.

109. S.M. de A.G.U. de Souza, K.A.S. Bonilla, A.A.U. de Souza, Removal of COD and color from hydrolyzed textile azo dye by combined ozonation and biological treatment, *J. Hazard. Mater.* (2010). https://doi.org/10.1016/j.jhazmat.2010.02.053.
110. K. Nadeem, G.T. Guyer, N. Dizge, Polishing of biologically treated textile wastewater through AOPs and recycling for wet processing, *J. Water Process Eng.* (2017). https://doi.org/10.1016/j.jwpe.2017.09.011.
111. U. Sathya, Keerthi, M. Nithya, N. Balasubramanian, Evaluation of advanced oxidation processes (AOPs) integrated membrane bioreactor (MBR) for the real textile wastewater treatment, *J. Environ. Manage.* (2019). https://doi.org/10.1016/j.jenvman.2019.06.039.
112. M. Klare, G. Waldner, R. Bauer, H. Jacobs, J.A.C. Broekaert, Degradation of nitrogen containing organic compounds by combined photocatalysis and ozonation, *Chemosphere.* (1999). https://doi.org/10.1016/S0045-6535(98)00414-7.
113. H. Fallmann, T. Krutzler, R. Bauer, S. Malato, J. Blanco, Applicability of the Photo-Fenton method for treating water containing pesticides, *Catal. Today.* (1999). https://doi.org/10.1016/S0920-5861(99)00192-3.
114. S. Malato, J. Blanco, J. Cáceres, A.R. Fernández-Alba, A. Agüera, A. Rodríguez, Photocatalytic treatment of water-soluble pesticides by photo-Fenton and TiO2 using solar energy, *Catal. Today.* (2002). https://doi.org/10.1016/S0920-5861(02)00220-1.
115. M.H. Pérez, G. Peñuela, M.I. Maldonado, O. Malato, P. Fernández-Ibáñez, I. Oller, W. Gernjak, S. Malato, Degradation of pesticides in water using solar advanced oxidation processes, *Appl. Catal. B Environ.* (2006). https://doi.org/10.1016/j.apcatb.2005.11.013.
116. W.K. Lafi, Z. Al-Qodah, Combined advanced oxidation and biological treatment processes for the removal of pesticides from aqueous solutions, *J. Hazard. Mater.* (2006). https://doi.org/10.1016/j.jhazmat.2006.02.027.
117. Y. Zhang, K. Pagilla, Treatment of malathion pesticide wastewater with nanofiltration and photo-Fenton oxidation, *Desalination.* (2010). https://doi.org/10.1016/j.desal.2010.06.031.
118. K. Roy, V.S. Moholkar, Sulfadiazine degradation using hybrid AOP of heterogeneous Fenton/persulfate system coupled with hydrodynamic cavitation, *Chem. Eng. J.* (2020). https://doi.org/10.1016/j.cej.2019.03.170.
119. N. Vela, J. Fenoll, I. Garrido, G. Pérez-Lucas, P. Flores, P. Hellín, S. Navarro, Reclamation of agro-wastewater polluted with pesticide residues using sunlight activated persulfate for agricultural reuse, *Sci. Total Environ.* (2019). https://doi.org/10.1016/j.scitotenv.2019.01.060.
120. M.Y. Kilic, W.H. Abdelraheem, X. He, K. Kestioglu, D.D. Dionysiou, Photochemical treatment of tyrosol, a model phenolic compound present in olive mill wastewater, by hydroxyl and sulfate radical-based advanced oxidation processes (AOPs), *J. Hazard. Mater.* (2019). https://doi.org/10.1016/j.jhazmat.2018.06.062.

121. G. Ferro, A. Fiorentino, M.C. Alferez, M.I. Polo-López, L. Rizzo, P. Fernández-Ibáñez, Urban wastewater disinfection for agricultural reuse: effect of solar driven AOPs in the inactivation of a multidrug resistant E. coli strain, *Appl. Catal. B Environ.* (2015). https://doi.org/10.1016/j.apcatb.2014.10.043.
122. M. Pidou, F.A. Mamon, T. Stephenson, B. Jefferson, P. Jeffrey, Greywater recycling: Treatment options and applications, *Proc. Inst. Civ. Eng. Eng. Sustain.* (2007). https://doi.org/10.1680/ensu.2007.160.3.119.
123. S. Paris, C. Schlapp, Greywater recycling in Vietnam - Application of the HUBER MBR process, *Desalination.* (2010). https://doi.org/10.1016/j.desal.2009.09.099.
124. L.D. Nghiem, N. Oschmann, A.I. Schäfer, Fouling in greywater recycling by direct ultrafiltration, *Desalination.* (2006). https://doi.org/10.1016/j.desal.2005.04.087.
125. S. Jabornig, E. Favero, Single household greywater treatment with a moving bed biofilm membrane reactor (MBBMR), *J. Memb. Sci.* (2013). https://doi.org/10.1016/j.memsci.2013.06.049.
126. G.P. Winward, L.M. Avery, T. Stephenson, B. Jefferson, Chlorine disinfection of grey water for reuse: Effect of organics and particles, *Water Res.* (2008). https://doi.org/10.1016/j.watres.2007.07.042.
127. Z. Ronen, A. Guerrero, A. Gross, Greywater disinfection with the environmentally friendly Hydrogen Peroxide Plus (HPP), *Chemosphere.* (2010). https://doi.org/10.1016/j.chemosphere.2009.09.067.
128. L.H. Leal, N. Vieno, H. Temmink, G. Zeeman, C.J.N. Buisman, Occurrence of xenobiotics in gray water and removal in three biological treatment systems, *Environ. Sci. Technol.* (2010). https://doi.org/10.1021/es101509e.
129. M.F. Sevimli, H.Z. Sarikaya, Ozone treatment of textile effluents and dyes: Effect of applied ozone dose, pH and dye concentration, *J. Chem. Technol. Biotechnol.* (2002). https://doi.org/10.1002/jctb.644.
130. W. Ding, W. Jin, S. Cao, X. Zhou, C. Wang, Q. Jiang, H. Huang, R. Tu, S.F. Han, Q. Wang, Ozone disinfection of chlorine-resistant bacteria in drinking water, *Water Res.* (2019). https://doi.org/10.1016/j.watres.2019.05.014.

Section II

NANOPARTICLE-BASED TREATMENT

5

Nanoparticles and Nanocomposite Materials for Water Treatment: Application in Fixed Bed Column Filter

Chhaya, Dibyanshu[1], Sneha Singh[1] and Trishikhi Raychoudhury[1,2]*

[1]Department of Civil and Environmental Engineering, Indian Institute of Technology Patna, Bihta, Patna, Bihar, India
[2]Department of Civil and Infrastructure Engineering, Indian Institute of Technology, Jodhpur, Rajasthan, India

Abstract

The contamination of water has become a major concern, where the source of contamination could either be geogenic or anthropogenic. The contaminants discussed here are broadly grouped as inorganic contaminants, organic contaminants, and emerging contaminants. Several nanoparticles (NPs) and nanocomposites (NCs) materials are explored for removal of different types of contaminants in recent times. Numerous studies have reported on performance evaluation of those NPs and NCs in removing different contaminants. In this chapter, first, a comprehensive review of efficiency and suitability of different NPs and NCs in removing various organic pollutants such as As, Cd, Cr, Cu, Mn, Ni, Pb and Zn, organic contaminants such as dyes, halogenated hydrocarbons, PAHs, etc., and emerging contaminants are discussed. Secondly, the process involved in removing contaminants in the fixed bed column filter and the performance of NPs and NCs as the filtering material in the column filter has been discussed. The parameters affecting the performance of the fixed bed column in removing the contaminants from water are also emphasized.

***Keywords*:** Emerging contaminants, filter, inorganic and organic contaminants, nano and composites, remediation

**Corresponding author*: trishikhi@iitj.ac.in

Siddhartha Moulik, Aditi Mullick and Anirban Roy (eds.) Sustainable Water Treatment: Advances and Technological Interventions, (173–244) © 2022 Scrivener Publishing LLC

5.1 Introduction

Water is an integral part of our life that fulfils our basic needs and maintains the reliability of the natural environment. However, the quality of water has become a significant concern in recent times. The presence of hazardous and toxic material causes water contaminations either from geogenic sources or from anthropogenic sources. The anthropogenic sources of water contamination could be from point sources such as discharge of municipal wastewater and industrial effluent, leakage from storage and septic tanks, and non-point sources such as runoff from the agricultural fields, leachate from landfill sites, storm water runoff, runoff from roads, urban areas, highways, etc. (Figure 5.1). Pollutant from various sources get partitioned in the different components of the environment such as air, soil, and water. Air and water pollution have a direct impact on human health, whereas soil pollution contributes to bioaccumulation to crops, which ultimately causes adverse effects on humans and other living organisms. There are various contaminants, which could broadly be grouped as inorganic contaminants, organic contaminants, and emerging contaminants. Different types of contaminants and their possible sources in the natural water are presented in Figure 5.1.

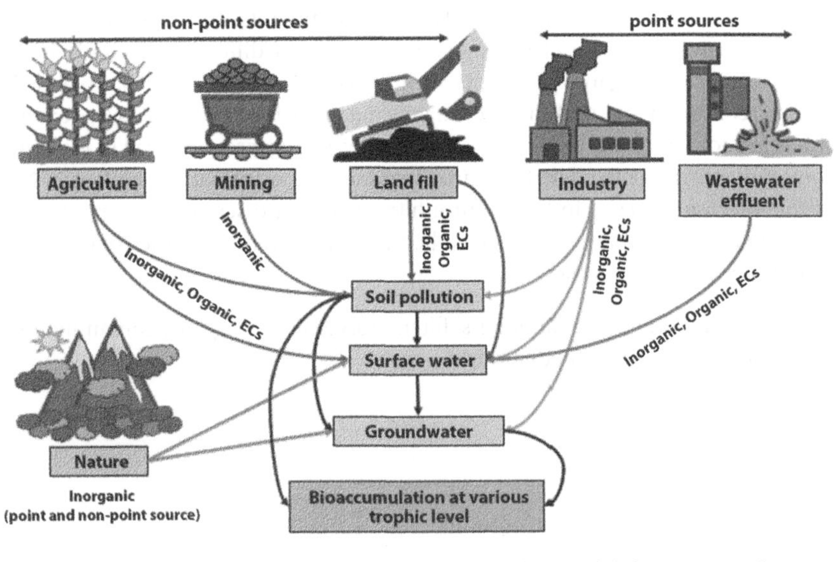

Figure 5.1 Schematic representation of sources and pathways of different types of contaminants in the aquatic system.

The inorganic contaminants include a diverse range of toxic heavy metals and various types of dissolved anions and cations. The heavy metals are defined by the group of metals and metalloids whose densities are greater than 5 g/cm^3 (Yang *et al.*, 2018). The most common heavy metal that acts as the contaminants are chromium (Cr), lead (Pb), mercury (Hg), cadmium (Cd), copper (Cu), nickel (Ni), cobalt (Co) and zinc (Zn). Arsenic (As) is also considered as heavy metal because of its environmental behaviour and chemical properties. Anthropogenic activities and natural processes have been proven to be a significant source of inorganic contamination. Dissolution of minerals and ions from rocks and sediments is the primary geogenic source of water contamination, whereas industrial and municipal waste effluents, leachate from landfill sites, from mining sites, etc., are the possible anthropogenic sources of this contamination. A high concentration of heavy metals such as Cr, Cu, iron (Fe), and Zn in different types of vegetables such as lettuce, radish, carrots, spinach, tomato, and eggplant have been reported (Qureshi *et al.*, 2016). Occurrences of Cd, Cu, Fe, manganese (Mn), Pb, Zn are frequently observed in contaminated groundwater (Chen *et al.*, 2016). Fluoride (F) and nitrate (NO_3) are other most common inorganic pollutants that are present in the groundwater (Rahman *et al.*, 2020; Ganyaglo *et al.*, 2019; Zhai *et al.*, 2017). Exposure to heavy metal contamination can cause carcinogenic risk, and their bioaccumulation may have adverse effects such as anaemia, brain damage, hepatitis, kidney failure, liver damage, skin diseases, ulcer, etc. (Wu *et al.*, 2020; Zhang *et al.*, 2018; Baby *et al.*, 2019). Consumption of F contaminated water can cause arthritis, osteoporosis, and dental fluorosis, where the intake of high NO_3 containing water may cause hypertension, and thyroid dysfunctions in adults, and methemoglobinemia in infants (Singh *et al.*, 2020).

Another form of contaminants in the water is organic pollutants, which have gained concern and raised alarm over the last several decades. The common organic compounds that are major concerns for water contamination are different types of dyes [i.e., methylene blue (MB), methyl orange (MO), methyl violet (MV), congo red (CR), malachite green (MG), rhodamine blue (RhB), etc.], halogenated hydrocarbons [i.e., chloroform (CF), trichloroethane (TCE), tetrachloroethane (PCE), etc.], polycyclic aromatic compounds (PAHs, i.e., polychlorinated benzene, phenyl, phenolic compounds, etc.). Some of these contaminants have very complex structures, high molecular weight, and different functional groups, making them more persistent in the different compartments of the environment. The typical characteristics of the organic compounds that make them more vulnerable are their partitioning behaviour, toxicity, non-biodegradability, bioaccumulation, and semi-volatile nature under wide environmental conditions. The primary sources of organic pollutants are from municipal sewage treatment plants and industrial effluents that are mostly released directly into the environment. A large number of organic contaminants such as PAHs, polychlorinated biphenyls (PCBs), etc., are released into the atmosphere as the secondary products of combustion, waste incineration, etc. The release of dye in the water is primarily from industrial effluent. These contaminants, when entering the water system, affect the aquatic life adversely, causing environmental degradation (Nagda *et al.*, 2007) and health hazards due to their mutagenic and carcinogenic nature (Gwenzi and Chaukura, 2018). The chlorinated hydrocarbons, phenolic compounds and their intermediates can penetrate the living cells and damage membranes, which have made them carcinogenic (Adeola, 2018).

There is a new group of contaminants named emerging contaminants (ECs), which generally refers to the synthetic or naturally occurring chemicals that are usually not monitored in the environment. The trace amount of pharmaceuticals such as clofibric acid, clotrimazole, dextropropoxyphene, diclofenac, ibuprofen, mefenamic acid, propranolol, tamoxifen, trimethoprim, carbamazepine, atenolol, metoprolol, sulfamethoxazole, gemfibrozil and propranolol, personal care products such as triclosan, triclocarban, 2-phenyl phenol, methylparaben, propylparaben, and pesticides such as trifluralin, atrazine, lindane, diazinon, acetochlor, and metolachlor are detected at wastewater effluent, and in the surface water at several locations (Ternes, 1998; Daughton and Ternes, 1999; Heberer, 2002; Székács *et al.*, 2015; Roberts *et al.*, 2016; Peng *et al.*, 2008). A few pharmaceuticals that have been identified in the groundwater include carbamazepine, sulfamethoxazole, ibuprofen, caffeine, diclofenac, and bisphenol A (Lapworth *et al.*, 2012). However, their fate in wastewater treatment plants and the environment is not yet completely understood. Generally, these ECs are released into the environment either through point sources, which include excretion in human and animal urine and feces, flushing of unused drugs, household uses, or through the non-point sources, which includes stormwater runoff, runoff from agricultural land, and leachate from landfill sites (Lapworth *et al.*, 2012). These contaminants, when released into the environment, may pose a potential threat to the ecosystem and human health even though they are present at a trace amount. They can cause several acute and long-term health hazards including endocrine destruction, immunotoxicity, cancers, and neurological disorders (Sanchez and Egea, 2018). The concentrations of some compounds have been raised during the removal of those contaminants, most likely due to the transformation of conjugates back to their parent compounds (Petrovic and Cristina, 2008).

There are several processes such as coagulation, flocculation, adsorption, advanced oxidation, photocatalytic degradation, membrane filtration, Fenton reaction, etc., that are used for the remediation of contaminated water. Several novel and emerging materials have been tried and applied to enhance the efficiency of these processes. In recent times different nanoparticles (NPs) and nanoparticle-based composite materials (NCs) are widely studied to remove different types of contaminants. The main advantage of NPs is their small size, which has resulted in large surface area (SSA) that are available for sorption and chemical degradation of contaminants (Anjum *et al.*, 2019; Qu *et al.*, 2013; Savage and Diallo, 2005; Theron *et al.*, 2008). Some of the NPs that are generally used for the remediation of contaminated water and wastewater includes graphene oxides (nGOs), carbon nanotubes (CNTs), iron oxide (nFe_xO_y), zero-valent iron (nZVI), titanium dioxide ($nTiO_2$), zinc oxide (nZnO) etc. (Crane and Scott, 2012; Kim *et al.*, 2018; Yu *et al.*, 2014; Raychoudhury and Scheytt, 2013). Though the efficiencies of several NPs are promising for removal of a range of contaminants, there are, however, some limitations for the application of those NPs that include their tendency to aggregate, difficulty in recovery, and reduction in reactivity, etc. To limit the aggregation and achieve the colloidal stability of NPs, different polymeric surface modifiers are commonly applied. Metals and metal oxides-based modifiers are responsible for the increase in the reactivity of the NPs, whereas magnetization of NPs can

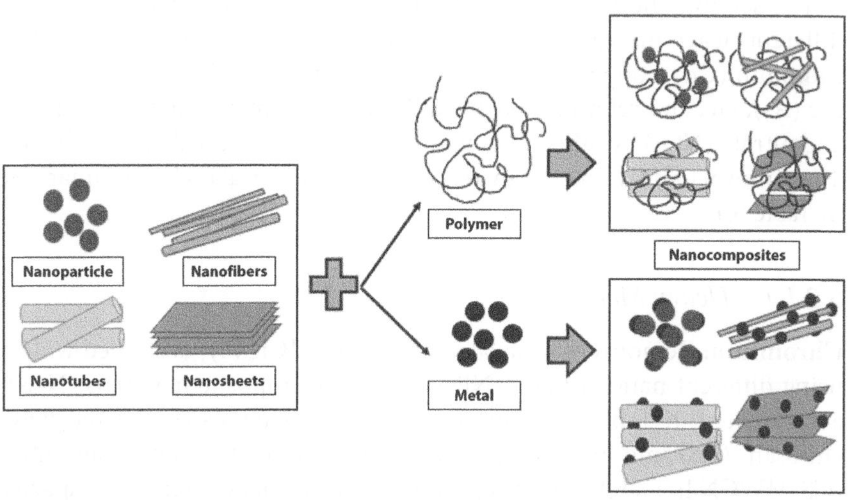

Figure 5.2 Conceptual diagram representing different types of nanoparticles and nanocomposites.

facilitate the recovery of NPs (Hu and Apblett, 2014; Kim *et al.*, 2018; Yu *et al.*, 2014). Some examples of NCs used for the treatment of pollutants are iron oxide (nFe_xO_y) modified graphene sheet, nGOs-chitosan, and etc. A conceptual schematic representation of different types of NPs and NCs are presented in Figure 5.2.

In recent years a significant amount of research has been focused on performance evaluation of NPs and NCs in removing different contaminants. In this chapter, first, a comprehensive review of efficiency and suitability of different NPs and NCs in removing various inorganic pollutants such as As, Cd, Cr, Cu, Mn, Ni, Pb and Zn, organic contaminants such as dyes, halogenated hydrocarbons, PAHs and some phenolic compounds, and emerging contaminants are discussed. Secondly, the process involved in removing contaminants in the fixed bed column filter and the performance of NPs and NCs as the filtering material in the column filter has been discussed. The parameters affecting the performance of the fixed bed column in removing the contaminants from water are also emphasized.

5.2 Target Contaminants: Performance of Nanoparticles and Nanocomposite Materials

5.2.1 Inorganic Contaminants

In this section, the removal of different inorganic contaminants including different heavy metals such as chromium (Cr), lead (Pb), mercury (Hg), cadmium (Cd), copper (Cu), nickel (Ni), cobalt (Co) and zinc (Zn), arsenic (As), etc, and relevant non-metallic contaminants such as fluoride (F) and nitrate (NO_3) by different emerging NPs and NCs are discussed below. The removal of heavy metals by relevant novel material is also summarized in Table 5.1.

5.2.1.1 Heavy Metals

Chromium: Removal of hexavalent chromium [Cr(VI)] is studied widely using different nanoparticles (NPs) and nanocomposite materials (NCs), as presented in Table 5.1. Amongst different materials considered here, chitosan (CS) modified carboxylated multi-walled carbon nanotubes (CS-MWCNTs-COOH) shows promising performance (removal capacity 164 ± 2 mg/g, equilibrium time- 30 min) in removing Cr(VI). However, the performance of the composite is more effective in an acidic aqueous solution of pH 2 (Huang *et al.*, 2018). The removal of Cr is predominantly governed by the synergic effect of sorption attributed to electrostatic interaction and chemical redox reaction. CS/clay NCs (CCN, Pandey and Mishra, 2011), exfoliated polypyrrole-organically modified montmorillonite clay NCs (PPy-OMMT, Setshedi *et al.*, 2013) can remove almost 100% Cr(VI) with a removal capacity of 357.1 mg/g and 209.6 mg/g, respectively. Positively charged (11.5 mV) cerium oxide NPs ($nCeO_2$) show promising performance (121.95 mg/g) in removing Cr(VI) at pH 7 (Recillas *et al.*, 2010). The adsorption follows the Freundlich isotherm model, which suggests heterogeneous surface active sites of the composite. Several studies have demonstrated that nanoscale zero-valent iron (nZVI) and iron oxide-based NPs perform reasonably well for removing Cr(VI). For example, nZVI, ultrasonic nZVI (US-nZVI), and pumice modified nZVI (P-nZVI) have Cr(VI) removal capacities of 47.2 mg/g (Nahuel Montesinos *et al.*, 2014), 66.7 mg/g (Zhou *et al.*, 2015) and 106.9 mg/g (Liu *et al.*, 2015), respectively. In the first case, the removal of Cr is mainly governed by the reduction of Cr(VI), and the removal efficiency decreases with an increase in the solution pH. For US-nZVI, ultrasound activates the surface of nZVI that induce new activate sites for chemical reaction, which eventually enhances the reduction of Cr(VI). Removal of Cr(VI) by iron oxide-based NCs such as polypyrrole/maghemite (PPy/γ-nFe_2O_3 MNC) and polyaniline/γ-nFe_2O_3 magnetic NC (PANI/γ-nFe_2O_3 MNCs) is very rapid (equilibrium reaches within 30 min) with a capacity of 209 mg/g and 170.7 mg/g, respectively (Chávez-Guajardo *et al.*, 2015). Functionalized nFe_3O_4 particles with graphene oxide and alginate bead (nFe_3O_4-nGO/beads) demonstrate moderate performance with the maximum removal efficiency of around 70% (Vu *et al.*, 2017). The advantage of magnetite NCs is these could be easily separated from contaminant solution due to their super magnetic property. The removal of Cr by graphene magnetic composite (nFe_3O_4-nGs, 17.29 mg/g at pH 6 to 7, Guo *et al.*, 2014) or CS grafted nGO composites (CS-nGO, 104.16 mg/g at pH 2, Samuel *et al.*, 2019) are assessed. The Cr removal by nFe_3O_4-nGs is mainly governed by chelation and ion exchange. The regeneration of the nano-microstructure of CS-nGO is possible, and the NCs could be recycled multiple times. The removal of Cr (VI) and Cr (III) by ZnO modified manganous-manganic oxide NCs (nMn_3O_4-nZnO) is attributed to the synergic effect of photocatalytic degradation and adsorption. The removal efficiency can be achieved up to 96.0% at pH 6 in 70 min (Li *et al.*, 2017). The removal efficiencies for Cr (VI) by polyacrylonitrile nanofibers functionalized with amine groups (PAN-NH_2, 156 mg/g at pH 2, Avila *et al.*, 2014) or PAN-PPy core-shell structure nanofibers (PAN-PPy nanofibers, 61.80 mg/g at pH 2, Wang *et al.*, 2013) or PPy-PANI nanofibers (227 mg/g, Bhaumik *et al.*, 2012) are promising. However, the working pH maintained in those cases is very low, and the removal efficiency of Cr (VI) decreases with an increase in the pH of the solution. For PAN-NH_2, surface protonation of amine group attributed to Cr removal. In contrast, for other cases, the synergic effect of anion exchange and reduction of Cr(VI) are the mechanisms involved in removing Cr (VI). The Cr (VI) removal capacity of PEI-nSi composites is as high as 183.7mg/g at pH 4 (Choi *et al.*, 2018). The amine groups present on the NCs surface is attributed to reduction reaction and formation of a hydrogen bond with chromium. Amongst different NPs and NCs referred here, CS/clay, PPy-OMMT, PPy/γ-nFe_2O_3 MNC, PANI/γ-nFe_2O_3 MNCs, nMn_3O_4-nZnOPPy-PANI and PEI-nSi types of NCs show the most promising performance in removing Cr (VI). The removal is mainly governed by adsorption. In most of the cases reported above, the removal of Cr (VI) is better fitted with the pseudo-second-order kinetic model.

Arsenic: Numerous studies have been done with a wide range of NPs and NCs for the removal of arsenic [arsenite As(III) and arsenate As(V)]. CNTs functionalized with poly amidoamine dendrimer (PAMAM/CNTs) exhibit excellent performance (432 mg/g) in removing As(III) at neutral pH (Hayati *et al.*, 2018). nFe_3O_4–nGO composite shows reasonably good performance (85 mg/g and 38 mg/g) for removal of As(III) and As(V) (Yoon *et al.*, 2016). The removal is mainly governed by electrostatic interaction between the anionic arsenic species and positively charged NCs. The removal capacities for As(III) and As(V) by nFe_3O_4-nGs macroscopic composites are around 400 mg/g and 280 mg/g, respectively (Guo *et al.*, 2015). The advantage of the NCs is, these could be easily separated from the aqueous solution by filtration. Spherical γ-nAl_2O_3 particles having a high specific surface area (231–497 m^2/g) can remove 64-86% arsenic (Patra *et al.*, 2012). The study suggests that the monolayer sorption of arsenic is predominant at the initial stage, followed by multilayer sorption for a prolonged duration. Removal of As(III) and As(V) by nZVI supported with reduced nGO (nZVI–nrGO) is very rapid (less than 30 min) with the removal capacity of 35.8 mg/g and 29 mg/g, respectively (Wang *et al.*, 2014). nFe_3O_4-nGO/beads can remove almost 100% As(V) at the optimum pH of 7 (Vu *et al.*, 2017). The nFe_3O_4-nGO/beads can be recycled and reused for at least five cycles. Covalent triazine framework encapsulated γ-nFe_2O_3 particles (γ-nFe_2O_3-CTF-1) shows good performance [98 mg/g for As(III) and 102.3 mg/g for As(V)] at pH 7 (Leus *et al.*, 2018). Regeneration of the adsorbent can be achieved for multiple cycles without compromising the performance. The immobilization of arsenic in red soil using manganese oxide modified-biochar composite (nMnO-BC) (14.36 mg/g) is assessed (Yu *et al.*, 2015). Goethite-silica NCs (Goethite-nSi) shows moderate performance (17.6 mg/g at pH 3.0) for the removal of As(V) (Attinti *et al.*, 2015). The adsorption of As(III) on hydrous titanium dioxide NPs ($nTiO_2{\cdot}xH_2O$, 96 mg/g at pH 9) is attributed to the high affinity of the surface hydroxyl group (Xu *et al.*, 2010). $nTiO_2$-impregnated CS bead (TICB) along with the exposure UV irritation shows rapid adsorption (equilibrium time 3 h) for both As(III), and As(V); however, the removal capacities are low (6.4 mg/g and 4.9 mg/g respectively) at an optimum pH of 7 (Miller and Zimmerman, 2010). The TICB surface area increases with UV light exposure and it causes photo-oxidation of As(III) to As(V). Significantly low performance (0.147 mg/g) for the removal of As(V) is observed at pH 6-7 by CS beads modified nFe_xO_y particles (INPB composite, Martínez-Cabanas *et al.*, 2016). Overall, PAMAM/CNT, nFe_3O_4-nGs macroscopic composites, nFe_3O_4-nGO/beads, and γ-nFe_2O_3-CTF-1 show promising performance in removing arsenic. The advantage of these NPs and NCs are that they work well at neutral pH.

Lead: Removal of lead [Pb(II)] by different NPs and NCs have been explored widely. Lignin grafted CNT (L-CNTs) shows promising performance (98.3%) in removing Pb(II) with the sorption capacity of 235 mg/g (Li *et al.*, 2017). The removal is facilitated by the interaction between positively charged Pb^{2+} and oxygen-functional groups present on the surface of the composite. The iron-based NPs such as superparamagnetic maghemite (γ-nFe_2O_3), nZVI, amino-functionalized magnetic nano-adsorbent (nFe_3O_4-NH_2), and hydrous ferric oxide NPs supported onto a porous poly-hydrogel (trans-aconitic acid/2-hydroxyethyl acrylate, HFO-P(TAA/HEA) show promising performance for the removal of Pb(II) from the aqueous solution. The adsorption by γ-nFe_2O_3 particles is primarily governed by electrostatic attractions with a moderate adsorption capacity (68.9 mg/g at 45 °C, pH 5.0) in slightly acidic condition (Rajput *et al.*, 2017). The nZVI can remove almost 100% of Pb(II) by the combined effect of chemical reduction, adsorption, and (co) precipitation method (Wang *et al.*, 2016). nFe_3O_4-NH_2 (Tan *et al.*, 2012) and HFO-P(TAA/HEA) (Zhang and Li, 2017) show moderate to high removal efficiency (40.1 mg/g and 304 mg/g, respectively) at pH 5. γ-nFe_2O_3 functionalized with organodisulfide polymer (γ-nFe_2O_3-PTMT) shows extremely high efficiency (603.16 mg/g) in removing Pb(II) (Huang *et al.*, 2018). The removal of Pb(II) by $Mg(OH)_2$ supported nZVI (nZVI-Mg-$(OH)_2$) is attributed to the synergic effect of reduction, adsorption, and precipitation, which shows significantly high removal efficiency (1986.6 mg/g, Liu *et al.*, 2015). nZVI supported by zeolite and montmorillonite [Z–nZVI (115.1 ± 11.0 mg/g) and Mt–nZVI (105.5 ± 9.0 mg/g)] can remove almost 99% of Pb(II) within 45 min with a reasonably good sorption capacity (Arancibia-Miranda *et al.*, 2016).

The 3D nGO based monoliths cross-linked with organic hydroxyl-containing agents (i.e., glycol with double hydroxyls), provides more active groups for Pb(II) binding and shows good removal capacity (146 mg/g at pH 3) of Pb (II) in acidic condition (Fang *et al.*, 2017). This composite has been reported to exhibit excellent recyclability EDTA functionalized magnetic nGO(EDTA- nFe_3O_4-nGO) shows promising performance (508.4 mg/g) for removal of Pb(II) (Cui *et al.*, 2015). nFe_3O_4-nGs demonstrated comparatively low performance (27.9 mg/g), where the removal of Pb (II) is attributed by chelation or ion exchange (Guo *et al.*, 2014). The removal efficiency of Pb(II) is moderate to high by amino-decorated Zr-based magnetic metal-organic frameworks composites (nFe_3O_4-$nSiO_2$) (Zr-MFCs, 102 mg/g, Huang *et al.*, 2018), hydrous Zr(IV) oxide-based NCs (NZP, 319 mg/g, Hua *et al.*, 2013) and $nMnO_2$/gelatin composites synthesized from protein-assisted synthesis process (319 mg/g, Wang *et al.*, 2018). In the first case, the removal is predominantly governed by electrostatic interaction between adsorbent and a metal cation. Pb(II) removal by nanosilica-immobilized-nanopolyaniline (Sil-Phy-NPANI) and cross-linked nanopolyaniline (Sil-Phy-CrossNPANI) composite is attributed to ion exchange with surface hydroxyl or amine group (Mahmoud *et al.*, 2016). Though the removal process was very fast where the equilibrium is achieved within 15–20 min; however, moderate to high sorption capacity (186.5-300.4 mg/g) is observed at the optimum pH range of 6.0-7.0. In general, $Mg(OH)_2$ supported nZVI shows the most promising performance at neutral pH. The EDTA-nFe_3O_4-nGO and functionalized Zr NCs have also shown reasonably good performance in removing Pb(II). In most of the cases reported above, the removal of Pb(II) is better fitted to the pseudo-second-order kinetic model. The sorption behaviour is better explained by the Langmuir isotherm model, which suggests a homogeneous active site of the NCs with monolayer sorption.

Mercury: Various NPs and NCs have been used for the removal of Hg (II). Among them, Silver NPs functionalized with mercaptosuccinic acid (nAg-MSA) demonstrate very high Hg (II) removal (800 mg/g) at a pH range of 5 to 6 (Sumesh *et al.*, 2011). The removal is primarily attributed to surface complexation and amalgamation. The Hg(II) removal by the CNTs functionalized with deep eutectic solvents (DES-CNTs) shows rapid removal (contact time of 28 min) with reasonably good adsorption capacity (186.9 mg/g, AlOmar *et al.*, 2017). Several iron-based NCs such as γ-nFe_2O_3-CTF-1 (165.8 mg/g, Leus *et al.*, 2018) pumice modified nZVI (P-nZVI, 107.1 mg/g, Liu *et al.*, 2015), fabricated poly(1-vinyl imidazole)-grafted nFe_3O_4-$nSiO_2$ magnetic particles (FSPV, 346 mg/g, Shan *et al.*, 2015), and organodisulfide polymer-modified magnetic NPs (nFe_2O_3-PTMT,

533.13 mg/g, Huang *et al.*, 2018), show moderate to very high performance in removing Hg(II). The γ-nFe_2O_3-CTF-1, nFe_2O_3-PTMT, and FSPV can be easily recovered and regenerated for multiple cycles. EDTA-nFe_3O_4-nGO performs reasonably well in removing Hg(II) (268.4 mg/g, Cui *et al.*, 2015) due to the presence of several surface functional groups and metal chelation property of the NCs. EDTA-nFe_3O_4-nGO also shows good magnetic performance, making it easy to separate from the liquid phase.nFe_3O_4-nGs (23.03 mg/g, Guo *et al.*, 2014) and Zr-based magnetic metal-organic (nFe_3O_4-$nSiO_2$-universiteteti) frameworks composites (nZr-MFC), (60 mg/g, Huang *et al.*, 2018) show comparatively low to moderate performance for removal of Hg(II). In some studies, cerium modified $nMnO_x$-based sorbents can remove almost 95% of elemental Hg (Liu *et al.*, 2017). Removal efficiency is reported to first increase and then decrease with the increase in the reaction temperature. The rapid removal of Hg(II) with two novel NCs such as Sil-Phy-NPANI, and Sil-Phy-CrossNPANI (160.5-210.6 mg/g) has been observed (Mahmoud *et al.*, 2016). Starch/$nSnO_2$ composites show promising performance (333 mg/g), which can remove almost 97% Hg(II) at a pH of 6 and 25°C. The rate of removal of Hg is also rapid, where the equilibration is reached in 60 min (Naushad *et al.*, 2016). In most of the cases reported above, the removal of Hg is better fitted with the pseudo-second-order kinetic model. In summary, nAg-MSA shows the best performance in removing Hg(II) at pH range 5-6. However, nFe_2O_3-PTMT, FSPV, and starch/$nSnO_2$ composites also show promising performance in removing Hg(II). Regeneration and reuse are an important advantage of these nFe_2O_3-PTMT and FSPV.

Cadmium: Removal of Cd (II) with a wide range of NPs and NCs is explored and nZVI shows promising performance [769.2 mg/g, (Boparai *et al.*, 2011) and 188 mg/g in the presence of NO_3 (Su *et al.*, 2014)] in removing Cd(II). Some of the other iron oxide-based NPs and NCs such as nFe_3O_4-nGs (27.83 mg/g, Guo *et al.*, 2014) and nFe_2O_3-PTMT (18.73 mg/g, Huang *et al.*, 2018) show moderate to low performance in removing Cd(II). The removal by those composites is dominated by sorption. HFO-P(TAA/HEA) performs reasonably good in removing Cd (149 mg/g, Zhang and Li, 2017). Cd (II) removal by $nMnO_2$/gelatin composite shows a moderate performance of 105.1 mg/g at pH 6.0 (Wang *et al.*, 2018). Active nMgO particles show excellent performance (2294 mg/g) in removing Cd (II) (Cai *et al.*, 2017). The equilibrium sorption is best fitted to the Langmuir isotherm model, suggesting monolayer sorption. Removal of Cd (II) is very rapid (equilibration time of 15–20 min) by Sil-Phy-NPANI and Sil-Phy-CrossNPANI, where the removal capacity is moderate (67.4-151.7 mg/g, Mahmoud *et al.*, 2016). The sorption of Cd(II) is accomplished by surface complexation and ion exchange with the surface amine or hydroxyl groups (at the pH range of 6.0 to 7.0, and at 25 ± 1 °C). Nanohybrid material such as $nZrO_2$-B_2O_3 shows reasonably good performance for Cd (II) removal (109.9 mg/g, Yalçinkaya *et al.*, 2011), where the material can be recycled for more than a hundred times without any major loss in metal (98 ± 4% recovery) content. Overall, active nMgO NPs and nZVI shows the most promising performance for the removal of Cd(II). In addition to that HFO-P(TAA/HEA), Sil-Phy-NPANI, and Sil-Phy-CrossNPANI also show good performance in removing Cd (II). The removal is dominated by the adsorption process.

Copper: Cu(II) removal has been explored extensively using various NPs and NCs. Magnetic multi-walled CNTs (MMWCNTs) having a high specific surface area (138.66 m^2/g) shows moderate (38.91 mg/g) Cu(II) removal capacity (Tang *et al.*, 2012). The regeneration of the MMWCNTs could be achieved for several cycles. Different iron oxide-based NPs and NCs such as γ-nFe_2O_3 particles (34.0 mg/g at optimum pH of 5.0, Rajput *et al.*, 2017), and nFe_2O_3–nAl_2O_3 mixed oxide NCs fiber (4.98 mg/g, Mahapatra *et al.*, 2013) show moderate to low performances in removing Cu(II). The sorption of Cu (II) by γ-nFe_2O_3 particles is governed by electrostatic interaction. HFO-P(TAA/HEA) shows a reasonable good performance of 107 mg/g for the removal of Cu (II) (Zhang and Li, 2017). Inner-sphere complexation is the primary mechanism responsible for the removal of Cu(II), and the removal efficiency increases in the presence of calcium ion. Some of the NCs with iron oxide-organic frameworks such as magnetic composite microspheres consisting of poly(acrylic acid) and nFe_3O_4 blended CS (CS/PAA-MCM, 174.0 mg/g, Yan *et al.*, 2012), PPY/γ-nFe_2O_3 MNC (196 mg/g, Chávez-Guajardo *et al.*, 2015) and PANI/γ-nFe_2O_3 MNCs (106.8 mg/g, Chávez-Guajardo *et al.*, 2015), show promising performance in removing Cu(II) under slightly acidic pH of 5.5. The second two adsorbents could be recycled without significant loss in adsorption capacities and can be used for four cycles. Few studies have demonstrated that engineered nGO functionalized with 4-amino thiophenol and 3-aminopropyltriethoxysilane (nGO-SH, 99.17 mg/g, and nGO-N, 103.28 mg/g, Chen *et al.*, 2016), EDTA- nFe_3O_4-nGO (301.2 mg/g, Cui *et al.*, 2015) show good to excellent performance in removing Cu(II) from aqueous solution. Sulfhydryl group and the amino group of nGO-SH and nGO-N attributed to an increase in the sorption active sites. The removal of Cu (II) by Sil-Phy-NPANI and Sil-Phy-CrossNPANI NCs is very rapid with moderate removal capacity (104.8-108 mg/g) at neutral pH (Mahmoud *et al.*, 2016). Zr-based magnetic metal-organic, nZr-MFCs (40-50 mg/g, Huang *et al.*, 2018), or $nZrO_2/B_2O_3$ (46.5 mg/g, Yalçinkaya *et al.*, 2011) show

Table 5.1 A comprehensive summary of the performance of NPs and NCs in removing targeted heavy metal contaminants.

Target contaminants / **Nanoparticles (NPs) & Nanocomposite (NCs)**	**Chromium (Cr)**	**Arsenic (As)**	**Lead (Pb)**	**Mercury (Hg)**	**Other heavy metals (Cu, Cd, Ni, Co, and Zn)**
Carbon Nanotubes (CNTs) [i.e., Chitosan (CS) modified carboxylated multi-walled CNTs (CS-MWCNTs-COOH), CNTs coated poly amidoamine dendrimer (PAMAM-CNTs), Lignin-grafted CNTs (L-CNTs), CNTs functionalized with deep eutectic solvents (DES-CNTs), Magnetic MWCNTs (MMWCNTs)]	- Removal of Cr (VI) with CS-MWCNTs-COOH is mainly governed by adsorption - The adsorption is dominated by electrostatic interaction - The adsorption capacity is reasonably high (143–164 mg/g) at a very low pH of 2	- Removal of As(III) with PAMAM-CNTs is mainly governed by adsorption - The adsorption capacity is high (432 mg/g) at a working pH of 7	- Removal of Pb(II) with L-CNTs is mainly governed by adsorption - The adsorption capacity is reasonably high (235 mg/g and 98.3%) at optimal pH of 7 - Removal of Pb(II) is governed by diffusion in the nano-pores, and interaction with the oxygen-containing groups on the surface of the composite	- Removal of Hg (II) by DES-CNTs is governed by adsorption - The removal efficiency is reasonably high (187 mg/g) at a pH of 5.5	- Removal of Co(II), Cu(II), and Zn(II) through PAMAM-CNTs and MMWCNTs is mainly governed by adsorption - Reasonably good removal efficiencies of 494 mg/g and 470 mg/g are observed for Co(II) and Zn(II) with PAMAM-CNTs - Removal of Cu (II) by MMWCNTs is comparatively low (38.9 mg/g) - The regeneration of MMWCNTs is possible for several cycles

(Continued)

Table 5.1 A comprehensive summary of the performance of NPs and NCs in removing targeted heavy metal contaminants. (*Continued*)

Target contaminants / **Nanoparticles (NPs) & Nanocomposite (NCs)**	**Chromium (Cr)**	**Arsenic (As)**	**Lead (Pb)**	**Mercury (Hg)**	**Other heavy metals (Cu, Cd, Ni, Co, and Zn)**
Graphene- metal framework [i.e., Graphene magnetic (nFe_3O_4-Gs), nFe_3O_4–graphene oxide (nFe_3O_4-nGO)]	- Removal of Cr(VI) using nFe_3O_4-nGs is governed by sorption that is accomplished by chelation or ion exchange - The adsorption capacity is moderate to low (17.3 mg/g) at the pH range of 1-3.5	- Removal of As(III) and As(V) by Fe_3O_4-nGO is governed by adsorption and electrostatic interaction - The removal efficiency is moderate to low (38 mg/g) at neutral pH	- Removal of Pb(II) by nFe_3O_4-nGs is governed by adsorption and accomplished with chelation or ion exchange - The removal capacity is moderate to low (27.9 mg/g) in the pH range of 6-7	- Removal of Hg(II) by nFe_3O_4-nGs is also governed by adsorption - The removal is moderate to low (23 mg/g) at neutral pH - The nFe_3O_4-nGs can be recovered from the water by magnetic separations	- Removal of Cd(II) and Ni(II) by nFe_3O_4-nGs is attributed to sorption -The removal of Cd and Ni is moderate to low (27.8 mg/g and 22 mg/g) in the pH range of 6-7

(*Continued*)

Table 5.1 A comprehensive summary of the performance of NPs and NCs in removing targeted heavy metal contaminants. (*Continued*)

Target contaminants / **Nanoparticles (NPs) & Nanocomposite (NCs)**	**Chromium (Cr)**	**Arsenic (As)**	**Lead (Pb)**	**Mercury (Hg)**	**Other heavy metals (Cu, Cd, Ni, Co, and Zn)**
Graphene oxide-organic framework [i.e., CS grafted nGO (CS-nGO), 3D nFe_3O_4-nGs macroscopic, nGO-based monoliths cross-linked by hydroxyl-containingagents, Ethylene diamine tetraacetic acid functionalized magnetic nGO composite (EDTA- nFe_3O_4-nGO), 4-amino thiophenol and 3-aminopropyltriethoxysilane functionalize nGO (nGO-SH and nGO-N)]	- Removal of Cr(VI) through CS-nGO is governed by adsorption - The adsorption capacity is moderate to high (104.2 mg/g) at the working pH of 2 - These NCs is recyclable for numerous cycles with the minimum loss in adsorption capacity	- Removal of As(III) and As(V) using 3D nFe_3O_4-nGs composite is governed by adsorption - The adsorption capacity is high (400 mg/g and 280 mg/g) at neutral pH - The NCs can be easily separated from aqueous solution by filtration	- Removal of Pb(II) by nGO-based monoliths cross-linked - hydroxyl-containingagents and EDTA- nFe_3O_4-nGO is dominated by adsorption - Both the adsorbent shows excellent reproducibility - The removal of the Pb is very high (508.4 mg/g) by EDTA-nGO at pH 4.2	- Removal of Hg(II) using EDTA-nFe_3O_4-nGO is governed by adsorption - Electrostatic interaction between metalions and oxygenic functional groups on nGO surface promote the removal of those contaminants - The removal efficiency is reasonably high (268.4 mg/g) at pH 4.1	- Removal of Cu (II) using EDTA- nFe_3O_4-nGO, nGO-SH, and nGO-Nis dominated by adsorption - High adsorption capacity (301.2 mg/g) of Cu (II) by EDTA- nFe_3O_4-nGO at pH 5.1 is reported

(*Continued*)

Table 5.1 A comprehensive summary of the performance of NPs and NCs in removing targeted heavy metal contaminants. (*Continued*)

Target contaminants / Nanoparticles (NPs) & Nanocomposite (NCs)	Chromium (Cr)	Arsenic (As)	Lead (Pb)	Mercury (Hg)	Other heavy metals (Cu, Cd, Ni, Co, and Zn)
Iron nanoparticle [i.e., Nanoscale zerovalent iron (nZVI), Ultrasonic modified nZVI (US-nZVI), Superparamagnetic maghemite (γ-nFe_2O_3), Covalent triazine framework encapsulated γ-nFe_2O_3 (γ-nFe_2O_3-CTF-1), Amino-functionalized nFe_3O_4 particles(nFe_3O_4-NH_2)]	- Removal of Cr(VI) using nZVI and modified-nZVI is due to the combined effect of chemical reduction and sorption - The removal capacity is moderate (66.7 mg/g), and the performance is best at pH 3	- Removal of As(III) and As(V) by γ-nFe_2O_3-CTF-1 is dominated by adsorption - The removal capacity is high (198 mg/g, 102.3 mg/g) for both As species at neutral pH - The composite can be recycled multiple time without compromising its performance	- Removal of Pb(II) by γ-nFe_2O_3 and nFe_3O_4-NH_2 is dominated by adsorption, - Removal of Pb(II) by nZVI is attributed to the synergic effect of adsorption, (co) precipitation, and chemical reduction - Moderate removal capacity (68.9 mg/g and 40.1 mg/g) by γ-nFe_2O_3 and nFe_3O_4-NH_2 is observed - Both the composites show optimal performance at pH 5	- Removal of Hg(II) by γ-nFe_2O_3-CTF-1 is attributed to adsorption - The removal efficiency is reasonably high (165.8 mg/g) at pH 7 - Regeneration of the adsorbent is possible for multiple cycles without compromising its performance	- Removal of Cu (II), Cd (II), and Zn (II) by nZVI and γ-n$Fe_2O_3$is dominated by sorption, chemical reduction, and (co)precipitation - The removal capacity is significantly high for Cd(II) (769.2 mg/g) by ZVI

(*Continued*)

Table 5.1 A comprehensive summary of the performance of NPs and NCs in removing targeted heavy metal contaminants. (*Continued*)

Target contaminants / **Nanoparticles (NPs) & Nanocomposite (NCs)**	**Chromium (Cr)**	**Arsenic (As)**	**Lead (Pb)**	**Mercury (Hg)**	**Other heavy metals (Cu, Cd, Ni, Co, and Zn)**
Iron-based metal framework [i.e., Pumice modified nanoscale zero-valent iron (P-nZVI), nZVI reduce graphite oxide (nZVI–nrGO), $Mg(OH)_2$ supported-nZVI (nZVI-Mg- **$(OH)_2$**), nFe_2O_3–nAl_2O_3 NCs fibers]	- Removal of Cr(VI) by P-nZVIis governed by the combined effect of chemical reduction and sorption - The removal capacity is reasonable high (106.9 mg/g) at a pH of 6.1 - The removal process is endothermic	- Removal of As(III) and As(V) using nZVI–nrGO is due to the combined effect of chemical reaction and adsorption, which is attributed by surface complexation and electrostatic interactions - The removal efficiency is moderate to low for both species of As (35.8 and 29 mg/g) - The optimal working pH is 4 to 10 for As(III) and pH 2 for As(V)	- Removal of Pb(II) using nZVI-Mg-$(OH)_2$is governed by the synergistic effect of adsorption, co-precipitation, and chemical reduction - Extremely high removal efficiency (1986 mg/g) at neutral pH is observed	- Removal of Hg(II) by P-nZVIis dominated by adsorption - The removal is reasonably high (107 mg/g) at pH 6.1	- Removal of Cu(II) and Ni(II) by nFe_2O_3–nAl_2O_3 NCs fiber is governed by adsorption - The removal capacity is on the lower side (4.9 and 32.4 mg/g) at pH 5.5

(*Continued*)

Table 5.1 A comprehensive summary of the performance of NPs and NCs in removing targeted heavy metal contaminants. (*Continued*)

Target contaminants / Nanoparticles (NPs) & Nanocomposite (NCs)	Chromium (Cr)	Arsenic (As)	Lead (Pb)	Mercury (Hg)	Other heavy metals (Cu, Cd, Ni, Co, and Zn)
Iron based-organic framework [i.e., Polypyrrole/ γ-nFe_2O_3(PPy/γ-nFe_2O_3 MNC), Polyaniline/ γ-nFe_2O_3(PANI/γ-nFe_2O_3 MNC), nFe_2O_3- nGO encapsulated in alginate beads (nFe_3O_4-nGO/ beads), CS beads modified nFe_xO_y particles (INPB), nFe_3O_4 modified by an organodisulfide polymer (nFe_3O_4-PTMT), Fabricated poly (1-vinylimidazole)-grafted nFe_3O_4-$nSiO_2$ magnetic NPs (FSPV), CS-poly acrylic acid magnetic composite microspheres (CS/PAA-MCM)]	- Removal of Cr(VI) using PPy/γ-nFe_2O_3 MNC, PANI/γ-nFe_2O_3 MNC and nFe_3O_4 -nGO/beads is governed by adsorption - The removal capacity is high (209 mg/g) under a wide range of pH between 2 to10	- Removal of As(V) by INPB composite and nFe_3O_4-nGO/beads is attributed by synergistic effects of sorption and ion exchange - The removal efficiency is low (0.1 mg/g) within a pH range of 6-8	- Removal of Pb (II) by nFe_3O_4-PTMT is dominated by adsorption - Very high removal efficiency (603.2 mg/g) of Pb (II) by nFe_3O_4-PTMT at neutral pH is reported	- Removal of Hg(II) by nFe_3O_4-PTMT and FSPV is governed by adsorption - The removal efficiency of Hg(II) is very high (533.1 mg/g) at neutral pH - The adsorbent is effectively regenerated	- Removal of Cd (II) and Cu (II) by nFe_3O_4-PTMT, PPY/γ-nFe_2O_3 MNC, PANI/γ-nFe_2O_3 MNC, and CS/PAA-MCM is dominated by adsorption - The removal capacities are reasonably high [216.5 mg/g for Cd (II) and 196 mg/g for Cu (II) by nFe_3O_4-PTMTand PPY/γ-nFe_2O_3 MNC, respectively]. - The performance of composites are good under low to neutral pH

(*Continued*)

Table 5.1 A comprehensive summary of the performance of NPs and NCs in removing targeted heavy metal contaminants. (*Continued*)

Target contaminants / Nanoparticles (NPs) & Nanocomposite (NCs)	Chromium (Cr)	Arsenic (As)	Lead (Pb)	Mercury (Hg)	Other heavy metals (Cu, Cd, Ni, Co, and Zn)
Manganese-metal framework [i.e., Zinc oxide modified Manganous-manganic oxide NCs (nMn_3O_4-nZnO, Manganese oxide modified-biochar composite (nMnO-BC), Manganese dioxide/gelatin composites ($nMnO_2$-gelatin), Ce modified nMnO($nMnO_x$)]	- Removal of Cr(III) and Cr(VI) by nMn_3O_4-nZnO is governed by the synergic effect of adsorption and photocatalytic reduction - The removal efficiency of Cr is high (96%) at pH of 6	- Removal of As(III) by nMnO-BC is dominated by adsorption - The removal capacity is in the lower range (4.4 mg/g) at neutral pH	- Removal of Pb(II) using $nMnO_2$/gelatin is governed by adsorption - The higher removal efficiency is high (318.7 mg/g) in the pH range of 2.0 to 7.0	- Removal of elemental Hg by $nMnO_x$ is dominated by chemisorption - Removal efficiency is reasonably high (94%) - The performance of the composites enhanced by doping with Ce	- Removal of Cd(II) using $nMnO_2$/gelatin is dominated by adsorption - The removal capacity is reasonably high (105.1 mg/g) at a pH range of 2.0 to 7.0

(*Continued*)

Table 5.1 A comprehensive summary of the performance of NPs and NCs in removing targeted heavy metal contaminants. (*Continued*)

Target contaminants / Nanoparticles (NPs) & Nanocomposite (NCs)	Chromium (Cr)	Arsenic (As)	Lead (Pb)	Mercury (Hg)	Other heavy metals (Cu, Cd, Ni, Co, and Zn)
Silica-metal framework [i.e., PEI-silica NCs (PEI-nSi), Goethite/silica NCs (Goethite-nSi), Nanosilica-immobilized-nanopolyaniline and cross-linked nanopolyaniline (Sil-Phy-NPANI) and (Sil-Phy-CrossNPANI)]	- Removal of Cr(VI) using PEI-nSiis governed by adsorption and reduction reaction - The sorption capacity is high (183.7mg/g) at pH 4	- Removal of As(V) by goethite-nSi is governed by sorption - The As removal capacity is in the lower range (17.6 mg/g) at pH 3	- Removal of Pb (II) by Sil-Phy-NPANI and Sil-Phy-CrossNPANI is dominated by sorption - The removal capacity is moderate (186.5-300.4 mg/g) at a pH range of 6.0-7.0 - The removal is fast where the equilibrium has reached in 15-20 min	- Removal of Hg (II) by Sil-Phy-NPANI and Sil-Phy-CrossNPANI is governed by sorption - The removal capacity is moderate (160.5-210.6 mg/g)at the pH range of 6.0-7.0	- Removal of Cd (II) and Cu (II) by Sil-Phy-NPANI and Sil-Phy-CrossNPANI is dominated by sorption - The removal efficiency is in the moderate range (104.8-108 mg/g for Cu(II) and 67.4-151.7 mg/g for Cd(II)) at the pH range 2.0 to 7.0

(*Continued*)

Table 5.1 A comprehensive summary of the performance of NPs and NCs in removing targeted heavy metal contaminants. (*Continued*)

Target contaminants / **Nanoparticles (NPs) & Nanocomposite (NCs)**	**Chromium (Cr)**	**Arsenic (As)**	**Lead (Pb)**	**Mercury (Hg)**	**Other heavy metals (Cu, Cd, Ni, Co, and Zn)**
Miscellaneous NPs [i.e., Cerium oxide ($nCeO_2$), Hydrous titanium dioxide NPs ($nTiO_2 \cdot xH_2O$), γ-nAl_2O_3 spherical particles, Magnesium oxide NPs (nMgO), Silver NPs functionalized with mercaptosuccinic acid (nAg- MSA), Hydrous ferric oxide NPs supported onto a porous poly(trans-Aconitic acid/2-hydroxyethyl acrylate) hydrogel (HFO-P(TAA/HEA))]	- Removal of Cr(VI) by $nCeO_2$is dominated by adsorption and photocatalytic degradation - The removal capacity is reasonably high (121.9 mg/g) by $nCeO_2$ at neutral pH	- Removal of As(III) using γ-nAl_2O_3 and $nTiO_2 \cdot xH_2O$ and As(V) using γ-nAl_2O_3 is governed by adsorption - The removal efficiency is low to moderate (64–86%) by γ-nAl_2O_3 at pH 6 while maximum removal capacity of 96 mg/g at pH 9.0	- Removal of Pb(II) by nMgO and HFO-P(TAA/HEA) is dominated by adsorption - The removal capacity is significantly high to moderate (2614 mg/g and 1.466 mmol/g) at pH 5 - The removal with nMgO rate is fast; equilibrium reaches within 60 min	- Removal of Hg(II) by nAg-MSA is dominated by adsorption - The maximum removal capacity is very high (800 mg/g) at optimum pH of 5.0-6.0	- Removal of Cd(II) nMgOis governed by adsorption - The Cd(II) removal capacity is extensively high (2294 mg/g) at working pH of 5 - The removal rate is fast, where the equilibrium reaches within 60 min

(*Continued*)

Table 5.1 A comprehensive summary of the performance of NPs and NCs in removing targeted heavy metal contaminants. (*Continued*)

Target contaminants / Nanoparticles (NPs) & Nanocomposite (NCs)	Chromium (Cr)	Arsenic (As)	Lead (Pb)	Mercury (Hg)	Other heavy metals (Cu, Cd, Ni, Co, and Zn)
Miscellaneous NCs [i.e., PPy -organically modified montmorillonite clay NCs (PPy-OMMT NC), CS/clay NCs (CCN), Polyacrylonitrile nanofibers functionalized with amine groups (PAN-NH_2), PAN/PPy nanofibers, PPy-PANI, Zr-based magnetic metal-organic frameworks composites(nFe_3O_4-nSiO_2-universiteteti) (nZr-MFC), nTiO_2-impregnated CS bead with UV irradiation (TICB), nZVI supported by zeolite and montmorillonite (Z–nZVI and Mt–nZVI), Hydrous Zr(IV) oxide-based NCs (NZP), Starch/nSnO_2, nZrO_2-B_2O_3]	- Removal of Cr(VI) by PPy-OMMT NC, CCN, PAN-NH_2, PAN/PPy nanofibers, PPy-PANI and Zr-MFC, is primarily governed by sorption and chemical reduction - The maximum removal capacity by CNN is reported as high as 357.14 mg/g	- Removal of As(III) and As(V) by TICB is dominated by adsorption - The removal capacities are in the lower range [6.4 mg/g for As(III) and 4.9 mg/g for As (V)] at pH 7 - The rate of removal is relatively slow where the equilibrium reaches at 185h - surface area of TICB increases after exposure to UV light.	- Removal of Pb (II) by Zr-MFC, Z–nZVI, Mt–nZVI and NZP is dominated by adsorption - The removal capacity is not very significant (66.17 mg/g) by NZP	- Removal of Hg (II) by starch/nSnO_2, Zr-MFC is dominated by adsorption - The removal capacity is high (333 mg/g) by starch/nSnO_2 with a maximum removal efficiency of 97% at an optimum pH of 6.0	- Removal of Cd(II), Cu (II), and Co(II) by Zr-MFC and nZrO_2-B_2O_3 is governed by adsorption - The removal capacity is high (109.9 mg/g) by nZrO_2-B_2O_3 for Cd(II) moderate (40-50 mg/g) to low (32.2 mg/g) for Cu(II) and Co (II) by Zr-MFC and nZrO_2-B_2O_3 respectively in the working pH range of 2.0 to 7.0

moderate efficiency for removal of Cu(II). In general, EDTA- nFe_3O_4-nGO, PPy/γ-nFe_2O_3 MNC, and CS/PAA-MCM show better performance for removing Cu(II). However, the other NCs such as PANI/γ-nFe_2O_3 MNCs, nGO-N, Sil-Phy-NPANI, and Sil-Phy-CrossNPANI NCs also demonstrate reasonably good performance.

Nickel: Mesoporous nMgO sheet shows promising performance (1684.25 mg/g) in removing Ni(II) (Feng *et al.*, 2015). Sorption is dominated by the formation of carbonate-complex. Removal of Ni(II) with HFO-P(TAA/HEA) shows moderate removal capacity (86 mg/g, Zhang and Li, 2017). The equilibrium isotherm is well-fitted to the Langmuir isotherm model, indicating monolayer sorption of Ni(II). The performance of iron-oxide based composites such as nFe_2O_3–nAl_2O_3 mixed oxide fiber (32.36 mg/g at pH6, Mahapatra *et al.*, 2013), nFe_3O_4-nGs (22.07 mg/g, Guo *et al.*, 2014) show moderate Ni(II) removal capacity.

Cobalt: PAMAM-CNTs shows promising performance in removing Co(II) (Hayati *et al.*, 2018) with the predicted maximum adsorption capacities of 494 mg/g. $nZrO_2$-B_2O_3 composites have demonstrated moderate performance (32.2 mg/g at pH 8, Yalçinkaya *et al.*, 2011) in removing Co. However, the recovery of the NCs is very high (96 ± 3%), which could be regenerated and reused for Co removal.

Zinc: The removal of Zn (II) by PAMAM-CNTs shows promising performance (470 mg/g) at neutral pH (Hayati *et al.*, 2018). nZVI can also remove 99% of Zn (II) from wastewater at pH 8 (Wang *et al.*, 2016). The removal of Zn by nZVI is governed by the synergistic effect of adsorption, chemical reduction, and co-precipitation.

5.2.1.2 Nonmetallic Contaminant

Nitrate: Microalgae immobilized cross-linked CS nanofiber shows moderate performance (87%) in removing NO_3 (Eroglu *et al.*, 2012). The removal of NO_3 is attributed to the combined effect of physicochemical sorption and biodegradation. Three NPs, namely nFe, nCu/Fe, and nMn/Fe supported on sodium Y zeolite, show 100% removal efficiency for NO_3 (Zeng *et al.*, 2017). At the initial stage, the first-order kinetic model fitted the data best. Chelating resin supported bimetallic iron–palladium NPs (DOW 3N-nFe/Pd, Shi *et al.*, 2016) and macroporous alginate substrate-supported nZVI (nZVI-MAS, Lee *et al.*, 2016) can remove more than 95% of NO_3. Removal of NO_3 by DOW-3N-nFe/Pd is primarily governed by sorption on the chelating resin (DOW-3N), while the removal of NO_3 by nZVI-MAS is attributed to the combined effect of reduction of NO_3 and sorption. The kaolinite supported bimetallic nFe/Ni particles have demonstrated comparatively low performance (40.4%) in removing NO_3 in the presence of Cu (II) as co-contaminants (Cai *et al.*, 2014). Iron NPs synthesized from green tea (GT-nZVI) can remove NO_3 up to 59.7% with the sorption capacity of 13.06 mg/g (Wang *et al.*, 2014). The removal is predominantly due to the synergic effect of adsorption, co-precipitation, and chemical reduction. In a similar study, nZVI in the presence of 0.2 % Ni (w/w) as the catalyst can effectively reduce NO_3 rapidly (within 1 min) under a wide range of concentration (10–10000 mg/l, Ryu *et al.*, 2011). The removal of NO_3 by Pd-Ag alloy NPs supported on amine-functionalized silicon dioxide (nPdAg-nSiO_2-NH_2) shows excellent performance (100% removal in 130 min) in acidic pH 3 (Ding *et al.*, 2017). The NO_3 removal efficiency increases with a decrease in pH. CS/Zeolite Y/ZrO composites show a moderate performance of 23.58 mg/g in removing NO_3 (Teimouri *et al.*, 2016). The elimination of NO_3 is dominated by the formation of chelate between NO_3 and the amine group present on the surface of the composite. CS-polystyrene-Zn composite performs reasonably good (removal efficiency 90% at pH 3) with very rapid sorption (Keshvardoostchokami *et al.*, 2017). Novel nFe_3O_4-ZrO_2-CS composites, having SSA of 212.9 m^2/g, have demonstrated excellentperformance (89.3 mg/g) in removing NO_3 (Jiang *et al.*, 2013). Overall, it could be said that zeolite supported multiple metals such as Fe, Ni, Mn-based NCs, nFe_3O_4-ZrO_2-CS composites, and multiple metals such as nPd-Ag alloy composites supported amine-functionalized nSiO_2 are most effective in removing NO_3. In addion to that nZVI shows rapid and complete removal of NO_3 at a very low concentration relevant for groundwater condition.

Fluoride: The removal of fluoride (F) by functionalized Zr (IV)-metalloporphyrin grafted nFe_3O_4 particles (nFe_3O_4-APTES-Zr, TCPP) is very rapid with reasonably good removal efficiency (92% in 20 min at pH 5.5) at slightly acidic pH (Poursaberi *et al.*, 2012). The advantage of this sorbent is it could be recycled and regenerated for multiple cycles. Nanosized sulfate-doped nFe_3O_4-nAl_2O_3 particles show good performance in removing F (70.4 mg/g) at neutral pH (Chai *et al.*, 2013). Ion exchange and formation of the inner-sphere surface complex are the governing mechanisms involved in the removal of F by this NPs. Removal capacity of F by calcium oxide (nCaO) NPs loaded mesoporous Al_2O_3 is reasoanbly good at pH 6.8 (137 mg/g, Dayananda *et al.*, 2014). nFeO(OH) particles extracted from natural clay show promising performance in removing F with a sorption capacity of 230 mg/g at pH 7.5 (Chaudhary *et al.*, 2016). Mg-Al-Zr, triple-metal composite demonstrated moderate performance (22.9 mg/g, Wang *et al.*, 2017). Zr-embedded polysulfone blends hollow fiber membrane (60.65 mg/g, He *et al.*, 2014) and needle-like Mg-La bimetal oxide composites (63.11 mg/g, Kong *et al.*, 2020) have demonstrated reasonable sorption capacity for F at neutral pH. The sorption in the bi- or tri-metalic NCs is mainly attributed to inner-sphere complexation and ion exchange with the carbonate and hydroxyl groups associated with the surface of the sorbent. This sorbent could be regenerated for four cycles with an efficiency of 80% of the initial adsorption capacities. Different metal such as Fe, Al, Ce impregnated AC based composites also show reasonably good performance (2.02 mg/g to 4.62 mg/g) in removing F (Kalidindi *et al.* 2017; Raychoudhury *et al.* 2017). Amongst different NPs and NCs, iron oxyhydroxide NPs, nAl_2O_3 coupled with other metal oxides such as nCaO and nFe_2O_3 and sulfate-doped nFe_3O_4/nAl_2O_3 particles shows promising performance in removing F at neutral pH, which is relevant for the natural condition.

5.2.2 Organic Contaminant

5.2.2.1 Organic Dyes

Methylene blue (MB): Amongst different NPs and NCs, graphene NPs (nGs) shows excellent (99% removal) performance in removing MB at pH 10 (Liu *et al.*, 2012). Other adsorbents like magnetite graphene sponge (nFe_3O_4-nGS), graphene carbon nanotubes hybrid (nGs-CNTs hybrid) and nZnO-nFe_3O_4/nrGO NCs shows reasonably good removal efficiency (97% within 2-3 hr) for MB (Yu *et al.*, 2015; Ai and Jiang, 2012; Ojha *et al.*, 2017). Moreover, nZnO-nFe_3O_4/nrGOcan also be regenerated and reused for five cycles (Ojha *et al.*, 2017). Different polymer/organic-based nGO particles such as chitosan (CS)-nGO (350 mg/g, Chen *et al.*, 2013), nGO-monolith (154 mg/g, Fang *et al.*, 2017), and nGO-alginate blue (3356 mg/g, Platero *et al.*, 2017) show good removal capacity for MB with a varying initial concentration of 100mg/l, 0.1mg/l, and 1600 mg/l, respectively. Complete removal (100%) of MB is observed using nGOs-Metal-nFe_2O_4 hybrids (where cobalt (Co), zinc (Zn), nickel (Ni), and manganese (Mn) are used as different hybrid metal) when the initial concentration of MB is kept at 5 mg/l and Co is used as the catalyst (Bai *et al.*, 2012). Several NPs such as nickel sulfide (nNiS), zinc oxide (nZnO), zinc sulfide copper (nZnS: Cu), and zinc hydroxide ($nZn(OH)_2$) is loaded on AC, which is capable of removing MB up to 96% within 5 min (except nZnO loaded AC, which takes 120 min for removal of MB) (Ghaedi *et al.*, 2014a; Nourmoradi *et al.*, 2014; Asfaram *et al.*, 2015; Ardekani *et al.*, 2017). Magnetic modified MWCNTs can remove a reasonably good (95%) amount of MB within 15 min with a very low dose of adsorbent (Madrakian *et al.*, 2011). Other studies indicate

MWCNTs-based NPs and NCs such as guar gum (GG)-MWCNTs-nFe_3O_4 (37 mg/g), magnetite (nFe_3O_4)-MWCNTs (42 mg/g), polymer calcium alginate (CA)-MWCNTs (182 mg/g) show moderate to high removal capacity of MB (Yan *et al.*, 2012; Qu *et al.*, 2008; Sui *et al.*, 2012). Iron oxide modified adsorbent such ascarboxymethyl cellulose (CMC) capped nFe_3O_4 (95%), tannic acid modified nFe_3O_4 (nFe_3O_4-TA, 97%), and xylan/poly magnetic hydrogels (90%) can remove high to moderate amount of MB (Anushree and Philip, 2019; Abkenar *et al.*, 2015; Sun *et al.*, 2015). Whereas excellent MB removal (99%) is observed by nFe_xO_y particles, where the removal is attributed to the synergic effect of Fenton reaction and sorption (Sawafta *et al.*, 2019). Wei *et al.* (2015) have shown that the removal of MB could be enhanced from 35% to 95% by hydroxyapatite (HAP) NPs in the presence of humic acid at pH 9. Copper oxide NPs (nCuO), which is synthesized from punica granatum leaf, shows good removal efficiency (98%) for MB at pH 6 (Vidovix *et al.*, 2019). In the presence of other contaminants such as congo red, almost 94% (at pH 11) of MB removal could be achieved within 30 secs by modified nFe_3O_4-nSiO_2 particles. The removal efficiency increases (up to 94%) with an increase in the initial concentration of MB up to 15 mg/l and then it decreases with further increase in concentration (Yimin *et al.*, 2018). Overall, nFe_xO_y and nGs particles and their composite show promising performances in removing MB, even in the presence of other contaminants.

Methyl/Crystal violet (MV): Significant amount (99%) of MV is removed by almond shell-nFe_3O_4 NPs (pH 8) and gum xantham/nFe_3O_4 based NCs hydrogels (pH 6.5) (Saeed *et al.*, 2016; Mittal *et al.*, 2016). Polyacrylic acid (PAA)-bound magnetic NPs (PAA-nFe_3O_4) show good (116 mg/g) and fast removal for MV, which reaches the equilibrium within 1 min. Furthermore, the PAA-nFe_3O_4 particles could also be regenerated and recycled several times without compromising their performance (Liao *et al.*, 2005). NPs derived from eggshell (91%, Foroutan *et al.*, 2019), polyamine-modified magnetic nGO (MMGO, 95%, Abdi *et al.*, 2019), halloysite nanotubes (nFe_2O_3-HNTs, 88%, Duan *et al.*, 2012), and nZVI in the presence of $NaBH_4$ (90%, Prema *et al.*, 2011) show moderate MV removal efficiency with a comparatively high initial concentration (in the range of 100 to 150 mg/l) of contaminant. nAg has shown very good (98% in 240 min, Sinha *et al.*, 2014) to moderate (85% in 20 min, Jishma *et al.*, 2018) removal efficiency for MV under varying synthesis and reaction conditions. Complete degradation of MV is observed using nSnO_2 modified AC and magnetite MWCNTs at pH 7 within 6 and 15 min, respectively (Li *et al.*, 2015; Madrakian *et al.*, 2011). Hydrolyzed polyacrylamide grafted xanthan gum incorporated in nSiO_2 (h-XG/nSiO_2) shows 99% removal within 20 min (Ghorai *et al.*, 2014). Magnetite (nFe_3O_4)/porous graphene NCs show excellent removal capacity (460 mg/g) within 5 min, which is facilitated by the porous nature of graphene and high magnetic property of nFe_3O_4 particles (Bharath *et al.*, 2017). Another study, using sodium dodecyl sulfate (SDS) coated nFe_3O_4 MNPs shows effective adsorption capacity (416 mg/g) for removal of MV (Keyhanian *et al.*, 2016). The good removal efficiency of MV (97%) by 3-D structuredpolylactic acid (PLA)-nGO/chitosan sponge (PLA-nGO/CSS) is reported, which is also dependent on the content of nGO particles in the composite (Zhou *et al.*, 2018). Some of the NPs and composites such as nZnS quantum dot particles, $nSnO_2$, Pd catalyst supported $nTiO_2$, monodispersed $nSiO_2$ particles show reasonably good performance (95%) in removing MV (Shamsipur *et al.*, 2014; Bhattacharjee *et al.*, 2016; Saeed *et al.*, 2017; Qin *et al.*, 2018).

Methyl orange (MO): Mn introduced ferrocene modified $nSiO_2$-nAl_2O_3 (Arshadi *et al.*, 2013), nZnO (Ghaderi *et al.*, 2015), MWCNTs (Ghaedi *et al.*, 2015), and bimetallic ferromagnetic NPs (Sarvari *et al.*, 2018) can remove more than 99% MO. Nickel oxide-based NPs and nZnO-nAl_2O_3 NCs can remove a high amount of MO (97%) within 30 min at pH 4 (Riaz *et al.*, 2020; Tajizadegan *et al.*, 2016). A few studies have demonstrated that removal of MO by AC modified with nCuS (99%, within 4 min) and $nSnO_2$ (98%, within 25 min) is excellent where the nAg particles (72% in 16 hr) shows moderate efficiency (Mokhtari *et al.*, 2016; Ghaedi *et al.*, 2016; Pal *et al.*, 2013). $nTiO_2$ particles based NCs such as sewage sludge-based AC-$nTiO_2$NCs ($nTiO_2$/ASS), $nTiO_2$/MIL-101(Cr) (Rashed *et al.*, 2017; Wu *et al.*, 2020) show reasonably good efficiency (93-94%) in removing MO. The removal of MO by $nTiO_2$/ASS NCs is attributed to both photocatalytic degradation and sorption. Graphene NPs and nickel-chromium layered double hydroxides modified by nrGO (nrGO–Ni–Cr–CO_3) shows promising performance (16 mg/g at 0.2 mg/l concentration and 160 mg/g at 100 mg/l concentration) in removing MO under varying concentration of the contaminant (Ruan *et al.*, 2016). Removal of MO by iron oxide NPs (79%, Debnath *et al.*, 2016) and bentonite supported nZVI (90%, Chen *et al.*, 2011) are in the moderate to good range. Ammonium polyethyleneimine (PEI) modified on $nSiO_2$ (QPEI/$nSiO_2$), nNiO (217 mg/g), and nCuO (370 mg/g) Nps exhibit good to excellent performance in removing MO, where the rate of removal is fastest (10 min) by QPEI/$nSiO_2$ (Liu *et al.*, 2013; Darwish *et al.*, 2019).

Congo red (CR): Iron oxide based NPs (Hao *et al.*, 2014), γ-nFe_2O_3 (Afkhami and Moosavi, 2010), MWCNTs (Sheibani *et al.*, 2015) and hollow Zn ferrite nanosphere (Rahimi *et al.*, 2011) show promising performance (>99%) for removal of congo red (CR) under a varying range of concentration

(0-100 mg/l). The hollow Zn ferrite nanosphere can remove a wide range (0-50 mg/l) of CR polluted wastewater by applying a little dose of NCs. The advantage of iron oxide NPs is that it could be regenerated and reused for several cycles with a slight decrease in its performance (Hao *et al.*, 2014). A few other NPs and NCs such as CS-nTiO_2 in the presence of Cu catalyst (Cu/CS-nTiO_2, Ali *et al.*, 2018), nGO (Debnath *et al.*, 2014), nZnO (Kataria and Garg, 2017), and silver decorated phosphorylated carbon nanotube/cyclodextrin nanosponge (pMWCNT/β-CD/nTiO_2-Ag, Taka *et al.*, 2020) show high removal efficiency (>97%) for CR. The rate of removal of CR is much faster by Cu/CS-nTiO_2 NCs (4 min) compared to that of pMWCNT/β-CD/nTiO_2-Ag (400 min). Different metal NPs loaded AC, such asZnO nanorods (nrZnO)-AC (90%), nPd-AC (94%), and nAg-AC (100%) show good to excellent CR removal efficiencies with a dose of 0.025 g of the NCs (Ghaedi *et al.*, 2012). The removal of CR by nrZnO-AC is much faster (7 min) than the other two adsorbents. The performance of nPd-AC and Ni-nZnS-AC in removing CR is very good (98% in 22 min at pH 2-3) at acidic pH conditions (Ahmadi *et al.*, 2015). Few studies demonstrated that removal capacities of CR by magnetic MWCNTs-Fe_3C (Konicki *et al.*, 2012), mixed iron and aluminum oxide (nFe_2O_3–nAl_2O_3, Mahapatra *et al.*, 2013), magnetic chitosan composite microparticles (MCCPs,Wang *et al.*, 2013), chitosan hydrogel beads (CSBs) modified on MWCNTs (Chatterjee *et al.*, 2010) and benzene tetracarboxylic acid-coated nFe_3O_4 particles (nFe_3O_4-BTCA, Chatterjee *et al.*, 2020) are 85 mg/g, 498 mg/g, 263 mg/g, 450 mg/g, and 630 mg/g, respectively. Amongst the above NCs, nFe_3O_4-BTCA shows the most promising performance (630 mg/g) for the removal of CR, and the NCs could also be recycled several times without compromising much of its performance. The Co and Cu modified-nTiO_3 ($Co_xCu_yTiO_3$, 92% in 30 min at pH 2), and quercetin (Qur) modified α-nFe_2O_3 particles (95% in 140 min at pH 6) shows reasonably good performance in removing CR (Hashemian and Foroghimoqhadam, 2014; Satheesh *et al.*, 2016).

Malachite green (MG): nrGO and Fe^{3+} ion-doped ZnS quantum dot show promising performance (>99%) in removing malachite green (MG) where the sorbent doses are kept at 0.5g/l and 0.08g/l, respectively (Gupta and Khatri, 2017; Rajabi *et al.*, 2013). Cerium oxide nanocubes (ncCeO_2) when synthesis with 0.5g urea can degrade 90% MG within 21 min (Sreekanth *et al.*, 2019). Removal of MG is moderate (80% to 86%) by La/Cu/Zr-trimetallic carbon dot (Sharma *et al.*, 2017), α-nFe_3O_4 (Dehbi *et al.*, 2020), nFe_3O_4-nNiO (Mohanta *et al.*, 2020), and nTiO_2 (Abou-Gamra and Ahmed, 2015) particles. Mohanta *et al.* (2020) have shown that the presence of chloride ions (Cl^-) has an adverse effect on the removal of MG. Different NPs loaded AC such as nrZnO-AC, nZnO-AC, nAu-AC, n$Zn(OH)_2$-AC, nZnS-AC, nAgOH-AC, ZnS:Cu-AC (Dastkhoon *et al.*, 2015; Azad *et al.*, 2015; Ghaedi *et al.*, 2014b; Roosta *et al.*, 2014; Bazrafshan *et al.*, 2015; Roosta *et al.*, 2015; Solaymani *et al.*, 2017) show excellent removal (99%) of MGat a very fast rate (within 4 to 0 min). Magnetic reduced graphene oxide/zeolitic imidazolate framework (MRGO/ZIF) NCs shows excellent removal capacities (3165 mg/g, Lin and Lee, 2016) of MG, whereas nGO-tannic acid (TA) (500 mg/g, Al-Kinani *et al.*, 2019), nFe_3O_4/β–CD/nGO (990 mg/g, Wang *et al.*, 2016), vinyl-modified mesoporous poly(acrylic acid)/$nSiO_2$ (PAA/$nSiO_2$) composite nanofiber membrane (220 mg/g, Raval *et al.*, 2017) and kappa-carrageenan-g-polyacrylic acid/$nTiO_2$–NH_2 hydrogel NC (K-Carrageenan-g-AA/$nTiO_2$–NH_2,666 mg/g, Pourjavadi *et al.*, 2014) have shown high to moderate range of removal capacities. The MRGO/ZIF could also be regenerated for four cycles. While MWCNTs-COOH and nrGO-nFe_3O_4 show relatively low performance (49mg/g and 22mg/g) in removing MG at neutral pH (Sadegh *et al.*, 2015; Sun *et al.*, 2011). Overall, NPs loaded AC are the most efficient composites, given that it could remove MGefficiently in a short duration.

Rhodamine B (RhB): Iron oxide modified NPs such as humic acid-modified iron oxide (Peng *et al.*, 2012), Cu doped cobalt iron oxide NPs ($Co_{1-x}Cu_xFe_2O_4$) (Sundararajan *et al.*, 2017), Co NPs implanted on magnetic ordered mesoporous carbon (nCo/OMC, Tang *et al.*, 2014) and magnetic nGO (MNPs-nGO, Junyong *et al.*, 2013) have shown very good (>96%) removal efficiency for rhodamine B (RhB). Amongst these NCsM NPs-nGO (Junyong *et al.*, 2013), shows excellent performance (98%) in removing RhB within a short duration of 5 min. However, the performance of the NCs decreases with an increase in contaminant concentration (99% and 67% removal an initial concentration of 10 mg/l and 100 mg/l respectively). L-Serine capped nFe_3O_4 particles (85%, Belachew *et al.*, 2017), $nCeO_2$ particles (87%, Calvache-Munoz, and Rodriquez-paez, 2020) and multiple nitrogens from terpyridine ligand (MNP-Tppy, 80%, Ojemaye *et al.*, 2019) have shown reasonably good performance in removing RhB. MNP-TppyNCs could be recycled for several cycles with insignificant loss in NCs. Palm kernel shell (PKSAC) synthesized AC coated nFe_3O_4 (Panneerselvam *et al.*, 2012), nZnO synthesized from Aervalanata and Aervajavanica flower extracts (Duraimurugan *et al.*, 2020), cadmium oxide (nCdO) synthesized from plant extract (Malini *et al.*, 2020) show reasonable performance (84% to 96%) in removing RhB. Several metal-based NCs such as Sn doped nCuO particles (Vomáčka *et al.*, 2016), nZnO-AC (Saini *et al.*, 2017), nrGO-nFe_3O_4 (Sun *et al.*, 2011), $nTiO_2$/MWCNTs (Zhou *et al.*, 2010) show excellent performance (>99%) in removing RhB within a 2 to 4 hr duration. TA-nFe_3O_4 mixed with nAg particles (Veisi *et al.*, 2019) shows an excellent performance (99%) for removal of RhB at very fast rate (within 1 min). Moreover, the NCs could be recycled for several cycles without compromising much in its performance. Amongst other NCs, nrGO-nFe_2O_4in the presence of Co (92% removal in 2 min, Bai *et al.*, 2012), CNTs and ZF-PANI (90% and 85% removal within 30 min, Kumar *et al.*, 2014; Rachna *et al.*, 2018) show reasonably good performance in removing RhB. The removal capacities of magnetic $ZnFe_2O_4$ NC (ZnFe-6mg/g), nrGO-hydrogel (29 mg/g) and nGs-TA (201 mg/g) for RhB varies from comparatively low to high ranges (Tiwari *et al.*, 2013; Liu *et al.*, 2015; Konicki *et al.*, 2017). Amongst different NPs and NCs, TA-nFe_3O_4mixed with nAg particles shows best performance with a fast rate of removal, where the NCs could also be regenerated.

5.2.2.2 *Halogenated Hydrocarbons*

Some of the most frequently occurring halogenated compounds that have a serious threat to human health, such as chloroform (CF), tetrachloroethylene (PCE), trichloroethylene (TCE), trichloroethane, etc., are discussed in this section. The removal of chlorinated hydrocarbons (CHCs) from water is controlled by dechlorination or degradation, or catalytic oxidation following different pathways. The toxic chlorinated compounds are transformed into less toxic CHCs or organic hydrocarbon after complete degradation. One of the limitations of the application of metal NPs and NCs based catalyst for degradation of CHCs is the formation of toxic by-products. Thus, adsorption along with the degradation process, often facilitates the complete removal.

A significant amount (98%) of TCE could be removed by Fe/FeS (Kim *et al.*, 2011), bimetallic NPs (nZVI/Pd, He and Zhao, 2005), pMWCNT/β-CD/nTiO_2-Ag (Taka *et al.*, 2020), polydopamine decorated nrGO supported Fe nanocatalyst (PDA-Fe/nrGO) (Farooq *et al.*, 2018), nZVI-nrGO (Ahmad *et al.*, 2015), nrGO-Fe/Ni (Sahu *et al.*, 2018). Amongst those NCs, nZVI-nrGO shows complete removal of TCE within 2 min. nAl_2O_3 modified MWCNT NCs shows good removal of TCE having an adsorption capacity of 19 mg/g (Liang *et al.*, 2015). nZVI, along with Degussa P-25 nTiO_2 (nZVI-nTiO_2) NCs, shows rapid (10 min) and complete removal of TCE in the presence of UV rays and Ni ions. They have shown good reusability of the composites for several cycles without much compromising in its efficiency (Parshetti and Doong, 2010). Different studies have also applied NPs supported granular composites for removal of the contaminants to limit the aggregation of NPs and to inhibit the release of NPs in the processed water. Bimetallic nZVI (nZVI/Pd, dose 50 g/l) anchored in alginate beads shows excellent removal (99%) of TCE within 4 hr (Kim *et al.*, 2010), where negligible amount (3%) of Fe is released in the processed water. The removal efficiency of TCE by CA encapsulated-nZVI (89-91%) and unmodified nZVI (88-90%) are reasonably good and in the same range (Bezbaruah *et al.*, 2011). However, the application of CA encapsulated-nZVI make the nZVI more stable, which can be stored for a longer time (up to 4 months) without much decrease in its performance. Agarose Fe NPs have a high removal efficiency (97%) for TCE, which decreases slightly (95%) in the presence of Cr(VI) ions (Luo *et al.*, 2016). Polymer modified-nZVI (Wang *et al.*, 2010), $nTiO_2$/AC (Salih *et al.*, 2012), nZVI/AC (Su *et al.*, 2013), nZVI supported with ordered mesoporous carbon (nZVI-CMK-3, Cheng *et al.*, 2020) and sulfide modified nZVI (Dong *et al.*, 2019) show moderate to reasonably good (60% to 90%) performance in removing TCE. With an increase in nZVI dose, complete removal of TCE is reported by polymer-modified-nZVI (Wang *et al.*, 2010). The Ni-dopping on nZVI can bring the TCE concentration below the detection limit (6 μg/l) within 2 hr (Schrick *et al.*, 2002). However, application of Pd as catalyst can even make the removal rate of TCE faster by degrading it completely within 5 to 10 min (Schrick *et al.*, 2002).

Amongst other CHCs, $nCeO_2$-AC shows excellent performance (>99%) in removing CF, dichloromethane (DCM), carbon tetrachloride within 10 min. The composite also shows good regeneration capacities and reusability for a few cycles (Alhooshani, 2019). Polyethylenimine (PEI) modified nZVI also shows high removal efficiency (96%) within 2 hr for 1, 2-DCP, PCE, and TCE from groundwater (Lin *et al.*, 2018). Reasonably good (92%) removal of CF within 45 min by Fe/Ni-ACF NCs is achieved (Jin *et al.*, 2019). nrGO-Ag also shows high removal efficiency (95%) in removing CF, DCM, and carbon tetrachloride. The composites could be regenerated and reused for several cycles without compromising its performance (Koushik *et al.*, 2016).

5.2.2.3 Polycyclic Aromatic Hydrocarbon (PAH)

Polycyclic aromatic hydrocarbons (PAHs) are other contaminants of concern that contain single or multiple aromatic rings in its structure. Removal of chlorinated benzene such as trichlorobenzene (TCB), dichlorobenzene (4,4'-DCB), chlorinated phenol such as chlorophenol (CP), dichlorophenol (2,4-DCP), dichlorophenoxyacetic acid(2,4-D), and chlorinated phenyls such as polychlorinated biphenyls (PCBs) by different NPs and NCs are summarized in Table 5.2 and discussed below.

Table 5.2 A comprehensive summary of the performance of NPs and NCs in removing targeted organic contaminants.

Target contaminants / **Nanoparticles and nanocomposites (NCs)**	**Organic dyes:** Congo red (CR), Methylene blue (MB), Malachite green (MG), Methyl orange (MO), Methyl violet (MV), Rhodamine B (RhB)	**Halogenated hydrocarbons:** Chloroform (2,4-CF), Carbon tetrachloride (CTC), Dichloromethane (DCM), Dichlorophenoxyacetic acid (2,4-D) Tetrachloroethylene (TCE), Trichloroethane (PCE)	**Polyhalogenated aromatic compounds:** Chlorophenol (4-CP) Dichlorobenzene (4,4'-DCB), Dichlorophenol (2,4-DCP), Dichlorodiphenyl, Polybrominated diphenyl esters (PBDEs), Polychlorinated biphenyls (PCBs)	**Miscellaneous aromatic pollutant:** Nitrophenol, Phenol
Graphene NPs [i.e., Graphene (nGs)/ Graphene oxide (nGO)/ reduces graphene oxide nanoparticles (nrGO), graphene nanosheets (nGSs), magnetite nGO (MNPs-nGO) nGOs-rhamnolopid (RL-nGO), polyamine-modified magnetic nGO (MMGO)]	- Different form of nGOs can remove more than 99% of CR, MB, MG, RhB - MNPs-nGO adsorbent also shows more than 98% removal of RhB in river water		- Magnetic nGO show 100% removal for PAH - nGSs show good (126-149 mg/g) adsorption capacity for benzene and toluene	- nrGO shows reasonably high adsorption capacity (238-266 mg/g) for dinitrobenzene, nitrobenzene, and p-nitrotoluene

(*Continued*)

Table 5.2 A comprehensive summary of the performance of NPs and NCs in removing targeted organic contaminants. (*Continued*)

Target contaminants / **Nanoparticles and nanocomposites (NCs)**	**Organic dyes:** Congo red (CR), Methylene blue (MB), Malachite green (MG), Methyl orange (MO), Methyl violet (MV), Rhodamine B (RhB),	**Halogenated hydrocarbons:** Chloroform (2,4-CF), Carbon tetrachloride (CTC), Dichloromethane (DCM), Dichlorophenoxy-acetic acid (2,4-D) Tetrachloroethylene (TCE), Trichloroethane (PCE)	**Polyhalogenated aromatic compounds:** Chlorophenol (4-CP) Dichlorobenzene (4,4'-DCB), Dichlorophenol (2,4-DCP),Dichlorodiphenyl, Polybrominated diphenyl esters (PBDEs), Polychlorinated biphenyls (PCBs)	**Miscellaneous aromatic pollutant:** Nitrophenol, Phenol
	- More than 95% MV, MO is removed by MMGO and unmodified nGO. - Removal of dyes are reasonably high with a different form of nGOs - Extent of removal of MB using RL-nGO (>95%) is comparatively high from the water taken from natural sources than that of dye spiked in ultrapure water.			

(*Continued*)

Table 5.2 A comprehensive summary of the performance of NPs and NCs in removing targeted organic contaminants. (*Continued*)

Target contaminants / **Nanoparticles and nanocomposites (NCs)**	**Organic dyes:** Congo red (CR), Methylene blue (MB), Malachite green (MG), Methyl orange (MO), Methyl violet (MV), Rhodamine B (RhB),	**Halogenated hydrocarbons:** Chloroform (2,4-CF), Carbon tetrachloride (CTC), Dichloromethane (DCM), Dichlorophenoxyacetic acid (2,4-D) Tetrachloroethylene (TCE), Trichloroethane (PCE)	**Polyhalogenated aromatic compounds:** Chlorophenol (4-CP) Dichlorobenzene (4,4'-DCB), Dichlorophenol (2,4-DCP),Dichlorodiphenyl, Polybrominated diphenyl esters (PBDEs), Polychlorinated biphenyls (PCBs)	**Miscellaneous aromatic pollutant:** Nitrophenol, Phenol
Carbon nanotubes (CNTs) [i.e., Multi-walled/single-walled carbon nanotubes (MWCNTs/ SWCNTs), pMWCNT/β-CD/nTiO_2-Ag[1], MWCNTs-COOH, graphitized CNTs, Pd-MWCNTs, CD-MWCNTs, CD-CNTs]	- MWCNTs is capable of removing more than 99% CR, MO, and MV (MV by magnetic MWCNTs) - Removal of MB (95%) and RhB (90%) by CNTs is reasonable - MWCNTs-COOH shows good removal of MG with an adsorption capacity of 49.45mg/g	- pMWCNT/β-CD/ nTiO_2-Ag) remove 98.5% of TCE - The removal decreases slightly(95.6%) under natural water condition	- Pd-MWCNTs removal 100% 4-CP within 30 min	- CD-CNTs shows 99% removal of p-NP

(*Continued*)

Table 5.2 A comprehensive summary of the performance of NPs and NCs in removing targeted organic contaminants. (*Continued*)

Target contaminants / **Nanoparticles and nanocomposites (NCs)**	**Organic dyes:** Congo red (CR), Methylene blue (MB), Malachite green (MG), Methyl orange (MO), Methyl violet (MV), Rhodamine B (RhB),	**Halogenated hydrocarbons:** Chloroform (2,4-CF), Carbon tetrachloride (CTC), Dichloromethane (DCM), Dichlorophenoxyacetic acid (2,4-D) Tetrachloroethylene (TCE), Trichloroethane (PCE)	**Polyhalogenated aromatic compounds:** Chlorophenol (4-CP) Dichlorobenzene (4,4'-DCB), Dichlorophenol (2,4-DCP),Dichlorodiphenyl, Polybrominated diphenyl esters (PBDEs), Polychlorinated biphenyls (PCBs)	**Miscellaneous aromatic pollutant:** Nitrophenol, Phenol
Iron-basedNPs [i.e., (nFe_xO_y), γ-nFe_2O_3, α-nFe_3O_4, nano zero-valent iron (nZVI), L-serine capped nFe_3O_4, humic acid (HA) modified nFe_xO_y (HA-nFe_xO_y), Fe/nNi, sodium dodecyl sulfate coated iron (SDS-MNPs), almond shell loaded on nFe_3O_4, Fe-Fe/S, nZVI-S, $CaCoO_3$-nFe_2O_3]	- More than 99% removal of MB, MV, and MO dye is observed using different types of modified nFe_xO_y particles - About 86% and 100% removal of MG and CR are observed using α-nFe_3O_4 and γ-nFe_2O_3 particles - Presence of HA enhances the removal efficiency of RhB (>98%)	- nZVI are largely used for the remediation of CHCs pollutants. - Removal of TCE is high - The specific surface area and content of elemental iron (Fe^0) is the main factor of remediation - Removal of TCE is very rapid (within 2 min) and efficient (100%) by Fe/FeS	- 97% removal of BTEX by nZVI is reported - 64% removal of 2,4-DCP by Fe NPs synthesized from leave extract - 100% oxidation of o-DCB using $CaCoO_3$-nFe_2O_3 is possible	

(*Continued*)

Table 5.2 A comprehensive summary of the performance of NPs and NCs in removing targeted organic contaminants. (*Continued*)

Target contaminants / **Nanoparticles and nanocomposites (NCs)**	**Organic dyes:** Congo red (CR), Methylene blue (MB), Malachite green (MG), Methyl orange (MO), Methyl violet (MV), Rhodamine B (RhB),	**Halogenated hydrocarbons:** Chloroform (2,4-CF), Carbon tetrachloride (CTC), Dichloromethane (DCM), Dichlorophenoxyacetic acid (2,4-D) Tetrachloroethylene (TCE), Trichloroethane (PCE)	**Polyhalogenated aromatic compounds:** Chlorophenol (4-CP) Dichlorobenzene (4,4'-DCB), Dichlorophenol (2,4-DCP),Dichlorodiphenyl, Polybrominated diphenyl esters (PBDEs), Polychlorinated biphenyls (PCBs)	**Miscellaneous aromatic pollutant:** Nitrophenol, Phenol
NPs modified AC [i.e., nZnO-AC, nAg-AC, $nSnO_2$-AC, nPd-AC, Ni-nZnS-AC, nNiS-AC, nZnS-Cu-AC, $nZn(OH)_2$-AC, nCuS-AC, nAu-AC, nAg-OH-AC, PKSAC-$nFe_3O_4^2$, nZnO-nanorods (NRs)-AC, $nTiO_2$-AC, nZVI-AC, $nTiO_2$-ACF, nFe_3O_4-AC, MWCNTs-AC]	- NPs loaded AC act as a good adsorbent for removing MB from water within a short span of contact time - nZnO-AC can remove 99% RhB in 120 min - nNiS and nZnS-Cu loaded AC can remove 99% MB within 7 min and 2 min	- 90% removal of TCE by nZVI-AC is observed		- $nTiO_2$-ACF degrade 98% phenol within 3 hr at very high temperature (500°C)

(*Continued*)

Table 5.2 A comprehensive summary of the performance of NPs and NCs in removing targeted organic contaminants. (*Continued*)

Target contaminants / **Nanoparticles and nanocomposites (NCs)**	**Organic dyes:** Congo red (CR), Methylene blue (MB), Malachite green (MG), Methyl orange (MO), Methyl violet (MV), Rhodamine B (RhB),	**Halogenated hydrocarbons:** Chloroform (2,4-CF), Carbon tetrachloride (CTC), Dichloromethane (DCM), Dichlorophenoxyacetic acid (2,4-D) Tetrachloroethylene (TCE), Trichloroethane (PCE)	**Polyhalogenated aromatic compounds:** Chlorophenol (4-CP) Dichlorobenzene (4,4'-DCB), Dichlorophenol (2,4-DCP),Dichlorodiphenyl, Polybrominated diphenyl esters (PBDEs), Polychlorinated biphenyls (PCBs)	**Miscellaneous aromatic pollutant:** Nitrophenol, Phenol
	- More than 99% removal of MV,MO, RhB, and CR is achieved by $nSnO_2$, nCuS, nZnO, and nZnO-NRs loaded AC - Several metal NPs loaded AC can remove more than 99% MG within 5 min			

(*Continued*)

Table 5.2 A comprehensive summary of the performance of NPs and NCs in removing targeted organic contaminants. (*Continued*)

Target contaminants / Nanoparticles and nanocomposites (NCs)	**Organic dyes:** Congo red (CR), Methylene blue (MB), Malachite green (MG), Methyl orange (MO), Methyl violet (MV), Rhodamine B (RhB),	**Halogenated hydrocarbons:** Chloroform (2,4-CF), Carbon tetrachloride (CTC), Dichloromethane (DCM), Dichlorophenoxy-acetic acid (2,4-D) Tetrachloroethylene (TCE), Trichloroethane (PCE)	**Polyhalogenated aromatic compounds:** Chlorophenol (4-CP) Dichlorobenzene (4,4'-DCB), Dichlorophenol (2,4-DCP),Dichlorodiphenyl, Polybrominated diphenyl esters (PBDEs), Polychlorinated biphenyls (PCBs)	**Miscellaneous aromatic pollutant:** Nitrophenol, Phenol
Graphene oxide (nGO)-organic/polymer framework [i.e., nGOs – chitosan (CS), Magnetic nGs – tannic acid (TA), nGOs-monoliths, nGOs sponge, nGOs – methylcellulose (MC), nGOs-alginate acid, PLA-nGO/CSS[3], PDA-Fe/nrGO[4], nrGO/polyHIPEs[5], nGO/PPy (polypyrrole)]	- Majority of the studies are done for MB removal - 100% removal of MB and RhB is observed by nGOs- and nGOs/PEI- hydrogels- within few hours - nGO-CS and nGO-sponge also shows high efficiency (>97%) for MV removal - 3-D filter PLA-nGO/CSS shows 97.8% removal for CV	- PDA-Fe/nrGO shows very high (>99%) degradation of TCE	- Highest removal is observed as nGO has a good affinity towards PAH - 3-D structure shows high efficiency in the presence of magnetite (nrGO/polyHIPEs) - It takes a few hours for complete degradation	

(*Continued*)

Table 5.2 A comprehensive summary of the performance of NPs and NCs in removing targeted organic contaminants. (*Continued*)

Target contaminants / **Nanoparticles and nanocomposites (NCs)**	**Organic dyes:** Congo red (CR), Methylene blue (MB), Malachite green (MG), Methyl orange (MO), Methyl violet (MV), Rhodamine B (RhB),	**Halogenated hydrocarbons:** Chloroform (2,4-CF), Carbon tetrachloride (CTC), Dichloromethane (DCM), Dichlorophenoxy-acetic acid (2,4-D) Tetrachloroethylene (TCE), Trichloroethane (PCE)	**Polyhalogenated aromatic compounds:** Chlorophenol (4-CP) Dichlorobenzene (4,4'-DCB), Dichlorophenol (2,4-DCP),Dichlorodiphenyl, Polybrominated diphenyl esters (PBDEs), Polychlorinated biphenyls (PCBs)	**Miscellaneous aromatic pollutant:** Nitrophenol, Phenol
Carbon nanotubes (CNTs)- metal framework [i.e., MWCNTs-nFe_3O_4, nTiO_2/MWCNTs, MWCNTs-Pd/Fe, Guar gum (GG)-MWCNTs-nFe_3O_4,nAl_2O_3/MWCNTs, MMWCNTs-Fe_3C, MWCNTs-$CoFe_2O_4$, nTiO_2/nSiO_2-CNTs, CNTs/nTiO_2]	- Magnetite used in NCs to enhance the reusability of CNTs - Rapid adsorption of MB byGG-MWCNTs-Fe_3O_4 is observed - nTiO_2/MWCNTs or nTiO_2/SWCNTs show 100% removal for RhB - MMWCNTs-Fe_3C shows high adsorption capacity (172.4 mg/g) for CR at pH 1	- nAl_2O_3/MWCNTs shows a moderate removal capacity of 19.84 mg/g for TCE under synthetic groundwater condition - Removal efficiency is higher than that by MWCNTs and functionalized MWCNTs	- More than 99% removal of DCP by MWCNTs-Pd/Fe is reported	- CNTs/nTiO_2 shows 98% removal of phenol - MWCNTs/nTiO_2 shows 100% removal of 2,4-DNP

(*Continued*)

Table 5.2 A comprehensive summary of the performance of NPs and NCs in removing targeted organic contaminants. (*Continued*)

Target contaminants / **Nanoparticles and nanocomposites (NCs)**	**Organic dyes:** Congo red (CR), Methylene blue (MB), Malachite green (MG), Methyl orange (MO), Methyl violet (MV), Rhodamine B (RhB),	**Halogenated hydrocarbons:** Chloroform (2,4-CF), Carbon tetrachloride (CTC), Dichloromethane (DCM), Dichlorophenoxy-acetic acid (2,4-D) Tetrachloroethylene (TCE), Trichloroethane (PCE)	**Polyhalogenated aromatic compounds:** Chlorophenol (4-CP) Dichlorobenzene (4,4'-DCB), Dichlorophenol (2,4-DCP),Dichlorodiphenyl, Polybrominated diphenyl esters (PBDEs), Polychlorinated biphenyls (PCBs)	**Miscellaneous aromatic pollutant:** Nitrophenol, Phenol
Carbon nanotubes (CNTs)- organic/polymer framework [i.e., MWCNTs-CSB[6], Calcium alginate (CA)/MWCNTs, MWCNTs-CS,MWCNTs-methyl methacrylate (MMA), CS modified MWCNTs, CNT-ACF[7]]	- Modification by the polymer of CNTs are done mainly to increase its colloidal stability - Amino and hydroxyl groups on the surface increases the efficiency of CNT - MWCNTs-CS hydrogel shows significantly (95%) removal of CR with an excellent adsorption capacity of 460 mg/g		- MWCNTs-MMA shows 95% removal for 4,4-DCP - CS-MWCNTs, when coated on polypropylene membrane, shows excellent (99-100%) removal for PCBs and PBDEs	- CNT-ACF shows high removal efficiency (99%)for phenol

(*Continued*)

Table 5.2 A comprehensive summary of the performance of NPs and NCs in removing targeted organic contaminants. (*Continued*)

Target contaminants / **Nanoparticles and nanocomposites (NCs)**	**Organic dyes:** Congo red (CR), Methylene blue (MB), Malachite green (MG), Methyl orange (MO), Methyl violet (MV), Rhodamine B (RhB),	**Halogenated hydrocarbons:** Chloroform (2,4-CF), Carbon tetrachloride (CTC), Dichloromethane (DCM), Dichlorophenoxyacetic acid (2,4-D) Tetrachloroethylene (TCE), Trichloroethane (PCE)	**Polyhalogenated aromatic compounds:** Chlorophenol (4-CP) Dichlorobenzene (4,4'-DCB), Dichlorophenol (2,4-DCP),Dichlorodiphenyl, Polybrominated diphenyl esters (PBDEs), Polychlorinated biphenyls (PCBs)	**Miscellaneous aromatic pollutant:** Nitrophenol, Phenol
	- CA/MWCNTs shows good adsorption capacity for MB and MO - More than 79% MB and MO can be desorbed at pH 2 and 13, which can facilitate the regeneration of CA/MWCNTs			

(*Continued*)

Table 5.2 A comprehensive summary of the performance of NPs and NCs in removing targeted organic contaminants. (*Continued*)

Target contaminants / **Nanoparticles and nanocomposites (NCs)**	**Organic dyes:** Congo red (CR), Methylene blue (MB), Malachite green (MG), Methyl orange (MO), Methyl violet (MV), Rhodamine B (RhB),	**Halogenated hydrocarbons:** Chloroform (2,4-CF), Carbon tetrachloride (CTC), Dichloromethane (DCM), Dichlorophenoxy-acetic acid (2,4-D) Tetrachloroethylene (TCE), Trichloroethane (PCE)	**Polyhalogenated aromatic compounds:** Chlorophenol (4-CP) Dichlorobenzene (4,4'-DCB), Dichlorophenol (2,4-DCP),Dichlorodiphenyl, Polybrominated diphenyl esters (PBDEs), Polychlorinated biphenyls (PCBs)	**Miscellaneous aromatic pollutant:** Nitrophenol, Phenol
Iron-based- metal framework **[i.e.,** $nZnO-nFe_3O_4$/nrGO, nFe_3O_4/Carbon, nFe_2O_3-HNTs[8], $nFe_2O_3-nAl_2O_3$, $nCuO-nFe_3O_4$, nFe_3O_4-nGO, nZVI-nrGO, nZVI/$nTiO_2$]	- $nFe_2O_3-nAl_2O_3$ shows excellent removal (100%) for CR - nFe_2O_3-HNTs removed 88.2% of MV - 97% removal of MB by $nZnO-nFe_3O_4$/nrGO can be possible - The composite can be regenerated for five cycles	- Complete degradation of TCE within 2 min using nZVI-nrGO - nZVI/$nTiO_2$ shows long-term efficiency for removal of TCE under UV and Ni ion	- $nCuO-nFe_3O_4$ shows 96.9% removal of 2,4-DCP - nFe_3O_4-nGO is used for the extraction of TCP	

(*Continued*)

Table 5.2 A comprehensive summary of the performance of NPs and NCs in removing targeted organic contaminants. (*Continued*)

Target contaminants / Nanoparticles and nanocomposites (NCs)	**Organic dyes:** Congo red (CR), Methylene blue (MB), Malachite green (MG), Methyl orange (MO), Methyl violet (MV), Rhodamine B (RhB),	**Halogenated hydrocarbons:** Chloroform (2,4-CF), Carbon tetrachloride (CTC), Dichloromethane (DCM), Dichlorophenoxyacetic acid (2,4-D) Tetrachloroethylene (TCE), Trichloroethane (PCE)	**Polyhalogenated aromatic compounds:** Chlorophenol (4-CP) Dichlorobenzene (4,4'-DCB), Dichlorophenol (2,4-DCP),Dichlorodiphenyl, Polybrominated diphenyl esters (PBDEs), Polychlorinated biphenyls (PCBs)	**Miscellaneous aromatic pollutant:** Nitrophenol, Phenol
Iron-based- organic/polymer framework [i.e., nFe_3O_4-CS,Polyacrylic acid (PAA)-nFe_xO_y, CMC-nFe_3O_4, agarose-nFe, nZVI-alginate beads, Xylan/Poly magnetic hydrogel, gum xantham/nFe_3O_4, nFe_3O_4-CTAB[10], clay/PANI/nFe_3O_4[11],nFe_3O_4-C/PB[12], BTCA-nFe_3O_4[13], nFe_3O_4-nGS[14], nFe_3O_4-β-CD[15], CS-poly(m-phenylenediamine)-nFe_3O_4, nZVI/CMK[16]-3, PEI-nZVI[17]]	- Surface modification is done mainly to limit aggregation - Decolorization efficiency for CR, MB, MV are reasonably good (>97%) by BTCA-nFe_3O_4, nFe_3O_4-GS and gum xantham/nFe_3O_4.	- nZVI-alginate beads show 99.8% removal of TCE - It releases very less Fe, thus making it environment friendly - PEI-nZVI can remove more than 97% 1,2-DCP and PCE	- nFe_3O_4-β-CD shows 100% removal of 4-CP, which is attributed to advanced oxidation process (AOPs) - nFe_3O_4-C/PB can remove 100% 2,4-DCP - CS-poly-nFe_3O_4 shows more than 94% removal of PCBs under varying initial concentrations	- TCF can remove 95% nitrophenol within 35 min

(*Continued*)

Table 5.2 A comprehensive summary of the performance of NPs and NCs in removing targeted organic contaminants. (*Continued*)

Target contaminants / Nanoparticles and nanocomposites (NCs)	**Organic dyes:** Congo red (CR), Methylene blue (MB), Malachite green (MG), Methyl orange (MO), Methyl violet (MV), Rhodamine B (RhB),	**Halogenated hydrocarbons:** Chloroform (2,4-CF), Carbon tetrachloride (CTC), Dichloromethane (DCM), Dichlorophenoxyacetic acid (2,4-D) Tetrachloroethylene (TCE), Trichloroethane (PCE)	**Polyhalogenated aromatic compounds:** Chlorophenol (4-CP) Dichlorobenzene (4,4'-DCB), Dichlorophenol (2,4-DCP),Dichlorodiphenyl, Polybrominated diphenyl esters (PBDEs), Polychlorinated biphenyls (PCBs)	**Miscellaneous aromatic pollutant:** Nitrophenol, Phenol
Metallic/Bimetallic/trimetallic NCs [i.e., nZVI-Ni/Fe, La/Cu/Zr-carbon dot trimetallic, DDS-Mg/Al LDH[18], nZVI-Pd, nrGO/Fe/Ni, nrGO-Ag, nZVI-Ni, Fe/Ni-ACF, Ni/Pd bimetallic NPs, La/Co/Ni-nGO, Ni-Zn nFe_2O_3, Pd/Fe-nFe_3O_4, Cu and Ag impregnated on MWCNTs-CD, catalyst Pt/Pd- $nTiO_2$]	- La/Cu/Zr-carbon dot can remove 86% MG - Synergic effect of sorption and photocatalytic degradation is the governing process	- Almost 100% dechlorination of TCEby nrGO/Fe/Ni is possible for multiple cycles - nrGO-Ag shows 95% removal for CF, DCM, CTC - Degradation efficiency of nZVI-Ni/Fe for TCE reduces significantly under groundwater condition compared to that of the DI water system	- Ni/Pd bimetallic NPs shows 100% removal for DCB - Ni-Zn nFe_2O_3tri-metallic NPs can remove almost 100% 4-DCP and 2,4-D	- Cu and Ag-MWCNTs-CD shows 88% removal for p-NP

(*Continued*)

Table 5.2 A comprehensive summary of the performance of NPs and NCs in removing targeted organic contaminants. (*Continued*)

Target contaminants / **Nanoparticles and nanocomposites (NCs)**	**Organic dyes:** Congo red (CR), Methylene blue (MB), Malachite green (MG), Methyl orange (MO), Methyl violet (MV), Rhodamine B (RhB),	**Halogenated hydrocarbons:** Chloroform (2,4-CF), Carbon tetrachloride (CTC), Dichloromethane (DCM), Dichlorophenoxyacetic acid (2,4-D) Tetrachloroethylene (TCE), Trichloroethane (PCE)	**Polyhalogenated aromatic compounds:** Chlorophenol (4-CP) Dichlorobenzene (4,4'-DCB), Dichlorophenol (2,4-DCP),Dichlorodiphenyl, Polybrominated diphenyl esters (PBDEs), Polychlorinated biphenyls (PCBs)	**Miscellaneous aromatic pollutant:** Nitrophenol, Phenol
Cerrium oxide ($nCeO_2$) and titanium oxide ($nTiO_2$) NCs [i.e., $nTiO_2$, $nCeO_2$, $nCeO_2$-AC, $nCeO_2$-$nTiO_2$, $nCeO_2$-nFe_2O_3,$nTiO_2$-ACF, Cu/CS-$nTiO_2$, $nTiO_2$/poly-(fluorine-co-thiophene) (PFT), Cu doped $CoTiO_3$, $nCeO_2$ NCs (nanocubes), $nTiO_2$ nanotube ferrite (TCF)]	- Cu/CS-$nTiO_2$ shows a high affinity towards MB dye in a matrix of MB, MO, CR, and nitrophenol solution - 99% removal of MO by $nTiO_2$-ACF is observed - The removal is dominated by photocatalytic degradation - About 90% MG is removed by $nCeO_2$ synthesized from urea	- 99.4% removal of CF is observed using $nCeO_2$-AC	- 90% combustion of o-DCB using $Ce_xTi_yO_2$is observed -$Ce_yFe_xO_2$ shows 100% removal of 4-CP, 2,4-DCP, and 2,4-D at different content of Ce, Fe	- Nitrophenols is transformed into less toxic aminophenols (AP) by Cu/CS-$nTiO_2$ - $nTiO_2$/poly-PFT shows 75% degradation of phenol

(*Continued*)

Table 5.2 A comprehensive summary of the performance of NPs and NCs in removing targeted organic contaminants. (*Continued*)

Target contaminants / **Nanoparticles and nanocomposites (NCs)**	**Organic dyes:** Congo red (CR), Methylene blue (MB), Malachite green (MG), Methyl orange (MO), Methyl violet (MV), Rhodamine B (RhB),	**Halogenated hydrocarbons:** Chloroform (2,4-CF), Carbon tetrachloride (CTC), Dichloromethane (DCM), Dichlorophenoxyacetic acid (2,4-D) Tetrachloroethylene (TCE), Trichloroethane (PCE)	**Polyhalogenated aromatic compounds:** Chlorophenol (4-CP) Dichlorobenzene (4,4'-DCB), Dichlorophenol (2,4-DCP),Dichlorodiphenyl, Polybrominated diphenyl esters (PBDEs), Polychlorinated biphenyls (PCBs)	**Miscellaneous aromatic pollutant:** Nitrophenol, Phenol
Miscellaneous NPs and composites: Fe^{3+} doped nZnS quantum dot, large pore carbon NPs, hydroxyapatite, ferrocene modified $nSnO_2$-nAl_2O_3, Cu doped $CoFe_2O_4$, MNP-Tppy[20], CA encapsulated nZVI, Fe-MWCNTs-CH_2COONa, $nCuO_2$ green synthesis,	- More than 99% removal of MG and MO by nZnS quantum dot in the presence of Fe^{3+} and $nSiO_2$-nAl_2O_3 is achieved - nCuO synthesized from punica granatum leaf can remove 97% MB - More than 95% removal of MV is observed using nZnS quantum dot in the presence of Fe^{3+} - The capsule of chitosan NPs can remove 99% CR within 1 min	- CA encapsulated nZVI can remove around 91% of TCE - It can be used for four months without any loss in its removal efficiency	- Fe-MWCNT-CH_2COONa remove 79%, 81%, and 83% of benzene, toluene, and dimethyl - 90% removal of 4-CP using $nTiO_2$	

(*Continued*)

Table 5.2 A comprehensive summary of the performance of NPs and NCs in removing targeted organic contaminants. (*Continued*)

Target contaminants / **Nanoparticles and nanocomposites (NCs)**	**Organic dyes:** Congo red (CR), Methylene blue (MB), Malachite green (MG), Methyl orange (MO), Methyl violet (MV), Rhodamine B (RhB),	**Halogenated hydrocarbons:** Chloroform (2,4-CF), Carbon tetrachloride (CTC), Dichloromethane (DCM), Dichlorophenoxyacetic acid (2,4-D) Tetrachloroethylene (TCE), Trichloroethane (PCE)	**Polyhalogenated aromatic compounds:** Chlorophenol (4-CP) Dichlorobenzene (4,4'-DCB), Dichlorophenol (2,4-DCP),Dichlorodiphenyl, Polybrominated diphenyl esters (PBDEs), Polychlorinated biphenyls (PCBs)	**Miscellaneous aromatic pollutant:** Nitrophenol, Phenol
nNiO, nZnO, Sn-nCuO, nAg, $nSnO_2$, monodispersed nSi particles, egg shell NPs, PEI-$nSnO_2$, nCdO, hollow Zn-nFe_2O_3 nanosphere, QPEI/$nSiO_2$[20]]	- nSi, $nSnO_2$, Zn-nFe_2O_4, nZnO, and Sn-nCuO can remove more than 99% removal of MB, MV, CR, MO, and RhB, respectively			

1) **pMWCNT/β-CD/$nTiO_2$-Ag**-silver decorated phosphorylated carbon nanotube/cyclodextrin (CD) nanosponge, 2) **PKSAC-nFe_3O_4** - palm kernel shellAC coated on nFe_3O_4, 3)**PLA-nGO/CSS** - polylactic acid-nGO/chitosan sponge, 4) **PDA-Fe/nrGO** - polydopamine decorated nrGO supported Fe nanocatalyst, 5) **nrGO/polyHIPEs** - nrGO-hybridized polymeric high interal phase emulsions, 6) **MWCNTs-CSB** - MWCNTs-chitosan hydrogel beads, 7) **ACF** - activated carbon fibres, 8) **nFe_2O_3-HNTs** - magnetic halloysite nanotubes NCs, 9) **CTAB** - cetrimonium bromide, 10) **clay/PANI/nFe_3O_4** - two-dimensional superparamagnetic clay/polyaniline/nFe_3O_4, 11) **nFe_3O_4-C/PB** - magnetic carbon supported by perssian blue, 12) **BTCA** - benzenetetracarboxylic acid 13) **nFe_3O_4-nGS** - magnetite graphene sponge, 14) **nFe_3O_4/β–CD/nGO** – magnetite β-cyclodextrin-nGO, 15) **nZVI/CMK** – ordered mesoporous carbon supported nZVI, 16) **PEI-nZVI** – Polyethylenimine-nZVI, 17) **DDS-Mg/Al LDH** – sodium dodecyl sulfate and magnesium–aluminum layered double hydroxides, 18) **MNP-Tppy** - multiple nitrogen from terpyridine ligand, 19) **QPEI/$nSiO_2$** - Quaternary ammonium polyethylenimine (PEI) modified silica.

Polychlorinated benzene: Ni/Pd bimetallic NPs (Jung *et al.*, 2018) and $CaCO_3/\alpha\text{-}nFe_2O_3$NRs (Ma *et al.*, 2016) show complete dechlorination of DCBs (1,2-DCB, o-DCB) within 30 min. In both cases, the removal of DCB pollutants depends on the amount of the catalyst loaded with the NPs and composites. $Ce_xTi_yO_2$is used as the catalyst for combustion of o-DCB, which shows moderate removal (90%) of the contaminants at 375 °C (Deng *et al.*, 2016). Graphitized CNTs show good removal capacity (28 mg/g) for DCB within 40 min (Peng *et al.*, 2003). Removal capacity of TCB and TCP by nGs is reasonably high (25 mg/g and 30 mg/g), where nGO shows comparatively lower removal capacity (2 mg/g and 5 mg/g) of those compounds (Pei *et al.*, 2013). The removal efficiency of benzene, toluene, ethylbenzene, and xylene (BTEX) by nZVI is high (97%) within a very short duration of 30 min and at a neutral pH (Mahmoud *et al.*, 2018). The Fe NPs functionalized MWCNTs (Fe-MWCNTs-CH_2COONa) show low to moderate range (79-83%) of removal efficiencies for benzene, toluene, dimethyl benzene (Jin *et al.*, 2007). The removal capacities for benzene and toluene are extremely high (126 mg/g and 149 mg/g) and reasonably fast (15 min to 145 min) by graphene nanosheets (nGSs) at neutral pH (Raad *et al.*, 2016).

Polychlorinated phenol: Complete dechlorination of 4-CP by Pd loaded MWCNTs in 30 min (Shu *et al.*, 2019), and by nFe_3O_4-β-cyclodextrin (CD) NCs in 90 min (Wang *et al.*, 2016) through catalytic degradation and advanced oxidation process has been reported. Kurian *et al.* (2017) have shown complete removal of 4-CP by $Ce_{0.75}Fe_{0.25}O_2$ (within 120 min), 2,4-D by $Ce_{0.25}Fe_{0.75}O_2$ (within 60 min) and 2,4-DCP by $Ce_{0.5}Fe_{0.5}O_2$ (within 45 min) at 70°C temperature, where the initial concentration of those contaminants are kept at 500 mg/l, 250 mg/l and 250 mg/l, respectively. A few studies have shown reasonable 4-CP removal efficiency (85%) by CD polymer-modified pMWCNTs (Mamba *et al.*, 2013) and by $nTiO_2$ (Hao *et al.*, 2007). Graphene oxide modified with La/Co/Ni trimetallic NCs shows moderate (71%) removal efficiency for 2-CP (Sharma *et al.*, 2019). Magnetic carbon supported by Persian blue (Guo *et al.*, 2019) and nickel-zinc ferrite NCs (Nair and Kurian, 2017) can degrade 2,4-DCP completely within 50 min and 75 min, respectively. The rate of removal of 2,4-DCP by magnetic carbon supported by Persian blue is dependent on the solution pH, where the nickel-zinc ferrite NCs could be regenerated and reused for multiple cycles. Fe NPs, synthesized from leaf extract can remove 2,4-DCP up to 64% when both sorption and Fenton reaction has been attributed to the removal of the contaminant (Guo *et al.*, 2017). The removal of 2,4-DCP by Pd/Fe-nFe_3O_4 NCs is reasonable (76%) at a nFe_3O_4 dose of 2g/l, while an increase in Pd loading (0.2% w/w) can enhance the 2,4 DCP removal efficiency significantly (98%). Around 97% removal of 2,4-DCP is observed when nCuO-nFe_3O_4 is used as an adsorbent (Sun *et al.*, 2015). Complete degradation of TCP by nCuO/nrGO in the persulfate system is reported, while negligible removal is observed when nCuO/nrGO and persulfate are applied separately (Du *et al.*, 2019).

Polychlorinated phenyl: Several NPs and NCs, such as magnetic carbon nitride NCs (Li *et al.*, 2017), nAu particles scattered on nGO sheet (Wu *et al.*, 2017), metal organic nanotube (nFe_3O_4-Co-MONT, Li *et al.*, 2016) can remove PCBs completely from aqueous solution. CS-poly (m-phenylenediamine)-nFe_3O_4 shows excellent removal efficiency (94-100%) for different types of PCBs under a very low range of concentration (i.e., 10ng/l to 50ng/l) that are generally found in the aqueous environment (Liao *et al.*, 2014). The removal efficiency of PCBs by carboxymethyl-β-CD modified nFe_3O_4-nTiO_2 particles is reasonably high (82%) with a fast rate of removal (30 min, Khammar *et al.*, 2020). Furthermore, the composite could be reused for several cycles. Amongst other NCs, methyl methacrylate (MMA) on MWCNTs (80%, Shao *et al.*, 2011a) and MWCNTs induced with chitosan (90%, Shao *et al.*, 2011b) show moderate to high removal of 4,4-dichlorinated phenyl and 4,4-dechlorinated biphenyl at pH 3.

5.2.2.4 *Miscellaneous Aromatic Pollutant*

Different NPs and NCs such as CD polymer-modified CNTs (Salipira *et al.*, 2007), CS-nTiO_2 fibre (Ali *et al.*, 2018) and MWCNTs/nTiO_2 NCs (Wang *et al.*, 2009) can remove nitrophenol and 2,4-dinitrophenol completely. Some of the NCs such as TiO_2 nanotubes-ferrite (TCF) NCs (94%, Ibrahim *et al.*, 2019), and montmorillonite nGO composite (MGC) (96%, Zhang *et al.*, 2019) shows reasonably good to excellent performance in removing nitrophenol. Significantly high removal capacity for nitrophenol (1548 mg/g) is reported by magnetic porous silica-nGO hybrid NCs at pH 8 (Liu *et al.*, 2016). Different NCs such as nTiO_2 particles on ACF (98%, Yao *et al.*, 2010), CNT-ACF (99%, Wang *et al.*, 2010), MWCNTs/nTiO_2 adsorbent (80%, Yao *et al.*, 2008) and nTiO_2/poly-(fluorene-co-thiophene) (PFT) (75%, Song *et al.*, 2007) show excellent to moderate performance in removing phenol. nrGO shows excellent removal capacity for dinitrobenzene (265 mg/g), nitrobenzene (260 mg/g), and para-nitrobenze (238 mg/g) (Chen and Chen, 2015). nFe_3O_4 (92% at pH 3, initial concentration of 50mg/l and in 2 hrs) and nZVI (92% within 24 hrs) NPs show reasonably good performance in removing DDT from water (Pushkar and Sevak, 2019).

5.2.3 Emerging Contaminants

5.2.3.1 Pharmaceuticals and Personal Care Products

Ibuprofen: Removal capacity of ibuprofen by $nZnO_2$ is significant (220mg/g in 50 min, Ulfa *et al.*, 2019), whereas degradation of ibuprofen by nZVI is moderate (66%) from the aqueous solution (Machado *et al.*, 2013). One of the studies shows that the removal efficiency of ibuprofen by single-walled carbon nanotubes (SWCNTs) with ultrasonication is much better (97%) than that of applying SWCNTs (48%) as a single entity. The study also suggests that high temperature and acidic conditions are favourable for the maximum degradation of ibuprofen (Al-Hamadani *et al.*, 2017). Another study in which $nTiO_2$ is modified with two compounds (*N*1-(2-aminobenzyliden)-*N*2,*N*2-bis(2-(2-aminobenzyliden) aminoethyl) ethane-1,2-diaminoruthenium(II) and *N*1, *N*2-bis(2aminobenziliden) ethane-1,2-diaminoruthenium(II)) show better performance in removing ibuprofen compared to that of unmodified $nTiO_2$ (Góngora *et al.*, 2017). nFe_3O_4 particles has reasonable good removal efficiency (92% at 0.6mg/l in 30 min at pH 7) with an adsorbent dose of 1.0 g/l (Ali *et al.*, 2016). Removal of ibuprofen by nCuO particles is moderate (74% at pH 4.5 in 60 min), and the removal kinetics follow the pseudo-second-order kinetics model (Husein *et al.*, 2019). Magnesium ferrite NPs ($nMgFe_2O_4$) can remove ibuprofen completely (100% in 40 min at pH 8 with the NPs dose of 0.5g/l, Ivanets *et al.*, 2020). CNTs are functionalized with nitric acid, sulphuric acid and with $nMnO_2$ and nFe_3O_4 (CNT-COOH/$nMnO_2$/nFe_3O_4) having SSA of 114.3 m^2/g show reasonable good performance (103.4 mg/g, Lung *et al.*, 2020) in removing ibuprofen.

Carbamazepine: The removal capacity of carbamazepine by granular activated carbon (AC), 191.5mg/g), multi-walled carbon nanotubes (MWCNTs, 225 mg/g, (Ncibi and Sillanpaa., 2017) and CNTs (242 mg/g, Álvarez-Torrellas *et al.*, 2017) are significant. However, the presence of different other organic contaminants such as bisphenol A and dissolved organic matter, has an adverse effect on removing carbamazepine from the aqueous solution (Lerman *et al.*, 2013). Bio-synthesized nFe_3O_4 has extremely high removal efficiency (100%) for carbamazepine, and the NPs could also be regenerated due to its magnetic property (Rajendran and Sen 2018). nZVI/nCuO prepared from Ficus Benjamina leaf extracts show promising efficiency (95% at pH 5 within 20 min) in removing carbamazepine (Abdel-Aziz *et al.*, 2019). Coconut shell–derived AC powder anchored $nTiO_2$ can degrade carbamazepine significantly (98% in 40 min, Khraisheh *et al.*, 2013). nFe_3O_4 and the same modified with methacrylic acid (MAA), $Al(OH)_3$, and SiO_2 show comparatively low removal efficiency (33, 48, 51, and 50% at pH 8 in 7 hrs) for carbamazepine (Yoon *et al.*, 2015). nGOs has also shown promising performance (99% removal at pH 6 in 120 min) in removing carbamazepine. Moreover, the NPs could be regenerated and reused for several cycles (Bhattacharya *et al.*, 2020).

Diclofenac: Removal capacity of diclofenac through nGO is promising (653.9 mg/g at a dose of 0.16 g/l within 15 min having an initial concentration of 400 mg/l at 40°C (Hiew *et al.*, 2018) and moderate with nGSs (74% at a dose of 0.25 g/l, at pH 6.2 within 300 min, Guerra *et al.*, 2019). The nZVI modified with sulfide can increase the removal efficiency (34%) of diclofenac two folds compared to that of unmodified nZVI (18%), furthermore, the sulfide-nZVI has a better recyclability (Su *et al.*, 2018). nCuO has shown remarkable removal efficiency (91% at 25°C at pH 4.5 and in 60 min) for diclofenac, where the removal rate follows the pseudo-second-order kinetics model (Husein *et al.*, 2019). Outstanding performances in removing diclofenac by cobalt ferrite NPs (nCoFe, 100% in 15 min at pH 5, Deng *et al.*, 2013), bio recovered palladium iron NPs (Pd/Fe-nFe_3O_4, 99% in 40 min, Wei *et al.*, 2020) and biogenic selenium nanoparticles (nSe) in the presence of hydrogen peroxide (H_2O_2, 97 % at pH 7.5, Ameri *et al.*, 2020) has been reported.

Paracetamol: $nTiO_2$ nanoparticles and lithium doped nZnO has shown significant removal efficiency for paracetamol through photodegradation (90% at pH 9) and (99% in 3h) respectively (Jallouli *et al.*, 2017; Thi and Lee 2017). $nTiO_2$ synthesized with lead sulfide (PbS/$nTiO_2$) NCs having a surface area of 38m^2/g can degrade up to 93% (Palma *et al.*, 2020). $nTiO_2$-Gs NCs can completely degrade paracetamol in 120 min (Vaiano *et al.*, 2018). CNT-COOH-nMnO-nFe_3O_4 NCs with a SSA of 114 m^2/g has a moderate capacity (81 mg/g) to treat paracetamol, and the rate of removal follows the pseudo-second-order kinetics model (Lung *et al.*, 2020).

Sulfamethoxazole: nAgO-nGO NCs has shown good performance (88% at pH 5 pH within 60 min) in removing sulfamethoxazole, where the maximum adsorption capacity reaches 250 mg/g (Keshvardoostchokami *et al.*, 2019) Tungsten trioxide (nWO_3, 43% at pH 4) has shown moderate removal efficiencies for sulfamethoxazolein acidic condition (Beheshti *et al.*, 2019). Manganese ferrite NPs ($nMnFe_3O_4$) synthesized on powder activated carbon PAC has shown good treatment capacity (159 mg/g at pH 7 and 25°C (Wan *et al.*, 2013). $nTiO_2$ NPs modified with different metals such as Ag, Au, and Cu have shown outstanding performance (100% at pH 5.5) in removing sulfamethoxazole (Zanella *et al.*, 2018). The rate of removal is much faster with UV radiation (completed in 180 min) compared to that of sunlight irradiation (completed in 360 min). The removal of sulfamethoxazole by $nTiO_2$ varies from moderate (61% at pH 3, Beheshti *et al.*, 2019) to good range (90% at pH 7 in 30 min, Alfred *et al.*, 2020). However, in the latter case the efficiency of the NPs decreases after its first use (58% at pH 7 in 30 min, Alfred *et al.*, 2020).

Amoxicillin: nZVI shows moderate removal efficiency (69% at pH 8 in 80min with initial concentration of 30mg/l), whereas Fe/Ni stabilized with CS shows reasonably good performance in removing amoxicillin (85% at pH 2 in 77 min with initial concentration of 30 mg/l, Yazdanbakhsh *et al.*, 2016). The removal efficiency (99% at pH 2, and in 200 min) for amoxicillin by nFe_3O_4 is significant, where the kinetics follows the pseudo-first-order model (Demirezen *et al.*, 2019). Bimetallic Fe/Ni NPs has shown promising result (98% in 60 min at pH 7) in removing amoxicillin (Weng *et al.*, 2018). Pistachio shell coated with nZnO has shown good removal capacity (132 mg/g at pH 5 in 120 min with initial concentration of 60mg/l, at a NCs dose of 0.1g/l) for amoxicillin (Mohammed *et al.*, 2020). The removal efficiency of amoxicillin by nZnOis reasonably high (92% in 60 min at pH 3) at acidic condition (Rahdar and Ahmadi, 2019). Exceptionally good removal of amoxicillin by nZnO coated carbon (100% at pH 7.5 in 60 min) is reported, where the NCs could be recycled for fifteen times (Chaba and Nomngongo, 2019). Carbon nano-fibres (CNFs) supported nCuO/nAgO NPs (100% in 30 min at pH 7) has shown excellent removal efficiency and the NCs could also be recycled multiple times (Zhang *et al.*, 2020). Zinc ferrite ($ZnFe_2O_4$) NPs and the same coated with graphene sand composite ($ZnFe_2O_4$-GSC) and bentonite ($ZnFe_2O_4$-BT) have shown good removal efficiency (up to 90% at pH 6 in 120 min) for amoxicillin (Singh *et al.*, 2017). AC NPs has shown average efficiency for removal of amoxicillin (50% at pH 2, 4g/L, 20mg/l, and in 8h) (Pouretedal and Sadegh, 2014).

Tetracycline: nFe_3O_4/nPd NPs has shown excellent efficiency for the removal (99.8% in 190 min having initial concentration 20mg/l, (Ravikumar *et al.*, 2020) of tetracycline, where nAC NPs has shown moderate performance (72% at pH 2 0.4g/l, 20mg/l and in 8h (Pouretedala and Sadegh, 2014). nCuO-nitrogen-doped porous carbon framework can degrade more than 94% (4h at pH 5.5) tetracycline (Su *et al.*, 2021). $nCeO_2$ perform significantly well (97% with an adsorption capacity of 58 mg/g in 60 min) for the removal of tetracycline (Nurhasanah *et al.*, 2020). nGO supported nZVI/ nCuO has shown excellent performance (100% with the capacity of adsorption 201.9mg/g in 15 min at pH 6.5 having an initial concentration of 100mg/l with the adsorbent dose 0.25g/l) in removing of tetracycline

(Tabrizian *et al.*, 2019). Cu doped zirconium metal frameworks NPs can be considered as a good adsorbent (adsorption capacity of 224.1mg/g within 10 min for the initial concentration of 20mg/l at a dose of 0.2g/l and pH 10) in removing tetracycline (Cao *et al.*, 2018). Unmodified clay and MnO NPs coated lightweight expanded clay aggregate have shown 52% and 99% efficiencies, respectively (at pH 3, 50mg/l, 2g/l, and 20^{0}C) for the removal of tetracycline (Sepehr *et al.*, 2019). MFe_2O_4 (M stands for Mn, Co, Zn) can remove more than 96% tetracycline in 5 min at pH 7 (NCs dose of 0.1g/l at 15°C) (Bao *et al.*, 2013). Zirconia (ZrO_2) NPs shows excellent capacity (526 mg/g at pH 6 in 15 min) in removing tetracycline (Debnath *et al.*, 2020).

Triclosan: nFe_3O_4 can be considered a good adsorbent (87.5% at pH 7, 30°C, 2g/l, and 120 min) for the removal of triclosan (Song *et al.*, 2020). Fe/Ni NPs can remove up to 76% triclosanduring competitive adsorption (Lin *et al.*, 2019). nAg_2O hydrogel has shown promising results (89% removal for 30 mg/l at pH 6 in 180 min) in removing triclosan (Sharma *et al.*, 2020). Modified nZnO with nGOisnot very efficient for triclosan (45%) at low concentration (8mg/l) (Cervantes Rincón *et al.*, 2018). CNTs shows reasonably good performance for the removal of triclosan in acidic solution (removal capacity 111.8mg/g at pH 2). The removal of triclosan bypolyvinylpyrrolidone (PVP) modified-$nFe_2O_4$is also in the higher range (90% in just 15 min at pH 7.5, Fard *et al.*, 2017).

5.2.3.2 *Miscellaneous Compounds*

$nTiO_2$ can degrade caffeine completely in 3h with an initial concentration of 15mg/l through photodegradation (Muangmora *et al.*, 2020). The removal capacity of $nTiO_2$ for caffeine is reasonable (55.3mg/g in 180 min with initial concentration of 7 mg/l) (León *et al.*, 2020). Silver ferrite NPs (nAg-FeO_2) shows promising performance in removing caffeine (99% removed in 15 min at pH 9 with an initial concentration of 50mg/l) (Muthukumar *et al.*, 2020). Caffeine removal efficiency is reasonable by nZVI/nCuO (86% at pH 5 in 45 min with NPs dose of 0.2g/l) (Abdel-Aziz *et al.*, 2020). nZnO loaded Fe/Ag metals shows moderate removal (70% at pH 6.5 and 23°C temperature with NCs dose of 0.3 g/l) for caffeine (Almeida *et al.*, 2020). nZnO modified with platinum (Pt), Ag and gold (Au) have shown excellent performance in degrading caffeine (100% in 3h,Vaiano *et al.*, 2019). The good performance of nZnO and $nTiO_2$ NPs (up to 90% at pH 7 with NPs dose of 1g/l) for removal of caffeine is also confirmed by different studies (Ghosh *et al.*, 2019). Degradation efficiencies of nFe_3O_4 for lindane is extremely high (100% in 12 h), whereas removal of aldrin and DDT are moderate (79%, and 81%, respectively in 12h (Shoiful *et al.*, 2016).

nFeS (94% in 8h at pH 5) and Ag reduced nGOs (99.9% in 120 min) show remarkably good performance in removing lindane (Paknikar *et al.*, 2005; Gupta *et al.*, 2015), whereas the removal efficiency by $nAl_2O_3$is moderate (80% at pH 4.5) (Derbalah *et al.*, 2015). $nTiO_2$ has shown excellent performance for removal of several contaminnats such as dicofol (100% in 2h, Yu *et al.*, 2008), a-, b-, c-, d-BHC, cypermethrin (99% in 10 min, Yu *et al.*, 2007), terbufos, monocrotophos, dichlorvos, acephate (100% in 4h, Wu *et al.*, 2009; Han *et al.*, 2009; Gomez *et al.*, 2015) and malathion (100% in 2h, Ramos-Delgadoa *et al.*, 2013). The removal of parathion methyl is reasonably good by $nCeO_2$ (90% in 1 min at pH 7) and $nMnO_2$ (90% in 2h at pH 7), where the rate of removal is much faster by $nCeO_2$ compared to that of $nMnO_2$ (Šťastný *et al.*, 2016; Janos *et al.*, 2014).

5.3 Application of Nanoparticles and Nanocomposite Materials in Fixed Bed Column Filter for Water Treatment

5.3.1 Fate and Transport Process of Contaminants in the Fixed Bed Column Filter

The fixed bed column could be considered as the packed bed reactor where the length and cross-sectional area of the packed bed is kept unchanged, and the flow occurs through the void spaces of the media. The contaminants are removed while passing through the novel NPs and NCs compacted porous media packed column. The removal of contaminants is mainly attributed to sorption, and chemical reaction (degradation or oxidation or elimination reaction) depending on the types of the contaminants and the NPs or NCs.

The fate and transport of the contaminants through the packed bed porous media could be represented by the solute transport model as shown in Equation 5.1. Equation 5.1 presents 1-D solute transport model, where the first and second terms in the left-hand side represents advection and dispersion, and the third and fourth terms represent the degradation and production of any contaminants due to chemical reaction. The retardation of contaminants due to sorption process is also incorporated in the equation. The detail of the derivation of retardation factor (*R*) in the equation is discussed later in this section. The schematic representing the conceptual diagram of contaminant transport mechanism is presented in Figure 5.3.

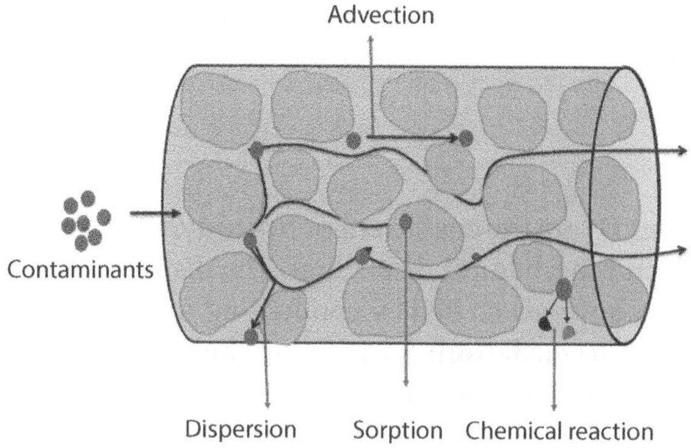

Figure 5.3 Conceptual diagram of the column showing transport and sorption of solute.

The 1-D solute transport equation is expressed as:

$$R\frac{\partial C}{\partial t} = -v_x\frac{\partial C}{\partial x} + D\frac{\partial^2 C}{\partial x^2} - k_1C + k_2C \tag{5.1}$$

where, R is retardation factor attributed to sorption, v_x is the Darcy's velocity, D is the hydrodynamic dispersion, k_1 and k_2 are the rate of reaction attributed to degradation and production of specific contaminants C is the concentration of the contaminants, t is the time and x is the length of the column.

Advective transport: Advectionor advective transport $\left(v_x\frac{\partial C}{\partial x}\right)$ is given in the Equation 5.1, refers to the transport of the contaminants along with the flow of water through the porous media.

Dispersive transport: Hydrodynamic dispersion or dispersive transport $\left(D\frac{\partial^2 C}{\partial x^2}\right)$ is the transport process of contaminant attributed to the combined effect of molecular diffusion and mechanical dispersion. Due to molecular diffusion the contaminant molecules and/ions get transported from a region of higher concentration to the lower concentration following the Fick's second law. The mechanical dispersion, on the other hand is primarily resulted from heterogeneous flow distribution within the porous media, which might be caused by variation in surface roughness, shape and size of packing material, etc. The mechanical dispersion process is analogous to the molecular diffusion; however, the earlier is a macroscale process and its extent in the porous media is much higher than the molecular diffusion.

$$D = D_{diff} + D_m$$

here, D, D_{diff} and D_m are the coefficient of hydrodynamic dispersion, molecular diffusion and mechanical dispersion. The dispersion coefficient could be estimated for any packed bed by conducting tracer test.

Sorption and Retardation: Sorption is the solid phase mass transfer of the contaminant molecules and/ ions from the liquid phase. The sorption could be chemisorption, physisorption or by electrostatic interactions between the solute-contaminants and the solid-surfaces. During the transport of the contaminants through porous media, rate of mass transfer from the liquid phase to solid phase due to sorption could be represented as $\frac{\partial S}{\partial t} = \frac{\rho_b}{\varepsilon}\frac{\partial C}{\partial t}$. Incorporating the mass transfer rate due to sorption in the 1-D transport equation, the retardation factor R, could be presented as $R = \left(1 + \frac{\rho_b}{\varepsilon}k_d\right)$. Here, S is the solid phase mass (mg/g), ρ_b(kg/m^3) is the dry bulk density of the porous medium, ε is the porosity and k_d (l/g) is the sorption coefficient. In absence of sorption, R becomes 1.

Chemical reaction: For removal of contaminants, different suitable material is used in the packed bed filter to facilitate chemical reaction and degradation of the contaminants. The NPs and NCs of interest, or other suitable packing material can degrade certain contaminants and at the same time specific compounds can also be produced. First order reaction kinetics for both chemical degradation $\left(\frac{\partial C}{\partial t} = -k_1 C\right)$, (or production of contaminants $\left(\frac{\partial C}{\partial t} = -k_2 C\right)$ has been considered and the same is incorporated in the contaminant transport equation (Equation 5.1).

5.3.2 Application of Nanoparticles and Nanocomposite Materials in Fixed Bed Column Filter

Fixed bed column filter could be considered as packed bed reactor for removal of contaminants from water. Granular porous media such as granular activated carbon (AC) or sand containing different NPs or NCs could be used as the packing material through which the contaminated water is passed. The interaction, fate and transport between the packed material and the solute contaminant is discussed above. The removal efficiency of the contaminant is very much dependent on the reaction or sorption kinetics of the contaminant by specific NPs and NCs as well as the resident time of the water in the column filter. The physical parameters that generally control the performance of the filter are the types of the contaminants and the packing material, their reaction kinetics, flow velocity, dimension of the column, etc. A few studies have been discussed below where different novel NPs and NCs are used in the fixed bed column for removal of various contaminants. In most of the studies, the contaminated water is injected in the one end of the column and the effluent is collected from the other end, and then the concentration of the contaminant is observed at the outlet to assess the performance of specific column filter. Figure 5.4 represents the packed bed reactor system and the tentative concentration at the effluent along with the volume of water treated. Two breakthrough curves (BTCs) under two conditions, where in one case the extent of sorption and chemical degradation is comparatively less (BTC_1) than that of other (BTC_2) are shown in Figure 5.4.

Performance of different NPs and NCs in the fixed bed filter for removing a range of contaminants are discussed here. CNTs-PAMAM is used as the packing material in fixed-bed column system for the removal of heavy metals such as Zn(II), Co(II) and As(III). The fixed bed column having a height (H) of 12 cm can treat 250 ml of Zn(II), Co(II) and As(III) when injected with the influent ion concentration of 100 mg/l, and at a flow rate of 0.5 ml/min. The treating capacity of the contaminated water increases with an increase in initial contaminant concentration, decrease in flow rate

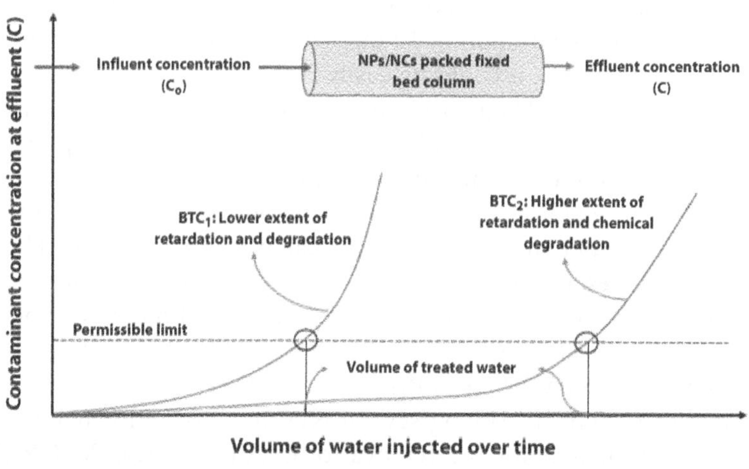

Figure 5.4 A diagram presenting the contaminant concentration over volume of water passed through the fixed bed column filter.

and increase in bed height (Hayati *et al.*, 2018). Fixed bed column of 35 cm height and 1cm internal diameter (ID) packed with 26.53 g of INPB can treat almost 3.6 L (flow rate at 2 ml/min) of As(V) in 30 hrs, where the influent concentration of As(V) is kept at 100 μg/L with an effluent concentration not exiting 10 μg/L (Martínez-Cabanas *et al.*, 2016). Micro-scale fixed bed column packed with 100 mg of Sil-Phy-NPANI or Sil-Phy-CrossNPANI show reasonable good performance in removing Cd(II) [(84.1%-98.7%)], Cu(II) [96.2%-97.0%], Hg(II)[77.7%] and Pb(II) [86.9%-99.2%] from the water collected from various sources (Mahmoud *et al.*, 2016). Polypyrrole/nFe_3O_4 (PPy/nFe_3O_4) composite is explored for Cr(VI) removal through fixed bed column system (ID – 2 cm and H- 30 cm). The column packed with 2g of PPy/nFe_3O_4 can process 5.0 L of water having initial ion concentration of 76.59 mg/l at acidic pH 2 (flow rate at 3ml/min). Studies with environmental water sample suggests that PPy/nFe_3O_4 shows promising performance for Cr removal in the presence of chlorine and sulphate ions (Bhaumik *et al.*, 2013). In similar condition, polypyrrole graphene oxide nanocubes (PPy–ncGO) is used as the packing material in fixed bed column (ID-1.6 cm and H-30 cm) for the removal of Cr(VI). Large volume of Cr (VI) (64.1 L) is successfully removed using 2g of PPy–ncGO and at a flow rate of 3 ml/min. The treating capacity of fixed bed filter increases with the decrease in initial concentration (Setshedi *et al.*, 2015). γ-nFe_2O_3/polyrhodanine composite shows moderate performance (63.88%) for the removal of Zn in fixed bed column system (H- 20 cm, influent concentration of 50 mg/l and at flow rate of 0.5 mL/min, Ghasemi *et al.*, 2015). Continuous fixed-bed column (ID- 4cm and H-20 cm) packed with 8 g of magnesia-pullulan composite (MgOP), can treat 6 L of fluoride with influent F concentration of 10 mg/l and flow rate of 16 ml/min. The advantage of MgOP is, it could be easily regenerated by calcination process (Ye *et al.*, 2018). Magnetized and quaternary amine-functionalized bio-resin derived from corn stalk (CS-MNB) is used as packing material in fixed bed column (H-2 cm) for the removal of NO_3. CS-MNB can treat 150ml of NO_3 when injected with the influent ion concentration of 50 mg/l, and at a flow rate of 8 ml/min. These NCs can be easily regenerated by two different processes, i.e., biological (89%) and chemical (93%) with reasonably good regeneration efficiency (Pan *et al.*, 2019). Both batch and column studies has been done with CNTs-PAMAM, INPB, [Sil-Phy-NPANI] or [Sil-Phy-CrossNPANI] and PPy–nGO NCs for the removal of various inorganic contaminant. Among several NPs and/ NCs PPy–nGO shows excellent performance for the removal of Cr(VI), whereas MgOP, PPy/nFe_3O_4, [Sil-Phy-NPANI] and [Sil-Phy-CrossNPANI] show reasonably good performance.

A few studies have been conducted on removal of organic pollutants and emerging contaminants using various NPs and NCs infixed-bed column. Zhuang *et al.* (2016) show that 2 gm of graphene/alginate double (GAD) and single (GAS) network NC beads packed fixed bed column (H – 4 cm, ID – 0.2 cm) can treat 20 ml MB of 50 mg/l concentration at a flow rate 0.5 ml/min. They have found that removal of MB through column is slightly low (GAS: 47%; GAD: 60%) compared to that of when the experiment is done in batch system (GAS: 1840 mg/g; GAD: 2300 mg/g). Alkali-activated PPy-based nGO particles (PACK) as the packing material in the column (H-25 cm, ID-1 cm) can treat 290 ml MO with a concentration of 30 mg/l, where the flow rate of 1ml/min is maintained (Alghamdi *et al.*, 2019). The column could be regenerated in 80 min, and the efficiency for MO removal of the recycled column is 73%. Zinc oxide NPs (nZnO) and groundnut shell powder (GSP) are packed in vertical bed and sequential bed columns (H-20 cm and ID-1.5 cm) to remove ciprofloxacin hydrochloride (CH). The removal capacity for ciprofloxacin by sequential bed (6.24-6.29 mg/g) is slightly better than vertical bed (5.59-6.10mg/g) system under the pH range of 4 to 6, with the initial ciprofloxacin concentration of 80mg/l and flow rate 5ml/min (Dhiman and Sharma, 2018). NiFe NPs have shown promising performance in treating tetracycline (1198 mg/g) in fixed bed column (ID-2.5 cm and H-30 cm), when injected with an initial concentration of 20mg/lat a flow rate of 1 ml/min (Ravikumar *et al.*, 2019). Silica-alumina (SA) NCs packed fixed bed column (H-10 cm and ID-0.6 cm) can treat 50 ml of metformin (removal capacity 45.7 mg/g) when injected with an initial concentration of 50mg/l and a flow rate of 2ml/min (Hethnawi *et al.*, 2020). The removal of sulfamethoxazole and sulfapyridine by CNTs packed fixed bed columns (ID-2.5cm and H-15cm) with the initial concentration of 20mg/l and at a flow rate of 2ml/min is reasonably high (84.9 mg/g and 123.4mg/g respectively, Tian *et al.*, 2013). Magnetic NPs coated with zeolite (MNCZ) can remove ibuprofen, naproxen and gemfibrozil up to 99% through fixed bed column (ID-1 cm and H-25cm) at a flow rate of 25ml/min (Attia *et al.*, 2013). Overall, it could be observed that different NPs and NCs shows reasonably good performance in fixed bed filter for removing targeted contaminants; however, the efficiency depends on dimension of the column, flow rate, concentration of contaminants and etc. Thus, it is important to optimize the filter parameters while designing the column filter.

References

Abdel-Aziz, H., Farag, R., Abdel-Gawad, S., 2019. *Int J Environ Res* 13, 843-852.

Abdel-Aziz, H.M., Farag, R.S., Abdel-Gawad, S.A., 2020. *Adsorpt. Sci. Technol.* 026361742094749.

Abkenar, S.D., Khoobi, M., Tarasi, R., Hosseini, M., Shafiee, A., Ganjali, M.R., 2015. *J. Environ. Eng.* 141, 4014049.

Abou-Gamra, Z.M., Ahmed, M.A., 2015. *Adv. Chem. Eng. Sci.* 5, 373.

Adeola, A.O., 2018. *Med. Anal. Chem. Int. J.* 2, 126.

Afkhami, A., Moosavi, R., 2010. J. Hazard. Mater. 174, 398–403.

Ahmad, A., Gu, X., Li, L., Lv, S., Xu, Y., Guo, X., 2015. *Environ. Sci. Pollut. Res.* 22, 17876–17885.

Ahmadi, K., Ghaedi, M., Ansari, A., 2015. Spectrochim. *Acta Part A Mol. Biomol. Spectrosc.* 136, 1441–1449.

Ai, L., Jiang, J., 2012. *Chem. Eng. J.* 192, 156–163.

Aksu Demirezen, D., Yıldız, Y.Ş., Demirezen Yılmaz, D., 2019. *Environ. Nanotechnology,* Monit. Manag. 11.

Alfred, M.O., Omorogie, M.O., Bodede, O., Moodley, R., Ogunlaja, A., Adeyemi, O.G., Günter, C., Taubert, A., Iermak, I., Eckert, H., Silva, I.D.A., de Camargo, A.S.S., de Jesus Motheo, A., Clarke, S.M., Unuabonah, E.I., 2020. *Chem. Eng. J.* 398, 125544.

Alghamdi, A.A., Al-Odayni, A.-B., Saeed, W.S., Almutairi, M.S., Alharthi, F.A., Aouak, T., Al-Kahtani, A., 2019. *Molecules* 24, 3685.

Al-Hamadani, Y.A.J., Jung, C., Im, J.K., Boateng, L.K., Flora, J.R.V., Jang, M., Heo, J., Park, C.M., Yoon, Y., 2017. *Chem. Eng. Sci.* 162, 300–308.

Alhooshani, K.R., 2019. *Arab. J. Chem.* 12, 2585–2596.

Ali, F., Khan, S.B., Kamal, T., Alamry, K.A., Asiri, A.M., 2018. *Sci. Rep.* 8, 1–18.

Ali, I., Al-Othman, Z.A., Alwarthan, A., 2016. *J. Mol. Liq.* 219, 858–864.

Al-Kinani, A., Gheibi, M., Eftekhari, M., 2019. *Model. Earth Syst. Environ.* 5, 1627–1633.

Almeida, L.N.B., Pietrobelli, J.M.T. A., Lenzi, G.G., Santos, O.A.A. *Braz. Arch. Biol. Technol.*, 63, 2020.

AlOmar, M.K., Alsaadi, M.A., Hayyan, M., Akib, S., Ibrahim, M., Hashim, M.A., 2017. *Chemosphere* 167, 44–52.

Álvarez-Torrellas, S., Peres, J. A., Gil-Álvarez, V., Ovejero, G., García, J. 2017. *Chem. Eng. J.* 320, 319–329.

Ameri, Atefeh, Shakibaie, M., Pournamdari, M., Ameri, Alieh, Foroutanfar, A., Doostmohammadi, M., Forootanfar, H., 2020. *J. Photochem. Photobiol. A Chem.* 392, 112382.

Anjum, M., Miandad, R., Waqas, M., Gehany, F., Barakat, M.A., 2019. *Arab. J. Chem.* 12, 4897–4919.

Anushree, C., Philip, J., 2019. *Colloids Surfaces A Physicochem. Eng. Asp.* 567, 193–204.

Arancibia-Miranda, N., Baltazar, S.E., García, A., Muñoz-Lira, D., Sepúlveda, P., Rubio, M.A., Altbir, D., 2016. *J. Hazard. Mater.* 301, 371–380.

Ardekani, P.S., Karimi, H., Ghaedi, M., Asfaram, A., Purkait, M.K., 2017. *J. Mol. Liq.* 229, 114–124.

Arshadi, M., Vahid, F.S., Salvacion, J.W.L., Soleymanzadeh, M., 2013. *Appl. Surf. Sci.* 280, 726–736.
Asfaram, A., Ghaedi, M., Hajati, S., Rezaeinejad, M., Goudarzi, A., Purkait, M.K., 2015. *J. Taiwan Inst. Chem. Eng.* 53, 80–91.
Attinti, R., Sarkar, D., Barrett, K.R., Datta, R., 2015. *Int. J. Environ. Sci. Technol.* 12, 3905–3914.
Avila, M., Burks, T., Akhtar, F., Göthelid, M., Lansåker, P.C., Toprak, M.S., Muhammed, M., Uheida, A., 2014. *Chem. Eng. J.* 245, 201–209.
Azad, F.N., Ghaedi, M., Dashtian, K., Hajati, S., Goudarzi, A., Jamshidi, M., 2015. *New J. Chem.* 39, 7998–8005.
Baby, R., Saifullah, B., Hussein, M.Z., 2019. *Sci. Rep.* 9, 1–11.
Bai, S., Shen, X., Zhong, X., Liu, Y., Zhu, G., Xu, X., Chen, K., 2012. Carbon N. Y. 50, 2337–2346.
Bao, X., Qiang, Z., Ling, W., Chang, J.H., 2013. *Sep. Purif. Technol.* 117, 104–110.
Bazrafshan, A.A., Hajati, S., Ghaedi, M., 2015. *Rsc Adv.* 5, 79119–79128.
Beheshti, F., M. A. Tehrani, R., Khadir, A., 2019. *Int. J. Environ. Sci. Technol.* 16, 7987–7996.
Belachew, N., Rama Devi, D., Basavaiah, K., 2017. *J. Exp. Nanosci.* 12, 114–128.
Bezbaruah, A.N., Shanbhogue, S.S., Simsek, S., Khan, E., 2011. J. Nanoparticle Res. 13, 6673–6681.
Bharath, G., Alhseinat, E., Ponpandian, N., Khan, M.A., Siddiqui, M.R., Ahmed, F., Alsharaeh, E.H., 2017. *Sep. Purif. Technol.* 188, 206–218.
Bhattacharjee, A., Ahmaruzzaman, M., Devi, T.B., Nath, J., 2016. *J. Photochem. Photobiol. A Chem.* 325, 116–124.
Bhattacharya, S., Banerjee, P., Das, P., Bhowal, A., Majumder, S.K., Ghosh, P., 2020. *Sustain. Environ. Res.* 30.
Bhaumik, M., Maity, A., Srinivasu, V. V., Onyango, M.S., 2012. *Chem. Eng. J.* 181–182, 323–333.
Bhaumik, M., Setshedi, K., Maity, A., Onyango, M.S., 2013. *Sep. Purif. Technol.* 110, 11–19.
Boparai, H.K., Joseph, M., O'Carroll, D.M., 2011. *J. Hazard. Mater.* 186, 458–465.
Cai, X., Gao, Y., Sun, Q., Chen, Z., Megharaj, M., Naidu, R., 2014. *Chem. Eng. J.* 244, 19–26.
Cai, Y., Li, C., Wu, D., Wang, W., Tan, F., Wang, X., Wong, P.K., Qiao, X., 2017. *Chem. Eng. J.* 312, 158–166.
Calvache-Muñoz, J., Rodríguez-Páez, J.E., 2020. *J. Environ. Chem. Eng.* 8, 103518.
Cao, J., Yang, Z. hui, Xiong, W. ping, Zhou, Y. yu, Peng, Y. rong, Li, X., Zhou, C. yun, Xu, R., Zhang, Y. ru, 2018. *Chem. Eng. J.* 353, 126–137.
Cervantes Rincón, N., Hammouda, S. Ben, Sillanpää, M., Escobar Barrios, V., 2018. *J. Environ. Chem. Eng.* 6, 6554–6567.
Chaba, J.M., Nomngongo, P.N., 2019. *Emerg. Contam.* 5, 143–149.
Chai, L., Wang, Y., Zhao, N., Yang, W., You, X., 2013. *Water Res.* 47, 4040–4049.
Chatterjee, S., Guha, N., Krishnan, S., Singh, A.K., Mathur, P., Rai, D.K., 2020. *Sci. Rep.* 10, 1–11.

Chatterjee, S., Lee, M.W., Woo, S.H., 2010. *Bioresour. Technol.* 101, 1800–1806.
Chaudhary, M., Bhattacharya, P., Maiti, A., 2016. *J. Environ. Chem. Eng.* 4, 4897–4903.
Chávez-Guajardo, A.E., Medina-Llamas, J.C., Maqueira, L., Andrade, C.A.S., Alves, K.G.B., de Melo, C.P., 2015. *Chem. Eng. J.* 281, 826–836.
Chen, D., Zhang, H., Yang, K., Wang, H., 2016. *J. Hazard. Mater.* 310, 179–187.
Chen, M., Qin, X., Zeng, G., Li, J., 2016. *Chemosphere* 152, 439–445.
Chen, X., Chen, B., 2015. *Environ. Sci. Technol.* 49, 6181–6189.
Chen, Y., Chen, L., Bai, H., Li, L., 2013. *J. Mater. Chem. A 1*, 1992–2001.
Chen, Zheng-xian, Jin, X., Chen, Zuliang, Megharaj, M., Naidu, R., 2011. *J. Colloid Interface Sci.* 363, 601–607.
Cheng, Y., Zhou, W., Zhu, L., 2020. *Sci. Total Environ.* 718, 137256.
Choi, K., Lee, S., Park, J.O., Park, J.A., Cho, S.H., Lee, S.Y., Lee, J.H., Choi, J.W., 2018. *Sci. Rep.* 8, 1–10.
Crane, R.A., Scott, T.B., 2012. *J. Hazard. Mater.* 211, 112–125.
Cui, L., Wang, Y., Gao, L., Hu, L., Yan, L., Wei, Q., Du, B., 2015. *Chem. Eng. J.* 281, 1–10.
Darwish, A.A.A., Rashad, M., AL-Aoh, H.A., 2019. *Dye. Pigment.* 160, 563–571.
Dastkhoon, M., Ghaedi, M., Asfaram, A., Goudarzi, A., Langroodi, S.M., Tyagi, I., Agarwal, S., Gupta, V.K., 2015. *Sep. Purif. Technol.* 156, 780–788.
Daughton, C.G., Ternes, T.A., n.d.
Dayananda, D., Sarva, V.R., Prasad, S. V., Arunachalam, J., Ghosh, N.N., 2014. *Chem. Eng. J.* 248, 430–439.
Debnath, A., Deb, K., Chattopadhyay, K.K., Saha, B., 2016. *Desalin. Water Treat.* 57, 13549–13560.
Debnath, B., Majumdar, M., Bhowmik, M., Bhowmik, K.L., Debnath, A., Roy, D.N., 2020. *J. Environ. Manage.* 261, 110235.
Debnath, S., Maity, A., Pillay, K., 2014. *J. Environ. Chem. Eng.* 2, 260–272.
Dehbi, A., Dehmani, Y., Omari, H., Lammini, A., Elazhari, K., Abdallaoui, A., 2020. *J. Environ. Chem. Eng.* 8, 103394.
Deng, J., Shao, Y., Gao, N., Tan, C., Zhou, S., Hu, X., 2013. *J. Hazard. Mater.* 262, 836–844.
Deng, W., Dai, Q., Lao, Y., Shi, B., Wang, X., 2016. *Appl. Catal. B Environ.* 181, 848–861.
Derbalah, A., El-Safty, S.A., Shenashen, M.A., Abdel Ghany, N.A., 2015. *Chempluschem* 80, 1119–1126.
Dhiman, N., Sharma, N., 2018. *J Environ Chem Eng* 6, 4391-4398.
Ding, Y., Sun, W., Yang, W., Li, Q., 2017. *Appl. Catal. B Environ.* 203, 372–380.
Dong, H., Hou, K., Qiao, W., Cheng, Y., Zhang, L., Wang, B., Li, L., Wang, Y., Ning, Q., Zeng, G., 2019. *Chem. Eng. J.* 359, 1046–1055.
Du, X., Zhang, Y., Si, F., Yao, C., Du, M., Hussain, I., Kim, H., Huang, S., Lin, Z., Hayat, W., 2019. *Chem. Eng. J.* 356, 178–189.
Duan, J., Liu, R., Chen, T., Zhang, B., Liu, J., 2012. *Desalination* 293, 46–52.

Duraimurugan, J., Shanavas, S., Ramesh, R., Acevedo, R., Anbarasan, P.M., Maadeswaran, P., 2020. *Optik (Stuttg).* 202, 163607.
Eroglu, E., Agarwal, V., Bradshaw, M., Chen, X., Smith, S.M., Raston, C.L., Swaminathan Iyer, K., 2012. *Green Chem.* 14, 2682–2685.
Fang, Q., Zhou, X., Deng, W., Liu, Z., 2017. *Chem. Eng. J.* 308, 1001–1009.
Farooq, U., Danish, M., Lu, S., Naqvi, M., Qiu, Z., Sui, Q., 2018. *Chem. Eng. J.* 347, 660–668.
Feng, J., Zou, L., Wang, Y., Li, B., He, X., Fan, Z., Ren, Y., Lv, Y., Zhang, M., Chen, D., 2015. *J. Colloid Interface Sci.* 438, 259–267.
Fard, M.A., Barkdoll, B., *J. Mol. Liq.* 249, 193-202, 2018.
Foroutan, R., Mohammadi, R., Farjadfard, S., Esmaeili, H., Ramavandi, B., Sorial, G.A., 2019. *Adv. Powder Technol.* 30, 2188–2199.
Ganyaglo, S.Y., Gibrilla, A., Teye, E.M., Owusu-Ansah, E.D.G.J., Tettey, S., Diabene, P.Y., Asimah, S., 2019. *Chemosphere* 233, 862–872.
Ghaderi, A., Abbasi, S., Farahbod, F., 2015. *Iran. J. Chem. Eng.* 12, 96–105.
Ghaedi, M., Ansari, A., Habibi, M.H., Asghari, A.R., 2014a. *J. Ind. Eng. Chem.* 20, 17–28.
Ghaedi, M., Biyareh, M.N., Kokhdan, S.N., Shamsaldini, S., Sahraei, R., Daneshfar, A., Shahriyar, S., 2012. *Mater. Sci. Eng. C 32*, 725–734.
Ghaedi, M., Pakniat, M., Mahmoudi, Z., Hajati, S., Sahraei, R., Daneshfar, A., 2014b. Spectrochim. Acta Part A Mol. *Biomol. Spectrosc.* 123, 402–409.
Ghaedi, M., reza Rahimi, M., Ghaedi, A.M., Tyagi, I., Agarwal, S., Gupta, V.K., 2016. *J. Colloid Interface Sci.* 461, 425–434.
Ghasemi, S., Ghorbani, M., Ghazi, M.M., 2015. *Appl. Surf. Sci.* 359, 602–608.
Ghorai, S., Sarkar, A., Raoufi, M., Panda, A.B., Schönherr, H., Pal, S., 2014. *ACS Appl. Mater. Interfaces* 6, 4766–4777.
Ghosh, M., Manoli, K., Shen, X., Wang, J., Ray, A.K., 2019. *J. Photochem. Photobiol. A Chem.* 377, 1–7.
Góngora, J.F., Elizondo, P., Hernández-Ramírez, A., 2017. *Photochem. Photobiol. Sci.* 16, 31–37.
Guerra, A.C.S., de Andrade, M.B., Tonial dos Santos, T.R., Bergamasco, R., 2019. *Environ. Technol.* (United Kingdom) 0, 1–11.
Guo, F., Wang, K., Lu, J., Chen, J., Dong, X., Xia, D., Zhang, A., Wang, Q., 2019. *Chemosphere* 218, 1071–1081.
Guo, L., Ye, P., Wang, J., Fu, F., Wu, Z., 2015. *J. Hazard. Mater.* 298, 28–35.
Guo, M., Weng, X., Wang, T., Chen, Z., 2017. *Sep. Purif. Technol.* 175, 222–228.
Guo, X., Du, B., Wei, Q., Yang, J., Hu, L., Yan, L., Xu, W., 2014. *J. Hazard. Mater.* 278, 211–220.
Gupta, K., Khatri, O.P., 2017. *J. Colloid Interface Sci.* 501, 11–21.
Gwenzi, W., Chaukura, N., 2018. *Sci. Total Environ.* 619, 1493–1514.
Hao, T., Yang, C., Rao, X., Wang, J., Niu, C., Su, X., 2014. *Appl. Surf. Sci.* 292, 174–180.
Hao, X.L., Zhou, M.H., 2007. *J. Hazard. Mater.* 141, 475–482.
Hashemian, S., Foroghimoqhadam, A., 2014. *Chem. Eng. J.* 235, 299–306.

Hayati, B., Maleki, A., Najafi, F., Gharibi, F., McKay, G., Gupta, V.K., Harikaranahalli Puttaiah, S., Marzban, N., 2018. *Chem. Eng. J.* 346, 258–270.
He, F., Zhao, D., 2005. *Environ. Sci. Technol.* 39, 3314–3320.
He, J., Siah, T.S., Paul Chen, J., 2014. *Water Res.* 56, 88–97.
Heberer, T., 2002. *J. Hydrol.* 266, 175–189.
Hethnawi, A., Alnajjar, M., Manasrah, A., Hassan, A., Vitale, G., Jeong, R., Nassar, N., 2020. *Colliod Surface A* 597, 124814.
Hiew, B.Y.Z., Lee, L.Y., Lee, X.J., Gan, S., Thangalazhy-Gopakumar, S., Lim, S.S., Pan, G.T. Yang, T.C.K. 2019. *J Taiwan Inst Chem. Eng.* 98, 150–162.
Hu, A., Apblett, A., 2014. *Nanotechnology for water treatment and purification.* Springer.
Hua, M., Jiang, Y., Wu, B., Pan, B., Zhao, X., Zhang, Q., 2013. *ACS Appl. Mater. Interfaces* 5, 12135–12142.
Huang, L., He, M., Chen, B., Hu, B., 2018. *Chemosphere* 199, 435–444.
Huang, X., Yang, J., Wang, J., Bi, J., Xie, C., Hao, H., 2018. *Chemosphere* 206, 513–521.
Huang, Y., Lee, X., Macazo, F.C., Grattieri, M., Cai, R., Minteer, S.D., 2018. *Chem. Eng. J.* 339, 259–267.
Husein, D.Z., Hassanien, R., Al-Hakkani, M.F., 2019. *Heliyon* 5, e02339.
Ibrahim, I., Athanasekou, C., Manolis, G., Kaltzoglou, A., Nasikas, N.K., Katsaros, F., Devlin, E., Kontos, A.G., Falaras, P., 2019. *J. Hazard. Mater.* 372, 37–44.
Ivanets, A., Prozorovich, V., Roshchina, M., Grigoraviciute-Puroniene, I., Zarkov, A., Kareiva, A., Wang, Z., Srivastava, V., Sillanpää, M. 2020. *J. Nanomater.*
Jallouli, N., Elghniji, K., Trabelsi, H., Ksibi, M., 2017. *Arab. J. Chem.* 10, S3640–S3645.
Janos, P., Kuran, P., Kormunda, M., Stengl, V., Grygar, T.M., Dosek, M., Stastny, M., Ederer, J., Pilarova, V. Vrtoch, L. 2014. *J. Rare Earths,* 32, 4, 360–370.
Jiang, H., Chen, P., Luo, S., Tu, X., Cao, Q., Shu, M., 2013. *Appl. Surf. Sci.* 284, 942–949.
Jin, J., Li, R., Wang, H., Chen, H., Liang, K., Ma, J., 2007. *Chem. Commun.* 386–388.
Jin, X., Chen, H., Yang, Q., 2019. *Environ. Eng. Sci.* 36, 681–689.
Jishma, P., Narayanan, R., Snigdha, S., Thomas, R., Radhakrishnan, E.K., 2018. *Biocatal. Agric. Biotechnol.* 14, 410–417.
Jung, H.J., Lee, S.J., Koutavarapu, R., Kim, S.K., Choi, H.C., Choi, M.Y., 2018. *Catalysts* 8, 390.
Junyong, C., Yongmei, H., Yan, L., Jiajia, G., 2013. *RSC Adv. 3*, 7254–7258.
Kalidindi, S., Vecha, M., Kar, A., Raychoudhury, T., 2017. *Water Sci. Technol. Water Supply* 17, 115–124.
Kataria, N., Garg, V.K., 2017. *J. Environ. Chem. Eng.* 5, 5420–5428.
Keshvardoostchokami, M., Babaei, S., Piri, F., Zamani, A., 2017. *Int. J. Biol. Macromol.* 101, 922–930.
Keshvardoostchokami, M., Rasooli, S., Zamani, A., Parizanganeh, A., Piri, F., 2019. *Environ. Monit. Assess.* 191.
Keyhanian, F., Shariati, S., Faraji, M., Hesabi, M., 2016. *Arab. J. Chem.* 9, S348–S354.

Khammar, S., Bahramifar, N., Younesi, H., 2020. *J. Hazard. Mater.* 122422.
Khraisheh, M., Kim, J., Campos, L., Al-Muhtaseb, A., Walker, G., AlGhouti, M., 2013. *Environ Eng Sci* 30, 515-526.
Kim, E.-J., Kim, J.-H., Azad, A.-M., Chang, Y.-S., 2011. *ACS Appl. Mater. Interfaces* 3, 1457–1462.
Kim, H., Hong, H.-J., Jung, J., Kim, S.-H., Yang, J.-W., 2010. *J. Hazard. Mater.* 176, 1038–1043.
Kim, S., Park, C.M., Jang, M., Son, A., Her, N., Yu, M., Snyder, S., Kim, D.-H., Yoon, Y., 2018. *Chemosphere* 212, 1104–1124.
Kong, L., Tian, Y., Pang, Z., Huang, X., Li, M., Li, N., Zhang, J., Zuo, W., Li, J., 2020. *Chem. Eng. J.* 382, 122963.
Konicki, W., Pełech, I., Mijowska, E., Jasińska, I., 2012. *Chem. Eng. J.* 210, 87–95.
Konicki, W., Siber, D., Narkiewicz, U., 2017. *Polish J. Chem. Technol.* 19, 65–74.
Koushik, D., Gupta, S. Sen, Maliyekkal, S.M., Pradeep, T., 2016. *J. Hazard. Mater.* 308, 192–198.
Kumar, S., Bhanjana, G., Jangra, K., Dilbaghi, N., Umar, A., 2014. *J. Nanosci. Nanotechnol.* 14, 4331–4336.
Kurian, M., Kunjachan, C., Sreevalsan, A., 2017. *Chem. Eng. J.* 308, 67–77.
Lapworth, D.J., Baran, N., Stuart, M.E., Ward, R.S., 2012. *Environ. Pollut.* 163, 287–303.
Lee, C.S., Gong, J., Huong, C.V., Oh, D.S., Chang, Y.S., 2016. *Chem. Eng. J.* 298, 206–213.
León, G.R., Aldás, M.B., Guerrero, V.H., Landázuri, A.C., Almeida-Naranjo, C.E., 2019. *MRS Adv.* 4, 3553–3567.
Lerman, I., Chen, Y., Xing, B., Chefetz, B., 2013. *Environ. Pollut.* 182, 169-176.
Leus, K., Folens, K., Nicomel, N.R., Perez, J.P.H., Filippousi, M., Meledina, M., Dîrtu, M.M., Turner, S., Van Tendeloo, G., Garcia, Y., Du Laing, G., Van Der Voort, P., 2018. *J. Hazard. Mater.* 353, 312–319.
Li, D., Zhu, J., Wang, M., Bi, W., Huang, X., Chen, D.D.Y., 2017. *J. Chromatogr. A* 1491, 27–35.
Li, H., Zhang, W., Zhang, Z., Zhang, X., 2017. *Sci. Total Environ.* 580, 1318–1326.
Li, J., Ng, D.H.L., Song, P., Kong, C., Song, Y., 2015. *Mater. Sci. Eng. B* 194, 1–8.
Li, N., Tian, Y., Zhao, J., Zhang, Jian, Zhang, Jun, Zuo, W., Ding, Y., 2017. *Appl. Catal. B Environ.* 214, 126–136.
Li, Q.-L., Wang, L.-L., Wang, X., Wang, M.-L., Zhao, R.-S., 2016. *J. Chromatogr. A* 1449, 39–47.
Li, Z., Chen, J., Ge, Y., 2017. *Chem. Eng. J.* 308, 809–817.
Liang, J., Liu, Junfeng, Yuan, X., Dong, H., Zeng, G., Wu, H., Wang, H., Liu, Jiayu, Hua, S., Zhang, S., 2015. *Chem. Eng. J.* 273, 101–110.
Liao, M., Wu, K., Chen, D., 2005. *Sep. Sci. Technol.* 39, 1563–1575.
Liao, Q.G., Wang, D.G., Luo, L.G., 2014. *Anal. Bioanal. Chem.* 406, 7571–7579.
Lin, K.-S., Mdlovu, N.V., Chen, C.-Y., Chiang, C.-L., Dehvari, K., 2018. *J. Clean. Prod.* 175, 456–466.
Lin, Y., Jin, X., Owens, G., Chen, Z., 2019. *Sci. Total Environ.* 695, 133878.

Liu, D., Zhou, W., Wu, J., 2017. *Fuel* 194, 115–122.
Liu, F., Wu, Z., Wang, D., Yu, J., Jiang, X., Chen, X., 2016. *Colloids Surfaces A Physicochem. Eng. Asp.* 490, 207–214.
Liu, J., Ma, S., Zang, L., 2013. *Appl. Surf. Sci.* 265, 393–398.
Liu, K., Li, H., Wang, Y., Gou, X., Duan, Y., 2015. *Colloids Surfaces A Physicochem. Eng. Asp.* 477, 35–41.
Liu, M., Wang, Y., Chen, L., Zhang, Y., Lin, Z., 2015. *ACS Appl. Mater. Interfaces* 7, 7961–7969.
Liu, T., Li, Y., Du, Q., Sun, J., Jiao, Y., Yang, G., Wang, Z., Xia, Y., Zhang, W., Wang, K., 2012. *Colloids Surfaces B Biointerfaces* 90, 197–203.
Liu, T., Wang, Z.L., Sun, Y., 2015. *Chem. Eng. J.* 263, 55–61.
Lung, I., Soran, M.L., Stegarescu, A., Opris, O., Gutoiu, S., Leostean, C., Lazar, M.D., Kacso, I., Silipas, T.D., Porav, A.S., 2021. *J. Hazard. Mater.* 403, 1–15.
Luo, F., Chen, Z., Megharaj, M., Naidu, R., 2016. *Chem. Eng. J.* 294, 290–297.
Ma, X., Zhao, M., Pang, Q., Zheng, M., Sun, H., Crittenden, J., Zhu, Y., Chen, Y., 2016. *Appl. Catal. A Gen.* 522, 70–79.
Machado, S., Stawiński, W., Slonina, P., Pinto, A.R., Grosso, J.P., Nouws, H.P.A., Albergaria, J.T., Delerue-Matos, C., 2013. *Sci. Total Environ.* 461–462, 323–329.
Madrakian, T., Afkhami, A., Ahmadi, M., Bagheri, H., 2011. *J. Hazard. Mater.* 196, 109–114.
Mahapatra, A., Mishra, B.G., Hota, G., 2013. *Ceram. Int.* 39, 5443–5451.
Mahapatra, A., Mishra, B.G., Hota, G., 2013. *J. Hazard. Mater.* 258–259, 116–123.
Mahmoud, A.S., Mostafa, M.K., Abdel-Gawad, S.A., 2018. *Water Supply* 18, 1650–1663.
Mahmoud, M.E., Fekry, N.A., El-Latif, M.M.A., 2016. *Chem. Eng. J.* 304, 679–691.
Malini, P.S.G., Durgadevi, P., Kumar, N.S., Rani, S., 2020. *Mater. Today Proc.*
Mamba, G., Mbianda, X.Y., Govender, P.P., 2013. *Carbohydr. Polym.* 98, 470–476.
Martínez-Cabanas, M., López-García, M., Barriada, J.L., Herrero, R., Sastre de Vicente, M.E., 2016. *Chem. Eng. J.* 301, 83–91.
Miller, S.M., Zimmerman, J.B., 2010. *Water Res.* 44, 5722–5729.
Misra, T., Mitra, S., Sen, S., 2018. *Nanotechnology Environ. Eng* 3.
Mittal, H., Kumar, V., Ray, S.S., 2016. *Int. J. Biol. Macromol.* 89, 1–11.
Mohammed, A.A., Al-Musawi, T.J., Kareem, S.L., Zarrabi, M., Al-Ma'abreh, A.M., 2019. *Arab. J. Chem.* 13, 4629–4643.
Mohanta, J., Dey, B., Dey, S., 2020. *ACS omega* 5, 16510–16520.
Mokhtari, P., Ghaedi, M., Dashtian, K., Rahimi, M.R., Purkait, M.K., 2016. *J. Mol. Liq.* 219, 299–305.
Mortazavi, K., Rajabi, H., Ansari, A., Ghaedi, M., Dashtian, K., 2016. *Synth. React. Inorganic, Met. Nano-Metal Chem.* 1553-3174.
Muangmora, R., Kemacheevakul, P., Punyapalakul, P., Chuangchote, S., 2020. *Catalysts* 10, 1–14.
Muthukumar, H., Shanmugam, M.K., Gummadi, S.N., 2020. *J. Water Process Eng.* 36, 101382.
Nagda, G.K., Diwan, A.M., Ghole, V.S., 2007. *Appl Ecol Env. Res* 5, 1–9.

Nahuel Montesinos, V., Quici, N., Beatriz Halac, E., Leyva, A.G., Custo, G., Bengio, S., Zampieri, G., Litter, M.I., 2014. *Chem. Eng. J.* 244, 569–575.
Nair, D.S., Kurian, M., 2017. *J. Water Process Eng.* 16, 69–80.
Naushad, M., Ahamad, T., Sharma, G., Al-Muhtaseb, A.H., Albadarin, A.B., Alam, M.M., ALOthman, Z.A., Alshehri, S.M., Ghfar, A.A., 2016. *Chem. Eng. J.* 300, 306–316.
Nourmoradi, H., Ghiasvand, A.R., Noorimotlagh, Z., 2015. *Desalin. Water Treat.* 55, 252–262.
Nurhasanah, I., Kadarisman, Gunawan, V., Sutanto, H., 2020. *J. Environ. Chem. Eng.* 8, 103613.
Paknikar, K.M., Nagpal, V., Pethkar, A. V., Rajwade, J.M., 2005. *Sci. Technol. Adv. Mater.* 6, 370–374.
Pal, A., Gin, K.Y.H., Lin, A.Y.C., Reinhard, M., 2010. *Sci. Total Environ.* 408, 6062–6069.
Pal, J., Deb, M.K., Deshmukh, D.K., Verma, D., 2013. *Appl. Water Sci.* 3, 367–374.
Palma, T.L., Vieira, B., Nunes, J., Lourenço, J.P., Monteiro, O.C., Costa, M.C., 2020. *J. Iran. Chem. Soc.* 17, 2013–2031.
Pan, J., Gao, B., Song, W., Xu, X., Jin, B., Yue, Q., 2019. *J. Clean. Prod.* 213, 762–775.
Pandey, S., Mishra, S.B., 2011. *J. Colloid Interface Sci.* 361, 509–520.
Panneerselvam, P., Morad, N., Tan, K.A., Mathiyarasi, R., 2012. *Sep. Sci. Technol.* 47, 742–752.
Parshetti, G.K., Doong, R., 2010. *Appl. Catal. B Environ.* 100, 116–123.
Patra, A.K., Dutta, A., Bhaumik, A., 2012. *J. Hazard. Mater.* 201–202, 170–177.
Pei, Z., Li, L., Sun, L., Zhang, S., Shan, X., Yang, S., Wen, B., 2013. *Carbon* N. Y. 51, 156–163.
Peng, L., Qin, P., Lei, M., Zeng, Q., Song, H., Yang, J., Shao, J., Liao, B., Gu, J., 2012. *J. Hazard. Mater.* 209, 193–198.
Peng, X., Li, Y., Luan, Z., Di, Z., Wang, H., Tian, B., Jia, Z., 2003. *Chem. Phys. Lett.* 376, 154–158.
Peng, X., Yu, Y., Tang, C., Tan, J., Huang, Q., Wang, Z. 2008. *Sci. Total Environ.* 397, 1-3, 158–166.
Petrovic, M., Cristina, R., 2008. 5, 1–35.
Platero, E., Fernandez, M.E., Bonelli, P.R., Cukierman, A.L., 2017. *J. Colloid Interface Sci.* 491, 1–12.
Pouretedal, H.R., Sadegh, N., 2014. *J. Water Process Eng.* 1, 64–73.
Pourjavadi, A., Doulabi, M., Doroudian, M., 2014. *J. Iran. Chem. Soc.* 11, 1057–1065.
Poursaberi, T., Hassanisadi, M., Torkestani, K., Zare, M., 2012. *Chem. Eng. J.* 189–190, 117–125.
Prema, P., Thangapandian, S., Selvarani, M., Subharanjani, S., Amutha, C., 2011. *Toxicol. Environ. Chem.* 93, 1908–1917.
Pushkar, B.K., Sevak, P., 2019.
Qin, P., Yang, Y., Zhang, X., Niu, J., Yang, H., Tian, S., Zhu, J., Lu, M., 2018. *Nanomaterials* 8, 4.
Qu, S., Huang, F., Yu, S., Chen, G., Kong, J., 2008. *J. Hazard. Mater.* 160, 643–647.

Qu, X., Alvarez, P.J.J., Li, Q., 2013. *Water Res.* 47, 3931-3946.

Qureshi, A.S., Hussain, M.I., Ismail, S., Khan, Q.M., 2016. *Chemosphere* 163, 54-61.

Raad, M.T., Behnejad, H., Jamal, M. El, 2016. *Surf. Interface Anal.* 48, 117–125.

Rachna, K., Agarwal, A., Singh, N.B., 2018. *Environ. Nanotechnology, Monit. Manag.* 9, 154–163.

Rahimi, R., Kerdari, H., Rabbani, M., Shafiee, M., 2011. *Desalination* 280, 412–418.

Rahman, M.M., Bodrud-Doza, M., Siddiqua, M.T., Zahid, A., Islam, A.R.M.T., 2020. *Sci. Total Environ.* 724, 138316.

Rajabi, H.R., Khani, O., Shamsipur, M., Vatanpour, V., 2013. *J. Hazard. Mater.* 250, 370–378.

Rajendran, K., Sen, S., 2018. *Environ. Nanotechnology, Monit. Manag.* 9, 122-127.

Rajput, S., Singh, L.P., Pittman, C.U., Mohan, D., 2017. *J. Colloid Interface Sci.* 492, 176–190.

Rani, M., Shanker, U., 2018. *Int. J. Environ. Sci. Technol.* 15, 1347–1380.

Rashed, M.N., Eltaher, M.A., Abdou, A.N.A., 2017. *R. Soc. open Sci.* 4, 170834.

Raval, N.P., Shah, P.U., Shah, N.K., 2017. *Appl. Water Sci.* 7, 3407–3445.

Ravikumar, K., Singh, A., Sikarwar, D., Gopal, G., Das, B., Mrudula, P., Natarajan, C., Mukherjee, A., 2019. *J Environ Manage* 236, 93-99.

Ravikumar, K.V.G., Kubendiran, H., Gupta, R., Gupta, A., Sharma, P., Alex, S.A., Natarajan, C., Das, B., Mukherjee, A., *J. Water Process. Eng.* 36, 101400, 2020.

Raychoudhury, T., Boindala, S.P., Kalidindi, S., 2017. *Water Sci. Technol. Water Supply* 17, 1377–1385.

Raychoudhury, T., Scheytt, T., 2013. Potential of Zerovalent iron nanoparticles for remediation of environmental organic contaminants in water: A review. *Water Sci. Technol.*

Recillas, S., Colón, J., Casals, E., González, E., Puntes, V., Sánchez, A., Font, X., 2010. *J. Hazard. Mater.* 184, 425–431.

Riaz, Q., Ahmed, M., Zafar, M.N., Zubair, M., Nazar, M.F., Sumrra, S.H., Ahmad, I., Hosseini-Bandegharaei, A., 2020. *Int. J. Environ. Anal. Chem.* 1–20.

Roosta, M., Ghaedi, M., Yousefi, F., 2015. *RSC Adv.* 5, 100129–100141.

Roosta, M.A., Ghaedi, M.A., Shokri, N., Daneshfar, A., Sahraei, R., Asghari, A., 2014. *Spectrochim. Acta Part A Mol. Biomol. Spectrosc.* 118, 55–65.

Ruan, X., Chen, Y., Chen, H., Qian, G., Frost, R.L., 2016. *Chem. Eng. J.* 297, 295–303.

Ryu, A., Jeong, S.W., Jang, A., Choi, H., 2011. *Appl. Catal. B Environ.* 105, 128–135.

Sadegh, H., Shahryari-ghoshekandi, R., Agarwal, S., Tyagi, I., Asif, M., Gupta, V.K., 2015. *J. Mol. Liq.* 206, 151–158.

Saeed, K., Ishaq, M., Sultan, S., Ahmad, I., 2016. *Desalin. Water Treat.* 57, 13484–13493.

Saeed, K., Khan, I., Gul, T., Sadiq, M., 2017. *Appl. Water Sci.* 7, 3841–3848.

Safiur Rahman, M., Khan, M.D.H., Jolly, Y.N., Kabir, J., Akter, S., Salam, A., 2019. *Sci. Total Environ.* 660, 1610–1622.

Sahu, R.S., Li, D.-L., Doong, R., 2018. *Chem. Eng. J.* 334, 30–40.

Saini, J., Garg, V.K., Gupta, R.K., Kataria, N., 2017. *J. Environ. Chem. Eng.* 5, 884–892.

Salem Attia, T., Hu, X., Yin, D., 2013. *Chemosphere* 93, 2076-2085.

Salih, H.H., Sorial, G.A., Patterson, C.L., Sinha, R., Krishnan, E.R., 2012. *Water, Air, Soil Pollut.* 223, 2837–2847.

Salipira, K.L., Mamba, B.B., Krause, R.W., Malefetse, T.J., Durbach, S.H., 2007. *Environ. Chem. Lett.* 5, 13–17.

Samuel, M.S., Bhattacharya, J., Raj, S., Santhanam, N., Singh, H., Pradeep Singh, N.D., 2019. *Int. J. Biol. Macromol.* 121, 285–292.

Sanchez, W., Egea, E., 2018. *Environ. Sci. Pollut. Res.* 25, 6085–6086.

Sarvari, H., Goharshadi, E.K., Samiee, S., Ashraf, N., 2018. *Phys. Chem. Res.* 6, 433–446.

Satheesh, R., Vignesh, K., Rajarajan, M., Suganthi, A., Sreekantan, S., Kang, M., Kwak, B.S., 2016. *Mater. Chem. Phys.* 180, 53–65.

Savage, N., Diallo, M.S., 2005. *J. Nanoparticle Res.* 7, 331–342.

Sawafta, R., Shahwan, T., 2019. *J. Mol. Liq.* 273, 274–281.

Schrick, B., Blough, J.L., Jones, A.D., Mallouk, T.E., 2002. *Chem. Mater.* 14, 5140–5147.

Sepehr, M.N., Allani, F., Zarrabi, M., Darvishmotevalli, M., Vasseghian, Y., Fadaei, S., Fazli, M.M., 2019. *Data Br.* 22, 676–686.

Setshedi, K.Z., Bhaumik, M., Onyango, M.S., Maity, A., 2015. *Chem. Eng. J.* 262, 921–931.

Setshedi, K.Z., Bhaumik, M., Songwane, S., Onyango, M.S., Maity, A., 2013. *Chem. Eng. J.* 222, 186–197.

Shamsipur, M., Rajabi, H.R., 2014. Spectrochim. *Acta Part A Mol. Biomol. Spectrosc.* 122, 260–267.

Shan, C., Ma, Z., Tong, M., Ni, J., 2015. *Water Res.* 69, 252–260.

Shao, D., Hu, J., Jiang, Z., Wang, X., 2011a. *Chemosphere* 82, 751–758.

Shao, D.D., Hu, J., Wang, X.K., Nagatsu, M., 2011b. *Chem. Eng. J.* 170, 498–504.

Sharma, G., Bhogal, S., Naushad, M., Kumar, A., Stadler, F.J., 2017. *J. Photochem. Photobiol. A Chem.* 347, 235–243.

Sharma, G., Kumar, A., Naushad, M., Sharma, S., Ghfar, A.A., Ahamad, T., Si, C., Stadler, F.J., 2019. *J. Mol. Liq.* 294, 111605.

Sharma, G., Thakur, B., Kumar, A., Naushad, M., Vo, D.V.N., Al-Misned, F.A., El-Serehy, H.A., Stadler, F.J., 2020. *Mater. Lett.* 276, 128169.

Sheibani, M., Ghaedi, M., Marahel, F., Ansari, A., 2015. *Desalin. Water Treat.* 53, 844–852.

Shi, J., Long, C., Li, A., 2016. *Chem. Eng. J.* 286, 408–415.

Shoiful, A., Ueda, Y., Nugroho, R., Honda, K. 2016. *J. Water Process. Eng.* 11, 110–117.

Shu, X., Yang, Q., Yao, F., Zhong, Y., Ren, W., Chen, F., Sun, J., Ma, Y., Fu, Z., Wang, D., 2019. *Chem. Eng. J.* 358, 903–911.

Singh, G., Rishi, M.S., Herojeet, R., Kaur, L., Sharma, K., 2020. *Environ. Geochem. Health* 42, 1833–1862.

Singh, P., Gautam, S., Shandilya, P., Priya, B., P. Singh, V., Raizada, P., 2017. *Adv. Mater. Lett.* 8, 229–238.

Sinha, T., Ahmaruzzaman, M., Bhattacharjee, A., 2014. *J. Environ. Chem. Eng.* 2, 2269–2279.

Solaymani, E., Ghaedi, M., Karimi, H., Ahmadi Azqhandi, M.H., Asfaram, A., 2017. *Appl. Organomet. Chem.* 31, e3857.

Song, L., Qiu, R., Mo, Y., Zhang, D., Wei, H., Xiong, Y., 2007. Catal. Commun. 8, 429–433.

Sreekanth, T.V.M., Nagajyothi, P.C., Reddy, G.R., Shim, J., Yoo, K., 2019. *Sci. Rep.* 9, 1–9.

Su, Y., Adeleye, A.S., Huang, Y., Sun, X., Dai, C., Zhou, X., Zhang, Y., Keller, A.A., 2014. *Water Res.* 63, 102–111.

Su, Y., Cheng, Y., Shih, Y., 2013. *J. Environ. Manage.* 129, 361–366.

Su, Y., Li, S., Jiang, G., Zheng, Z., Wang, C., Zhao, S., Cui, D., Liu, Y., Zhang, B., Zhang, Z., 2021. *J. Colloid Interface Sci.* 581, 350–361.

Sui, K., Li, Y., Liu, R., Zhang, Y., Zhao, X., Liang, H., Xia, Y., 2012. *Carbohydr. Polym.* 90, 399–406.

Sumesh, E., Bootharaju, M.S., Anshup, Pradeep, T., 2011. *J. Hazard. Mater.* 189, 450–457.

Sun, C., Zhou, R., Jianan, E., Sun, J., Ren, H., 2015. *RSC Adv.* 5, 57058–57066.

Sun, H., Cao, L., Lu, L., 2011. *Nano Res.* 4, 550–562.

Sun, X.-F., Liu, B., Jing, Z., Wang, H., 2015. *Carbohydr. Polym.* 118, 16–23.

Sundararajan, M., Kennedy, L.J., 2017. *J. Environ. Chem. Eng.* 5, 4075–4092.

Székács, A., Mörtl, M., Darvas, B., *J Chem.*, 2015.

Šťastný, M., Štengl, V., Henych, J., Tolasz, J., Vomáčka, P., Ederer, J. 2016. *J. Mater. Sci.* 51, 5, 2634–2642.

Tabrizian, P., Ma, W., Bakr, A., Rahaman, M.S., 2019. *J. Colloid Interface Sci.* 534, 549–562.

Tajizadegan, H., Torabi, O., Heidary, A., Golabgir, M.H., Jamshidi, A., 2016. Desalin. *Water Treat.* 57, 12324–12334.

Taka, A.L., Fosso-Kankeu, E., Pillay, K., Mbianda, X.Y., 2020. *J. Environ. Chem. Eng.* 8, 103602.

Tan, Y., Chen, M., Hao, Y., 2012. *Chem. Eng. J.* 191, 104–111.

Tang, J., Wang, J., 2018. *Chem. Eng. J.* 351, 1085–1094.

Tang, L., Cai, Y., Yang, G., Liu, Y., Zeng, G., Zhou, Y., Li, S., Wang, J., Zhang, S., Fang, Y., 2014. *Appl. Surf. Sci.* 314, 746–753.

Tang, W.W., Zeng, G.M., Gong, J.L., Liu, Y., Wang, X.Y., Liu, Y.Y., Liu, Z.F., Chen, L., Zhang, X.R., Tu, D.Z., 2012. *Chem. Eng. J.* 211–212, 470–478.

Teimouri, A., Nasab, S.G., Vahdatpoor, N., Habibollahi, S., Salavati, H., Chermahini, A.N., 2016. *Int. J. Biol. Macromol.* 93, 254–266.

Ternes, T.A., 1998. *Water Res.* 32, 3245–3260.

Theron, J., Walker, J.A., Cloete, T.E., 2008. *Crit. Rev. Microbiol.* 34, 43–69.

Thi, V.H.T., Lee, B.K., 2017. *Mater. Res. Bull.* 96, 171–182.

Tian, Y., Gao, B., Morales, V., Chen, H., Wang, Y., Li, H., 2013. *Chemosphere* 90, 2597-2605.

Tiwari, J.N., Mahesh, K., Le, N.H., Kemp, K.C., Timilsina, R., Tiwari, R.N., Kim, K.S., 2013. *Carbon* N. Y. 56, 173–182.
Ulfa, M., Worabay, N. S., Pradipta, M. F., Prasetyoko, D., 2019. In *AIP Conference Proceedings*. 2194, 1, 020131.
Vaiano, V., Jaramillo-Paez, C.A., Matarangolo, M., Navío, J.A., del Carmen Hidalgo, M., 2019. *Mater. Res. Bull.* 112, 251–260.
Vaiano, V., Sacco, O., Matarangolo, M., 2018. *Catal. Today* 315, 230–236.
Veisi, H., Moradi, S.B., Saljooqi, A., Safarimehr, P., 2019. *Mater. Sci. Eng. C* 100, 445–452.
Vidovix, T.B., Quesada, H.B., Januário, E.F.D., Bergamasco, R., Vieira, A.M.S., 2019. *Mater. Lett.* 257, 126685.
Vomáčka, P., Štengl, V., Henych, J., Kormunda, M., 2016. *J. Colloid Interface Sci.* 481, 28–38.
Vu, H.C., Dwivedi, A.D., Le, T.T., Seo, S.H., Kim, E.J., Chang, Y.S., 2017. *Chem. Eng. J.* 307, 220–229.
Wan, J., Deng, H.P., Shi, J., Zhou, L., Su, T., 2014. *Clean - Soil, Air, Water* 42, 1199–1207.
Wang, C., Luo, H., Zhang, Z., Wu, Y., Zhang, J., Chen, S., 2014. *J. Hazard. Mater.* 268, 124–131.
Wang, H., Wang, H.-L., Jiang, W.-F., Li, Z.-Q., 2009. *Water Res.* 43, 204–210.
Wang, J., Pan, K., He, Q., Cao, B., 2013. *J. Hazard. Mater.* 244–245, 121–129.
Wang, J.-P., Yang, H.-C., Hsieh, C.-T., 2010. *Sep. Sci. Technol.* 46, 340–348.
Wang, M., Fang, G., Liu, P., Zhou, D., Ma, C., Zhang, D., Zhan, J., 2016. *Appl. Catal. B Environ.* 188, 113–122.
Wang, M., Yu, X., Yang, C., Yang, X., Lin, M., Guan, L., Ge, M., 2017. *Chem. Eng. J.* 322, 246–253.
Wang, P., Yan, T., Wang, L., 2013. BioResources 8, 6026–6043.
Wang, T., Lin, J., Chen, Z., Megharaj, M., Naidu, R., 2014. *J. Clean. Prod.* 83, 413–419.
Wang, W., Hua, Y., Li, S., Yan, W., Zhang, W. xian, 2016. *Chem. Eng. J.* 304, 79–88.
Wang, W., Zhou, M., 2010. Colloids Surfaces A Physicochem. *Eng. Asp.* 369, 232–239.
Wang, X., Huang, K., Chen, Y., Liu, J., Chen, S., Cao, J., Mei, S., Zhou, Y., Jing, T., 2018. *J. Hazard. Mater.* 350, 46–54.
Wei, W., Yang, L., Zhong, W.H., Li, S.Y., Cui, J., Wei, Z.G., 2015. Dig. J. Nanomater. *Biostruct.* 19, 1343–1363.
Wei, X., Zhu, N., Huang, X., Kang, N., Wu, P., Dang, Z., 2020. *J. Environ. Manage.* 260, 110072.
Weng, X., Cai, W., Lan, R., Sun, Q., Chen, Z., 2018. *Environ. Pollut.* 236, 562–569.
Wu, H., Liu, Q., Ma, J., Liu, L., Qu, Y., Gong, Y., Yang, S., Luo, T., 2020. *Chemosphere* 245, 125591.
Wu, Lidong, Lu, X., Fu, X., Wu, Lingxia, Liu, H., 2017. *Sci. Rep.* 7, 1–7.
Wu, W., Yao, T., Xiang, Y., Zou, H., Zhou, Y., 2020. *Dalt. Trans.* 49, 5722–5729.
Xu, Z., Li, Q., Gao, S., Shang, J.K., 2010. *Water Res.* 44, 5713–5721.

Yalçinkaya, Ö., Kalfa, O.M., Türker, A.R., 2011. *J. Hazard. Mater.* 195, 332–339.
Yan, H., Yang, L., Yang, Z., Yang, H., Li, A., Cheng, R., 2012. *J. Hazard. Mater.* 229–230, 371–380.
Yan, L., Chang, P.R., Zheng, P., Ma, X., 2012. *Carbohydr. Polym.* 87, 1919–1924.
Yang, Q., Li, Z., Lu, X., Duan, Q., Huang, L., Bi, J., 2018. A review of soil heavy metal pollution from industrial and agricultural regions in China: Pollution and risk assessment. *Sci. Total Environ.*
Yao, S., Li, J., Shi, Z., 2010. *Particuology* 8, 272–278.
Yao, Y., Li, G., Ciston, S., Lueptow, R.M., Gray, K.A., 2008. *Environ. Sci. Technol.* 42, 4952–4957.
Yazdanbakhsh, A.R., Daraei, H., Rafiee, M., Kamali, H., 2016. *Water Sci. Technol.* 73, 2998–3007.
Ye, Y., Yang, J., Jiang, W., Kang, J., Hu, Y., Ngo, H.H., Guo, W., Liu, Y., 2018. *J. Environ. Manage.* 206, 929–937.
Yimin, D., Jiaqi, Z., Danyang, L., Lanli, N., Liling, Z., Yi, Z., Xiaohong, Z., 2018. Colloids Surfaces A Physicochem. *Eng. Asp.* 550, 90–98.
Yoon, S., Mahanty, B., Kim, C., 2015. *Desalin Water Treat* 57, 7789-7800.
Yoon, Y., Park, W.K., Hwang, T.M., Yoon, D.H., Yang, W.S., Kang, J.W., 2016. *J. Hazard. Mater.* 304, 196–204.
Yu, B., Zhang, X., Xie, J., Wu, R., Liu, X., Li, H., Chen, F., Yang, H., Ming, Z., Yang, S.-T., 2015. *Appl. Surf. Sci.* 351, 765–771.
Yu, J.-G., Zhao, X.-H., Yang, H., Chen, X.-H., Yang, Q., Yu, L.-Y., Jiang, J.-H., Chen, X.-Q., 2014. *Sci. Total Environ.* 482, 241–251.
Yu, Z., Zhou, L., Huang, Y., Song, Z., Qiu, W., 2015. *J. Environ. Manage.* 163, 155–162.
Zanella, R., Avella, E., Ramírez-Zamora, R.M., Castillón-Barraza, F., Durán-Álvarez, J.C., 2018. Environ. Technol. (United Kingdom) 39, 2353–2364.
Zeng, Y., Walker, H., Zhu, Q., 2017. *J. Hazard. Mater.* 324, 605–616.
Zhai, Y., Lei, Y., Wu, J., Teng, Y., Wang, J., Zhao, X., Pan, X., 2017. *Environ. Sci. Pollut. Res.* 24, 3640–3653.
Zhang, C., Luan, J., Yu, X., Chen, W., 2019. *J. Hazard. Mater.* 378, 120739.
Zhang, Y., Li, Z., 2017. *Sci. Total Environ.* 580, 776–786.
Zhang, Y., Yin, C., Cao, S., Cheng, L., Wu, G., Guo, J., 2018. *Sci. Total Environ.* 622–623, 1499–1508.
Zhang, Y., Zhang, B.T., Teng, Y., Zhao, J., Kuang, L., Sun, X., 2020. *J. Hazard. Mater.* 400, 123290.
Zhao, L., Deng, J., Sun, P., Liu, J., Ji, Y., Nakada, N., Qiao, Z., Tanaka, H., Yang, Y., 2018. *Sci. Total Environ.* 627, 1253–1263.
Zhou, G., Wang, K.P., Liu, H.W., Wang, L., Xiao, X.F., Dou, D.D., Fan, Y.B., 2018. *Int. J. Biol.* Macromol. 113, 792–803.
Zhou, W., Pan, K., Qu, Y., Sun, F., Tian, C., Ren, Z., Tian, G., Fu, H., 2010. *Chemosphere* 81, 555–561.
Zhou, X., Lv, B., Zhou, Z., Li, W., Jing, G., 2015. *Chem. Eng. J.* 281, 155–163.
Zhuang, Y., Yu, F., Chen, J., Ma, J., 2016. *J. Environ. Chem. Eng.* 4, 147–156.

6

Nanomaterials for Wastewater Treatment: Potential and Barriers in Industrialization

Snehasis Bhakta

Department of Chemistry, Cooch Behar College, WB, India

Abstract

Recent advancements in nanotechnology towards wastewater treatment especially in industries has drawn huge attention. Numerous studies have shown that nanomaterials can effectively remove various pollutants in water and thus have been successfully applied in water and wastewater treatment. This chapter will mainly focus on the application of various nanoparticles including some new smart materials which are based on molecular imprinting technology for wastewater treatment and the potential application in industry. Drawbacks associated with the nanotechnology are also discussed and the major problems associated with industrialization are also summarized.

***Keywords*:** Nanoparticles, wastewater, molecular imprinting, nanotechnology, micropollutants

6.1 Introduction

Pure water is the most important natural source to the human body as well as to all living organisms in the world. Without clean water, the world cannot survive [1, 2]. According to the survey, the world's population will reach 9 billion by 2050 [3]. Thus, demand for clean and pure water will rise with time for the survival of the earth. With ongoing industrial and technological development around the globe, one of the biggest problems is contamination of the water due to several factors including industrial waste disposal to the waterbody, improper and insufficient

Email: Snehasis.bhakta1989@gmail.com; snehasisbhakta@coochbeharcollege.org.in

Siddhartha Moulik, Aditi Mullick and Anirban Roy (eds.) Sustainable Water Treatment: Advances and Technological Interventions, (245–270) © 2022 Scrivener Publishing LLC

sewage management systems, havoc usage of pesticides in agricultures, radioactive waste disposal, etc. These all directly lead to water pollution, and indirectly cause soil and air pollution which ultimately severely affect human health [4]. Besides, water pollution also causes huge loss to economic growth, and as a result it also has an impact on the societies around us. Thus, there is huge demand for better solution with the help of technological advancement to treat wastewater efficiently and make it useful to the world.

Contaminant of water occurs when undesired chemicals, materials mix into water and make it unsuitable as well as harmful for normal uses. Heavy metals, organic molecules, inorganic molecules, pathogens, all are responsible for water pollution and can cause serious damage to the aquatic life also. Examples of a few pollutants and their impact on human health are summarized in Table 6.1. Researchers are looking for affordable alternative solutions for the treatment of polluted water. Conventionally, wastewater can be treated with some physical, chemical, mechanical, and some biological methods; however, there are some major drawbacks associated with those techniques. The cost of implementation at the industrial site sometimes is huge and most of the time the removal efficiency of those existing methods is not enough to fulfill the demands at industrial scale. Thus, there is an urgent need for a low-cost alternative to treat wastewater to meet the ongoing demand for pure water. Recently, nanotechnology has shown its potential to clean water due to its higher binding capacity, selective adsorption capacity, and even the ability to degrade pollutants more efficiently than the existing methods at much lower cost. This new emerging technology has already been applied successfully in various medical industry, energy sectors, etc. Nanomaterials could be a better option to treat wastewater due to some of the interesting properties linked with these materials such as higher surface area, variable surface functionalities, hydrophobicity, hydrophilicity, faster mobility, strong mechanical properties, reusability, etc. [5, 6].

The concept of nanotechnology started in the late 19th century; however, it was well recognized by famous physicist Prof. Richard Feynman in 1959. Later, around 1981, when IBM launched the scanning tunneling microscope (STM), the real application and importance of this new technology started booming [17]. Nanoparticles are generally referred to as particles in the range of 1-100 nm in diameter, thus they have high surface-to-volume ratio which makes better choice for faster chemical reactions. The nanomaterials have some special characteristics which are completely absent in the micron-sized particles. They have unique optical, electrical,

Table 6.1 Examples of a few pollutants found in wastewater and the health hazards associated with those chemicals.

Pollutants	Source	Health issue	Reference
Bacteria	Untreated wastewater	Diarrhea, severe vomiting	[7]
Viruses	Untreated wastewater	Fever, sever headache	[8]
Pesticides	Agricultural waste	Diarrhea, waterbody damage	[9]
Radioactive materials	Different isotopes	Organ damage	[10]
Arsenic	Pesticides, fungicides	Affect skin, lung, brain	[11]
Mercury	Paper industries, batteries	Affect lung, reproductive organs	[12]
Lead	Paint, batteries	Affect bones, kidney	[13]
Cadmium	Welding, electroplating	Affect bones, lung, kidney	[14]
Organic chemicals	Textile, paint industries	Directly affect waterbody	[15]
Microplastics	Plastic products	Oxidative stress, inflammatory lesions	[16]

catalytical, magnetic properties which are suitable for wastewater treatment [18]. Several types of nanomaterials are being introduced for water treatment; however, there are mainly three categories where this technology can be applied successfully: (i) detection of pollutants at ultralow concentration, (ii) removal of contaminants, and (iii) degradation of the pollutants to less toxic or eco-friendly products [19]. Although there are lot of research publications comprising the possibility of application for nanomaterials in wastewater treatment, a lot of work needs to be done to make it economically friendly materials for industries. This chapter will mainly focus on the application of different types of nanomaterials including some smart nanoparticles, and natural polymer-based nanomaterials in wastewater treatment. Some of the major drawbacks associated with the nanotechnology are also discussed herein.

6.2 Nanomaterials in Wastewater Treatment

Present research on nanotechnology offers a sustainable, inexpensive alternative for wastewater treatment. Water can be cleaned and usable with the help of nanotechnology which can be categorized into the following:

i. Removal of pollutants using nanomaterials
ii. Degradation of pollutants using nanomaterials
iii. Filtration of contaminant water
iv. Disinfection of polluted water

All of these methods involve various types of nanoparticles with required modification. Different types of inorganic zerovalent nanoparticles, metal oxide nanoparticles, polymeric nanoparticles, etc., can be employed as per the requirement for effective removal or detection of harmful pollutants present in wastewater. Figure 6.1 shows examples of a few nanoparticles along with the main functions associated with nanotechnology.

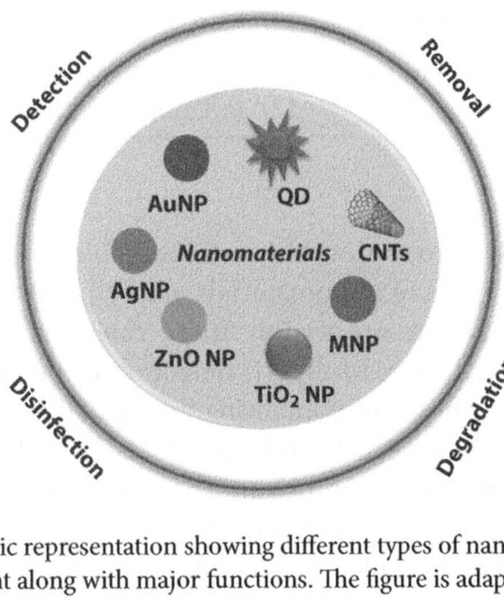

Figure 6.1 Schematic representation showing different types of nanomaterials used in wastewater treatment along with major functions. The figure is adapted from reference [20].

6.2.1 Nanotechnological Processes for Wastewater Treatment

6.2.1.1 Nanofiltration

Filtration using a membrane is a well-known technique to remove unwanted materials from a mixture. However, the biggest problem associated with the wide implementation of filtration was the huge investment cost. With more research and new innovation in nanotechnology, the overall cost is reduced, and because filtration technique has great efficiency for removal of pollutants, it is now becoming popular in various industries [21]. Likewise, reverse osmosis, nanofiltration is also a high-pressure filtration method with much less energy consumption. Organic, inorganic pollutants, viruses, and bacteria can be removed from contaminant water.

6.2.1.2 Adsorption

Nanoparticles possess some unique characteristics for which these are very good adsorbents for toxic chemicals or other pollutants. The higher specific surface area and the surface functionalities allow to bind more pollutants. The surface can be easily modified with other functional groups, thus can be used for selective removal of certain chemicals. Besides, the added functionalities sometimes improve the stability of the system, allowing it to function in harsh chemical condition without degradation. Additionally, the incorporated surface groups sometimes help the adsorption-desorption process to go much faster by just changing some physical conditions such as pH of the media [22].

6.2.1.3 Photocatalysis

Metal and metal oxide nanoparticles can be used as catalytic agents. Generally, the surface when irradiated with light generates highly reactive radicals which oxidizes the pollutants. As a result, either much less reactive chemicals are formed or the pollutants are totally degraded to eco-friendly chemicals. The smaller size of the particles leads to high surface-to-volume ratio, and, thus make these very reactive towards pollutants. However, the production cost for this technology is very high and it is thus not economically friendly yet [23].

6.2.1.4 *Disinfection*

In addition to pollutant removal, filtration, and photocatalytic behavior, nanoparticles can also be used as disinfectant materials. Many inorganic metal oxide nanoparticles such as copper oxide, zinc oxide nanoparticles, silver, carbon nanotubes and copper nanoparticles possess inherent antibacterial properties, thus can be used as a disinfectant system as well. However, these nanoparticles should come in direct contact with the cell surfaces of bacteria or viruses to be effective herein [24].

6.2.2 Different Nanomaterials for Wastewater Treatment

6.2.2.1 *Zerovalent Metal Nanoparticles*

In general, zerovalent metal nanoparticles refer to iron nanoparticle in its neutral form. However, it could be any metals with zero charge. These materials are being used as reducing agents which facilitate conversion of oxidized pollutants in wastewater. Most of the zerovalent metal nanoparticles, such as silver, iron, zinc, and gold have excellent optical, mechanical, electrical, and catalytic properties for which these are excellent materials for wastewater treatment [25].

Zerovalent zinc nanoparticles are considered to be a better option than the zerovalent iron due to high negative reduction potential. Yan *et al.* [26] found that the zinc could degrade the pollutant in wastewater at a much faster rate than the iron. Other study also revealed that zinc could slow down the bacterial growth in wastewater substantially [27]. However, zerovalent zinc is not very effective in halogenated water, thus, application in large scale can be a tricky one.

Zerovalent iron nanoparticles has already shown the potential to degrade pollutants effectively at industry level. Crane *et al.* reported that these nanoparticles could act very effectively on halogenated chemicals such as hexachlorocyclohexane (lindane), vinyl chloride, carbon tetrachloride, chlorinated organic compounds (COCs), and trichloroethane [28]. Besides, in anaerobic condition, zerovalent iron nanoparticles exhibited as a potential reducing agent and could react to various inorganic, organic pollutants in wastewater. However, there are some drawbacks such as aggregation, leaching, and separation of zerovalent iron nanoparticles which need to be addressed for future consideration [29].

Other important zerovalent nanoparticles are silver nanoparticles which possess antibacterial properties; thus, these are extensively used to decontaminate water from pathogens like bacteria, viruses, etc., from wastewater. However, the aggregation problems sometimes hindered the on-site application at large scale [30].

6.2.2.2 *Metal Oxide Nanoparticles*

Metal oxide nanoparticles have already shown immense potential as a low-cost, sustainable, and user-friendly choice for wastewater treatment. The higher surface area and functionalities onto the surface made these better adsorbent as well as selective one for pollutants. Metal oxide nanoparticles are extremely helpful to remove heavy metals, and simply by changing the pH, these can be easily regenerated and reused for further use [31].

One of the widely used metal oxide nanoparticles in wastewater treatment is titanium dioxide (TiO_2) nanoparticles which act as an extremely good photocatalyst due to formation of reactive oxygen species with minimal toxicity to humans [32]. Panahi *et al.* [33] reported the use of titanium dioxide nanoparticles as a catalyst in ozonation process and could totally degrade the organic pollutants in wastewater. Besides, these nanoparticles can also kill pathogens like viruses, and can be a very effective disinfectant material when it is combined with silver nanoparticles.

Similar to previously discussed titanium dioxide nanoparticles, zinc oxide (ZnO) nanoparticles also showed photocatalytic properties. These act as an oxidizing agent with very wide band gap in UV-VIS region, thus making it extremely helpful for degradation of pollutants in wastewater. Besides, these also possess antibacterial properties. However, when zinc oxide nanoparticles are coupled with some conducting oxides such as reduced graphene oxide, titanium oxide, CeO, SnO_2 etc., the photocatalytic properties enhance substantially [34].

Iron oxide nanoparticles, being superparamagnetic in nature with surface-to-volume ration, have been already applied in laboratory and industry scale for wastewater treatment. The magnetic nature definitely helped for a faster and easy separation process. Mainly, these nanoparticles act as an adsorbent for pollutants due to the interaction with the functional groups and electrostatic interactions onto the surface [45, 46]. However, agglomeration of these nanoparticles remains a problem which needs to be addressed properly for future application. Table 6.2

Table 6.2 Example of various metal oxide nanoparticles used in wastewater treatment.

Nanoparticles	Pollutants	Application	References
AgO, NiO CuO	Malachite green	Degradation of dye	[35]
Cu_2O, ZnO, NiO	Bacterial cells	Disinfection of pathogens	[36]
TiO_2	Heavy metal ions	Photocatalytic degradation	[37]
TiO_2	Methylene blue	Photocatalytic degradation	[38]
Iron oxide	Heavy metal ions	Removal of heavy metal ions	[39]
Iron oxide	Mercury	Removal of mercury	[40]
Iron oxide	Heavy metals	Remediation of heavy metals	[41]
ZnO	Metal ions	Selective adsorption of metals	[42]
ZnO	Methylene blue	Cleaning of water	[43]
Alumina-silica	Cadmium ions	Adsorption of ions	[44]

lists examples of some metal oxide nanoparticles used in wastewater treatment.

6.2.2.3 *Other Nanoparticles*

Carbon nanotubes (CNTs) have been used extensively for entrapment of inorganic, organic pollutants from wastewater. These materials have an excellent mechanical, electronic, optical, and hollow-like structure which also helps to act as a photocatalytic agent. CNTs can also kill bacteria cells when the surfaces are properly functionalized. However, extremely high production cost for CNTs in pilot-scale set up is a barrier for widespread applicability [47].

Development of nanocomposites or mixture of nanomaterials opens up smart technologies to use materials for multifunction purposes. Mixing of materials (CNTs with silver nanoparticles, titanium dioxide with bimetallic nanoparticles, nanoparticles embedded in polymeric nanocarriers) make them useful for disinfection of wastewater, and improve the binding efficiency of the pollutants [48, 49].

Quantum dots also play a crucial role in wastewater treatment. These unique nanoparticles help for selective detection of pollutants present in contaminant water. Although there are many advanced technologies available for the detection and removal of pollutants, sometimes it is extremely difficult to detect them when present in ultralow concentration. Quantum dots with their strong fluorescence properties can act as a nanosensor for the same [50].

6.3 Smart Nanomaterials: Molecularly Imprinted Polymers (MIP)

Different types of new and smart technologies based on nanoparticles have emerged recently to clean wastewater effectively at minimal cost [51, 52]. Researchers have been using physical adsorption [53, 54], biological degradation [55], and oxidation [56] to treat wastewater from various resources. All of these methods have been used extensively over the past decades; however, sometimes the above-mentioned methods are not effective to remove pollutants present at low concentration from wastewater [57]. There are numerous numbers of micropollutants such as endocrine-disrupting compounds (EDC) and selective serotonin reuptake inhibitors (SSRI) which are present in wastewater at very low concentration and these need to be removed selectively [58, 59]. Those existing methods sometimes fail to selectively detect and separate these micropollutants. Besides, these methods are often very expensive and non-selective against pollutants [60, 61]. Even at very low concentration, EDCs and SSRIs can cause human health problems [62, 63]. Thus, there is a need for alternate solutions which are inexpensive as well as highly selective to detect and remove pollutants from wastewater [64]. More recently, molecularly imprinted polymers (MIP) have drawn huge attention due to its high selectivity and binding capacity toward the target analytes.

6.3.1 Molecularly Imprinted Polymers (MIP)

MIPs are made of polymeric materials surrounding a target analyte, similar to "lock-key", either on the surface of nanoparticles or in bulk solvent media. Removal of the analyte creates a cavity or pocket-like structure which is similar in size and shape to the template. Addition of monomeric units with various functionality further adds up the possibility of several binding interactions while rebinding which makes the synthetic binding sites highly selective [65]. In the last few decades, there were significant efforts to create MIP-based binding sites for small molecules, proteins, and peptides as well as microorganisms. The major advantages of this technique includes better thermal stability than natural bioreceptor, low production cost, and reusability [66, 67].

The concept of MIPs was first introduced in 1930 by Polyakov. Then, in 1940, the concept was modified and Pauling introduced the term molecular imprinting due to the similarities with the "lock-key" mechanism [68]. Initially, imprinting was carried out in a bulk solution where target analyte along with the polymerizable monomers were mixed together. Finally, the binding cavity resembling the size and shape of the analyte was prepared by mechanically crushing the polymers. Although this imprinting strategy was used broadly back in the 2000s due to its simple synthetic strategy, this method has some serious problems such as heterogenous binding sites, time-consuming, and laborious [69]. The drawbacks were resolved by introducing a relatively new and modified version of the concept, namely, surface molecular imprinting. Herein, the target or template molecule is first attached onto a surface of nanoparticles, then different types of monomers along with cross-linkers are mixed together and allowed to polymerize. This creates highly selective homogeneous binding sites [70, 71].

6.3.2 Application of MIP-Based Nanomaterials in Wastewater Treatment

Recently, MIP-based nanomaterials are being used extensively to treat wastewater by researchers all over the world. This unique polymer-coated nanoparticle can be applied in contaminated water for various purposes: (i) selectively detect pollutants at ultralow concentration in wastewater, (ii) removal of pollutants from wastewater, (iii) improve the catalytic degradation of organic matters such as dyes present in wastewater. Due to advantages like low-cost production, reusability, high-selectivity, and better stability, MIP-based technologies are becoming more demanding for recognition of micropollutants and separation of the same from contaminant water.

6.3.2.1 Recognition of Pollutants

Since the emergence of surface molecular imprinting concept, scientists have looked for cheap alternative materials for wastewater treatment.

MIP-based materials have already been introduced in various analytical tools such as in sensors, solid phase extraction, and chromatography due to well-oriented molecular recognition and stability. Surface molecular imprinting comprises different types of non-covalent interactions such as H-bonding, hydrophobic, and van der Waals along with the shape and size of the target molecules. This enables very effective recognition and adsorption of the pollutants. In recent years, several publications have shown the potential of the materials for the detection and extraction of a spectrum of compounds from water including carbaryl, estradiol, catechol, bisphenol A, organic dyes, cyanide, etc.

Xu *et al.* (2009) [72] studied the recognition of low concentration estrone. They coupled high performance liquid chromatography (HPLC) with molecularly imprinted polymer for the separation and achieved detection limit of 5.7 ng L^{-1} and could extract 0.5 μg L^{-1} from wastewater. Small nanoparticles with their high surface-to-volume are an obvious choice for researchers; however, the separation could be a tedious process with some loss of materials. This problem was overcome via introduction of magnetic onto the nanoparticles, which made the separation process much faster. Lin *et al.* [73] applied magnetic nanoparticles for detection and separation of bisphenol A from water. The materials could remove approximately 90% of the analyte from the contaminant water and detection limit was 2.50 pg mL^{-1}. Khan *et al.* introduced a biomimetic sensor for sensitive detection of methyl green dye using electrochemical platform onto carbon paste coupled with magnetic molecularly imprinted polymers [74]. This showed satisfactory and very stable sensor for the recognition of any pollutants from wastewater.

6.3.2.2 *Removal of Pollutants*

Adsorption of pollutants onto the surface of materials is an effective and well-known process for the removal of pollutants from water. To date, a huge number of materials, namely carbon black, graphite, magnetic nanoparticles, and silica nanoparticles have been applied to isolate the pollutants from wastewater and make the water usable again. Addition of imprinting techniques could improve the adsorption capacity dramatically due to the high selective nature towards the target molecule. However, when it comes to the application of any materials in a real sample, one must consider the cost and handling of the same. Thus, MIP-based nanomaterials add up all the above-mentioned benefits to treat wastewater. Pan *et al.* [75] prepared imprinted nanomaterials for the separation of 2,4-dichloropheol (2,4-DCP). The magnetic materials showed high adsorption capacity of 145.79 mg g^{-1} and the inherent magnetism made the separation process very fast. Hu *et al.* [76] synthesized magnetic MIP using β-cyclodextrin as monomer to remove pollutants from water. They showed rapid removal of contaminants such as PA, MB, BO_2, RhB, Cr(III), Pb(II), Zn(II), and Cu(II) from wastewater. More examples are given in Table 6.3.

6.3.2.3 *Catalytic Degradation of Organic Molecules*

Along with the diverse application of MIPs for recognition and separation, enormous attention was also given for the potential uses in catalytic

Table 6.3 Recent literature on wastewater treatment using MIP.

Nanomaterials	Target analytes	Application	Reference
Fe_3O_4	Sulphonamides	Detection of sulphonamides	[77]
Fe_3O_4	EDCs	Extraction of endocrine disruptors	[78]
Multiwalled carbon natubes	2, 4-dichlorophe-noxya Ceticatid	Recognition of the analyte	[79]
Ag_2S-TiO_2	ethyl p-hydroxybenzoate	Photocatalytic degradation	[80]
MIL100	phthalate esters	Catalytic degradation of phthalate esters	[81]
Graphene QD	Chloroyrifos	Ultralow detection of chloropyrifos	[82]
TiO_2	O-phenylphenol	Selective degradation of analyte	[83]
Fe_3O_4	Quinoline	Selective adsorption of quinoline	[84]
AgNP	Caffeine	Nanosensor for caffeine	[85]
SiNP	Tannin	Selective removal of tannin	[86]

degradation of organic molecules using this smart nanomaterial. Since this technique creates enzyme-like binding cavities, and the materials could tolerate harsh conditions such as high temperature, pH, organic solvent, MIP-based materials are becoming an obvious alternate choice instead of natural enzyme, catalytic antibodies. Shen *et al.* [77] showed the photocatalytic degradation of 2-nitrophenol and 4-nitrophenol onto imprinted TiO_2 nanomaterials. The imprinted nanomaterials enhanced photocatalytic efficiency and improved the selectivity compared to only TiO_2.

6.3.3 Barriers in Industrialization

In spite of many promising alternatives using different types of nanoparticles such as carbon nanotube, carbon quantum dots, gold nanoparticles, silica nanoparticles, zinc oxide nanoparticles, magnetic nanoparticles and many more, the imprinted nanoparticles are yet to get success in the industrial market for wastewater treatment. With all the advantages, there are still some challenges to make it widely useful in industry. Firstly, leakage of the template during adsorption process remains a problem. Sometimes, multilayer adsorption causes difficulty in removing the analyte effectively. Secondly, most of the synthesis methods are employed in organic solvent, and thus it is not easy to use in aqueous media. Lastly, sometimes, the soft polymer coating tends to degrade with time and thus creates inhomogeneous binding pockets. However, with proper choice of monomers and nanoparticle surface, these obstacles can be minimized and MIP-based nanomaterials could be a cheap alternative choice for industries.

6.4 Cheap Alternative Nanomaterials

Nanotechnology has been widely applied in treating wastewater effectively. The high surface-to-volume ratio helps improve adsorption capacity of any pollutants, thus reducing the use of material for the same. In this regard, various types of inorganic, polymeric, organic coated nanoparticles are being introduced to clean wastewater and make it usable. Overall, the use of nanoparticles has shown great potential for wastewater treatment to reduce the cost and time. There are a huge number of applications based on nanomaterials such as zinc oxide (ZnO), iron oxide magnetic nanoparticles (IONPs), titanium dioxide, carbon nanotubes, etc. [88, 89]. However, the use of such nanomaterials also draws attention to the toxicity effect associated with them. Thus, researchers are always looking for alternative solutions which are eco-friendly, non-toxic to the environment, less-expensive, renewable, recyclable. In this regard, two naturally materials nanocellulose and nanoclay play a crucial role in fulfilling the growing demands [90].

6.4.1 Nanoclay for Wastewater Treatment

Nanoclays are excellent adsorbents for pollutants from wastewater due to higher surface area, easy tunable surface chemistry and sustainability. Nanoclays are generally modified with suitable surface chemistry or coupling with other functional materials according to the demand. A few advantages of these nanoclays are abundant occurrence in nature, can be reused with some treatment and is very cheap. There are a sufficient number of research articles describing the usage of these naturally occurring nanomaterials to detect and isolate pollutants. Bentonite was used to adsorb amoxicillin antibiotics from contaminant water [91]; similarly, montmorillonite clay can be used to adsorb trimethoprim [92]. In a similar fashion, halloysite clay was widely used to adsorb heavy metal ions, dyes, and other organic pollutants from wastewater to make it useful for drinking purposes [93, 94]. Recent advancement comprising magnetic nanoparticles and clay composites has shown excellent potential to remove contaminants from wastewater easily via magnetic separation [95]. These nanomaterials have also shown their application to clean oils and toxic pollutants, and thus can be an alternative cleaning material in textile and oil industries.

6.4.1.1 Water Filtration by Nanoclays

Halloysite clay showed promising results for the filtration of Rhodamine 6G and Chrome azurol S dyes due to its relatively high surface area. When compared with kaolinite, halloysite exhibited higher removal capacity. Additionally, both these nanoclays can be easily regenerated and reused for multiple times [96]. Incorporation of halloysite nanoclays with polyacrylonitrile revealed created nanofibrous-like structure and could be useful as an excellent filtration membrane. While only polyacrylonitrile showed ~4% oil removal capacity, the addition of the clay improved the capacity to 31% [97].

6.4.1.2 Water Treatment by Hybrid Gel

Hybrid gel formed by naturally occurring clays are eco-friendly and can be used for heavy metal ions removal from wastewater. Alginate/pectin gel beads was used to isolate heavy metal ions such as cadmium (II), copper (II) very effectively [98]. Similarly, halloysite/alginate gel beads can be prepared by dropping technique and this type of gel is excellent material to remove organic with more than 90% efficiency. Likewise, chitosan can also be combined with halloysite and could be an alternative, cheap, eco-friendly material for removal of dyes in textile industries [99, 100].

6.4.1.3 Nanosponges

An interesting biocompatible material with very high adsorption capacity is nanosponge. This material can be formed by mixing halloysite and ucurbituril at appropriate quantity. The nanosponge showed ~50 times more adsorption efficiency when compared with the unmodified halloysite. This functional halloysite can be an exceptional entrapping material for aliphatic and aromatic hydrocarbons present in wastewater [101]. Additionally, various types of hybrid nanomaterials comprising nanoclays can be formed for treating wastewater.

6.4.2 Nanocellulose for Wastewater Treatment

Another largely accepted renewable, biocompatible nanomaterial for wastewater treatment in recent years is nanocellulose. Cellulose is one of the highly abundant biomolecules found on earth and the nanocellulose can be obtained from the forestry-based lignocellulose resources with the help of new technologies. With modifications and adopting various synthetic strategies, one can get different types of low-cost, high adsorbing materials. Without doubt, this type of renewable material possesses a huge potential to replace the existing materials which are generally obtained from fossil resources for wastewater treatment [102].

The main sources for cellulose are lignocellulosic materials such as woods (hard woods, soft woods), grasses (bamboo, bagasse), annual plants (jute) and also animal or bacteria. Recent technological advancement allowed the exploration the structure of the cellulose at nanoscale level with much better mechanical strength and higher surface area. Nanocellulose is generally categorized into two main groups: cellulose nanocrystals (CNCs) and cellulose nanofibrils (CNFs). These nanocelluloses along with some other forms such as nanofibers, nanocrystals or a combination of these can be applied for adsorption, remediation and separation of pollutants from wastewater at very low cost.

6.4.2.1 *Adsorption of Heavy Metals by Nanocellulose*

Consumption of heavy metals cause adverse effects on human health. High surface area and the presence of surface functional groups enhance the adsorption capacity for heavy metals for cellulose nanoparticles. The surface groups onto nanocellulose can easily be modified according to the need and thus can be used for selective adsorption of metal ions from wastewater. Phosphorylation of the nanocellulose was used to remove lanthanide ions (La^{3+}, Sm^{3+}, Ho^{3+}) [103]. Chemical treatment of rice straw with a mixture of hydrochloric and sulfuric acid produced CNCs which was used to remove toxic metal ions from contaminant water. The removal efficiency was 8.55 mg g^{-1} for Ni^{2+}, 9.42 mg g^{-1} for Pb^{2+} and 9.7 mg g^{-1} for Cd^{2+} from a mixture of solution at pH 6 [104]. Titanate aerogel was used for metal ions adsorption from polluted water. When the 3D-titanate aerogel was mixed with NFC, the adsorption efficiency for metal ions (Pb^{2+}, Sr^{2+}, Cu^{2+}, Cd^{2+}, Ra^{2+}) was improved dramatically to 122% from 70% [105]. With proper modification and mixing with other material, user-friendly materials thus can be formed and these materials have appealing potential for application in wastewater and sewage water treatment.

6.4.2.2 *Adsorption of Dyes by Nanocellulose*

Chemical dyes used in the textile, paper, pharmaceutical, and food industries generate an enormous number of pollutants to the aquatic body. Nanocellulose with some chemical modifications can play a crucial role for the isolation of these harmful chemicals from water. By incorporation of negatively charged sulfate groups onto CNC, the material was used to remove methylene blue with very promising binding capacity of ~120 mg g^{-1} at pH 9 [106]. The material was reused by adsorption-desorption mechanism simply by changing the pH of the medium. Upon carboxylation of the nanocellulose, the dye removal efficiency increased to 6-fold and these materials were used to separate cationic dyes such as crystal violet, basic fuchsin, and malachite green [107].

6.4.2.3 *Barriers in Industrialization*

The major advantages for nanocellulose are natural resources and their biocompatibility. In spite of these, more intensive research is needed before industrialization of these materials. Because it is biocompatible, when it is used as filtration or separating materials, the bacteria present in the wastewater can degrade this naturally occurring materials. However, proper surface modification or coating with some polymers could overcome the problems. Next, the cost of production of the nanocellulose still remains high and special attention should be given. Sometimes, it is difficult to form homogeneous nanocellulose with the existing synthetic methodology. Often, it forms inhomogeneous mixture with variant mechanical strength [108]. In future, if these challenges are addressed properly, these materials can replace most existing materials for wastewater treatment.

6.5 Toxicity Associated with Nanotechnology in Wastewater Treatment

Nanotechnology brings a new era to modern society. At present, nanoparticles have been used in almost every field starting from medicinal industries, oil and paint industries, textile industries, automobile industries, cosmetic industries, etc. [109, 110]. Most of the nanoparticles ultimately enter into a waterbody or landfill. As discussed earlier, due to high adsorption capacity and other functionalities, nanomaterials are also extensively applied for wastewater treatment. However, many nanomaterials are also associated with toxicity to human health which sometimes stops the nanoparticles for application in wastewater. Thus, special attention should be given to address the hazardous effect with various nanoparticles [111].

Most of the untreated nanomaterials generally end up piling up in the soil and can cause heavy pollution in the longer run. The nanomaterials can be found in the sludge formed in the wastewater treatment plant. Most of the treatment plants apply polyelectrolytes for faster aggregation and sedimentation purposes which eventually affects the properties of the nanomaterials. Electrolytes, in general, reduce the reactivity of the nanomaterials, lead to flocculation, and finally get consumed in the sludge [112]. However, extensive research works need to be done to understand the effects of these nanomaterials on human health. Otherwise, these may be a huge obstacle to apply nanomaterials in wastewater treatment expansively.

Toxicity of the nanomaterials is dependent on the concentration of the same. Sometimes, the presence of very low concentration could lead to serious human health issues. Carbon nanotubes (CNTs) are widely used to remove pollutants, degradation of chemicals from wastewater, and the toxicity effect associated with CNTs depends on the size, shape, initial concentration, and aggregation degree of the materials. In general, single-walled CNTs are less toxic than multiwalled CNTs. Major health issues associated with these materials are pulmonary disorder, basic inflammation, apoptosis, oxidative stress, etc. [113, 114]. Thus, serious attention should be given before applying nanomaterials for wastewater treatment. Nanoparticles tend to aggregate with time due to several factors and this causes toxicity to the environment. To avoid the aggregation, different types of polymeric or organic capping agents are being used; however, when these nanoparticles come into biological media, there is always some aggregation. Generally, silver nanoparticles with size less than 100 nm can enter bacterial cell, then react with the enzyme, DNA in the bacterial cell and thus inhibit growth of the pathogens. The size of silver nanoparticles, for example, in the presence of bacterial cell, increases 10-15-fold and thus the overall antibacterial nature decreases consequently. The accumulated silver nanoparticles in the sludge or biomass could slow down the denitrification process of the soil. The presence of ZnO and TiO_2 are relatively stable nanoparticles, but still these particles also aggregate with time due to gravitation and ultimately get consumed in nature [115, 116]. Besides the size effect, the toxicity during wastewater treatment by nanoparticles is also associated with the reaction conditions. In case of aerobic oxidation, metal or metal oxides generally release reactive oxygen species, thus, the toxicity is associated with the metal ions. However, in case of anaerobic oxidation, no reactive oxygen species is formed, and as a result the toxicity is associated with the release of metal ions. Iron oxide and carbon nanotube composites are found to be very toxic to DNA even at very low concentration [117]. Similarly, other metal ions can also affect the environment as well as human health. Thus, special scientific research should be conducted to understand the toxicity effect of these nanoparticle.

6.6 Barriers in Industrialization

Nanotechnology has already shown its potential in wastewater treatment. Most of the nanomaterials can easily be employed to detect, remove, and degrade pollutants present in wastewater. However, most of these studies are conducted in laboratory scale, and it is not easy to implement nanotechnology in industries to clean contaminant water from various sources due to several reasons. First of all, the cost of production of nanomaterials still remains very high while considering its applicability in industrial level. Robust research is necessary to synthesize nanomaterials at minimal cost so that it becomes sounding for industries. Next, the toxicity associated with various nanomaterials is another alarming concern to the environment and human health. Most of the nanomaterials are accumulated into the sludge generated by wastewater and finally the sludge is mixed with soil. Thus, proper attention should be given to overcome the toxicity effect associated with the usage of nanomaterials if it is planned to implement [118]. Besides, more research should be conducted to understand the compatibility of the nanotechnology with the existing methods used in wastewater treatment. In developed countries where the wastewater management and sewage treatment facilities are robust, it is not recommended to entirely change the existing system. However, modifications are necessary with very specific nanotechnology for an easy, fast and effective way to treat wastewater and make it useful. However, in underdeveloped or developing countries, this new nanotechnology-based system could be adopted for wastewater treatment. Thus, with proper management and risk assessment studies, the nanotechnology could be a blessing to the whole world as this new technology can supply clean water.

6.7 Future Aspect and Conclusions

The demand for fresh and clean water is growing rapidly. The economic and societal growth of a nation is somehow indirectly dependent on the availability and supply of clean water. With the progression of industries, water pollution also increases astonishingly due to lack of management and availability of robust, cost-effective wastewater treatment technologies. Nanotechnology offers a wide range of solutions to treat wastewater effectively. However, when it comes to implementation of practical application, many problems are arising. Except for one technology, zerovalent iron nanoparticles, there is not much success yet. Still, nanotechnology holds the possibility due to its unique optical, electrical, and mechanical properties. Nanomaterials can be applied to remove pollutants from wastewater, to detect harmful chemicals from contaminant water, to degrade organic chemicals present in wastewater, and sometimes to act as disinfectants. Extensive research on nanotechnology can bring a solution to the whole world. We also need to look into the cost associated with the nanotechnology to be implemented in the real world. Besides, the health risk associated with accumulation of nanoparticles has to be carefully monitored. The new synthetic nanomaterials have already shown their applicability in wastewater treatment at much lower cost. Natural biopolymer-based nanomaterials could be an ideal choice for treating wastewater since these are biodegradable and available everywhere in the world. There is a tremendous potential in nanotechnology to fulfill the supply of clean water to the whole world.

References

1. A. Ahmad, S.H. Mohd-Setapar, C.S. Chuong, A. Khatoon, W.A. Wani, R. Kumar, and M. Rafatullah, *RSC Adv.* Vol. 5, p. 30801–30818, 2015.
2. V. Gitis, and N. J. Hankins, *Water Process Eng.*, Vol. 25, p. 34–38, 2018.
3. D. Bhatia, N.R. Sharma, J. Singh, and R.S. Kanwar, *Critcal Rev. Environ. Sci. Technol.* Vol. 47, 1836–1876, 2017.
4. Hodges, B.C.; Cates, E.L.; Kim, *J. Nat. Nanotechnol.*, Vol. 13, p. 642–650, 2018.
5. S. Baruah, and M.N. Dutta, *J. Environ. Chem. Lett.*, Vol. 14, p. 1–14, 2016.
6. Y. Wu, H. Pang, Y. Liu, X. Wang, S. Yu, D. Fu, J. Chen, and X. Wang, *Environ. Pollut.*, Vol. 246, p. 608–620, 2019.
7. M.A. Bahlaoui, B. Baleux, and M. Troussellier, *Water Research*, Vol. 31, p. 630-638, 1997.
8. C.P. Gerba, J.L. Melnick, C. Wallis, *Journal of the Irrigation and Drainage Division*, Vol. 101, p. 157-174, 1975.
9. F.C. Moreira, V.J. Vilar, A.C. Ferreira, F.R. dos Santos, M. Dezotti, M.A. Sousa, C. Gonçalves, R.A. Boaventura, and M.F. Alpendurada, *Chemical Engineering Journal*, Vol. 209, p. 429-441, 2012.
10. X. Zhang, and Y. Liu, *Environmental Science: Nano*, Vol. 7, p. 1008-1040, 2020.
11. E. Fogarassy, I. Galambos, E. Bekassy-Molnar, and G. Vatai, *Desalination*, Vol. 240, p. 270-273, 2009.
12. H. Von Canstein, Y. Li, K.N. Timmis, W.D. Deckwer, and I. Wagner-Döbler, *Applied and environmental microbiology*, Vol. 65, p. 5279-5284, 1999.
13. D. Singh, A. Tiwari, and R. Gupta, *J Agric Technol*, Vol. 8, p. 1-11, 2012.
14. K.K. Singh, A.K. Singh, and S.H. Hasan, *Bioresource Technology*, Vol. 97, p. 994-1001, 2006.
15. C. Fernández, M.S. Larrechi, and M.P. Callao, *TrAC Trends in Analytical Chemistry*, Vol. 29, p. 1202-1211, 2010.
16. S. Ziajahromi, P.A. Neale, L. Rintoul, and F.D. Leusch, *Water Research*, Vol. 112, p. 93-99, 2017.
17. I.J. El Saliby, H.K. Shon, J. Kandasamy, and S. Vigneswaran, *Encyclopedia of life support system (EOLSS)*, Vol. 7, 2008.
18. O.S. Ayanda, and L.F. Petrik, *International Journal of Chemical, Material and Enviromnmental Research*, Vol. 1, p. 1-2, 2014.
19. K. Watlington, National network for environmental management studies fellow, North Carolina State University, 2005.
20. J. Chen, S.M. Andler, J.M. Goddard, S.R. Nugen, and V.M. Rotello, *Chemical Society Reviews*, Vol. 46, p. 1272-1283, 2017.
21. G. Tchobanoglous, H.D. Stensel, R. Tsuchihashi, and F. Burton, *Wastewater engineering: Treatment and resource recovery*, Fifth Ed., New York, McGraw-Hill, 2014.

22. X. Ren, C. Chen, M. Nagatsu, and X. Wang, *Chemical Engineering Journal*, Vol. 170, p. 395-410, 2011.
23. P.N.L. Lens, J. Virkutye, V. Jegatheesan, S.h. Kim, and S. Al-Abed, *Nanotechnology for water and wastewater treatment*, IWA Publishing, 2013.
24. Q. Li, S. Mahendra, D.Y. Lyon, L. Brunet, M.V. Liga, D. Li, and P.J.J. Alvarez, *Water research*, Vol. 42, p. 4591-4602, 2008.
25. V.K. Sharma, J. Filip, R. Zboril, and R.S. Varma, *Chem Soc Rev.*, Vol. 44, p. 8410–8423, 2015.
26. F. Fu, D.D. Dionysiou, and H. Liu, *J Hazard Mater* Vol. 267, p. 194–205, 2014.
27. G. Liu, D. Wang, J. Wang, and C. Mendoza, *Sci Total Environ*, Vol. 409, p. 2852–2857, 2011.
28. R.A. Crane, and T.B. Scott, *J Hazard Mater*, Vol. 211, p. 112–125, 2012.
29. D. Dutta, D.Thakur, and D. Bahadur, *Chem Eng J*, Vol. 281, p. 482–490, 2015.
30. R. Prasad, V. Kumar, K.S. Prasad, *Afr J Biotechnol*, Vol. 13, p. 705–713, 2014.
31. I. Corsi, M. Winther-Nielse, R. Sethi, C. Punta, C. Della Torre, G. Libralato, and F. Cinuzzi, *Ecotoxicol Environ Saf*, Vol. 154, p. 237–244, 2018.
32. X. Qu, P.J. Alvarez, and Q. Li, *Water Res*, Vol. 7, p. 3931–3946, 2013.
33. Y. Panahi, H. Mellatyar, M. Farshbaf, Z. Sabet, T. Fattahi, and A. Akbarzadehe, *Mater Today Proc*, Vol. 5, p. 15550–15558, 2018.
34. N. Rana, K.S. Ghosh, S. Chand, and A.K. Gathania, *Indian J Pure Appl Phys*, Vol. 56, p. 19–25, 2018.
35. N. Sapawe, M.A. Rustam, M.H.H. Mahadzir, M.K.E.M. Lani, A. Raidin, and M.F. Hanafi, *Materials Today: Proceedings*, Vol. 19, p. 1441-1445, 2019.
36. J. Hrenovic, J. Milenkovic, N. Daneu, R.M. Kepcija, and N. Rajic, *Chemosphere*, Vol. 88, p. 1103-1107, 2012.
37. K. Zhang, K.C. Kemp, and V. Chandra, *Materials Letters*, Vol. 81, p. 127-130, 2012.
38. A. Najafidoust, S. Allahyari, N. Rahemi, and M. Tasbihi, *Ceramics International*, Vol. 46, p. 4707-4719, 2020.
39. Z. Cheng, A.L.K. Tan, Y. Tao, D. Shan, K.E. Ting, and X.J. Yin, *International Journal of Photoenergy, 2012.*
40. E. Vélez, G.E. Campillo, G. Morales, C. Hincapié, J. Osorio, O. Arnache, J.I. Uribe, and F. Jaramillo, In *Journal of Physics: Conference Series*, Vol. 687, p. 012050, IOP Publishing, 2016.
41. V.K. Yadav, D. Ali, S.H. Khan, G. Gnanamoorthy, N. Choudhary, K.K. Yadav, V.N. Thai, S.A. Hussain, and S. Manhrdas, *Nanomaterials*, Vol. 20, p. 1551, 2020.
42. M. Gu, L. Hao, Y. Wang, X. Li, Y. Chen, W. Li, and L. Jiang, *Chemical Physics*, Vol. 534, p. 110750, 2020.
43. T.M. Tamer, W.M. Abou-Taleb, G.D. Roston, M.S. Mohyeldin, A.M. Omer, R.E. Khalifa, and A.M. Hafez, *Environmental Nanotechnology, Monitoring & Management*, Vol. 10, p.112-121, 2018.

44. S. Pacheco, J. Tapia, M. Medina, and R. Rodriguez, *Journal of Non-Crystalline Solids*, Vol. 352, p. 5475-5481, 2006.
45. A.M. Gutierrez, T. D. Dziubla, and J. Z. Hilt, *Rev Environ Health*, Vol. 32, p. 111–117, 2017.
46. R. Dinali, A. Ebrahiminezhad, M. Manley-Harris Ghasemi, and Y. Berenjian, *A Crit Rev Microbiol*, Vol. 43, p. 493–507, 2017.
47. O.G. Apul, and T. Karanfil, *Water Res*, Vol. 68, p. 34–55, 2015.
48. Rodrigues SM, Demokritou P, Dokoozlian N, Hendren CO, Karn B, Mauter MS, Sadik OA, Safarpour M, Unrine JM, Viers J, Welle P (2017). *Environ Sci Nano* 4:767–781.
49. P. Thoniyot, M.J. Tan, A.A. Karim, D.J. Young, and X.J. Loh, *Adv Sci*, Vol. 2, p. 1400010, 2015.
50. J. Yu, X. Wang, Q. Kang, J. Li, D. Shen, and L Chen, *Environ Sci Nano* Vol. 4, p. 493–502, 2017.
51. J.K.H. Wong, H.K. Tan, S.Y. Lau, P.S. Yap, and M.K. Danquah, *Journal of Environmental Chemical Engineering*, Vol. 7, p. 103261, 2019.
52. S.A. Younis, H.A. Maitlo, J. Lee, and K.H., Kim, *Advances in Colloid and Interface Science*, Vol. 275, p. 102071, 2020.
53. S.A. Snyder, S. Adham, and A. M. Redding et al., *Desalination*, Vol. 202, p. 156–181, 2007.
54. D.L. Huang, G. M. Zeng, X. Y. Jiang et al., *J Hazard Mater*, Vol. 134, p. 268–276, 2006
55. Y. Lee, B. I. Escher, and U. Von Gunten, *Environ Sci Technol*, Vol. 42, p. 6333–6339, 2008.
56. H.J. Guo, S.L. Luo, L. Chen et al., *Bioresource Technol*, Vol. 101, p. 8599–8605, 2010.
57. B. Okutucu, In *Waste in Textile and Leather Sectors*, London, IntechOpen, 2020.
58. W. Mnif, A.I.H. Hassine, A. Bouaziz, A. Bartegi, O. Thomas, and B. Roig, *International journal of Environmental Research and Public Health*, Vol. 8, p. 2265-2303, 2011.
59. J.C. Jakobsen, K.K. Katakam, A. Schou, S.G. Hellmuth, S.E. Stallknecht, K. Leth-Møller, M. Iversen, M.B. Banke, I.J. Petersen, S.L. Klingenberg, and J. Krogh, *BMC Psychiatry*, Vol. 17, pp.1-28, 2017.
60. M. Bilal, M. Adeel, T. Rasheed, Y. Zhao, and H.M. Iqbal, *Environment International*, Vol. 124, p. 336-353, 2019.
61. S. Krishnan, H. Rawindran, C.M. Sinnathambi, and J.W. Lim, In *IOP Conference Series: Materials Science and Engineering*, Vol. 206, p. 012089, IOP Publishing, 2017.
62. M. Beszterda, and R. Frański, *Pediatric Endocrinology, Diabetes & Metabolism*, Vol. 24, 2018.
63. J.C. Jakobsen, K.K. Katakam, A. Schou, S.G. Hellmuth, S.E. Stallknecht, K. Leth-Møller, M. Iversen, M.B. Banke, I.J. Petersen, S.L. Klingenberg, and J. Krogh, *BMC Psychiatry*, Vol. 17, p.1-28, 2017.

64. R. Fiorenza, A. Di Mauro, M. Cantarella, A. Gulino, L. Spitaleri, V. Privitera, and G. Impellizzeri, *Materials Science in Semiconductor Processing*, Vol. 112, p. 105019, 2020.
65. G. Goyal, S. Bhakta, and P. Mishra, *ACS Applied Nano Materials*, Vol. *2*, p. 6747-6756, 2019.
66. L. Chen, S. Xu, and J. Li, *Chemical Society Reviews*, Vol. 40, p. 2922-2942, 2011.
67. S. Bhakta, M.S.I. Seraji, S.L. Suib, and J.F. Rusling, *ACS Applied Materials & Interfaces*, Vol. 7, p. 28197-28206, 2015.
68. K. Mosbach, *Trends in Biochemical Sciences*, Vol. 19, p. 9-14, 1994.
69. M. Komiyama, T. Takeuchi, T. Mukawa, and H. Asanuma, *Molecular imprinting*. Wiely, Mörlenbach, Germany, 2003.
70. L. Chen, X. Wang, W. Lu, X. Wu, and J. Li, 2016. *Chemical Society Reviews*, Vol. 45, p. 2137-2211, 2016.
71. S. Bhakta, C.K. Dixit, I. Bist, J. Macharia, M. Shen, K. Kadimisetty, J. He, B. Dutta, S.L. Suib, and J.F. Rusling, 2017. *Chemical Communications*, Vol. 53, p. 9254-9257, 2017.
72. P. Xu, G. M. Zeng, D.L. Huang *et al.*, *Sci Total Environ*, Vol. 424, p. 1–10, 2012.
73. Z.K. Lin, W.J. Cheng, Y.Y. Li et al., *Anal Chim Acta*, Vol. 720, p. 71–76, 2012.
74. S. Khan, A. Wong, A.V.B. Zanoni, M.P.T. Sotomayor, *Materials Science & Engineering C.*, Vol. 103, p. 109825-109832, 2019.
75. J. Pan, L. C. Xu, J. D. Dai, et al., *Chem Eng J*, Vol. 174, p. 68–75, 2011.
76. X. Hu, Y. Hu, G. Xu, M. Lia, Y. Zhu, L. Jiang, et al., *Environmental Research*, Vol. 180, p. 108796-108807, 2020.
77. X.T. Shen, L.H. Zhu, G.X. Liu, et al., *Environ Sci Technol*, Vol. 42, 1687–1692, 2008.
78. H.Y. Chen, Y.Q. Zhang, B. Gao, et al., *Environ Sci Pollut R*, Vol. 20, p. 8567–8578, 2013.
79. X. Xia, E.P.C. Lai, B. Örmeci, *Enviro Sci Pollut R*, Vol. 20, p. 3331–3339, 2013.
80. W.J. Yang, F.P. Jiao, L. Zhou, et al., *Appl Surf Sci.*, Vol. 284, p. 692–699, 2013.
81. X. Liu, L. Zhu, X. Wang, and X. Meng, *Scientific reports*, Vol. 10, p.1-10, 2020.
82. X. Li, J. Wan, Y. Wang, H. Chi, Z. Yan, and S. Ding, *Chemosphere*, Vol. 240, p. 124875, 2020.
83. M.L. Yola, and N. Atar, *Journal of the Electrochemical Society*, Vol. *164*, p. B223, 2017.
84. R. Fiorenza, A. Di Mauro, M. Cantarella, A. Gulino, L. Spitaleri, V. Privitera, and Impellizzeri, G., *Materials Science in Semiconductor Processing*, Vol. *112*, p. 105019, 2020.
85. Y. Cui, W. Kang, L. Qin, J. Ma, X. Liu, and Y. Yang, *Chemical Engineering Journal*, Vol. 397, p. 125480, 2020.
86. R. Hu, R. Tang, J. Xu, and F. Lu, *Analytica chimica acta*, Vol. 1034, p. 176-183, 2018.
87. J. Benvenuti, L.B. Capeletti, M. Gutterres, and J.H.Z. dos Santos, *Journal of Sol-Gel Science and Technology*, Vol. 85, p. 446-457, 2018.

88. C. Burda, Y. Lou, X. Chen, A. C. S. Samia, J. Stout, and J. L. Gole, *Nano Lett.* Vol. 3, p. 1049–1051, 2003.
89. V. K. Gupta, and T. A. Saleh, *Environ Sci. Pollut. Res.*, Vol. 20, p. 2828–2843, 2013.
90. K. G. Satyanarayana, G. G. C. Arizaga, and F. Wypych, *Prog. Polym. Sci.*, 34, 982–1021, 2009.
91. E. K. Putra, R. Pranowo, J. Sunarso, N. Indraswati, and S. Ismadji, *Water Res.*, Vol. 43, p. 2419–2430, 2009.
92. Z. Bekci, Y. Seki, and M. K. Yurdakoc, *J. Hazard. Mater.*, Vol. 133, p. 233–242, 2006.
93. Z. Wang, S. S. Zhao, F. Liu, L. Y. Yang, Y. Song, X. Y. Wang, and X. J. Xi, *Desalination*, Vol. 259, p. 235–242, 2010.
94. Z. Shu, Y. Chen, J. Zhou, T. T. Li, Z. Sheng, C. W. Tao, and Y. X. Wang, *Appl. Clay Sci.*, Vol. 132, p. 114–121, 2016.
95. L. H. Ai, H. Y. Huang, Z. L. Chen, X. Wei, and J. Jiang, *Chem. Eng. J.*, Vol. 156, p. 243–249. 2010.
96. Y. F. Zhao, E. Abdullayev, A. Vasiliev, and Y. Lvov, *J. Colloid Interface Sci.*, Vol. 406, p. 121–129, 2013.
97. M. Makaremi, R. T. De Silva, and P. Pasbakhsh, *J. Phys. Chem. C*, Vol. 119, p. 7949–7958, 2015.
98. S. Cataldo, G. Cavallaro, A. Gianguzza, G. Lazzara, A. Pettignano, D. Piazzese, and I. Villaescusa, *J. Environ. Chem. Eng.*, Vol. 1, p. 1252–1260, 2013.
99. L. Liu, Y. Wan, Y. Xie, R. Zhai, B. Zhang, and J. Liu, *Chem. Eng. J.*, Vol. 187, p. 210–216, 2012.
100. Q. Peng, M. Liu, J. Zheng, and C. Zhou, *Microporous Mesoporous Mater.*, Vol. 201, p. 190–201, 2015.
101. M. Kryuchkova, and R. Fakhrullin, *Environ. Sci. Technol. Lett.*, Vol. 5, p. 295–300, 2018.
102. G. Yuan, and L. Wu, *Science and Technology of Advanced Materials*, Vol. 8, p. 60, 2007.
103. Oshima, T.; Kondo, K.; Ohto, K.; Inoue, K.; Baba, Y. React. Funct. Polym. 2008, 68, 376–383.
104. A. Kardam, K. R. Raj, S. Srivastava, and M. M. Srivastava, *Clean Technol. Environ. Policy*, Vol. 16, p. 385–393, 2014.
105. Y. Xiong, C. Wang, H. Wang, Q. Yao, B. Fan, Y. Chen, Q. Sun, C. Jin, and X. Xu, *J. Mater. Chem. A*, Vol. 5 (12), p. 5813-5819, 2017.
106. R. Batmaz, N. Mohammed, M. Zaman, G. Minhas, R. M. Berry, and K. C. Tam, *Cellulose* Vol. 21, p. 1655–1665, 2014.
107. H. Qiao, Y. Zhou, F. Yu, E. Wang, Y. Min, Q. Huang, L. Pang, and T. Ma, *Chemosphere*, Vol. 141, p. 297–303, 2015.
108. B. Yuan, L. Li, V. Murugadoss, S. Vupputuri, J. Wang, N. Alikhani, and Z Guo, *ES Food Agrofor.*, Vol. 1, p. 41-52, 2020.

109. S. Ripp, and T.B. Henry, Biotechnology and nanotechnology risk assessment: minding and managing the potential threats around us, *American Chemical Society*, Vol. 1079, p. 0, 2011.
110. D. E. Meyer, M. A. Curran, and M. A. Gonzalez, *Environ. Sci. Technol.*, Vol. 43, p. 1256–1263, 2009.
111. Y.H. Zhang, C.Z. Hu, F. Liu, Y. Yuan, H. Wu, and A. Li, *Sci. Total Environ.*, Vol. 646, p. 265–279, 2019.
112. P.N.L. Lens, J. Virkutye, V. Jegatheesan, S.h. Kim, and S. Al-Abed, *Nanotechnology for water and wastewater treatment*, IWA Publishing, 2013.
113. D. Jones, S. Caballero, and G.D. Pardo, *Adv. Food Nutr. Res.*, Vol. 1, p. 2–9, 2019.
114. S. Wang, W. Lu, O. Tovmachenko, U.S. Rai, H. Yu, and P.C. Ray, *Chem. Phys. Lett.*, Vol. 463, p. 145–149, 2008.
115. M. Auffan, J. Rose, M. R. Wiesner, and J.-Y. Bottero, *Environ. Pollut.*, Vol. 157, p. 1127–1133, 2009.
116. R. J. Barnes, O. Riba, M. N. Gardner, A. C. Singer, S. A. Jackman, and I. P. Thompson, *Chemosphere*, Vol. 80, p. 554–562, 2010.
117. H.L. Karlsson, P. Cronholm, J. Gustafsson, and L. Möller, *Chem. Res. Toxicol.*, Vol. 21, p. 1726–1732, 2008.
118. F. Lu, and D. Astruc, *Coord. Chem. Rev.*, Vol. 408, p. 213180, 2020.

Section III

MEMBRANE-BASED TREATMENT

7

Microbial Fuel Cell Technology for Wastewater Treatment

Nilesh Vijay Rane[1], Alka Kumari[2*], Chandrakant Holkar[1], Dipak V. Pinjari[1] and Aniruddha B. Pandit[1†]

[1]Department of Chemical Engineering, Institute of Chemical Technology, Nathalal Parekh Marg, Matunga, Mumbai, India
[2]Process Engineering and Technology Transfer (PETT) Division, CSIR-Indian Institute of Chemical Technology, Tarnaka, Hyderabad, India

Abstract

Improving energy recovery from wastewater is a sustainable approach for its treatment and also one of the interests of an effluent treatment plant (ETP) in protecting and improving the water environment in India. This chapter describes in detail microbial fuel cell (MFC) systems and sustainable technologies that are the need of the hour. MFC is one such technology of wastewater treatment (WWT), by which energy can be extracted out of the effluent in terms of electricity or hydrogen. WWT is one of the most complex chemical industry processes due to the varied nature of effluent. Microbes, i.e., bacteria, are used to oxidize bio-degradable matter. They are sustainable and cheap catalysts to decrease the activation energy of oxidation reaction. The working principle and the performance parameters, along with its technique of measurement of MFC, is discussed in great detail. Various materials and reactor configurations of MFC are also reported. The incorporation of MFC into conventional WWT and its effects are also demonstrated with a case study.

***Keywords*:** Microbial fuel cell, wastewater, electricity, low-cost electrodes, efficient and cheap membranes

**Corresponding author*: alka@iict.res.in; alka.bits@gmail.com
†*Corresponding author*: ab.pandit@ictmumbai.edu.in; dr.pandit@gmail.com

Siddhartha Moulik, Aditi Mullick and Anirban Roy (eds.) Sustainable Water Treatment: Advances and Technological Interventions, (273–324) © 2022 Scrivener Publishing LLC

7.1 Introduction

India is a country with more than 1.2 billion people (2016) and expected to be the most populous country by 2022, surpassing China with a staggering figure of 1.6 billion. It is one of the main reasons that we are facing a crunch in terms of resources. In addition to this, India, being a developing country, requires more resources to develop technology and infrastructure. The two significant resources being referred to here are energy and water.

In the last six decades, India's energy application has increased by 16 times and the installed electricity capacity by 84 times. In 2008, India's energy use was the fifth highest in the world. Nevertheless, the country suffers from significant energy poverty and pervasive electricity deficits. In recent years, India's energy consumption has increased faster due to population growth and economic development. With an economy projected to grow at 8-9% per annum, rapid urbanization, and improving living standards for millions of Indian households, the demand is likely to increase significantly. We are heavily dependent on fossil fuels as a source of energy. Coal is the most essential and abundant fossil fuel used in India and accounts for 55% of its energy needs, followed by crude oil and natural gas [1]. These fossil fuels have been used indiscriminately in the past years, considering their low cost and relatively high energy efficiencies. But in the past few years, there has been a growing concern about the social and environmental impacts of using these fossil fuels. Global warming due to excessive CO_2 emissions has led to severe climate change all around the world. Figure 7.1 shows the emission levels of different energy resources.

These alarming figures depict the harm done to the environment and the health of people in society. Alternative, efficient and economical energy sources need to be found and applied in the years to come.

A similar decline has occurred in the case of other resources also, mostly water. India accounts for around 16% of the population and only 4% of water resources globally. The per capita water availability in India is approximately 1,170 m^3/person/year [3]. Thus, the increasing water demand coupled with the decline in water resources due to climate change is a serious issue. People cannot stop water usage entirely, and there is a need to recycle water as much as possible. In earlier days, not much attention was paid to the treatment of the wastewater generated from various sources. Given the environmental and health impacts of the release of wastewater without treatment, stringent rules have been made by the government of India regarding wastewater treatment. It is a complex process as it contains too many chemicals and doesn't have a standard formulation. Hence, multiple

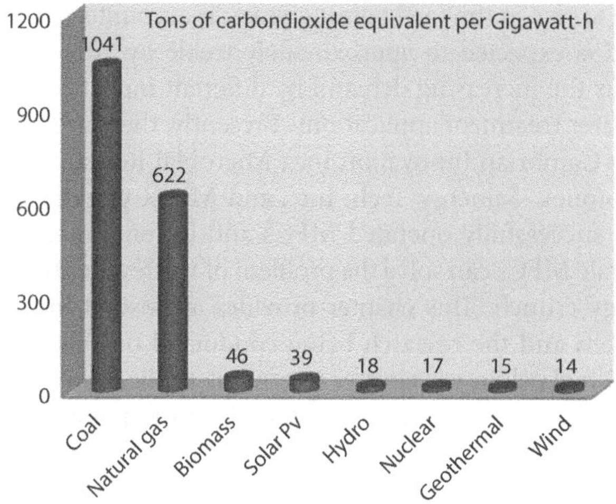

Figure 7.1 Comparison of life cycle emissions of various resources [2].

stages (primary, secondary and tertiary) are involved in the treatment, and it is highly energy-intensive and time-consuming. Better ways are being found to improve the treatment of wastewater.

Nowadays, research is directed towards sustainable use of resources, discovering more and more renewable resources, and developing new, green, and clean technologies to address such global issues. Tapping energy from wastewater is one such idea. Wastewater contains high organic content, thereby, a high energy content. In the USA, each year, 3% (1.5×1010 W) of all electrical power is consumed for the treatment of organic-rich wastewater [4]. A considerable amount of sludge is produced after wastewater treatment, disposal of such waste is energy-intensive and uneconomical [5]. Moreover, greenhouse gases such as carbon dioxide, nitrous oxide, etc., are generated and released into the atmosphere. If efficiently extracted, this technology will lead to simultaneous energy generation and wastewater treatment, like killing two birds with one stone. It is estimated that effluent from domestic, animal, and food processing industries contains 17 GW of energy. This energy, if recovered, can make water infrastructure self-sufficient. Microbial fuel cell (MFC) technology can tap power from biodegradable organic matter in the form of electricity. This idea dates back to 1911 when M. Potter thought of using microbes to generate electricity, but his work did not gain much coverage [6]. MFCs are being researched heavily due to the various benefits, and efforts are being made to make them as efficient as possible. With efficient treatment of the wastewater, its release in the environment, sanitation, etc., is ruled out. The global market of MFCs is predicted to rise in the forthcoming years; it accounted for $8.60 million in 2017 and is expected to approximately treble by 2026. The reason for the growth is the increasing demand by different industries for biosensors and wastewater treatment applications. Presently, there are various companies such as Cambrian Innovation Inc., Microbial Robotics, ElectroChem, Emefcy, Protonex, Sainergy Tech, Inc., and MICR Organic Technologies, which have successfully operated MFCs and become market players [7]. The large-scale MFCs can solve the problem of wastewater treatment as well as the energy crunch. This chapter provides an insight into the basics of these fuel cells and the research being conducted on them. The potential of microbial fuel cell in wastewater treatment application and the material of construction, design modification and effect of operating parameters is covered in this chapter.

7.2 Microbial Fuel Cell

Microbial fuel cells (MFCs) are bio-electrochemical devices that use bacteria as catalysts to oxidize organic and inorganic matter and produces electricity while treating wastewaters. These systems mainly depend on electron-rich bacteria, which metabolises and transfers an electron out of its cell wall in the presence of electrode. These unique systems rely on the ability of certain electroactive bacteria to transfer their electrons extracellulary as a part of their metabolism when placed in proximity to a solid electrode [1]. This electricity is termed as bio-electricity as it is produced by microbes. The greatest advantage of the MFC is that any kind of wastewater can be used as a fuel and energy, in a usable form, can be harnessed from it.

7.2.1 Working Principle

Before understanding the principle of MFC, it is necessary to know all the components involved and their functions.

a. Fuel: Bio-degradable organic matter
b. Microbes: They oxidize the fuel an aerobically to release electrons and protons.
c. Anode: Electrons are transferred to the anode with or without the use of exo-electrogens.

d. Cathode: Electrons are further transferred to the cathode (negative electrode) by a conducting wire operated under load.
e. Proton exchange membrane: It is the most crucial part of the MFC as it allows selective permeation of protons to the cathode chamber from the anode chamber.
f. Terminal electron acceptor (TEA): Chemical oxidizing agents (generally, oxygen) get reduced on reacting with the transferred electrons and protons transferred from the membrane at cathode. The working of the MFC is demonstrated in Figure 7.2.

An MFC contains three major sections, i.e., an aerobic cathode section, an anaerobic anode section, and a partition linking the two chambers made up of proton exchange membrane. The anode section gives the required environment to grow microorganism which then degrade (oxidise) the organic matter in the wastewater to convert chemical bond energy of organic compounds into electrons. The oxidative metabolic reactions in the anode chamber are not permitted to finish and electrons are withdrawn intermittently from the chamber to generate electrical work. Microbes transfer electrons to the anode and are carried to the cathode by a resistor generating electricity. The electrolyte solution acts as a transfer medium for proton towards the cathode to carry out oxygen reduction reaction (ORR). A separating media, i.e., membrane, divides the MFCs into two compartments of anode (where oxidation occurs) and cathode (where reduction

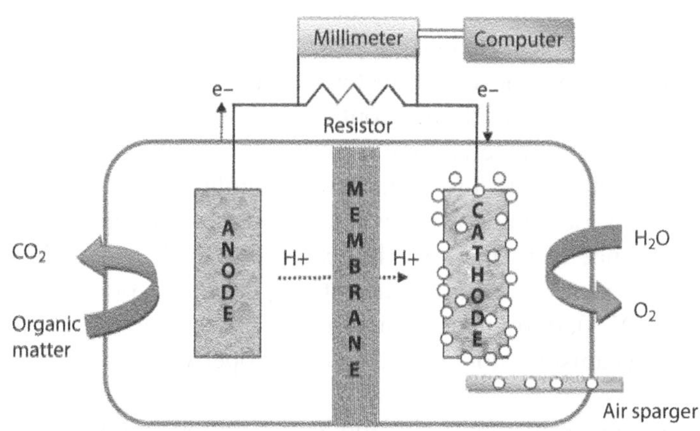

Figure 7.2 Working principle of microbial fuel cell.

occurs). An external resistor positioned between the anode and cathode to harvest energy produced by MFCs.

Membrane: The anode and cathode chambers are separated by a membrane. This is done so as to prevent the catholyte (TEA) from entering the anode chamber, as this inhibits electricity generation. Also, the membrane allows diffusion of only protons from the anode chamber to the cathode chamber.
Fuel and microbes: The microbes (generally bacteria) grow on the anode, oxidize (degrade) the organic content of the wastewater and release electrons to the anode and protons to the solution. A variety of biodegradable materials is used in MFC, such as,volatile acids, carbohydrates, proteins, cellulose, alcohols, etc. Either pure culture or mixed culture of bacteria is used. The self-replicating nature of the bacteria imparts a sustainable nature to the operation of MFC.
Transfer of electrons to anode: The transfer of electrons to the anode is the next stage. Electron mediators, direct membrane electron transfer or nanowires produced by bacteria is used for the transmission of electrons from cathode to anode. To increase the rate of electron transfer to the anode, chemical mediators are used in the system [8]. Following are the mechanisms of electron transfer:

a. Electron shuttling via self-produced mediators: These kinds of mediators are called endogenous mediators. Pycocyanin is a mediator produced by Pseudomonas aeruginosa [9].
b. Nanowires: The measurement of conductivity of bio-films of bacteria by conductive scanning tunnelling microscopy (STM) indicates the formation of nano-wires on the surface of anode. Nanowires of Geobacter and Shewanella species have been reported [9].
c. Cell-surface electron transfer: Surface blebs, i.e., lumps on the surface that do not contain nanowires and also act as conductive material for electrons.
d. Mediators: This mechanism was mainly studied for E.coli and yeast as they are known not to produce electricity in the absence of mediators. The exogenous mediators include, for example, neutral red, anthraquinone-2, 6-disulfonate (AQDS), thionin, potassium ferricyanide [9]. However, this mechanism is not yet understood properly.

The reactions occurring at the anode and cathode are very important and determine the performance of the microbial fuel cell. Positive current travels from positive to negative terminal, a heading inverse to the flow of electrons traditionally. As mentioned earlier, electrons are transported to cathode through a conducting wire. At the cathode, reaction between terminal electron acceptor (TEA) and the protons that have travelled across the membrane and electrons occurs. The cathode is sparged with air to provide dissolved oxygen (TEA) for the reaction with electrons and protons, thereby producing water. Oxygen has usually been a final electron acceptor, due to its easy availability, intense oxidation potential and formation of no poisonous end products. Nitrates, sulphates, ferricyanide or Mn (IV), have been used as chemical oxidizers and therefore it needs to replaced or regenerated [6]. The concentration and species of the oxidant (electron acceptor), proton availability and electrode structure and its catalytic ability affect the reaction at the cathode. Catalysts are very important in both cathodic and anodic reaction to decrease the activation energy of the reaction and thus, increase the rate of reaction. The current is obtained based on measuring the voltage drop across the resistor using a multimeter connected to a data acquisition system.

In the early 1900s, it was discovered that MFCs generate electricity, and M. Potter in 1911 invented the idea to produce electricity using E. Coli [10], although M. J. Allen and Peter Benetto are credited as the forerunners in this field. They started their work in the early 1980s, and since then many advances have been done in this field and new materials developed. However, MFCs have very limited commercial applications due to the high capital cost associated with setting up an MFC and its relatively low energy density output.

$$C_{12}H_{22}O_{11} + 13H_2O \rightarrow 12CO_2 + 48H^+ + 48e^- \qquad (7.1)$$

7.2.2 Role of MFC Components

The role of different parts of MFCS is described below

a. Anode: It is the chamber where oxidation of substrate takes place with the help of microbes. Typical reaction at anode is mentioned here:

$$CH_3COO^- + 2H_2O \rightarrow 2CO_2 + 7H^+ + 8e^- \qquad (7.2)$$

b. Cathode: It is the electrode where reduction takes place with the help of electrons from anode, which allows continuous flow of electrons and generates electricity. Cathode is necessary for generation of electricity; however, a cathodic chamber is not necessary.

$$O_2 + 4e^- + 4H^+ \rightarrow 2H_2 \tag{7.3}$$

c. Membranes: It is an important component of MFC, which transports the excess protons formed in anode chamber towards cathode. Generally, proton exchange membrane (PEM) is used in MFC, while using cation exchange membrane (CEM) it should be able to permeate oxygen, ferricyanide, other ions, or organic matter applied as the substrate. The market for ion exchange membranes is constantly growing, and more systematic studies are required to evaluate the effect of the membrane on the performance and its long-term stability.

7.2.3 Performance Indicator of MFC

The performance of MFC is determined by electrical indicators such as current and power densities, voltage difference and electrical resistance and its efficacy in terms of organic pollutant removal. Coulombic efficiency (CE) is one of the indicators of MFC. It is a combination of the amount of electrons produced as current and total charge available [11]. The performance of an MFC depends on the voltage of reaction between substrate and water in anode at low potential and reaction of electrons and protons using oxygen in cathode at higher potential. However, it is difficult to measure the exact voltage because of the fact that electrons are transferred to anode from substrate via microbes. Microbes complete this process by complex respiratory chains; the exact mechanism is yet unknown. Moreover, the actual performance of an MFC is always less than its ideal one due to irreversible losses, and is given by:

$V_{cell} = E_{cathode} - E_{anode} - iR_i$ – (concentration polarization losses in cathode + anode) – (activation polarization losses in cathode + anode) – Ohmic losses (7.4) where R_i is the internal resistance of MFC, Ohmic losses occur in MFC due to resistance of flow of ions in electrolyte and resistance of flow of electrons in electrode [11].

Voltage or potential of an electrode: It is determined by measuring the potential of an electrode against standard electrodes. Electrode potential is affected by pH, therefore pH of the solution is to be reported as different solutions have different potentials at different pH [12]. Current is measured directly by using Ohm's law by measuring voltage of MFC and dividing it by the external resistance.

$$I = E_{cell}/R_{ext} \tag{7.5}$$

Power: It is calculated by measuring the voltage across a fixed external resistor (R_{ext}), however maximum power is estimated using polarization curves plotted as cell voltage against current density [11]:

$$P = E_{cell}^2 / R_{ext} \tag{7.6}$$

Power density: it is defined as power (P) per unit area as reference area which entirely depends on an individual. In many cases, anode reaction limits power generation and the efficacy of anode is found out by taking the reference area as the area of anode.

$$\text{Power density} = \text{P/Area of anode} \tag{7.7}$$

However, cathode reaction limits power generation in certain cases. It may also happen that the anode surface area is difficult to measure, and then the cathode area is used for the power density. Moreover, in some cases, to have an idea on the efficacy of entire MFC, the total volume of reactor is considered for power density calculation as this is the best estimate to see how much power is generated per unit volume, although there is a limitation to compare power generated by a single-chambered and a two-chambered MFC. In such cases, anode area or cathode area is used [11].

Coulombic efficiency: It is the ratio of total coulombs actually transferred to the anode from the substrate, to maximum possible coulombs if all substrate removal produced current. Under steady-state conditions, coulumbic efficiency is calculated as:

$$E_{ce} = M \times I/Fbq\Delta COD \tag{7.8}$$

M is the molecular weight of electron acceptor, I is current under steady-state conditions, q is the volumetric influent flow rate and ΔCOD is the difference in the influent and effluent chemical oxygen demand (COD).

Treatment efficiency: This evaluates the overall performance in terms of biochemical oxygen demand (BOD), chemical oxygen demand (COD), or total organic carbon (TOC) removal. The performance in terms of COD removal is indicated as it is a common measure for WWT efficiency, and even required in the Coulombic and energy calculations. This parameter measures how much of the available "fuel" has been converted in the MFC, either into electrical current (via the Coulombic efficiency) or biomass (via the growth yield) or through competitive reactions with alternative electron acceptors [10].

Growth Yield: It is computed in terms of substrate utilization for growth as:

$$Y = X/\Delta\ COD \tag{7.9}$$

where Y represents the growth yield, X represents the biomass produced over time.

Energy efficiency: It is the most important factor for assessing the performance of an MFC for generating electricity, which evaluates the system in terms of the energy recovery. The overall energy efficiency is calculated as the ratio of power produced by the cell over a time interval t to the heat of combustion of the organic substrate added in that time:

$$E_e = \int E_{cell}\ Idt/\Delta HM_{added} \tag{7.10}$$

where ΔH is the heat of combustion (J mol^{-1}) and M_{added} is the amount (mol) of substrate added. This is calculated only for influents with known composition (i.e., for synthetic wastewaters) as ΔH is not known for actual wastewaters. In MFCs, energy efficiencies range from 2% to 50% or more when easily biodegradable substrates are used.

7.2.4 Design Parameters

The design of microbial fuel cell with respect to reactor size, shape and configuration mainly depends on the designer and there is no said protocol to follow. However, reactor volume, supply of oxygen, type of membrane and electrode spacing can affect the overall performance of MFC. It is also considered that the hydraulic retention time (HRT), anode and cathode surface areas and the reactor volume are the primary design parameters to develop a MFC system. Generation of power and effluent quality is mainly affected by HRT. The longer the HRT, the less organic matter is present in the system effluent which increases reactor size and cost [13]. The efficiency of MFC depends on the bio-film present, mixing and mass transfer in reactor, organic loading rate, etc. Other than the above parameters, over potentials at anode and cathode electrode is a very important parameter measured with respect to open circuit potential and depends on various factors, such as electrode surface area, electrode potential and electrochemical characteristics of electrodes. Cathode displays the potential loss in the system. Finally, another important factor that affects the performance of MFC is the resistance between electrolyte of two electrodes and membrane to favor transfer of proton. To reduce this resistance, the electrodes need to be placed as close to each other as possible, facilitating the proper mixing of the reactor substrate. Following are the different types of MFCs used in the treatment of wastewater:

7.2.5 Types of Microbial Fuel Cell

Since the MFCs have been a topic of interest for wastewater treatment for two decades, various designs of MFCs have been evolved such as single chamber, double chamber, stack MFC and many others. All these different types of MFC are discussed in the subsequent sections.

Single-Chambered MFC:

A typical single-chamber MFC is shown in Figure 7.3. This type of MFC is easier to design, containing an anodic chamber and typically no

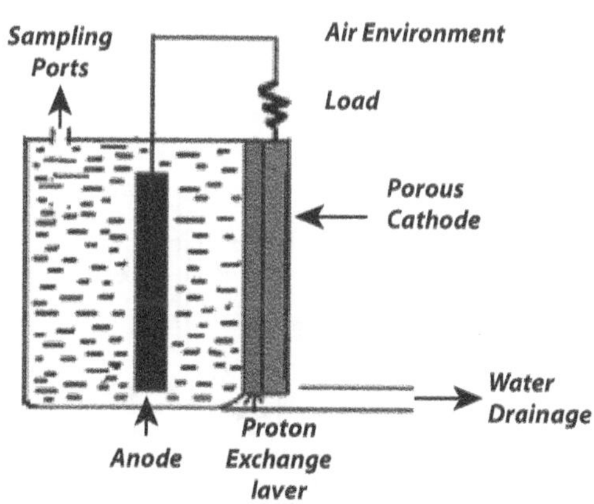

Figure 7.3 Schematic diagram of single chamber MFC. (Figure reproduced with permission from authors) [14, 15].

separate chamber for cathode. The porous electrode present in the cathode allows oxygen from the atmosphere to come in contact and protons diffuse through it. Due to simplicity of design, it is easier to scale up than double-chambered fuel cells, occupy less area and this is a subject of current research. Traditional carbon electrodes are used as an anode while for the cathodes either porous carbon electrodes or PEM bonded with flexible carbon cloth electrodes are used. Cathodes are always coated with graphite and dipped in electrolytes which act as catholytes and prevent the membrane and cathode from drying. Therefore water management or better fluid management is a critical step in such single-chambered fuel cells [15].

Two-Chambered MFC:
In this type of MFC (Figure 7.4), cathode and anode are placed in different compartments separated by a proton exchange membrane or a salt bridge which acts as a medium for transfer of protons to complete the electrical circuit. The membrane also prevents the anode from coming in direct contact with oxygen and other oxidizers. This type of MFC is generally operated in batch mode. These are used only for laboratory-scale purposes as their scale-up requires large volumes of fuel cells, which further reduce the efficiency of MFC, and their design is complex. Several modifications need to be made to this type of MFC before it can be used in a large-scale production of electricity and purification of wastewater. A double-chamber cathode is advantageous because cathode efficiency can be increased by adjusting the pH, purging pure oxygen, maintaining flow rate and also by adding electron mediators in the cathode chamber [15, 16].

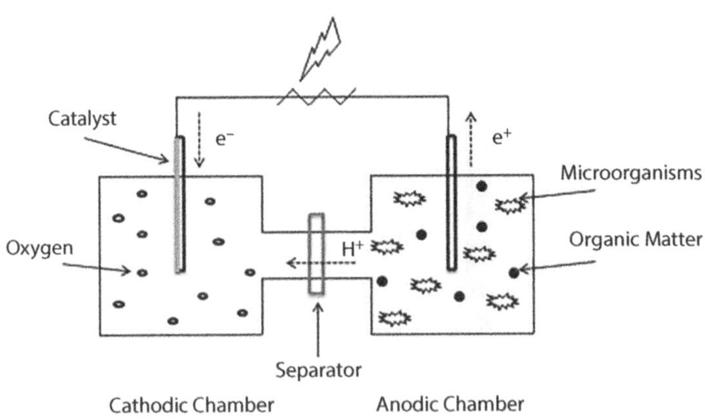

Figure 7.4 Schematic diagram of two-chambered MFC [16].

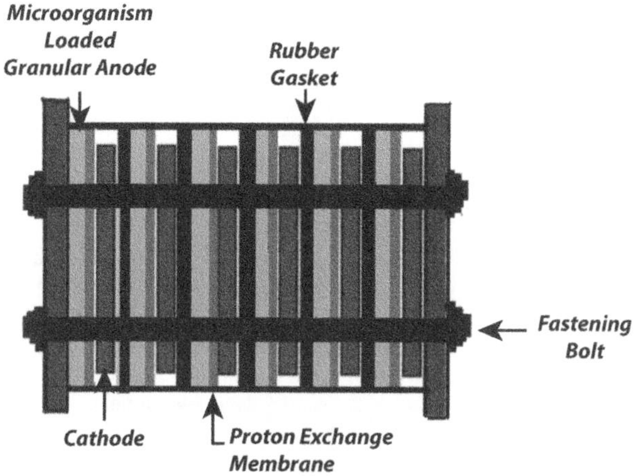

Figure 7.5 Schematic diagram of stack MFC. (Figure reproduced with permission from authors) [15].

Stack MFC:

These is another type of construction, wherein microbial fuel cells are stacked to form a battery of fuel cells as shown in Figure 7.5. This type of construction doesn't affect each cell's individual Coulombic efficiency; however, altogether it increases the output of overall battery to be comparable to normal power sources. This can be either stacked in series or parallel; both have their importance due to high power efficiency and can be practically utilized as a power source. Coulombic efficiencies differ greatly by the type of connection. When connected in parallel, efficiency is 6 times as compared to the series at the same flow rate. Moreover, in parallel arrangement, higher chemical reaction rate is allowed due to higher short circuit current. Thus, it is better to operate in parallel to maximize COD removal [17].

Upflow Mode MFC Systems:

It is a modification of a double-chambered MFC in such a way that continuous operation of MFC is possible and schematically shown in Figure 7.6. In fact, it is midway between a double- and single-chamber MFC. They are mediatorless and sometimes even membraneless. Its scale-up is easy and is used for generation of large quantities of electricity. Anode and cathode are at the bottom and top, respectively. The feed stream is allowed to enter the cell from the bottom and passes to the cathode chamber at the top through membrane. There are no separate anolyte and catholyte and the diffusion barriers between the anode and cathode provide a concentration gradient for proper operation of the MFCs.

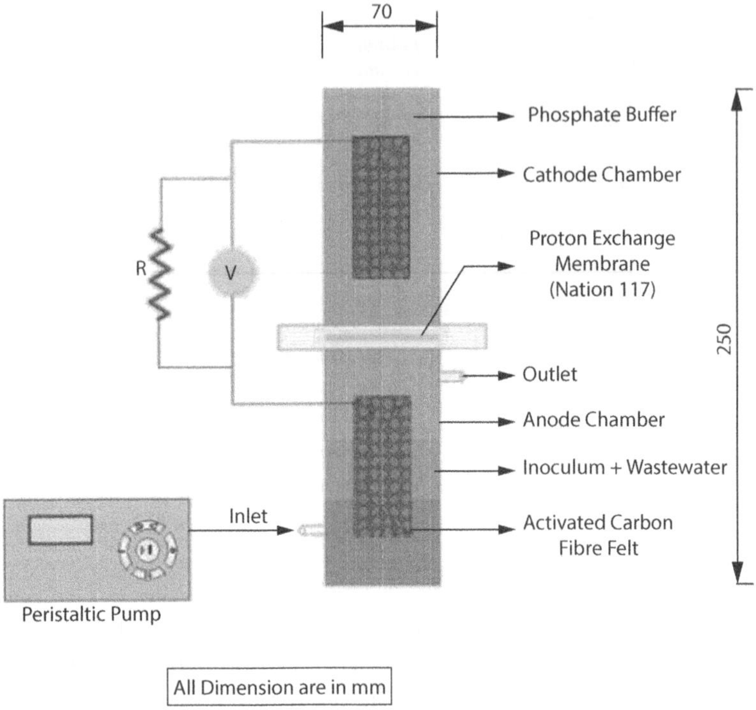

Figure 7.6 Schematic diagram of upflow MFC. (Figure reproduced with permission from authors) [18].

7.3 Recent Development in MFC Component

7.3.1 Recent Development in Cathode Used in MFC

In MFC, oxygen is converted into water or hydrogen peroxide at the intersection of the three phases, such as air (gas), electrolyte (liquid) and cathode electrode (solid). Low concentrations of oxygen in MFC at pH 7 and low temperature conditions mark high over potential, and was one of the limitations in MFC performance. To increase the reduction reaction, rate the catalyst is required to overcome the lower activity of carbon materials used in cathode. After an increase in catalytic activity at the cathode, a decrease in activation energy barrier is observed. Oxygen reduction reaction (ORR) can follow two different pathways: one is the 4-electron pathway which is a more preferred pathway than the other 2-electron pathway. The ORR in two different electron pathways in acidic medium can be expressed as:

$$O_2 + 4H^+ + 4e^- \rightarrow 2H_2O \tag{7.11}$$

$$O_2 + 2H^+ + 2e^- \rightarrow H_2O_2 \tag{7.12}$$

$$H_2O_2 + 2H^+ + 2e^- \rightarrow 2H_2O \tag{7.13}$$

Different materials which have been used for cathode preparation are discussed in Table 7.1 along with their advantages and disadvantages.

The performance of the electrodes profoundly depends on the kind of materials it is made of. The table indicates that different researchers have

Table 7.1 Materials used for the preparation of cathode.

Sr. no	Material	Advantage	Disadvantage	Ref.
1	Carbon paper	Increase in Coulombic efficiency with current density	Energy recovery was low	[19]
2	Carbon cloth	Increase in power production, energy recovery was greater	Poor bacteria adhesion and low EET efficiency, limits the MFC's power output	[20]
3	Fe-N-C	Open circuit potential was higher, it has better electro chemical stability, higher power density, potential much higher than AC	Resistance of fe-n-c cathode is decreased when compared to AC	[21]
4	Graphite plates	Excellent microbial adhesion performance, and good electrochemical reversibility	Low specific conductivity, r, bulk graphite is mechanically rigid and is difficult to get into specific shapes.	[22]

(*Continued*)

Table 7.1 Materials used for the preparation of cathode. (*Continued*)

Sr. no	Material	Advantage	Disadvantage	Ref.
5	Air cathode	Resistance was higher, better power output than Pt basing on oxygen reduction reaction	Lower power density than fe-n-c, potential was lower than fe-n-c, high cost, huge labor consuming and less accuracy	[21]
6	Carbon brush	Excellent electrochemical stability, low cost, and commercial availability	Poor bacteria adhesion and low EET efficiency, limits the MFC's power output	[23]
7	Platinum-carbon	pH, polarization is further reduced by embedding the electrodes	Decreases the ORR performance	[24]
8	Stainless steel	High electric conductivity and good inoxidizability, inexpensive and easy to process and connect	-	[22]
9	Graphite rod	Excellent electrochemical stability, low cost, and commercial availability	Poor bacteria adhesion and low EET efficiency, limits the MFC's power output	[23]
10	Platinum	High efficiency towards oxygen reduction	Expensive	[25]

(*Continued*)

Table 7.1 Materials used for the preparation of cathode. (*Continued*)

Sr. no	Material	Advantage	Disadvantage	Ref.
11	Iron ethylenediaminetetraacetic acid cathode	Significantly improved power density, 10% greater than plain activated carbon cathode, and also the power density is comparable to platinum catalyst-based cathode.	-	[25]
12	Granular activated carbon based packed bed air cathode	Highest power density, potential cost effective alternates to platinum air cathodes	Power generation was significantly dropped	[25]
13	Mesoporous nitrogen	Rich carbon cathode, less power production than platinum cathode	Less reduction compared to platinum cathode	[25]
14	Nitrogen doped ionothermal carbon aerogel as air cathode	Higher power density than usual platinum electrode	Efficient cost effective	[25]
15	VITO AC	High activity towards oxygen reduction.	-	[26]
16	Ferricyanide	-	Use of ferricyanide is not sustainable for long-term operation of MFC	[27]

(*Continued*)

Table 7.1 Materials used for the preparation of cathode. (*Continued*)

Sr. no	Material	Advantage	Disadvantage	Ref.
17	Abiotic cathodes	Highly favorable in the presence of a catalyst.		[27]
18	Mn_2O_3, Fe_2O_3, and heteroatom doped carbon	These are low cost and replaced by platinum cathode	-	[27]

explored a wide variety of materials for cathode to enhance the performance of MFCs. Platinum coated on carbon electrodes was the most common material used for cathodes in the microbial fuel cell. However, a major disadvantage of this material is the high cost associated with platinum. It comprises 50% of the total cost of MFC, therefore up-scaling is not economically feasible. Hence, other alternatives are being explored for cathodes. Mainly, carbon-based materials have been used for producing cathodes with the main importance being laid on biocompatibility and high specific surface area along with good conductivity. Air should be able to percolate through the pores of cathode to ensure oxygen is available for reduction. Many other modifiers such as Co-naphthalocyanine (CoNPc) have also been used to study the effect of modifiers [28] but none of them have produced power density higher than Pt/C MFC under ordinary conditions. CoTTP-based cathode gives better performance than Pt/C with respect to COD removal and power generation only in case of high concentration of methanol, which is detrimental to the stability of Pt/C cathode [29]. Another cathode material examined is stainless steel material (SSM) and on addition of Co_3O_4, higher proton transfer rate and a little bit lower power density are obtained as compared to Pt/C [30]. Graphene can also be added to carbon cloth, which forms a porous 3D structure around cloth with high surface area conducive to oxygen reduction. It has been a major advancement as it provides a much higher surface area than planar structure. Power density is 60% that of Pt/C cathode and it is relatively stable in the long run due to its porosity [31]. PANI/V_2O_5 catalyst on carbon electrode produces a power density higher than Pt/C but had low Coulombic efficiencies. PANI/Prussian Blue catalyst also has high power densities but poor Coulombic efficiencies and low current densities [32]. The application of CuPc/C as cathode is also promising having low internal resistance and high power and current densities along with a Coulombic efficiency of 29.3% which is quite high [33]. A blend of AC/CB has also been examined as a cathode catalyst which showed promising signs due to its stability and current density which is even greater than that produced by Pt/C at a potential of -0.1V [34]. This way, many catalysts have been added to carbon polymorphs (carbon cloth, graphite, activated carbon, carbon brush and carbon rods). Graphite and activated carbon have a higher specific surface area as compared to others. However, no electrode produced all characteristics similar to Pt/C or better than that.

Different transition metal oxides have emerged as alternative materials to manufacture electrodes over existing expensive material for oxygen reduction reaction. Recent discoveries also focus on use of nanofibers/nanotubes of carbon to increase surface area and enhance biocompatibility. Co3O4/nanocarbon composite was examined by Song *et al.* (2015) which almost matched the performance of Pt/C in all aspects like Coulombic efficiency, current at reduction peak and power density [35]. As of now, nanofibers of carbon can be expected to match the performance of Pt/C electrode or even perform better if suitable blending catalysts are found. Therefore, more work needs to be done for modification of carbon nanofiber-based cathode using a more efficient and cheaper catalyst for MFC. Application of nanofiber composites along with a cheap catalyst like cobalt, iron and manganese dioxide, silver and palladium looks like a promising cathode material, although more detailed work needs to be carried out before any claims can be made.

7.3.2 Recent Development in Anode Used in MFC

Choosing suitable electron material is the main and important parameter of MFC in order to carry out electron transfer and increase electrochemical efficiency. Anode is the heart of any MFC since it is the most crucial part of any MFC and as electricity generates. Important features such as high conductivity, high photochemical stability, and low resistance should be considered while selecting the anode material. The wastewater is treated by reducing the chemical oxygen demand (COD)/organic compounds along with the electricity and hydrogen production, and bio-film forms at the anode chamber. The anode material significantly impacts the biofilm formation and the electron transfer between the microorganism and the electron acceptor. Generally, the utilized non-metallic anode materials in an MFC involve carbon rods, carbon cloth, carbon fiber, carbon foam, graphite rod/plate, stainless steel mesh etc. [20, 24, 36, 37]. It was also traditionally made using platinum electrode coated on carbon; however, this is not viable for scale-up due to the huge cost of platinum. Metals cannot be used as anode materials as their smooth surfaces do not facilitate the growth of bacteria. Growth of bacteria facilitates creation of nanowires and direct electron transfer which help in conduction of electrons. Hence, almost all anodes are made up of polymorphs of carbon. Apart from anode material, other parameters such as substrate applied, microbial species, pH and mediators affect the overall performance of the MFC. A key concern in the designing of MFC is the very weak interaction of conventional carbon electrode. To overcome this and implement the MFC at commercial scale, cost-effective and reliable and modified anode materials are recommended. Modified material is considered to provide a relatively higher number of active sites which finally promotes the electron transfer. Table 7.2 shows the different kind of materials used for the preparation anode.

As indicated in Table 7.2, various modifications have been done and different catalysts have been added to examine the performance of anodes. Paul *et al.* found approx. 3.6 times enhanced efficiency when carbon felt anode modified with the grapheme oxide zeolite [41]. Another researcher utilized the modified anodic macroporous graphite carbon foam (MGCF) having polydopamine coating, resulting in increased electron transfer due to the improved interactions of microbe [42]. Zhou *et al.* (2012) used electrolytes as catalysts on carbon mesh, which increased stability and power density as well as COD removal efficiency [43]. Use of graphene on carbon cloth was studied by Liu *et al.* (2012) and Xiao *et al.* (2012) which decreased polarization resistance [31, 44]. Moreover, they found that crumpled graphene particles performed better than sheets due to development of 3D porous structure with higher surface area. Further (Yang-Chun Yong, Xiao-Chen Dong, Mary B. Chan-Park, Hao Song, 2012) added PANI to graphene foam which had much less resistance than carbon cloth. Moreover, a power density of 768 mW/m^2 is obtained which is 4 times that of carbon cloth [45]. Addition of chitosan to carbon improves biocompatibility of anode with bacteria as demonstrated by He *et al.* (2012) [46]. Current density is improved and charge transfer resistance decreases, consequently polarization resistance. Other catalysts such as Nitric acid/EDTA on AC electrode also were examined. All of them had higher power densities as compared with plain AC electrode [47, 48]. Chen *et al.* (2015) made a carbon felt anode with the pore size of anode according to bacteria which helps in better contact between bacteria and anode resulting in better performance and faster performance. Here, peak current was observed at an earlier stage [49].

Recently, the anode material is also being utilized made of modified carbon nanofibers. Activated carbon nanofibers have a high COD removal

Table 7.2 Materials used for the preparation of anode.

Sr. no.	Material	Advantage	Disadvantage	Ref.
1	Graphite rod	Good electrical conductivity and chemical stability, relatively cheap, and easy to get	Difficult to increase the surface area	[36]
2	Graphite fiber brush	Higher specific surface areas, easy to construct	Clogging	[37]
3	Carbon cloth	Large relative porosity, chemical stability	Expensive, limited surface area, limited active reaction sites	[20]
4	Carbon paper	Easy to connect wiring, low cost	Lack of durability, fragile	[19]
5	Carbon felt	Large aperture, good conductivity	Large resistance	[38]
6	RVC	Good electrical conductivity and plasticity	Large resistance, fragile	[39]
7	prepared from platinum-nickel alloy	pH, polarization is further reduced by embedding the electrodes	Decreases the ORR performance	[24]
8	Graphite plates	Excellent microbial adhesion performance, and good electrochemical reversibility	Low specific conductivity, r, bulk graphite is mechanically rigid and is difficult to get into specific shapes.	[22]

(Continued)

Table 7.2 Materials used for the preparation of anode. (*Continued*)

Sr. no.	Material	Advantage	Disadvantage	Ref.
9	3D-Nitrogen-doped porous carbon on carbon cloth	Electrocatalytic performance, good biocompatibility, unique electron properties	-	[40]
10	Binder-free 3D-carbon black/ stainless steel mesh composite	Higher current density, robust and suitable for upscaling ,low cost	-	[22]
11	Binder-free reduced graphene oxide & manganese oxide coated on carbon felt	Enhance the enrichment and growth of electrochemically active bacteria and facilitate the extracellular electron transfer	-	[23]
12	manganese oxide (MnO2) coated on carbon felt	Improve power out put	Reduced surface area, restricting bacterial attachment	[23]

efficiency and power density. However, they have a high internal resistance as well which may be due to film formation or large pore size. Power can further improve if this limitation is surpassed in future. Multi-walled carbon nanotubes (MWCNT), another modified electrode, have shown promising electricity generation of approx 49% by growth of microbial biofilm over unmodified electrode [50]. MWCNT-SnO2 coating on granular carbon electrode (GCE) was used which showcased that nano tubes improve stability, power and reduce charge transfer resistance [51–53]. Thus, the majority of materials used for electrodes are made up of carbon which have high resistivity. This resistivity can be minimized by applying edged metal collectors like Cu TiO2, Ni, Si, etc. Therefore the use of carbon nanofiber composites along with cheap catalyst Cu, TiO2, Ni, Si might be a promising anode material in MFC; however, more detailed research needs to be done in this respect.

7.3.3 Recent Developments in Membranes Used in MFC

The scope of MFC for commercial use is highly dependent on membranes which have low resistance, high selectivity and are cheap with long-term stability. Membranes are used in MFC to ensure transport of ions from one chamber to the other. Table 7.3 depicts the various membranes which have been investigated for their use in MFC.

Table 7.3 indicates that Cation exchange membranes (CEMs) are most widely used for this purpose with nafion being the most commonly used CEM. It has various advantages like enhanced proton transfer and high-power density due to low internal resistance. However, its major disadvantage is its high cost and low selectivity for H^+ ions. Moreover, pH control in the case of CEM is also an issue and the performance of CEM deteriorates with time. Specifically, the Polymer backbone associated with $-SO_3H$ group is stable, while that with -COOH group decomposed completely. Hence, the options to stabilize the membrane are the addition of ionomers or some of nano additives which can reduce the degradation. High ionic conductivities (1 S cm^{-1}) associated with the liquid KOH, phosphoric acid have been used along with the thick sheets of membranes such as Aromatic Sulphonic Acid sulfonated poly(sulfones), sulfonated poly (ether ketones) [62].

Biodegradable natural rubbers and earthen pots are also good candidates due to their high-power densities but they have their own limitations. For porous, non-woven cloth, the main problem is sustainable and constant production of electricity which is difficult due to the non-uniform size of fibres. Hence, the focus should be on the uniform production of current using non-woven cloth of uniform pore size. Moreover, electricity production has also been investigated by using materials like ceramic and earthen pot which provide less resistance [63]. This can be a major breakthrough if above low-cost membranes can be utilized to transfer of ions due to size difference. But the issue of selectivity still remains as it is. One can do this by modifying membranes, which is permeable only to protons. Furthermore, K^+, Na^+, Ca_2^+, NH_4^+, and Mg^{2+}cation species from waste water will cross the Nafion membrane like protons. As, the concentrations of these cation species is higher in MFCs than the proton concentration, these species get accumulated and pH of the solution into cathode chamber increases. Due to which efficiency of the MFCs decreases with reduced

Table 7.3 Materials used for the preparation of membrane.

Sr. no.	Material	Advantage	Disadvantage	Ref.
1	PVA- borosilicate	High oxygen diffusion	Lower proton conductivity than nafion borosilicate, low power density	[54]
2	PVA-Nafion-borosilicate (MPN)	Higher proton conductivity, maximum power density	Low oxygen diffusion, expensive	[54]
3	Microfiltration membrane	Reduce PH gradient across membrane, High internal resistance, power density & high percentage of COD removal	-	[55]
4	Proton exchange membrane	High internal resistance, power density & high percentage of COD removal	High cost	[55]
5	J- cloth	High Coulombic efficiency, using reactors with different configurations	Initially reduced the volumetric power density	[56]

(*Continued*)

Table 7.3 Materials used for the preparation of membrane. (*Continued*)

Sr. no.	Material	Advantage	Disadvantage	Ref.
6	Polyvinyl alcohol	Coulombic efficiency was 10% - 15% higher than glass fiber separators	Power was reduced if a PVA separator was used in higher conductivity solutions with either Pt or activated carbon catalysts	[57]
7	Monmorittonite	Had higher open circuit voltage, electrochemical performance	Reduced oxygen permeability	[58]
8	Triaminopyrimidine	The oxygen permeability, proton conductivity and hydrophilic behaviour	-	[59]
9	Activated carbon	Higher proton conductivity & coulombic efficiency	High operational cost	[60]
10	Sulfonated poly ether ether ketone (SPEEK)	Good hydrophilic properties and proton conductivity	Low power production	[61]
11	Nafion 117	High power density than SPEEK	Non-commercial due to its high cost	[61]

microorganism activity and decrease in the thermodynamic cell potential is observed. Many strategies were projected to solve the problem of the pH gradient in both the cathodic and anodic chambers, such as: (i) the use of pH-static control or other membrane materials such as cation exchange membranes (CEM), anion exchange membranes (AEM). The immobilization of ionic liquid (IL) in membranes could open up this field of improvement in MFCs.

Disinfectant, such as catholyte (liquid inside the cathode compartment) solution can be synthesized by using ion exchange membranes with ceramic chassis. The antagonism of the established biofilm on the anode electrode against any non-electroactive pathogenic organism results in the elimination of pathogens simply by exposure to the MFC environment.

7.4 MFC for Wastewater Treatment

Wastewater treatment using MFC was adapted in early 1991 [64]. As municipal wastewater has high organic molecules such as acetate, propionate, and butyrate, etc., it can be systematically converted into CO_2 and H_2O. The more organics in wastewater can be biodegraded incorporating hybrid of both anodophiles and electrophiles. MFCs can remove sulphides employing certain microbes due to the obligation of reducing sulphides in the wastewater [65, 66]. The treatment of wastewater by MFCs promotes the growth of bio-electrochemical active microbes, which makes the entire system operation smooth and stable. Concerning the scale-up of MFCs, either membraneless or single compartment in continuous flow MFCs are preferred for the wastewater treatment [67, 68]. The reactions at cathode and anode that take place during the treatment of wastewater are mentioned as:

Oxidation Reaction at Cathode:

$$\text{Glucose: } C_6H_{12}O_6 + 12H_2O \rightarrow 6HCO_3^- + 30H^+ + 24e^- E^0 = -0.429V \text{ vs. SHE} \tag{7.14}$$

$$\text{Glycerol: } C_3H_8O_3 + 6H_2O \rightarrow 3HCO_3^- + 17H^+ + 14e^- E^0 = -0.289V \text{ vs. SHE} \tag{7.15}$$

$$\text{Malate: } C_4H_5O_5^- + 7H_2O \rightarrow 4H_2CO_3 + 11H^+ + 12e^- E^0 = -0.289 \text{ V vs. SHE} \quad (7.16)$$

$$\text{Sulfur: } HS^- \rightarrow S^o + H^+ + 2e^- \; E^0 = -0.230 \text{ V vs. SHE} \quad (7.17)$$

Reduction Reaction at Anode:

$$O_2 + 4H^+ + 4e^- \rightarrow 2H_2O \; E^0 = +1.230 \text{ V vs. SHE} \quad (7.18)$$

$$O_2 + 2H^+ + 2e^- \rightarrow H_2O_2 E^0 = +0.269 \text{ V vs. SHE} \quad (7.19)$$

$$NO_3^- + 2e^- + 2H^+ \rightarrow NO_2^- + H_2O \quad E^0 = +0.433 \text{ V vs. SHE} \quad (7.20)$$

$$NO_2^- + e^- + 2H^+ \rightarrow NO + H_2O \quad E^0 = +0.350 \text{ V vs. SHE} \quad (7.21)$$

$$NO + e^- + H^+ \rightarrow 1/2N_2O + 1/2H_2O \; E^0 = +1.175 \text{ V vs. SHE} \quad (7.22)$$

$$1/2H_2O + e^- + H^+ \rightarrow 1/2N_2 + 1/2H_2O \; E^0 = +1.355 \text{ V vs. SHE} \quad (7.23)$$

$$2NO_3^- + 12H^+ + 10e^- \rightarrow N_2 + 6H_2O \; E^0 = +0.734 \text{ V vs. SHE} \quad (7.24)$$

$$Fe^{3+} + e^- + H^+ \rightarrow Fe^{2+} + 1/2H_2O \; E^0 = +0.773 \text{ V vs. SHE} \quad (7.25)$$

$$MnO_2 + 4H^+ + 3e^- \rightarrow Mn^2 + 2H_2O \; E^0 = +0.602 \text{ V vs. SHE} \quad (7.26)$$

7.4.1 Advantages of MFC Over Conventional Treatment

MFC is a promising technology realizing its unique advantages and sustainability in the wastewater treatment. MFC is capable of directly producing electricity from the organic matter in wastewater with a high energy conversion rate with low energy consumption, whereas the conversion of biogas (e.g., CH4 or H2) into electricity causes a significant energy loss of more than 60% [69, 70]. It was estimated that approx. 0.024 kW or 0.076 kWh/kg-COD (mainly) is consumed in MFCs especially for reactor feeding and mixing than the energy approx 0.3 kW or 0.6 kWh/kg-COD consumed during the activated sludge aerobic treatment of wastewater [5, 71]. Additionally, a broad variety of organic molecules available in the wastewaters can be used as substrates in MFC to produce electricity and pay off the treatment cost [72]. It is an energy- saving technology, which is not due to the need for aeration or temperature maintenance, and it generates 50-90% less sludge to be disposed off compared to conventional activated sludge processes [73–75]. This reduces the pollution and if implemented commercially, provides the dual benefit of safe and clean energy for people and the effective treatment of the wastewaters with negligible noise and emissions. Figure 7.7 below depicts the advantages of MFC over conventional processes.

7.4.2 Challenges in the Wastewater Treatment Using MFC

Despite the theoretical advancement in MFCs, its implementation at commercial scale is far from reality, which is mainly due to the relative low power production level of MFCs at larger scales. All the challenges in wastewater treatment using MFC are shown in Figure 7.8. As the milliliter-scale fuel cell size increased to liter-scale, the power density decreased substantially < 35 W/m^3, even though the 30 mL MFC generates approximately 2.87 kW/m^3 with a cloth electrode assembly configuration, making it very difficult to realize the energy and economic revenues [76, 77]. For real-world application, MFCs should produce 400 W/m^3 to be competitive with conventional anaerobic digestion [78] and to have an output of 1 kW/m^3 at an organic loading rate of 10 kg COD/(m^3 ·d) to achieve self-sufficient

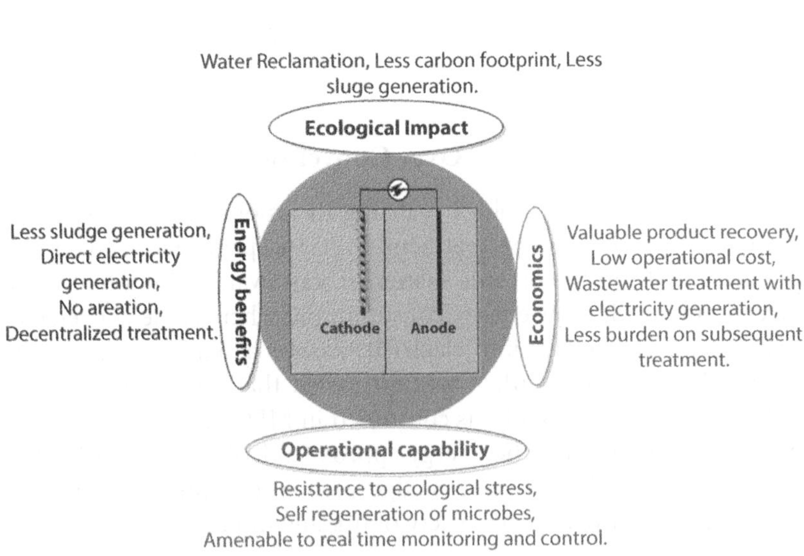

Figure 7.7 Advantages of MFC in wastewater treatment.

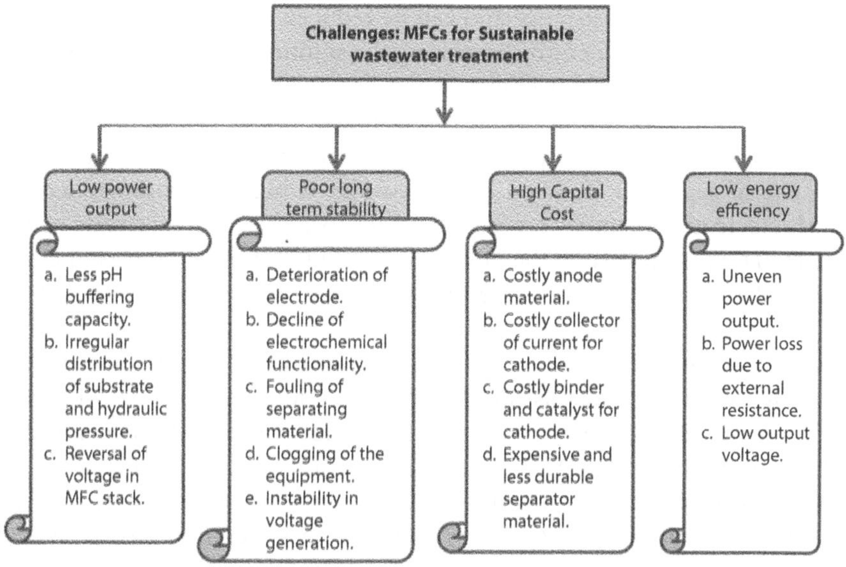

Figure 7.8 Challenges in the up-scaling of MFCs in wastewater treatment.

energy [79]. It is recommended to improve the power production level of MFC as there is much scope of research for improvement in technology. Other factors, such as high capital cost, low power harvesting efficiency and poor long-term system stability are also challenging the real-world application of MFC. In addition to low power generation, there are additional factors such as poor performance of air cathode, life cycle impact, low pollutant degradation rates and huge cost involved in electrodes and membrane material of MFCs in wastewater treatment as compared to conventional processes. The following sections will discuss the current status, recent developments and the desired attributes of an ideal MFC to make this intriguing technology an exciting reality.

7.5 Different Ways for Increasing the Throughput of MFC

7.5.1 Big Reactor Size

During the scale-up of MFC from laboratory scale to higher volume, the power output of MFC is reduced considerably 2-4 orders of magnitude as compared to laboratory scale [80–82]. This reduction in power is observed due to the internal resistance and can be reduced by decreasing the space between two electrodes or by increasing the solution conductivity [80, 83]. However, reduction in power output can also be observed when electrodes are placed too closed which increases changes of short circuit and oxygen diffusion towards anode. To avoid this and to keep internal resistant low, the use of separator between two electrodes is highly recommended. A cloth electrode assembly can also be used to reduce the electrode spacing by 0.6 mm increasing the power generation. But over the period cloth electrodes can be degraded due to microorganism present in the system affecting the stability of the system. To overcome this, a non-biodegradable low-cost separator needs to be developed with low oxygen permeability and high electron transmission. Power generation in large-size MFC can be increased due to increase in solution conductivity and balancing the pH of solution. In 20L MFC, cathode performance can be increased by aerating catholyte solution with pure oxygen instead of air and increasing the flow rate resulting in the increase in power generation by 144 W/m^3 [77]. Domestic and industrial wastewater are generally found to be poor conductive and pH buffered; this makes it critical to operate and maintain the high performance of MFC. To overcome this, additives such as NaCl, bicarbonate and phosphate buffered saline can be added to MFC and are found to work efficiently in lab scale setup. However, in large-scale MFC addition of chemicals will not be sustainable for real-world wastewater treatment. Use of close electrode assembly to reduce high internal resistant caused by low conductivity of solution and increasing the flow rate by stirring is highly recommended to offset the drawbacks while using actual wastewater [4].

7.5.2 Stacking

There are different types of MFCs based on geometry, the electrode separator and cathode type. Nowadays many MFC designs that have been investigated for better power generation have their own advantages and disadvantages. For large-scale energy generation by treating wastewater, the MFC model used to be placed in parallel or in series should be simple and use cost-effective materials. The MFCs give higher current and higher voltage when designed in parallel and in series, respectively [4]. Aelterman *et al.* have designed the MFC system by connecting six individual MFCs in parallel. The output current is equal to the sum of all the six individual MFCs while output voltage is the same as individual MFC. Ieropoulos *et al.* and Wang and Han *et al.* also reported that the maximum power density is higher when MFCs are connected in parallel. When MFCs are connected in series, the output voltage is equal to the sum of all the individual MFC connected in series in ideal case; however, efficiency of MFC is experienced to be as low as 38-41% due to voltage reversal and ionic short circuits [84, 85]. Incorporating air cathode of high performance in MFC can help to reduce voltage reversal. A substantial development in electrical array to design MFCs in parallel or in series is required to obtain maximum power output and constant stability of stacking system [86].

7.5.3 Cathode

The cost involved in the cathode electrode contributes approx 45-75% of the capital investment in MFC [73]. The economic viable cathode material is to use promptly available oxygen in air as air cathode or terminal electron acceptor and doesn't require solution aeration. The cost of air cathode can be reduced by developing low-cost modern collection materials, catalysts, dispersion sheet materials and binders. Presently, a cost-effective and highly conductive collection material is the metal mesh, like nickel foam and stainless steel mesh. Poly-dimethylsiloxane (PDMS) is an affordable catalyst binder over costly Nafion and polytetrafluoroethylene (PTFE) is used as low-cost oxygen distribution sheets. Instead of using highly expensive Pt metal, cheaply available catalysts comprising affordable metals, viz. Co/Fe/N/CNT, Co-OMS-2, β-MnO2, MnPc, MnOx, CoTMPP can be recommended, with special emphasis on the low-cost activated carbon having more catalytic activity [87]. Nevertheless, a detailed study is required to understand the mechanism of oxygen reduction catalyzed by activated carbon [88]. The choice and design of cathode material is very tricky as the complex three-phase (gas-liquid-solid) interface involved for oxygen reduction, while difficulty in the reduction of the cost of cathode.

7.5.4 Anode

Graphite fiber brush is one of the best suitable structures as an anode material, which is produced by graphite fibers containing non-corrosive metals such as stainless steel, titanium, etc., as the core material. An activated carbon (AC) granule is the best alternative anode material when combined with metal mesh. In anode chamber, AC is stacked and provides higher specific surface area for electricity production.

7.5.5 Separating Material

The separator is the second expensive component that needs to be taken care of while designing MFC. Liu and Logan *et al.* (2004) have shown in their study that small-size MFC without a separator lowers the internal resistant and reduces the cost. Also, for large-size MFC, it is necessary to use a separator for close electrode assembly thereby reducing internal resistant and blocking the transfer of oxygen to the anode [69]. Expensive membranes such as ion exchange membrane, ultra filtration membranes and forward osmosis membranes should be avoided for the construction of large-size MFC. Non-woven cloth is a favorable inexpensive separator material [76, 89] but its mechanical strength and long-term stability should be improved. From future perspective, various studies need to be conducted for separating material to increase its mechanical strength, long-term stability and it should be cost effective for large-size application of MFC.

7.5.6 Harnessing Output Energy

Efficient harvesting of energy is the main challenge in scaling up of MFCs. Due to the low voltage and current levels, there is a difficulty in handling the practical load in MFC [90]. Therefore a power management system (PMS) needs to be combined with MFC to make energy available for energy devices such as wireless sensors to monitor the environment [91, 92]. DC/DC voltage boost convertors have been proved to harvest energy from MFCs by high frequency switching action. The optimization of convertors and development of efficient power output tracking devices is highly recommended for future perspective [93, 94]. Capacitors used in the design of MFC can be able to extract and store energy unless its desired amount is stored and then released for further applications. Donovan *et al.* (2011) has reported that capacitor connected to sediment microbial fuel cell (SMFC) can be able to extract 2.5 W of energy, enough to charge the wireless sensors. Kim *et al.* (2011) showed that voltage reversal and increasing internal resistance created drawbacks when MFCs connected in parallel, which can be nullified by connecting capacitors to MFC. Capacitors connected to MFCs can affect the stability of the system, as it passively captures the energy. It is beneficial to develop the best suitable capacitors, optimization of electrode materials and suitable combination of capacitor for MFC.

7.5.7 Increasing Long-Term Stability

Long-term stability is an important factor for constant energy production and for an economically viable system. Many studies have reported that the performance of MFC decreases along the duration of operation [92, 95–97]. This can be observed due to reasons such as fouling, degradation of separating material, decrease in cathodic performance, loss of electrochemical activity of anodic biofilm and clogging due to the solid particles present in wastewater.

7.5.8 Coupling of MFC with Other Techniques

There are many factors affecting the performance and scale-up activity of MFCs as mentioned above. To overcome these challenges, it is advisable to combine MFC with other processes to have a synergistic and sustainable system. The output of this combination can be of high effluent quality with low COD, phosphate, nitrogen, nutrient and salinity content, whereas another output can be a value-added chemical with great demand in the market. In order to recycle and reuse water from effluent treatment process, the untreated effluent should have low organic matter, low salinity and low inorganic ions. MFC is a good choice to treat this kind of low- to medium-strength wastewater; however, wastewater containing high organic matter can be non-biodegradable by microorganisms present in MFC system. Therefore, depending upon the properties of recycle water required the MFC system can be combined with the other processes. MFC in combination with anaerobic digestion (AD) system is used to treat the high organic wastes from brewery and winery industries [98–100]. Anaerobic digestion systems hydrolyses and ferments the substrates in high-strength wastewater into simple volatile fatty acids with methane and hydrogen gas as by-products and further MFC treat these volatiles and produce electricity [100–102]. MFC coupled with electro dialysis can be used to treat high salinity industrial wastewater together called as micro desalination cells (MDCs). An MDCs system generates energy by removing the ions from wastewater. Apart from this MFCs can be combined with electrochemical technology to produce high valuable chemicals such as H_2, CH_4, H_2O_2, acetate, and ethanol along with electricity generated during the process [103–106].

7.6 Different Case Studies Indicating Commercial Use of MFC

Recently many researchers have focused on the scale-up studies of MFCs and are still counting [107]. Wu *et al.* (2016) in their study reported that a 72-L pilot-scale stacked MFC can be successfully demonstrated with granular activated carbon packed bed electrodes to get higher efficiency of COD removal with synthetic wastewater. Also, Ge *et al.* (2015) developed a 200-L MFC system for the treatment of real municipal wastewater [108, 109]. Zhang *et al.* (2013) incorporated 2 L tubular MFCs to the aeration tank for the wastewater treatment in order to reduce aeration and infrastructure while Feng *et al.* (2014) utilized a 250-L MFC to treat municipal wastewater for one year as a stand-alone system [71, 109]. Also, (Hiegemann *et al.*, 2016) integrated a 45-L pilot MFC system for nine months to treat organic wastes from primary clarifier in a municipal wastewater treatment plant. In all the above cases higher COD removal is observed; however, it also experiences the low power generation and higher cost of MFC electrodes which opens opportunities for development and scale-up studies of MFC in future perspective [110–112].

An important criterion for the sustainable wastewater treatment is good effluent quality. The effluent from the treatment process should be low in organic matter, salinity, and inorganic nutrient ions to satisfy water reclamation and reuse criteria. In general, MFCs are excellent at handling the relatively simple composition of medium- and low-strength wastewater. However, some real wastewater can contain high concentrations of organic matter, many of which are not degradable by microorganisms in MFCs. In certain cases, stringent discharge laws require a very low concentration of phosphate and nitrogen ions in wastewater effluent, which is difficult to achieve by MFC alone. MFCs can be combined with different conventional processes and integrated with different traditional processes depending on the properties of wastewater and the recovery requirements. For high-strength wastewater, such as brewery and winery wastewater, the combination of MFC with anaerobic digestion (AD) technique should be an attractive synergistic mode [98, 99]. First, when producing methane or hydrogen gas, the AD reactors hydrolyze and ferment the complex substrates and particles in high-strength wastewater into more functional substrates (mainly volatile fatty acids). In order to further polish the consistency of water and generate electrical energy, MFCs subsequently degrade the remaining organics in AD effluent. This synergistic mechanism of AD-MFC has been successfully demonstrated in several studies [102]. Sharma and Li (2010), for example, combined MFCs with an anaerobic biofermentor producing hydrogen (HPB) to produce hydrogen and electricity from glucose wastewater simultaneously. This synergistic scheme could achieve a maximum of 2.85 mol H2/mole glucose hydrogen output and a 4200 mW/m3 MFC power density, with a total energy recovery efficiency of 29% (559 J/L) and a 97% COD removal efficiency. It would be interesting to combine MFCs with the electrodialysis technique to form microbial desalination cells (MDCs) for industrial wastewater with high salinity [101, 103–105], which can simultaneously eliminate ionic species and produce electricity from wastewater. By inserting one or more pairs of anion exchange membrane (AEM) (next to the anode) and cation exchange membrane (CEM) (by the cathode) into an MFCC, an MDC is built [103].

Wastewater for food production is abundant in readily biodegradable carbohydrates and organic acids with relatively low organic nitrogen concentrations (e.g., proteins) [113, 114]. Depending on the BOD and the amount of water used in the process, MFCs can generate electricity within the range of 2e260 kWh/ton of the wastewater product used to process food products. A total of 46 MW of electricity can theoretically be generated from wastewater from low BOD wastewater from milk dairy farms in the United States, while a maximum of 1960 MW of electricity can be generated from high BOD wastewater from the milk industry [115]. Animal wastewater from the livestock-related industry is also especially high in organic matter content (approximately 100,000 mg/L COD for animal waste) and can contain high levels of nitrogen-containing components, such as proteins, and organic materials, such as cellulose, that are more difficult to degrade [116]. Because of its high carbohydrate content (high energy density) and low ammonium nitrogen concentration, it may also be an ideal substrate for MFCs. Feng *et al.* (2008) studied the treatment of beer brewery wastewater using air cathode MFC and obtained a maximum power density of 528 mW/m^2 when the wastewater was applied to the 50 mM phosphate buffer. In this case, when both wastewaters were compared at comparable strengths, the maximum power produced by brewery wastewater was lower than that obtained using domestic wastewater [98].

Municipal wastewater typically has lower BOD concentrations of less than 300 mg/L, known as low-energy density carriers or MFC feedstocks. However, due to the anaerobic conditions in the anode chamber, MFCs are also capable of handling high strength wastewater (high energy density) with BOD concentrations above 2000 mg/L. From the food processing industry, breweries, dairy farms and animal feeding operations and other industrial waste streams, these high-strength wastewater sources are

Table 7.4 Different sources of wastewater used in MFC.

Sr. no.	MFC details (type, anode, cathode and membrane)	Type of wastewater	%degradation (COD or TOC reduction)	Power density or current density (W/m2)	Ref.
1	Single chamber, Air-Cathode, Ammonia treated graphite anode	Domestic wastewater	2580.00%	422 mW/ m2, 12.8 W/ m3	[37]
4	Dual Chamber	Chemical wastewater	62	65.82W/m^2	[117]
5	Single chamber, Air-Cathode	Starch wastewater	98%	2239 W/m^2	[118]
6	Single chamber, Air-Cathode	Swine manure slurry (i) non-diluted	27	261 W/m^2	[119]
		(ii) diluted manure slurry	92	110 W/m^2	
7	Single-chamber down flow	Landfill leacheat	57 (BOD)	0.19 W/m^2	[120]
8	Two-chamber MFC, Cathode & Anode : Granular Graphite	Ago food waste water	80	27.3 W/m^3	[121]
10	Single-chamber mediatorless MFC	Composite vegetable waste	62.86	26.41 W/m^2	[122]

(*Continued*)

Table 7.4 Different sources of wastewater used in MFC. (*Continued*)

Sr. No.	MFC details (type, anode, cathode and membrane)	Type of wastewater	%degradation (COD or TOC reduction)	Power density or current density (W/m2)	Ref.
12	Single-chamber air cathode MFC, Cathode: Carbon cloth with platinum, Anode: Carbon brush with titanium	Sea food wastewater	90	570 mW/m^2	[123]
13	Single-chamber MFC, graphite anode and carbon cloth	Dairy wastewater	91	20.2 W/m^2	[124]
15	Single-chamber air cathode and anode graphite fiber MFC	Olive mill wastewater	65	-	[125]
16	Single cell MFC, Cathode carbon cloth	Brewery wastewater	72.84	124.35 mW/m^2	[126]
18	Two-chamber MFC, Cathode & anode: graphite plates	Distillery wastewater	63.5	63.8 mW/m^2	[127]
19	Two-chamber MFC, Cathode: Carbon paper, Anode: Nickel coated sponge	Acids from Fermentation	75.59	543.75 mW/m^2	[128]

created. Table 7.4 shows the efficiency of current/power density in MFCs using different sources of wastewater as fuel in the anode section.

Table 7.4 indicates that treating industrial waste streams such as brewery and agrochemical, domestic wastewater, with higher concentrations of organic matter, could be more suitable for MFC technology and required to study scale-up opportunities to treat such waste streams [129]. Another study tested a low-cost MFC system treating brewery wastewater for almost 1 year showed up to 98% COD removal [42] which might be associated to low flow rate and high hydraulic retention time (HRT).

7.7 Other Applications of MFC

Figure 7.9 shows the different applications of MFC's currently used in particular sector. MFC is a fantastic technology used mostly for bio-electricity generation from bio-degradable matter using bacteria and in the production of bio-hydrogen from electrolysis of the organic matter by the bacteria. The basic idea is that the potential generated at the anode, in a typical MFC, can be augmented with an additional voltage to generate hydrogen gas at the cathode. MFC technology is used as biosensor for analyzing pollutant and monitoring the processes [130]. The toxic content in water can be easily detected by monitoring the perturbations in the electric current generated by MFCs. It is also applicable that MFCs can be used for powering electrochemical sensors as batteries having limited lifetime. It is

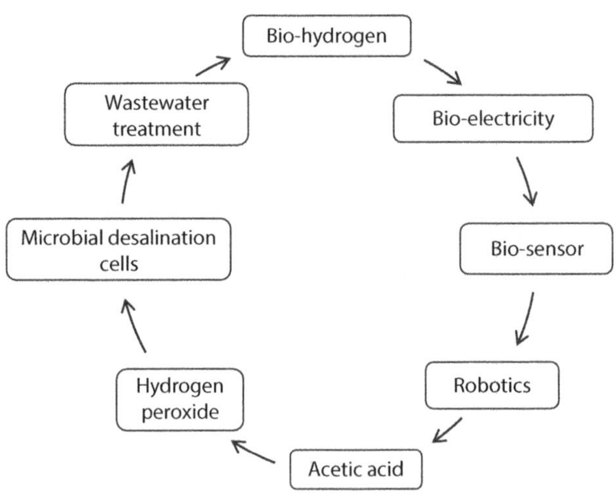

Figure 7.9 Application of MFCs.

probable to use MFCs as biological oxygen demand (BOD) sensors and it is exhibited that this type of BOD sensor has an excellent operational sustainability and reproducibility and can be kept operating for 5 years [51]. Another major application of MFC is the wastewater treatment, which is a different approach for wastewater treatment as it attempts at extracting energy out of wastewater, rather than expending energy to treat it. The desalination of wastewater can also be performed. It has also emerged as the secondary fuel production, i.e., the production of H_2 and methane by microbial electrolysis in modified MFC with better exterior potential at the cathode terminal. An environment benign solvent, hydrogen peroxide (H_2O_2) can be produced economically at cathode and even acetic acid co-produced with electricity using MFC technology. MFC is being applied in robotics as the main source of energy in the diverse generations of robots. The chemical fuel cell was used to charge Ni-Cd batteries for the world's first robot, i.e., Wilkinson's Gastronome (Chew-chew) in 2000. MFC is used to produce biomass such as an algae Chlorella vulgaris, acts as the biological electron acceptor at the cathode while reducing CO_2 to biomass using a mediator resulting in the generation of bioelectricity. It is also being explored for bioremediation (electrosorption, electrodeposition, and electrocoagulation). MFCs engage microbial catabolic activities to produce electricity from a wide range of complex organic waste such as domestic wastewater, lignocellulosic biomass, brewery wastewater, starch processing wastewater, and landfill leachates.

7.8 Conclusions and Recommendations (Future Work)

Microbial fuel cell technology has a large scope for comprehending the sustainability aspect of wastewater treatment. At instance many large-scale applications were under development for real-world wastewater. Scientists and engineers around the world are making hard efforts to develop large-scale MFCs and to deal with the limitations of scaling up of MFCs, such as the low power output, high capital cost, low energy harvesting efficiency, and poor system stability. MFCs in combination with other techniques for good quality effluent and higher energy recovery is under investigation and the challenges in the electricity production efficiency, efficiency of wastewater treatment, and economic feasibility requires more detailed studies for its practical application. Combined efforts in the area of reactor design, materials, and optimization of systems are required to develop sustainable wastewater treatment method using MFCs.

In the MFC sector, numerous challenges for successful deployment in real environments remain unresolved, although some initiatives have been reported. The great challenges that need to be addressed include low power performance, high capital cost, and other device limitations. With synthetic substrates and model microbial catalysts or microbial consortia developed in the laboratory, remarkable progress has been achieved starting from the anode. However, implementation study problems are inherently more difficult when it comes to operating in more dynamic settings, such as actual industrial waste or natural ecosystems. As an example, many forms of organic waste have been used as microbial anode fuel, but the kinetics of electroactive bacteria remain poor and the interaction between electrodes and bacteria has not yet been completely understood. In addition, interaction and/or co-existence between bacteria and solid electrodes in electron transfer mechanisms is not well explained, especially in complex environments, where a multitude of microbial species (electroactive or not) can be found on the electrodes. Finally, the electrode attraction of microbial cells, bio-film formation and anode surface development, contact and inter-species cooperation. Another problem is the low energy provided by MFCs, which is currently lower in magnitude compared to chemical fuel cells. By combining MFCs with external off-the-shelf harvesting systems based primarily on supercapacitors, the harvesting and management of the low power produced by MFCs has given rise to new hybrid systems which partially address this issue.

In particular, we address the difficulties and challenges of scaling up the MFC systems. This is the most important issue for realizing this technique's functional implementation. MFCs may also be combined with other techniques to achieve high effluent efficiency or to extract more economic value (i.e., by producing energy-rich or high-value chemicals) from wastewater in order to achieve sustainability. However, research in this field is still ongoing, and before real-world implementation, many issues need to be solved. Progress in terms of performance, economic viability, system stability and reliability are required. For the economical generation of electricity, the high surface area electrode and cheap construction material of electrodes should be preferred in MFC. The treatment of actual wastewater has been realized to be more focused on process development studies in the future research activities. A proper attention is sought for understanding electron transfer losses and mechanism of the electron release transfer acceptance by emergent novel configurations in cathode-anode building. Focus should also be given to more in-depth analysis on the life cycle of MFCs to find out the critical regions of new developments. The urgent need of the hour is large-scale system development and demonstration and forthcoming challenges and restriction should be addressed in a systematic manner. Use of carbon nanofiber composites along with cheap catalyst Cu, TiO2, Ni, Si looks like a promising anode material. The use of expensive catalysts for the cathode must also be avoided. For the sustainable development of MFC, algae-based cathode electrode can be economically practiced. Oil-algae-based electrodes can be grown to have an extra advantage of oil recovery and greenhouse gas sequestration. Therefore, design and optimization of process parameters should be performed for the performance enhancement of MFC reactor. Moreover, the application of algae species can increase the advantages such as bio-diesel production of MFC. A different decisive feature is the separation of non-carbon-based substrates from the waste streams: nitrogen, sulfur, and phosphorus-containing compounds often cannot be discharged into the environment at influent concentrations. Similarly, even particulate organic compounds will need to be removed and converted to easily biodegradable compounds, as part of an effective wastewater treatment operation. The future of MFC technologies is full of exciting opportunities and challenges and demands greater discoveries and advancements in science and engineering aspects to overcome several barriers and to develop sustainable energy harvesting wastewater treatment technology based on bioelectrochemical principles.

References

1. Garg P. Energy Scenario and Vision 2020 in India. *J Sustain Energy Environ* 2012;3:7–17.
2. Meier PJ. Life-cycle assessment of electricity generation systems and applications for climate change policy analysis 2002.
3. Unicef P. Prakash , Anjal , Medhavi Sharma and Jayati Chourey. 2013. Water in India: Situation and 2015.
4. Wei-feng Liu and Shao-an Cheng. Microbial fuel cells for energy production from wastewaters: the way toward practical application. *J Zhejiang Univ Sci A* 2014;15:841–61. https://doi.org/10.1631/jzus.A1400277.
5. McCarty PL, Bae J, Kim J. Domestic wastewater treatment as a net energy producer-can this be achieved? *Environ Sci Technol* 2011;45:7100–6. https://doi.org/10.1021/es2014264.
6. Logan Bruce. *Microbial Fuel Cells.* Wiley; 2007. https://doi.org/10.1002/9780470258590.
7. Dr. Priyom Bose. Recent Developments in Microbial Fuel Cells (MFCs). AZO Cleantech n.d. https://www.azocleantech.com/article.aspx?ArticleID=1085#:~:text=Global Market Outlook of MFCs&text=It accounted for %248.60 million,to approximately treble by 2026.

8. Verstraete W, Rabaey K. Critical Review Microbial Fuel Cells: Methodology and Technology † 2006;40:5181–92. https://doi.org/10.1021/es0605016.
9. B E Logan. *Microbial Fuel Cells*. Wiley & Sons, New Jersey; 2007.
10. Potter MC. Electrical effects accompanying the decomposition of organic compounds. *Proc R Soc London Ser B, Contain Pap a Biol Character* 1911;84:260–76. https://doi.org/10.1098/rspb.1911.0073.
11. Stoll Z, Dolfing J, Xu P. Minimum Performance Requirements for Microbial Fuel Cells to Achieve Energy-Neutral Wastewater Treatment. *Water* 2018; 10:243. https://doi.org/10.3390/w10030243.
12. Bullen RA, Arnot TC, Lakeman JB, Walsh FC. Biofuel cells and their development. *Biosens Bioelectron* 2006;21:2015–45. https://doi.org/10.1016/j.bios.2006.01.030.
13. Akman D, Cirik K, Ozdemir S, Ozkaya B, Cinar O. Bioelectricity generation in continuously-fed microbial fuel cell: Effects of anode electrode material and hydraulic retention time. *Bioresour Technol* 2013;149:459–64. https://doi.org/10.1016/j.biortech.2013.09.102.
14. Parkash A. Microbial Fuel Cells: A Source of Bioenergy. *J Microb Biochem Technol* 2016;8. https://doi.org/10.4172/1948-5948.1000293.
15. Shaikh S, Patil NP. Microbial Fuel Cell : Design and Operation Research and Reviews : *Journal of Microbiology and Biotechnology Microbial Fuel Cell: Design and Operation. J Microbiol Biotechnol* 2016:1–8.
16. Ortiz-Martínez VM, Salar-García MJ, de los Ríos AP, Hernández-Fernández FJ, Egea JA, Lozano LJ. Developments in microbial fuel cell modeling. *Chem Eng* J 2015;271:50–60. https://doi.org/10.1016/j.cej.2015.02.076.
17. Nevin KP, Richter H, Covalla SF, Johnson JP, Woodard TL, Orloff AL, *et al.* Power output and columbic efficiencies from biofilms of Geobacter sulfurreducens comparable to mixed community microbial fuel cells. *Environ Microbiol* 2008;10:2505–14. https://doi.org/10.1111/j.1462-2920.2008.01675.x.
18. Tamilarasan K, Banu JR, Jayashree C, Yogalakshmi KN, Gokulakrishnan K. Effect of organic loading rate on electricity generating potential of upflow anaerobic microbial fuel cell treating surgical cotton industry wastewater. *J Environ Chem Eng* 2017;5:1021–6. https://doi.org/10.1016/j.jece.2017.01.025.
19. Kim J, Jung S, Regan J, Logan B. Electricity generation and microbial community analysis of alcohol powered microbial fuel cells. *Bioresour Technol* 2007;98:2568–77. https://doi.org/10.1016/j.biortech.2006.09.036.
20. Ishii S, Watanabe K, Yabuki S, Logan BE, Sekiguchi Y. Comparison of Electrode Reduction Activities of Geobacter sulfurreducens and an Enriched Consortium in an Air-Cathode Microbial Fuel Cell. *Appl Environ Microbiol* 2008;74:7348–55. https://doi.org/10.1128/AEM.01639-08.
21. Pan Y, Mo X, Li K, Pu L, Liu D, Yang T. Iron–nitrogen–activated carbon as cathode catalyst to improve the power generation of single-chamber air-cathode microbial fuel cells. *Bioresour Technol* 2016;206:285–9. https://doi.org/10.1016/j.biortech.2016.01.112.

22. Zheng S, Yang F, Chen S, Liu L, Xiong Q, Yu T, *et al.* Binder-free carbon black/stainless steel mesh composite electrode for high-performance anode in microbial fuel cells. *J Power Sources* 2015;284:252–7. https://doi.org/10.1016/j.jpowsour.2015.03.014.
23. Zhang C, Liang P, Yang X, Jiang Y, Bian Y, Chen C, *et al.* Binder-free graphene and manganese oxide coated carbon felt anode for high-performance microbial fuel cell. *Biosens Bioelectron* 2016;81:32–8. https://doi.org/10.1016/j.bios.2016.02.051.
24. Rao JR, Richter GJ, Von Sturm F, Weidlich E. The performance of glucose electrodes and the characteristics of different biofuel cell constructions. *Bioelectrochemistry Bioenerg* 1976;3:139–50. https://doi.org/10.1016/0302-4598(76)85014-3.
25. Bhargavi G, Venu V, Renganathan S. Microbial fuel cells: recent developments in design and materials. *IOP Conf Ser Mater Sci Eng* 2018;330:012034. https://doi.org/10.1088/1757-899X/330/1/012034.
26. Pant D, Van Bogaert G, Porto-Carrero C, Diels L, Vanbroekhoven K. Anode and cathode materials characterization for a microbial fuel cell in half cell configuration. *Water Sci Technol* 2011;63:2457–61. https://doi.org/10.2166/wst.2011.217.
27. Kalathil S, Patil SA, Pant D. Microbial Fuel Cells: Electrode Materials. *Encycl. Interfacial Chem.*, Elsevier; 2018, p. 309–18. https://doi.org/10.1016/B978-0-12-409547-2.13459-6.
28. Kim JR, Kim J-Y, Han S-B, Park K-W, Saratale GD, Oh S-E. Application of Co-naphthalocyanine (CoNPc) as alternative cathode catalyst and support structure for microbial fuel cells. *Bioresour Technol* 2011;102:342–7. https://doi.org/10.1016/j.biortech.2010.07.005.
29. Liu B, Brückner C, Lei Y, Cheng Y, Santoro C, Li B. Cobalt porphyrin-based material as methanol tolerant cathode in single chamber microbial fuel cells (SCMFCs). *J Power Sources* 2014;257:246–53. https://doi.org/10.1016/j.jpowsour.2014.01.117.
30. Gong X-B, You S-J, Wang X-H, Zhang J-N, Gan Y, Ren N-Q. A novel stainless steel mesh/cobalt oxide hybrid electrode for efficient catalysis of oxygen reduction in a microbial fuel cell. *Biosens Bioelectron* 2014;55:237–41. https://doi.org/10.1016/j.bios.2013.12.015.
31. Xiao L, Damien J, Luo J, Jang HD, Huang J, He Z. Crumpled graphene particles for microbial fuel cell electrodes. *J Power Sources* 2012;208:187–92. https://doi.org/10.1016/j.jpowsour.2012.02.036.
32. Fu L, You S-J, Zhang G-Q, Yang F-L, Fang X-H, Gong Z. PB/PANI-modified electrode used as a novel oxygen reduction cathode in microbial fuel cell. *Biosens Bioelectron* 2011;26:1975–9. https://doi.org/10.1016/j.bios.2010.08.061.
33. Ghasemi M, Daud WRW, Rahimnejad M, Rezayi M, Fatemi A, Jafari Y, *et al.* Copper-phthalocyanine and nickel nanoparticles as novel cathode catalysts

in microbial fuel cells. *Int J Hydrogen Energy* 2013;38:9533–40. https://doi.org/10.1016/j.ijhydene.2013.01.177.

34. Zhang X, Xia X, Ivanov I, Huang X, Logan BE. Enhanced Activated Carbon Cathode Performance for Microbial Fuel Cell by Blending Carbon Black. *Environ Sci Technol* 2014;48:2075–81. https://doi.org/10.1021/es405029y.
35. Song T-S, Wang D-B, Wang H, Li X, Liang Y, Xie J. Cobalt oxide/nanocarbon hybrid materials as alternative cathode catalyst for oxygen reduction in microbial fuel cell. *Int J Hydrogen Energy* 2015;40:3868–74. https://doi.org/10.1016/j.ijhydene.2015.01.119.
36. Liu H, Cheng S, Logan BE. Power Generation in Fed-Batch Microbial Fuel Cells as a Function of Ionic Strength, Temperature, and Reactor Configuration. *Environ Sci Technol* 2005;39:5488–93. https://doi.org/10.1021/es050316c.
37. Ahn Y, Logan BE. Effectiveness of domestic wastewater treatment using microbial fuel cells at ambient and mesophilic temperatures. *Bioresour Technol* 2010;101:469–75. https://doi.org/10.1016/j.biortech.2009.07.039.
38. Kim HJ, Park HS, Hyun MS, Chang IS, Kim M, Kim BH. A mediator-less microbial fuel cell using a metal reducing bacterium, Shewanella putrefaciens. *Enzyme Microb Technol* 2002;30:145–52. https://doi.org/10.1016/S0141-0229(01)00478-1.
39. He Z, Minteer SD, Angenent LT. Electricity Generation from Artificial Wastewater Using an Upflow Microbial Fuel Cell. *Environ Sci Technol* 2005;39:5262–7. https://doi.org/10.1021/es0502876.
40. Bi L, Ci S, Cai P, Li H, Wen Z. One-step pyrolysis route to three dimensional nitrogen-doped porous carbon as anode materials for microbial fuel cells. *Appl Surf Sci* 2018;427:10–6. https://doi.org/10.1016/j.apsusc.2017.08.030.
41. Paul D, Noori MT, Rajesh PP, Ghangrekar MM, Mitra A. Modification of carbon felt anode with graphene oxide-zeolite composite for enhancing the performance of microbial fuel cell. *Sustain Energy Technol Assessments* 2018;26:77–82. https://doi.org/10.1016/j.seta.2017.10.001.
42. Jiang H, Yang L, Deng W, Tan Y, Xie Q. Macroporous graphitic carbon foam decorated with polydopamine as a high-performance anode for microbial fuel cell. *J Power Sources* 2017;363:27–33. https://doi.org/10.1016/j.jpowsour.2017.07.064.
43. Zhou M, Chi M, Wang H, Jin T. Anode modification by electrochemical oxidation: A new practical method to improve the performance of microbial fuel cells. *Biochem Eng J* 2012;60:151–5. https://doi.org/10.1016/j.bej.2011.10.014.
44. Liu J, Qiao Y, Guo CX, Lim S, Song H, Li CM. Graphene/carbon cloth anode for high-performance mediatorless microbial fuel cells. *Bioresour Technol* 2012;114:275–80. https://doi.org/10.1016/j.biortech.2012.02.116.
45. Yong Y-C, Dong X-C, Chan-Park MB, Song H, Chen P. Macroporous and Monolithic Anode Based on Polyaniline Hybridized Three-Dimensional

Graphene for High-Performance Microbial Fuel Cells. *ACS Nano* 2012;6:2394–400. https://doi.org/10.1021/nn204656d.
46. He Z, Liu J, Qiao Y, Li CM, Tan TTY. Architecture Engineering of Hierarchically Porous Chitosan/Vacuum-Stripped Graphene Scaffold as Bioanode for High Performance Microbial Fuel Cell. *Nano Lett* 2012;12:4738–41. https://doi.org/10.1021/nl302175j.
47. Zhu N, Chen X, Zhang T, Wu P, Li P, Wu J. Improved performance of membrane free single-chamber air-cathode microbial fuel cells with nitric acid and ethylenediamine surface modified activated carbon fiber felt anodes. *Bioresour Technol* 2011;102:422–6. https://doi.org/10.1016/j.biortech.2010.06.046.
48. Zhang Y, Mo G, Li X, Zhang W, Zhang J, Ye J, *et al.* A graphene modified anode to improve the performance of microbial fuel cells. *J Power Sources* 2011;196:5402–7. https://doi.org/10.1016/j.jpowsour.2011.02.067.
49. Chen X, Cui D, Wang X, Wang X, Li W. Porous carbon with defined pore size as anode of microbial fuel cell. *Biosens Bioelectron* 2015;69:135–41. https://doi.org/10.1016/j.bios.2015.02.014.
50. Fan M, Zhang W, Sun J, Chen L, Li P, Chen Y, *et al.* Different modified multi-walled carbon nanotube–based anodes to improve the performance of microbial fuel cells. *Int J Hydrogen Energy* 2017;42:22786–95. https://doi.org/10.1016/j.ijhydene.2017.07.151.
51. Karra U, Troop E, Curtis M, Scheible K, Tenaglier C, Patel N, *et al.* Performance of plug flow microbial fuel cell (PF-MFC) and complete mixing microbial fuel cell (CM-MFC) for wastewater treatment and power generation. *Int J Hydrogen Energy* 2013;38:5383–8. https://doi.org/10.1016/j.ijhydene.2013.02.092.
52. Mehdinia A, Ziaei E, Jabbari A. Multi-walled carbon nanotube/SnO2 nanocomposite: a novel anode material for microbial fuel cells. *Electrochim Acta* 2014;130:512–8. https://doi.org/10.1016/j.electacta.2014.03.011.
53. Mink JE, Rojas JP, Logan BE, Hussain MM. Vertically Grown Multiwalled Carbon Nanotube Anode and Nickel Silicide Integrated High Performance Microsized (1.25 μL) Microbial Fuel Cell. *Nano Lett* 2012;12:791–5. https://doi.org/10.1021/nl203801h.
54. Tiwari BR, Noori MT, Ghangrekar MM. A novel low cost polyvinyl alcohol-Nafion-borosilicate membrane separator for microbial fuel cell. *Mater Chem Phys* 2016;182:86–93. https://doi.org/10.1016/j.matchemphys.2016.07.008.
55. Tang X, Guo K, Li H, Du Z, Tian J. Microfiltration membrane performance in two-chamber microbial fuel cells. *Biochem Eng J* 2010;52:194–8. https://doi.org/10.1016/j.bej.2010.08.007.
56. Zhang X, Cheng S, Wang X, Huang X, Logan BE. Separator Characteristics for Increasing Performance of Microbial Fuel Cells. *Environ Sci Technol* 2009;43:8456–61. https://doi.org/10.1021/es901631p.
57. Hoskins DL, Zhang X, Hickner MA, Logan BE. Spray-on polyvinyl alcohol separators and impact on power production in air-cathode microbial fuel

cells with different solution conductivities. *Bioresour Technol* 2014;172:156–61. https://doi.org/10.1016/j.biortech.2014.09.004.
58. Hasani-Sadrabadi MM, Dashtimoghadam E, Sarikhani K, Majedi FS, Khanbabaei G. Electrochemical investigation of sulfonated poly(ether ether ketone)/clay nanocomposite membranes for moderate temperature fuel cell applications. *J Power Sources* 2010;195:2450–6. https://doi.org/10.1016/j.jpowsour.2009.11.090.
59. Ilbeygi H, Ghasemi M, Emadzadeh D, Ismail AF, Zaidi SMJ, Aljlil SA, *et al.* Power generation and wastewater treatment using a novel SPEEK nanocomposite membrane in a dual chamber microbial fuel cell. *Int J Hydrogen Energy* 2015;40:477–87. https://doi.org/10.1016/j.ijhydene.2014.10.026.
60. Dharmalingam S, Kugarajah V, Sugumar M. Membranes for Microbial Fuel Cells. Microb. *Electrochem. Technol.*, Elsevier; 2019, p. 143–94. https://doi.org/10.1016/B978-0-444-64052-9.00007-8.
61. Ghasemi M, Daud WRW, Ismail AF, Jafari Y, Ismail M, Mayahi A, *et al.* Simultaneous wastewater treatment and electricity generation by microbial fuel cell: Performance comparison and cost investigation of using Nafion 117 and SPEEK as separators. *Desalination* 2013;325:1–6. https://doi.org/10.1016/j.desal.2013.06.013.
62. Ayyaru S, Dharmalingam S. Enhanced response of microbial fuel cell using sulfonated poly ether ether ketone membrane as a biochemical oxygen demand sensor. *Anal Chim Acta* 2014;818:15–22. https://doi.org/10.1016/j.aca.2014.01.059.
63. Daud SM, Kim BH, Ghasemi M, Daud WRW. Separators used in microbial electrochemical technologies: Current status and future prospects. *Bioresour Technol* 2015;195:170–9. https://doi.org/10.1016/j.biortech.2015.06.105.
64. Habermann W, Pommer EH. Biological fuel cells with sulphide storage capacity. *Appl Microbiol Biotechnol* 1991;35. https://doi.org/10.1007/BF00180650.
65. Liu H, Ramnarayanan R, Logan BE. Production of Electricity during Wastewater Treatment Using a Single Chamber Microbial Fuel Cell. *Environ Sci Technol* 2004;38:2281–5. https://doi.org/10.1021/es034923g.
66. Rabaey K, Van de Sompel K, Maignien L, Boon N, Aelterman P, Clauwaert P, *et al.* Microbial Fuel Cells for Sulfide Removal †. *Environ Sci Technol* 2006;40:5218–24. https://doi.org/10.1021/es060382u.
67. Jang JK, Pham TH, Chang IS, Kang KH, Moon H, Cho KS, *et al.* Construction and operation of a novel mediator- and membrane-less microbial fuel cell. *Process Biochem* 2004;39:1007–12. https://doi.org/10.1016/S0032-9592(03)00203-6.
68. Moon H, Chang IS, Kim BH. Continuous electricity production from artificial wastewater using a mediator-less microbial fuel cell. *Bioresour Technol* 2006;97:621–7. https://doi.org/10.1016/j.biortech.2005.03.027.
69. Liu H, Logan BE. Electricity Generation Using an Air-Cathode Single Chamber Microbial Fuel Cell in the Presence and Absence of a Proton

Exchange Membrane. *Environ Sci Technol* 2004;38:4040–6. https://doi.org/10.1021/es0499344.
70. Rittmann BE. Opportunities for renewable bioenergy using microorganisms. *Biotechnol Bioeng* 2008;100:203–12. https://doi.org/10.1002/bit.21875.
71. Zhang F, Ge Z, Grimaud J, Hurst J, He Z. Long-Term Performance of Liter-Scale Microbial Fuel Cells Treating Primary Effluent Installed in a Municipal Wastewater Treatment Facility. *Environ Sci Technol* 2013;47:4941–8. https://doi.org/10.1021/es400631r.
72. D Davila, J P Esquivel, N Vigues, O Sanchez, L Garrido, N Tomas, N Sabate, F J del Campo, F J Munoz and JM. Development and Optimization of Microbial Fuel Cells. *J New Mater Electrochem Systems* 2008;11:99–103.
73. Rozendal RA, Hamelers HVM, Rabaey K, Keller J, Buisman CJN. Towards practical implementation of bioelectrochemical wastewater treatment. *Trends Biotechnol* 2008;26:450–9. https://doi.org/10.1016/j.tibtech.2008.04.008.
74. Oh ST, Kim JR, Premier GC, Lee TH, Kim C, Sloan WT. Sustainable wastewater treatment: How might microbial fuel cells contribute. *Biotechnol Adv* 2010;28:871–81. https://doi.org/10.1016/j.biotechadv.2010.07.008.
75. He Z. Microbial Fuel Cells: Now Let us Talk about Energy. *Environ Sci Technol* 2013;47:332–3. https://doi.org/10.1021/es304937e.
76. Fan Y, Han S-K, Liu H. Improved performance of CEA microbial fuel cells with increased reactor size. *Energy Environ Sci* 2012;5:8273. https://doi.org/10.1039/c2ee21964f.
77. Dekker A, Heijne A Ter, Saakes M, Hamelers HVM, Buisman CJN. Analysis and Improvement of a Scaled-Up and Stacked Microbial Fuel Cell. *Environ Sci Technol* 2009;43:9038–42. https://doi.org/10.1021/es901939r.
78. Pham TH, Rabaey K, Aelterman P, Clauwaert P, De Schamphelaire L, Boon N, *et al.* Microbial Fuel Cells in Relation to Conventional Anaerobic Digestion Technology. *Eng Life Sci* 2006;6:285–92. https://doi.org/10.1002/elsc.200620121.
79. Rabaey K, Verstraete W. Microbial fuel cells: novel biotechnology for energy generation. *Trends Biotechnol* 2005;23:291–8. https://doi.org/10.1016/j.tibtech.2005.04.008.
80. Liu H, Cheng S, Huang L, Logan BE. Scale-up of membrane-free single-chamber microbial fuel cells. *J Power Sources* 2008;179:274–9. https://doi.org/10.1016/j.jpowsour.2007.12.120.
81. Logan BE. Scaling up microbial fuel cells and other bioelectrochemical systems. *Appl Microbiol Biotechnol* 2010;85:1665–71. https://doi.org/10.1007/s00253-009-2378-9.
82. Cheng S, Logan BE. Increasing power generation for scaling up single-chamber air cathode microbial fuel cells. *Bioresour Technol* 2011;102:4468–73. https://doi.org/10.1016/j.biortech.2010.12.104.
83. Clauwaert P, Aelterman P, Pham TH, De Schamphelaire L, Carballa M, Rabaey K, *et al.* Minimizing losses in bio-electrochemical systems: the

road to applications. *Appl Microbiol Biotechnol* 2008;79:901–13. https://doi.org/10.1007/s00253-008-1522-2.
84. Ieropoulos I, Greenman J, Melhuish C. Microbial fuel cells based on carbon veil electrodes: Stack configuration and scalability. *Int J Energy Res* 2008;32:1228–40. https://doi.org/10.1002/er.1419.
85. Wang B, Han J-I. A single chamber stackable microbial fuel cell with air cathode. *Biotechnol Lett* 2009;31:387–93. https://doi.org/10.1007/s10529-008-9877-0.
86. Roy S, Marzorati S, Schievano A, Pant D. Microbial Fuel Cells. *Encycl. Sustain. Technol.*, Elsevier; 2017, p. 245–59. https://doi.org/10.1016/B978-0-12-409548-9.10122-8.
87. Zhou M, Chi M, Luo J, He H, Jin T. An overview of electrode materials in microbial fuel cells. *J Power Sources* 2011;196:4427–35. https://doi.org/10.1016/j.jpowsour.2011.01.012.
88. Watson VJ, Nieto Delgado C, Logan BE. Influence of Chemical and Physical Properties of Activated Carbon Powders on Oxygen Reduction and Microbial Fuel Cell Performance. *Environ Sci Technol* 2013;47:6704–10. https://doi.org/10.1021/es401722j.
89. Wang Y-K, Sheng G-P, Shi B-J, Li W-W, Yu H-Q. A Novel Electrochemical Membrane Bioreactor as a Potential Net Energy Producer for Sustainable Wastewater Treatment. *Sci Rep* 2013;3:1864. https://doi.org/10.1038/srep01864.
90. Kim Y, Hatzell MC, Hutchinson AJ, Logan BE. Capturing power at higher voltages from arrays of microbial fuel cells without voltage reversal. *Energy Environ Sci* 2011;4:4662. https://doi.org/10.1039/c1ee02451e.
91. Donovan C, Dewan A, Peng H, Heo D, Beyenal H. Power management system for a 2.5W remote sensor powered by a sediment microbial fuel cell. *J Power Sources* 2011;196:1171–7. https://doi.org/10.1016/j.jpowsour.2010.08.099.
92. Zhang F, Tian L, He Z. Powering a wireless temperature sensor using sediment microbial fuel cells with vertical arrangement of electrodes. *J Power Sources* 2011;196:9568–73. https://doi.org/10.1016/j.jpowsour.2011.07.037.
93. Park J-D, Ren Z. High efficiency energy harvesting from microbial fuel cells using a synchronous boost converter. *J Power Sources* 2012;208:322–7. https://doi.org/10.1016/j.jpowsour.2012.02.035.
94. Park J-D, Ren Z. Hysteresis controller based maximum power point tracking energy harvesting system for microbial fuel cells. *J Power Sources* 2012;205:151–6. https://doi.org/10.1016/j.jpowsour.2012.01.053.
95. Min B, Logan BE. Continuous Electricity Generation from Domestic Wastewater and Organic Substrates in a Flat Plate Microbial Fuel Cell. *Environ Sci Technol* 2004;38:5809–14. https://doi.org/10.1021/es0491026.
96. Chung K, Fujiki I, Okabe S. Effect of formation of biofilms and chemical scale on the cathode electrode on the performance of a continuous two-chamber microbial fuel cell. *Bioresour Technol* 2011;102:355–60. https://doi.org/10.1016/j.biortech.2010.04.091.

97. Zhuang L, Yuan Y, Wang Y, Zhou S. Long-term evaluation of a 10-liter serpentine-type microbial fuel cell stack treating brewery wastewater. *Bioresour Technol* 2012;123:406–12. https://doi.org/10.1016/j.biortech.2012.07.038.
98. Feng Y, Wang X, Logan BE, Lee H. Brewery wastewater treatment using air-cathode microbial fuel cells. *Appl Microbiol Biotechnol* 2008;78:873–80. https://doi.org/10.1007/s00253-008-1360-2.
99. Cusick RD, Kiely PD, Logan BE. A monetary comparison of energy recovered from microbial fuel cells and microbial electrolysis cells fed winery or domestic wastewaters. *Int J Hydrogen Energy* 2010;35:8855–61. https://doi.org/10.1016/j.ijhydene.2010.06.077.
100. Zhang B, Zhao H, Zhou S, Shi C, Wang C, Ni J. A novel UASB–MFC–BAF integrated system for high strength molasses wastewater treatment and bioelectricity generation. *Bioresour Technol* 2009;100:5687–93. https://doi.org/10.1016/j.biortech.2009.06.045.
101. Sharma Y, Li B. Optimizing energy harvest in wastewater treatment by combining anaerobic hydrogen producing biofermentor (HPB) and microbial fuel cell (MFC). *Int J Hydrogen Energy* 2010;35:3789–97. https://doi.org/10.1016/j.ijhydene.2010.01.042.
102. Durruty I, Bonanni PS, González JF, Busalmen JP. Evaluation of potato-processing wastewater treatment in a microbial fuel cell. *Bioresour Technol* 2012;105:81–7. https://doi.org/10.1016/j.biortech.2011.11.095.
103. Cao X, Huang X, Liang P, Xiao K, Zhou Y, Zhang X, *et al.* A New Method for Water Desalination Using Microbial Desalination Cells. *Environ Sci Technol* 2009;43:7148–52. https://doi.org/10.1021/es901950j.
104. Chen X, Xia X, Liang P, Cao X, Sun H, Huang X. Stacked Microbial Desalination Cells to Enhance Water Desalination Efficiency. *Environ Sci Technol* 2011;45:2465–70. https://doi.org/10.1021/es103406m.
105. Jacobson KS, Drew DM, He Z. Efficient salt removal in a continuously operated upflow microbial desalination cell with an air cathode. *Bioresour Technol* 2011;102:376–80. https://doi.org/10.1016/j.biortech.2010.06.030.
106. Harnisch F, Schröder U. From MFC to MXC: chemical and biological cathodes and their potential for microbial bioelectrochemical systems. *Chem Soc Rev* 2010;39:4433. https://doi.org/10.1039/c003068f.
107. Anthony Janicek YF and HL. Design of Microbial fuel cells for practical application:a review and analysis of scale-up studies. *Biofuels* 2014:79–92. https://doi.org/https://doi.org/10.4155/bfs.13.69.
108. Shijia Wu, Hui Li, Xuechen Zhou, Peng Liang, Xiaoyuan Zhang, Yong Jiang XH. Anovel pilot-scale stacked microbial fuel cell for efficient electricity generation and wastewater treatment. *Water Res* 2016;98:396–403. https://doi.org/https://doi.org/10.1016/j.watres.2016.04.043.
109. Zheng Ge, Liao Wu, Fei Zhang ZH. Energy extraction from a large-scale microbial fuel cell system treating municipal wastewater. *J Power Sources*

2015;297:260–4. https://doi.org/https://doi.org/10.1016/j.jpowsour.2015.07.105.
110. Heinz Hiegemann, Daniel Herzer, Edith Nettmann, Manfred Lübken, Patrick Schulte, Karl-Georg Schmelz, Sylvia Gredigk-Hoffmann MW. An integrated 45L pilot microbial fuel cell system at a full scle wastewater tratment plant. *Bioresour Technol* 2016;218:115–22. https://doi.org/10.1016/j.biortech.2016.06.052.
111. Logan BE, Wallack MJ, Kim K-Y, He W, Feng Y, Saikaly PE. Assessment of Microbial Fuel Cell Configurations and Power Densities. *Environ Sci Technol Lett* 2015;2:206–14. https://doi.org/10.1021/acs.estlett.5b00180.
112. Dong Y, Qu Y, He W, Du Y, Liu J, Han X, *et al.* A 90-liter stackable baffled microbial fuel cell for brewery wastewater treatment based on energy self-sufficient mode. *Bioresour Technol* 2015;195:66–72. https://doi.org/10.1016/j.biortech.2015.06.026.
113. Fuchs W, Binder H, Mavrias G, Braun R. Anaerobic treatment of wastewater with high organic content using a stirred tank reactor coupled with a membrane filtration unit. *Water Res* 2003;37:902–8. https://doi.org/10.1016/S0043-1354(02)00246-4.
114. Lepistö SS, Rintala JA. Start-up and Operation of Laboratory-Scale Thermophilic Upflow Anaerobic Sludge Blanket Reactors Treating Vegetable Processing Wastewaters. *J Chem Technol Biotechnol* 1997;68:331–9. https://doi.org/10.1002/(SICI)1097-4660(199703)68:3<331::AID-JCTB657>3.0.CO;2-Z.
115. Borole AP, Hamilton CY. Energy Production from Food Industry Wastewaters Using Bioelectrochemical Cells. *Emerg. Environ. Technol.* Vol. II, Dordrecht: Springer Netherlands; 2010, p. 97–113. https://doi.org/10.1007/978-90-481-3352-9_5.
116. Hoffmann RA, Garcia ML, Veskivar M, Karim K, Al-Dahhan MH, Angenent LT. Effect of shear on performance and microbial ecology of continuously stirred anaerobic digesters treating animal manure. *Biotechnol Bioeng* 2008;100:38–48. https://doi.org/10.1002/bit.21730.
117. Raghavulu SV, Mohan SV, Goud RK, Sarma PN. Effect of anodic pH microenvironment on microbial fuel cell (MFC) performance in concurrence with aerated and ferricyanide catholytes. *Electrochem Commun* 2009;11:371–5. https://doi.org/10.1016/j.elecom.2008.11.038.
118. Lu N, Zhou S, Zhuang L, Zhang J, Ni J. Electricity generation from starch processing wastewater using microbial fuel cell technology. *Biochem Eng J* 2009;43:246–51. https://doi.org/10.1016/j.bej.2008.10.005.
119. Min B, Kim J, Oh S, Regan JM, Logan BE. Electricity generation from swine wastewater using microbial fuel cells. *Water Res* 2005;39:4961–8. https://doi.org/10.1016/j.watres.2005.09.039.
120. Greenman J, Gálvez A, Giusti L, Ieropoulos I. Electricity from landfill leachate using microbial fuel cells: Comparison with a biological aerated

filter. *Enzyme Microb Technol* 2009;44:112–9. https://doi.org/10.1016/j.enzmictec.2008.09.012.

121. Cecconet D, Molognoni D, Callegari A, Capodaglio AG. Agro-food industry wastewater treatment with microbial fuel cells: Energetic recovery issues. *Int J Hydrogen Energy* 2018;43:500–11. https://doi.org/10.1016/j.ijhydene.2017.07.231.
122. Venkata Mohan S, Mohanakrishna G, Sarma PN. Composite vegetable waste as renewable resource for bioelectricity generation through non-catalyzed open-air cathode microbial fuel cell. *Bioresour Technol* 2010;101:970–6. https://doi.org/10.1016/j.biortech.2009.09.005.
123. Jamal MT, Pugazhendi A, Jeyakumar RB. Application of halophiles in air cathode MFC for seafood industrial wastewater treatment and energy production under high saline condition. *Environ Technol Innov* 2020;20:101119. https://doi.org/10.1016/j.eti.2020.101119.
124. Mahdi Mardanpour M, Nasr Esfahany M, Behzad T, Sedaqatvand R. Single chamber microbial fuel cell with spiral anode for dairy wastewater treatment. *Biosens Bioelectron* 2012;38:264–9. https://doi.org/10.1016/j.bios.2012.05.046.
125. Sciarria TP, Tenca A, D'Epifanio A, Mecheri B, Merlino G, Barbato M, *et al.* Using olive mill wastewater to improve performance in producing electricity from domestic wastewater by using single-chamber microbial fuel cell. *Bioresour Technol* 2013;147:246–53. https://doi.org/10.1016/j.biortech.2013.08.033.
126. Mohanakrishna G, Venkata Mohan S, Sarma PN. Bio-electrochemical treatment of distillery wastewater in microbial fuel cell facilitating decolorization and desalination along with power generation. *J Hazard Mater* 2010;177:487–94. https://doi.org/10.1016/j.jhazmat.2009.12.059.
127. Samsudeen N, Radhakrishnan TK, Matheswaran M. Bioelectricity production from microbial fuel cell using mixed bacterial culture isolated from distillery wastewater. *Bioresour Technol* 2015;195:242–7. https://doi.org/10.1016/j.biortech.2015.07.023.
128. Xia T, Zhang X, Wang H, Zhang Y, Gao Y, Bian C, *et al.* Power generation and microbial community analysis in microbial fuel cells: A promising system to treat organic acid fermentation wastewater. *Bioresour Technol* 2019;284:72–9. https://doi.org/10.1016/j.biortech.2019.03.119.
129. Gude VG. Microbial fuel cells for wastewater treatment and energy generation. *Microb. Electrochem. Fuel Cells*, Elsevier; 2016, p. 247–85. https://doi.org/10.1016/B978-1-78242-375-1.00008-3.
130. Cui Y, Lai B, Tang X. Microbial Fuel Cell-Based Biosensors. *Biosensors* 2019;9:92. https://doi.org/10.3390/bios9030092.

8

Ceramic Membranes in Water Treatment: Potential and Challenges for Technology Development

Debarati Mukherjee[1] and Sourja Ghosh[2]*

[1]Department of Chemical Engineering, Indian Institute of Technology Guwahati, Guwahati, Assam, India
[2]Water Technology Division, CSIR-Central Glass and Ceramic Research Institute, Raja S.C. Mullick Road, Jadavpur, Kolkata, India

Abstract

Increased water consumption, the gradual depletion of the fresh water reserves and increased contamination of the surface water and groundwater resources due to the over growth of population, rapid industrialization and urbanization has led to significant water scarcity across the world. Accordingly, sustainable technology development for treatment of the contaminated water has been the major focus of researchers. The technological development in the field of water and wastewater treatment involves potable drinking water production from the contaminated surface/groundwater streams, treatment and recycling of industrial effluent, as well as domestic and municipal wastewater. While there are a number of technologies involving various conventional processes such as biological processes, adsorption, chemical coagulation-flocculation, oxidation and ion-exchange, etc., membrane-based technologies are becoming significant due to their clean and sustainable nature, high efficiency, modular design, ease in operation and cost-effectiveness. Although the major portion of the membranes used in industrial and domestic water treatment are polymeric, the ceramic membranes are becoming increasingly popular due to their inherent material stability, prolonged life and low fouling and therefore are largely preferred in the harsh and corrosive industrial effluent treatment and high fouling feed with bio-solids, as well as water/wastewater systems with high organic and inorganic loading compared to their polymeric counterparts. This chapter emphasizes the application of various types of ceramic membranes in the field of water and wastewater treatment processes, involving the

**Corresponding author*: sourja@cgcri.res.in; sourja.g@gmail.com

Siddhartha Moulik, Aditi Mullick and Anirban Roy (eds.) Sustainable Water Treatment: Advances and Technological Interventions, (325–382) © 2022 Scrivener Publishing LLC

traditionally used inorganic oxide and non-oxides, such as alumina (Al_2O_3), titania (TiO_2), zirconia (ZrO_2) and silicon carbide (SiC), etc., as well as new materials like clay, metal oxide nanoparticle and graphine oxide–coated ceramic membranes. In addition, with the pressure-driven membrane processes like micro, ultra and nano filtration, ceramic membrane–assisted hybrid processes, such as membrane bioreactor, photocatalytic membrane reactor and membrane photobioreactor processes are also covered. Performance of the ceramic membrane–based pilot plants is discussed which is critical in view of the technological acceptability of such processes. Finally, the various challenges in ceramic membrane–based processes are outlined which need to be addressed towards development of cost-effective, robust and sustainable technologies for water and wastewater treatment.

***Keywords*:** Water treatment, ceramics, membranes, sustainability, pilot scale implementation

8.1 Introduction

Water is the primary source to maintain the sustainability of the natural ecosystem. Survival of human beings, flora and fauna is strongly dependent upon the availability of pure water. However, the present era has witnessed a gradual increase in industrialization and urbanization which results in discharge of various toxic contaminants in surface water and groundwater. The hazardous environmental impacts of these contaminants limit the available useable water resources. This calls for design and implementation of advanced water treatment technologies which can fulfil the regulatory standards as determined by US Environmental Protection Agency (EPA) [1]. In this regard, research is being performed in this area to promote human welfare and economic growth.

8.1.1 Background and Current State-of-the-Art

Conventional processes such as adsorption, biodegradation, coagulation, ion-exchange, etc., are accompanied by certain disadvantages, such as rapid saturation of active sites, handling of toxic components, low efficiency, toxic by-product production, etc. [2]. Membrane-based processes appear to be an emerging option in water and wastewater treatment applications due to their low footprint area, cost-effectiveness, long-term operations, and high performance efficiency [2]. Membranes can be broadly classified into ceramic and polymeric category based on the material of fabrication. Although polymeric membranes are easy to fabricate and are comparatively cost-effective, they are accompanied by certain limitations, which restricts their industrial usage [3]. The typical advantages of ceramic membranes include (i) better control of porosity and pore size distribution resulting in enhanced contaminants separation efficiency from water and higher permeate flux, (ii) higher mechanical stability due to which they can be operated at high pressure, (iii) higher thermal stability which makes their applications viable at high temperature, and (iv) higher longevity as they are resistant to acid and alkaline environment due to their enhanced chemical stability [3, 4]. These characteristics of ceramic membranes assure their long-term, industrial-scale applications. Moreover, fouling, which is a detrimental factor for membrane processes, can be easily mitigated by application of backpulsing or cleaning agents, ensuring consistent permeate flux throughout the process. Ceramic microfiltration (MF), ultrafiltration (UF), and nanofiltration (NF) membranes are usually fabricated by deposition of slurry consisting of a suitable inorganic oxide, plasticizer, binder, and pore former over a substrate followed by sintering [5].

Ceramic membranes have been widely used for treatment of wastewater generated from various sectors, such as textile industries, tannery industries, distilleries, hospitals, diary industries, municipalities, etc. [6–10]. This wastewater contains a wide spectrum of contaminants which includes dyes, heavy metals, oil and grease, natural organic matter (NOM), ions, microorganisms, pesticides, pharmaceuticals and personal care products (PPCPs), value added products, etc. Moreover, ceramic membranes show high efficiency for treatment of groundwater and drinking water contaminated with arsenic, fluoride, nitrate, various PPCPs, etc. [11–14]. Numerous research studies report high separation of the toxic contaminants from water and wastewater, making the water suitable for disposal and reuse. A detailed picture of successful application and execution of these membranes has been presented in sections 8.2 and 8.3.

In spite of the numerous aforementioned benefits of ceramic membranes, their comparatively higher cost limits their pilot-scale industrial applications. To overcome this limitation, the recent focus has been shifted to use of cost-effective, widely available precursors like bauxite, kaolin for membrane preparation [15]. Different industrial wastes have also been used for this purpose which reduces the waste generation by providing a technique for waste inertization. The wastes were immobilized in the ceramic membrane matrix via crystalline phase formation or vitrification.

8.1.2 Ceramic Membranes: An Approach to Trade-Off the Bridge Between Theoretical Research and Industrial Applications

As mentioned in the previous section, ceramic membranes find a place in industries treating wastewater at harsh conditions of temperature, chemicals, and pressure. Improved stability of ceramic membranes in these harsh conditions makes them an ideal replacement of other processes including polymeric membrane systems. Some typically used inorganic oxide and non-oxides used for ceramic membrane preparation are alumina (Al_2O_3), titania (TiO_2), zirconia (ZrO_2), silicon carbide (SiC), etc. In recent years, industries have been employing ceramic membranes to fulfill the target of zero liquid discharge' in an environmental friendly way. Integration of low-cost raw materials/fillers such as clay can considerably reduce the high cost of ceramic membrane. In CSIR-CGCRI, porous ceramic membranes with pore size of ~1 μm and porosity of 40% have been prepared using an optimum composition of alumina (80%) and clay (20%) in tubular single and multichannel configuration. Further, the greater longevity of the ceramic membranes nullifies the higher installment costs [16].

A ceramic UF membrane-based hybrid system was developed by Veolia Water Ltd. for providing clean and safe drinking water [17]. The system encompasses coagulation and adsorption as pre-filtration stages for filtration using the UF membrane. Provision for disinfection using chlorination and ultraviolet oxidation are also present. The throughput of the process per day was about 360 m^3 of drinking water.

Metawater Co., Ltd., has been involved in drinking water treatment for 15 years using a ceramic MF membrane system integrated with coagulation and having high throughput [18]. The higher longevity of ceramic membranes was the key factor leading to replacement of polymeric membranes by the same in another drinking water treatment plant. Small-scale ceramic membrane water treatment systems have been installed in various schools and airports by Likuid Nanotek [19]. They work on the principle of submerged membrane bioreactor (MBR) process in cross-flow mode. The main advantage of these processes lies in the minimum requirement of maintenance and cleaning. Industrial wastewaters are also treated by ceramic MBRs. Olive mill, brewery wastewater has been effectively treated by a MBR comprising tubular ceramic membrane. The system has been installed by Likuid Nanotek and effectively treats about 3-300 m^3/day of wastewater. Meidensha Corporation has developed a system integrating up-flow anaerobic reactor with flat-sheet Al_2O_3 membrane with 100 nm pore diameter. Veolia Water, Ltd., has developed and installed ceramic monolith membranes–based systems at about 60 locations for oily wastewater treatment [20]. The membrane permeate produced showed >98% turbidity removal and oil and grease was less than 1%. Pilot plants consisting of ion exchange, ozonation, and ceramic membrane filtration have been installed in Australia, the UK, and Singapore [21].

The overwhelming advantages of ceramic membranes lead to the anticipation of the spike in their commercialization globally in recent years. Research efforts relating to engineering of the membrane structures to achieve the desired application are being performed. Due to high surface area and ease of module design using hollow fibre membranes, they are specifically expected to become an integral part of this water treatment industry in the upcoming years. Use of cost-effective precursors is being targeted to circumvent the issue of higher cost. MBRs are also expected to meet the demands and be used for large-scale wastewater treatment applications due to their environmental friendly nature. In spite of all these advantages, the commercial market demands designing of unique membrane modules with low footprint area to minimize the installation cost. Moreover, membrane cleaning produces a contaminants-laden waste stream, which needs proper disposal. Thus designing of low-fouling membranes needs to be achieved. Study of interactions between the membrane surface and contaminants and tailoring the surface properties of the membrane should be done to achieve the desired properties. This would be a trade-off of the gap between the theoretical research and practical application of ceramic membranes in water treatment and pave a way for their commercial use.

8.1.3 Industrial Wastewater Treatment

Due to the robustness of the inorganic material, ceramic membranes have increasingly been preferred in recent days for a wide range of industrial wastewater treatments. Some of the major industrial applications of the ceramic membranes are described here.

Textile Wastewater Treatment

Textile manufacturing processes result in the highest water pollution among all industrial sectors, both in terms of volume of wastewater produced and their toxic composition [6]. Indiscriminate, unregulated discharge of various unreacted dyes, chromium, and salts results in the accumulation of such hazardous contaminants in surface water [6]. This produces wastewater with high chemical oxygen demand (COD), dissolved solids, suspended solids, turbidity, colour, and pH. Research results show that such textile effluents when subjected to ceramic membrane processes undergo efficient removal of COD, dyes, heavy metals, salts, and hence, produce water that can be reused or safely discharged [22, 23].

Low-cost tubular ceramic MF membranes were fabricated from local Moroccan Perlite and applied for textile wastewater treatment. The active layer, having a pore diameter of 0.27 μm, showed greater than 50% removal of COD and completely decolorized the wastewater at pressure of 1 bar [22]. The membrane showed better performance than commercial alumina membrane having a pore diameter of 0.2μm. Disc-shaped MF membrane was prepared by mixing silica-rich drinking water treatment sludge with clay [23]. The membrane efficiently removed turbidity and COD from textile wastewater, below the allowable limit. The probable mechanism of separation in MF membranes might be their surface properties, such as hydrophobicity/hydrophillicity, charge, and presence of adsorptive sites. Ceramic tubular 19-channeled UF membrane with molecular weight cut-off (MWCO) of 8800 was found to show high rejection (>98%) of dyes, namely, Reactive Brilliant Blue, Reactive Black, Reactive Red, Cation Yellow due to charge separation [24]. The membrane showed better performance compared to a commercial polymeric membrane in terms of the permeate flux and separation efficiency. An interesting observation of this study was enhanced fractionation of dyes and salts (i.e., sodium chloride and sodium sulphate) by the ceramic membrane, which concludes its application for desalination as the future research scope. Effluent from a textile industry was subjected to successive stages of UF and NF membrane filtration [25]. The UF layer was based on alumina while the active NF layer was composed of alumina/zirconia mixture. The wastewater underwent ten cycles of UF+NF treatment followed by four cycles of NF treatment to produce water suitable for reuse [25]. This pilot scale system resulted in more than 80% removal of colour, COD, and total organic carbon (TOC) from the textile effluent. Development of single-channel and 19-channel tubular ceramic membranes was carried out at CSIR-CGCRI Kolkata from Tunisian apatite as the precursor [26]. The lab-scale filtration unit showed about 99% removal of reactive red dye from synthetic solution after two hours of operation. The membrane could be easily regenerated using hydrogen peroxide, hence reducing the fouling effects and rendering higher longevity. The efficiency of dye removal by the apatite-based membrane was found to be higher compared to adsorptive removal by apatite powder. This proved that synergistic effect of adsorption and membrane filtration was responsible for the improved efficiency. Figure 8.1 gives a comparative pictorial representation of colour removal from textile wastewater by various membranes.

Application of the ceramic MF and UF membranes can enable separation of different types of dyes from textile wastewater based on various interactions, the process being energy-efficient as it operates under low pressure. The system can provide high flux by effective controlling of the membrane fouling by mechanical (back washing/back pulsing), as well as chemical (harsh cleaning agents) means. Development of tight UF

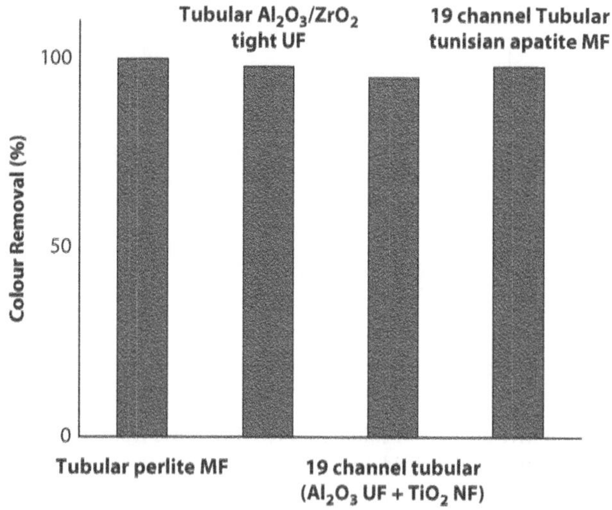

Figure 8.1 Comparative analysis of colour removal from textile wastewater by ceramic membranes.

(MWCO 10-20 kDa) and NF membranes (MWCO <1 kDa) are required for efficient removal of such dyes at high pressure due to their smaller pore size. Further, total dissolved solids (TDS) reduction is another crucial aspect in textile effluent since the typical dye bath effluent usually has high TDS value (~10 g/L). Thus, a thorough analysis of the wastewater composition is essential since it determines the type of membrane and the process to be chosen to achieve the desired product, i.e., treated water suitable for recycling at various intermediate steps of textile processing with minimum operating costs.

Apart from single-stage membrane filtration, ceramic membranes are often used in a hybrid process for more effective rejection efficiency and flux recovery, as well as attaining the quality of treated effluent suitable for recycling [27–31]. Zirconia-based ultrafiltration membrane (pore size 50 nm) was integrated with electrodialysis process for treating textile wastewater (primary treated). The process resulted in 94% of TDS removal [27]. Clay- and alumina-based ceramic microfiltration membrane (pore size 1 μm) was used in combination with biosorbent prepared from sugar cane bagasse for enhanced removal of dyes (99%) and COD (91%) from concentrated dyebath effluent. The recycling potential of the treated effluent was studied in dyeing of cotton fabric wherein the membrane-treated water showed higher dye uptake for the light and medium shade dyeing compared to the fresh water system [28]. In another study γ-alumina coated

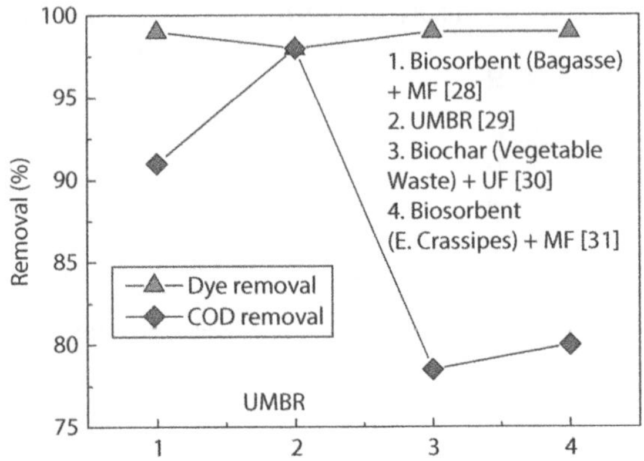

Figure 8.2 Comparative analysis of colour and COD removal from textile wastewater by ceramic membrane integrated hybrid processes.

ultrafiltration membrane prepared on porous clay-alumina support was used in a side-stream membrane bioreactor process for detoxification of textile effluent. The integrated process involving an isolated bacterial consortium was responsible for effective degradation of the dyes in effluent (>98%) and COD reduction (>98%), the treated water was found safe on a living system, ensuring an effective detoxification of the industrial effluent [29]. Further, in a recent study biochar prepared from vegetable wastes was combined with titania UF membrane [Tami Industry] having pore size of 200 nm for treatment of reactive dye containing effluent. The hybrid process could achieve >99% of dye removal and 77-80% removal of COD [30]. Figure 8.2 shows efficiency of different ceramic membrane–based hybrid processes for dye and COD removal in textile wastewater.

Tannery Wastewater Treatment

Tannery industries is another major sector which produces an enormous amount of hazardous solid and liquid wastes which adversely affects the aquatic and terrestrial environment. The effluent produced from a typical tannery industry contains organic and inorganic contaminants, high value of BOD, COD, oil and grease, surfactants, ammonia, chromium, arsenic, suspended solids, unreacted salts used in tanning process, such as sulfates, chlorides etc. [7]. Step-by-step processing of 1 ton (approx. 2200 lbs) of skin/hide of animals, via soaking, liming, bating, pickling, skin degassing, and tanning, produces about 7920 gallons of wastewater [7]. For decades, discharge of tannery effluent has been a problem in India, especially Kolkata, which is a hub of tanning industries. Discharge of these untreated/insufficiently treated effluents in water bodies results in pollution.

Tannery effluents are usually subjected to conventional treatment methods such as chemical coagulation, activated sludge, adsorption, etc., in common effluent treatment plants (CETP) [32]. However, the treated water obtained from these CETPs is still unsuitable for reuse. Membrane processes emerge as a promising alternative to treatment of such wastewater. Studies on application of ceramic MF and UF membranes for removal of various contaminants from textile wastewater have been reported. Macroporous supports of Al_2O_3 and corderitre were coated with ZrO_2 for fabrication of a ceramic MF membrane which was efficient in terms of removal of COD and suspended solids from tannery effluent [32]. At CSIR-CGCRI, Kolkata several lab-scale, as well as pilot-scale research studies have been performed on wastewater from various collection points of tannery industries such as untreated composite effluent, primary and secondary clarifier, equalization tank, etc. Porous 19-channel MF membrane, prepared from cost-effective composition of clay-alumina, having pore size of 0.5 μm was applied for treatment of wastewater collected from secondary clarifier of a CETP in Kolkata [33]. Initially, studies were conducted on a lab scale for assessment of membrane performance and optimization of process parameters to maximize the efficiency. This was followed by scale-up and subsequent onsite installation of the pilot plant near the leather complex. COD reduction of >80% was achieved for both lab scale and pilot plant. The various other parameters, such as BOD, chromium content and turbidity were reduced to below their detection limit. The membrane-treated effluent could be successfully reused in the leather tanning process and provided comparable performance with that of the fresh water system, proving the efficiency of the developed membrane process in water conservation. Steps were taken to prevent the membrane fouling and maintain uniform flux by equipping the system with *in situ* cleaning techniques. In another study conducted by the same research group at CSIR-CGCRI, Kolkata, the efficiency of two different membrane processes, *viz.*, (i) single-stage microfiltration and (ii) two-stage route involving ceramic MF membrane followed by polymeric reverse osmosis (RO) membrane, for treatment of tannery effluent was analyzed and compared [34]. The tubular single-channel and 19-channel membrane used are shown in Figure 8.3. The industrial effluent of about 400 tannery industries in Kolkata are collectively treated in a CETP, wastewater sample was collected from the inlet of this CETP. The treated was devoid of most of the organic and inorganic contaminants, suspended solids and oil and grease in both the processes. However, owing to the micron range pore sizes of

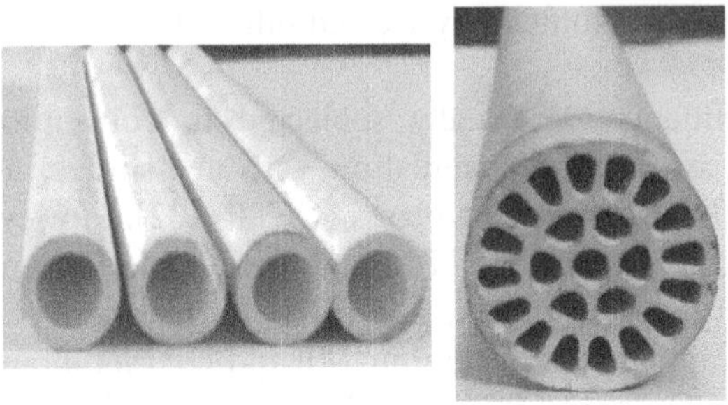

(a) (b)

Figure 8.3 (a) Single-channel and (b) 19-channel MF supports used for tannery wastewater treatment.

the MF process, negligible reduction of the TDS was found in this case, whereas, >97% removal occurred in the two-stage process. Toxicity studies of the treated effluent on aquatic snail showed no significant adverse effects, justifying their reuse for aquaculture farming. Figure 8.4 shows the surface microstructure of clay-alumina-based ceramic microfiltration membrane before and after treatment of tannery wastewater [34].

The analytical characteristics of tannery effluent vary widely based on the differences in the type of end product, plant capacity, and the chemicals applied. This directs us to wisely choose a membrane that will comply with our target. The effects of membrane pore size and operating pressure was studied on the flux and COD removal from tannery wastewater [35]. Single-channel tubular Al_2O_3 membranes with pore size of 10, 50, and 200 nm were used. COD removal increased from 58 to 90% with

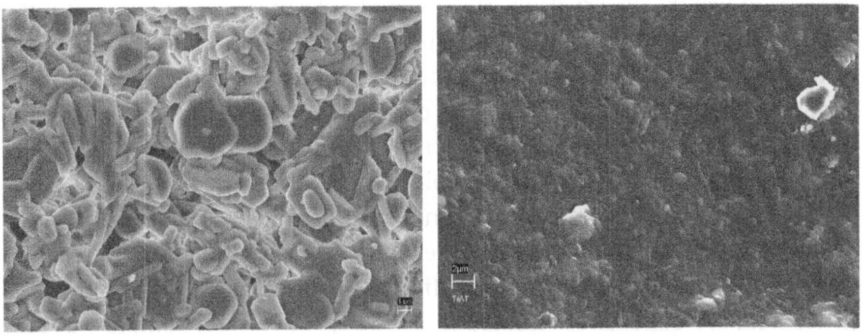

(a) (b)

Figure 8.4 Surface microstructure of clay-alumina based ceramic microfiltration membrane: (a) before and (b) after tannery wastewater treatment.

increase in operating pressure from 1 to 4 bar for the 10 nm UF membrane. However, using 200 nm MF membrane, COD removal of only 20% was obtained. It may be concluded that the removal of contaminants is largely dependent on the membrane properties, as well as filtration parameters. Low-cost ceramic MF membrane was developed from natural pozzolan, collected from a mountain in Morocco, and micronized phosphate [36]. The membrane facilitated about 97% turbidity removal for tannery wastewater.

In view of the complex nature of the tannery wastewater, ceramic membrane technology can play a crucial role in effluent treatment processes. An optimum design of effluent treatment strategy is required to enable adequate reduction of various contaminants in a cost-effective and sustainable manner. Inclusion of ceramic MF membrane (0.5-1 μm) can result in considerable reduction of the organic loading (COD and BOD), suspended and colloidal solids (turbidity and TSS), and partial removal of dissolved solids (TDS) and thus can be an effective choice as a pretreatment strategy. So far the majority of the studies conducted indicates the prevalence of ceramic MF and UF membranes for treatment of tannery wastewater. Stability of ceramic membranes upon exposure to harsh chemicals makes them suitable for use in tannery effluent treatment unlike polymeric membranes. Significantly, integration of the ceramic-based MF pretreatment system can enhance the performance of the successive polymer-based RO process in terms of the output (process flux and permeate quality) and thus can be considered an important pathway from technology development perspectives.

Food Industry Wastewater Treatment

Distillery Wastewater Treatment

Typically ethanol is produced by fermentation of sugar-based (e.g., sugarcane), starch-based (e.g., wheat) sources followed by distillation [8]. The spent liquid residue of the fermentation is termed as distillery wastewater [8]. The volume of wastewater produced is about 30 times greater than the desired product. The wastewater has very high content of BOD, COD, odor, color, and organics, such as polymers, tannic acid, humic acid, etc., inorganic components, such as calcium, sulphate, nitrogen, phosphate., etc., which restricts their direct discharge into land or water bodies. The composition of the distillery wastewater is highly dependent on the type of source used for extraction of the biofuel.

Application of membrane separation process for distillery wastewater treatment efficiently removes the contaminants. Additionally, value-added chemicals, such as amino acids and proteins, can be extracted from the retentate and the retentate can also be used as source of fertilizer or biogas production depending upon their composition [8]. Ceramic membranes are being applied in treatment of such wastewater. Wastewater produced from potato fermentation was subjected to treatment of three channelled ceramic membranes [37]. The process efficiency was compared among three different membranes having pore size varying from 200 nm to 50 kDa and arranged in different combinations. A two-stage treatment process involving MF membrane of 200 nm followed by UF membrane of 50 kDa, connected in series, was found to be most efficient in terms of reduction of COD, solids, and nitrogen content. Separation efficiency was dependent upon membrane alignment and operating conditions. Starch-based distillery wastewater was subjected to microfiltration treatment by using membranes of pore size 800 nm, 450 nm, and 200 nm [38]. The 200 nm membrane showed highest efficiency in terms of removal and permeate flux. The BOD, COD, organic content, and total nitrogen were sufficiently reduced in the membrane treated permeate. Ceramic MF and UF membranes were used for treating distillery wastewater and the permeate produced was found suitable for reuse in the bioethanol industry [39]. The multi-channeled MF membrane resulted in about 100%, 78%, 80%, 80% removal of suspended solids, dry matter, COD, and total nitrogen, respectively, from the wastewater at transmembrane pressure of 2 bar. Further, the removal of these components increased to 99%, 90%, and 99%, respectively, for the last three contaminants upon UF membrane treatment at 4 bar operating pressure. Research studies assessing the reusability of the treated water in the fermentation and biofuel yield and the economic viability of the entire process need to be explored. A UF membrane composed of the mixture of zirconia and titania was applied for treatment of wine vines, i.e., wastewater obtained by fermentation of grapes, and recovery of the phenolic compounds simultaneously from the waste stream [40]. Chloride, phosphate, sulphate and total solids were sufficiently reduced in the permeate stream. Flux recovery of 96% by periodic cleaning with water and detergent ensured higher longevity of the UF membrane. The wastewater upon centrifugation and membrane filtration resulted in product containing 45% gallic acid equivalent. The toxicity of membrane-treated water was significantly reduced compared to the untreated effluent, demonstrating that it could be reused or safely discharged.

It may be noted that for high organic loaded distillery wastewater with high values of the BOD, COD total organic carbon (TOC) and total kjeldahl nitrogen (TKN) content, ceramic MF membrane-based technology can be an effective choice involving operating pressures < 3 bar along with significant reduction of the odor and colour. Moreover, various value-added products can be derived from the retentate which is a further motivation to adopt such membrane-based processes for distillery wastewater treatment.

Dairy Wastewater Treatment

Ceramic membranes find significant applications in the dairy industry which tops among various food processing industries in consumption of freshwater and production of wastewater [41]. The various associated units, such as sanitization, milk processing and cleaning produces about 2.5 L of wastewater for 1 L of milk processed [41]. The wastewater is rich in fats, lactose, suspended solids, ammonia, phosphate, etc. [42]. The presence of such organics and nutrients increases the BOD and COD of the wastewater, hence posing detrimental effects to the natural ecosystem.

MF membrane was fabricated from low-cost local clays such as, kaolin, quartz, ball clay, pyrophyllite, feldspar and calcium carbonate as precursor to circumvent the limitation of high membrane preparation cost [41]. The membrane efficiency was assessed in terms of the COD reduction from wastewater collected from a local dairy industry in Guwahati. The process could achieve reduction of COD value below the safe discharge limit in the treated water; i.e., 135 mg/L from an initial feed COD value of 1462 mg/L at low transmembrane pressure of 207 KPa. Membrane fouling studies revealed prevalence of reversible cake filtration, thus ensuing higher longevity. Comparative analysis of the membrane performance with other polymeric NF and UF membranes, shows its better efficiency in dairy wastewater treatment. Ceramic MF and UF membranes composed of TiO_2-ZrO_2, were used as in a single-stage filtration and double-stage combination process [43]. MF+UF system showed COD removal of 99% followed by UF (95%) and MF (89%), respectively. About 3-18% COD removal was due to the adsorption of the organics on the membrane surface in addition to the size exclusion effect. More than 95% of removal of color, turbidity, and suspended solids were found in all the systems. The flux in MF+UF process was about 80% higher than UF, concluding MF+UF as the most efficient process to fulfil the target. This is due to decrease in UF membrane resistance, owing to pretreatment by MF process, thus reducing the pore clogging and fouling effect. Wastewater in a dairy plant rich in whey was treated adequately for use as a drinking water source at Algiers [44]. In this regard, two different UF membranes, i.e., a commercial alumina membrane and a synthesized TiO_2 membrane over clay support were applied for treatment of this wastewater. Considerable reduction of lactose, proteins, BOD, and COD were obtained. Using the commercial membrane the BOD and COD values were reduced from 4000 mg/L and 1800 mg/L to 90 mg/L and 62 mg/L, respectively, which were below the safe limit. Although the commercial membrane showed good rejection efficiency, it was accompanied with high membrane cost. Thus, the aforementioned UF membrane was synthesized which showed BOD and COD reduction to 60 mg/L and 42 mg/L, respectively, due to the combined effect of the surface adsorption and size exclusion. The fouling effects in the synthesized membrane were also less compared to the commercial one. Additionally, recovery of valuable whey was undertaken from the concentrated retentate. ISOFLUX™ MF membranes are composed of TiO_2 and uniquely designed with thickness decreasing from inlet to outlet to overcome the pressure drop effect [45]. These ISOFLUX™ membranes show high efficiency in dairy industry in terms of reduction of bacteria and spores from milk. Another aspect in dairy industries is production of large volumes of spent alkaline cleaning solutions. Studies target purification of these cleaning solutions by separating the milk compounds using ceramic membranes, so that it can be reused for the same cleaning purpose [46]. This would minimize the freshwater consumption. 19-channeled, ZrO_2 membranes with pore size of 20 nm were used in the study. The composition, properties and temperature of the cleaning solution were found to affect the flux and removal.

Thus ceramic membranes, especially MF and UF, found a wide application in dairy industries for wastewater treatment as well as product recovery. TiO_2, ZrO_2 are the typically used precursors for membrane preparation. Recent studies report use of locally available low-cost precursors such as, kaolin, quartz, etc., to reduce the overall process cost.

Poultry Wastewater treatment

Increasing population has led to increased demand in broiler consumption. The poultry slaughterhouse industry requires a large amount of wastewater at various stages of processing. The wastewater produced from this industry is rich in BOD, COD, solid particles, due to the presence of carbohydrates, proteins, and fats from the skin, meat, blood, feathers, etc. The COD, BOD, and total solids content may rise up to 10000 mg/L, 2000 mg/L, and 800 mg/L, respectively, making them toxic to the flora and fauna. In order to address this alarming issue, inorganic membranes can be applied judiciously for waste minimization and recycling of the treated water.

Cost-effective tubular ceramic membranes were prepared from waste source i.e., flyash (which constituted 75% of the total composition) [9]. The membrane had pore diameter and porosity of 0.133 μm and 40.17%, respectively. Almost complete removal of turbidity and COD was achieved by the membrane, directing reuse of wastewater for further processing in the poultry industry. MF membranes prepared from natural aluminosilicates could efficiently remove >85% of COD, microorganisms, insoluble components and total carbon content [47]. Apart from production of the clean permeate, a membrane-based process could facilitate extraction of proteins from the concentrated retentate streams [48].

The primary limitation of using membranes in food industries is formation of spores and bacteria on membrane surface, if not properly cleaned after every cycle [49]. Ceramic membranes being resistant to the cleaning agents with extreme pH, can overcome this roadblock unlike polymeric membranes. Thus, ceramic membranes can play a dominant role in food industries by efficient removal of the contaminants and generation of treated water suitable for recycle and reuse.

Petrochemical Wastewater

Petroleum and gas processing industries generate huge amounts of wastewater in the process of petrochemical production. The wastewater is highly contaminated with various organic and inorganic components, oil and grease, suspended and dissolved solid particles [49]. Efficient treatment methods must be applied to this wastewater so that they can be safely discharged in the surroundings following strict environmental guidelines. γ-Al_2O_3 MF membranes of 200 nm pore size was used for removal of coke (100%) and COD (70%) from a petrochemical plant wastewater at 15 bar pressure, cross flow velocity of 2 m/sec, and temperature 20-80 °C [50]. However, the membrane was ineffective in removing volatile organic components. Flux recovery was performed using sodium hydroxide as the cleaning agent. Based on the positive findings of the aforementioned study, a cost analysis of the wastewater treatment process comprising 7-channeled and 19-channeled γ-Al_2O_3 MF membranes was performed to judge the economic viability of the cross-flow filtration process and scope for full-scale pilot plant implementation [51]. Accordingly, the payback period and breakeven point was found to be 2 years and 3%, respectively, with a capital investment of $535,300, concluding the membrane process for efficient coke removal from the wastewater. The economic and technical feasibility of a pilot scale process involving ZrO_2- Al_2O_3 membrane with MWCO of 300 KDa for treatment of oily wastewater was performed by a group of researchers [52]. The membrane possessed permeate flux of 100 $Lm^{-2}hr^{-1}$ and efficiently removed 97% of hydrocarbons (treated water has less than 1 ppm of hydrocarbons). This ceramic membrane unit has gained many accreditations and has been installed by ferries, thus reducing the volume of wastewater to be treated onshore. α-Al_2O_3 membrane with pore size 200 nm could remove up to 100%, 98%, and 85% of TSS, turbidity, and oil and grease, respectively, from the wastewater of a refinery in

Iran [53]. The temperature, pressure, and flow velocity of the process for obtaining maximum separation efficiency were 32.5°C, 1.25 bar and 2.25 m/sec, respectively. The membrane fouling was minimized by application of backwashing. Al_2O_3 MF membrane was coated by kaolin-MnO_2 bilayer dynamic membrane by circulating kaolin and $KMnO_4$ solution, each for 30 min successively [54]. More than 98% of oil removal was achieved in the membrane process at concentration of 0.1 g/L to 1 g/L at alkaline conditions. However, the main limitation of this membrane was its poor efficiency at low pH due to the corrosion of MnO_2 particles. In order to produce cost-effective membranes and low cost, abundantly available sources such as quartz, feldspar, kaolin, etc., have been used for membrane fabrication [55]. The membrane showed >98% oil removal and flux of 5.36×10^{-6} m^3/m^2 s for an initial oil concentration of 250 mg/L at a pressure of 68.95 kPa, within 60 min of filtration. Fouling analyses of the study showed that cake filtration was the dominant mechanism. Ceramic membranes have been installed at 60 different locations by Veolia Technologies to treat wastewater from petrochemical plants and reuse them in fracking water in shale oil operations [49].

However, in spite of the advancements in ceramic membranes for petrochemical wastewater treatment, the limiting factors are the formation of concentration polarization layer and membrane fouling due to deposition of oil droplets and other components on the membrane surface. Research efforts have contributed to reduce this fouling effect by applying shear at membrane surface, which can be achieved by turbulence promoter, pulsed flow and ultrasound cleaning, etc. [49].

Pharmaceutical Wastewater

The pharmaceutical industry requires water for manufacturing various medicines, and in the process generates wastewater. This wastewater contains inorganic and organic compounds (pharmaceutically active compounds, PhACs), which are the contaminants of emerging concern. They pose serious threats to flora, fauna, and even humans where they might cause endocrine disruption [49]. Discharge of such untreated wastewater may result in accumulation of PhACs in the human body resulting in adverse effects. Membrane technology, being a robust and efficient water treatment, is expected to remove such contaminants from wastewater.

The efficiency of ceramic MF membrane composed of clay-alumina mixture was found to efficiently treat herbal pharmaceutical wastewater [56]. The filtration process could remove about 85%, 98% and 99% of COD, TSS and turbidity, respectively, from the wastewater upon 2 h of filtration when used with biosorption as the pretreatment step. Commercially available disc-shaped ceramic UF membranes made up of ZrO_2 were modified by coating of TiO_2-graphene oxide (GO) nanocomposites [57]. The membrane showed >80% removal of ibuprofen and diclofenac while lower removal of carbamazepine (30%) was attained. The removal mechanism was primarily size exclusion and electrostatic repulsion. Carbamazepin, being a neutral molecule, was therefore poorly removed by the negatively charged membrane unlike the other pharmaceuticals which are negatively charged. The membrane surface was highly hydrophilic, thus reducing the fouling effect and maintaining a stable flux throughout the filtration process. TiO_2 membrane was modified by GO via vacuum coating for removal of the pharmaceuticals, natural organic matter (NOM), humic acid, tannic acid, and salts from simulated wastewater [58]. About 51%, 93.5% and 31.4% removal of the pharmaceuticals, NOM, and salts were attained by the membrane which was much higher compared to the unmodified pristine TiO_2 membrane. Hydrophobic membranes were synthesized by surface modification of clay-alumina-based MF membranes using poly-dimethylsiloxane (PDMS) [59]. The membrane showed >90% removal of ciprofloxacin, a widely used antibiotic from simulated water. Here hydrophobic interaction and π-π interaction effects between the PDMS-coated membrane surface and contaminants were responsible for the removal of ciprofloxacin. Membrane fouling was mitigated by applying pneumatic backpulsing. Apart from the bench scale studies, pilot-scale installation of the ceramic membranes has been reported. Inopor GmbH has installed ceramic UF membrane for water treatment in the pharmaceutical industry [49]. The Inopor membranes are resistive towards bacterial growth and steam sterilization unlike polymeric membranes, making them suitable alternatives in pharmaceutical wastewater treatment.

Although a lot of ceramic membranes are being prepared for separation of drugs from pharmaceutical wastewater, an insight into the removal mechanism and establishment of appropriate explanation is significantly necessary. Moreover, most of the membranes are drug-specific. Since a wide range of drugs with various structural, functional, and chemical properties are being manufactured and consumed daily, development of a generic ceramic membrane with tailored properties for removal of a wide range of drugs needs to be investigated.

Table 8.1 gives an overall glimpse of ceramic membrane application for treatment of various types of industrial wastewater.

8.1.4 Domestic Wastewater Treatment

The increasing population and industrialization have led to depletion of water resources in order to fulfil the growing demands of civilization.

Table 8.1 Application of ceramic membranes for industrial wastewater treatment.

Type of membrane	Membrane material	Contaminants removed	Reference
Textile Wastewater			
MF	Moroccan Perlite	Dyes, COD	22
MF	Mixture of silica rich drinking water treatment sludge and clay	Turbidity, COD	23
UF	TiO_2/ZrO_2	Dyes, Salts	24
UF + NF	UF- Al_2O_3 NF - Al_2O_3/ZrO_2	Dyes, COD, salts, turbidity	25
MF	Clay/α-Al_2O_3	Dyes	26
MF	Tunisian apatite	Dyes	26
Tannery Wastewater			
MF	ZrO_2	COD, suspended solids	27
MF	Clay/α-Al_2O_3	COD, Turbidity, BOD (below detection limit), TDS	7
MF	Al_2O_3	COD	28
MF	Pozzolan	Turbidity	29
Distillery Wastewater			
MF + UF	$Al_2O_3 + ZrO_2 + TiO_2$	COD, solids, nitrogen content	30
MF	TAMI membrane	BOD, COD, total nitrogen, organic matter	31

(*Continued*)

Table 8.1 Application of ceramic membranes for industrial wastewater treatment. (*Continued*)

Type of membrane	Membrane material	Contaminants removed	Reference
MF + UF	Multi-channel, NOVASEP	TDS, dry matter, COD, and total nitrogen respectively	32
UF	ZrO_2+ TiO_2	Chloride, phosphate, sulfate, TDS	33
Dairy Wastewater			
MF	Kaolin, quartz, ball clay, pyrophyllite, feldspar and calcium carbonate	COD	34
MF+UF	ZrO_2+ TiO_2	COD, color, turbidity, and suspended solids	36
UF1+UF2	UF1- Al_2O_3 UF2 - TiO_2	Lactose, proteins, BOD, and COD	37
MF	TiO_2	Bacteria and spores	38
Poultry Wastewater			
MF	Flyash	COD, turbidity	9
MF	Natural aluminosilicates	COD, microorganisms, insoluble components, total carbon content	40
Petrochemical Wastewater			
MF	γ-Al_2O_3	COD, coke	43
MF	ZrO_2- Al_2O_3	Hydrocarbons	44

(*Continued*)

Table 8.1 Application of ceramic membranes for industrial wastewater treatment. (*Continued*)

Type of membrane	Membrane material	Contaminants removed	Reference
MF	α-Al_2O_3	TSS, turbidity, and oil and grease	45
MF	Kaolin-MnO_2	Oil	47
MF	Quartz, feldspar, kaolin	Oil	48
Pharmaceutical Wastewater			
MF	Clay- Al_2O_3	COD, TSS, turbidity	49
UF	ZrO_2/TiO_2-GO	ibuprofen, diclofenac, carbamazepine	50
UF	TiO_2-GO	Pharmaceuticals, natural organic matter (NOM), humic acid, tannic acid, and salts	51
MF	Clay- Al_2O_3 modified by PDMS	Ciprofloxacin	52

In India, the volume of domestic wastewater produced is almost twice that of industrial wastewater [60]. The wastewater contains large amounts of organic and inorganic components, such as pharmaceuticals and personal care products (PPCPs), pesticides, suspended solids, humic acid, salts, nitrogen, phosphorous, etc. Use of day-to-day household ingredients, cookwares, drugs, cosmetics, pesticides, etc., are the main sources of accumulation of such toxic contaminants in domestic water via household and agricultural run-off. However, there is a lack of proper treatment facilities for such wastewater compared to industrial wastewater [60]. The membrane separation process appears to be appealing for the treatment of domestic wastewater as well as valuable resource recovery.

Flat sheet ceramic membranes have been used for treatment of raw domestic wastewater collected from a sewage treatment plant in Hong Kong

[61]. The primary advantage of a flat sheet membrane is its ease of installation and upgradation in any water treatment plants. The membranes were composed of α-Al_2O_3 and had pore diameter of about 100 nm. Coagulation was applied as a pretreatment step to minimize membrane fouling effects. About 99% of TSS and phosphorous removal and 90% COD removal was attained by this process. Long-term filtration studies showed stable permeate flux of 41.7 L/m^{2}h. High flux recovery upon cleaning with acid, alkali or hydrogen peroxide for just 10 min after a span of 3-5 days proves the low-fouling nature of the membrane surface. In continuation with the previous work, the same group of researchers studied the fouling mechanism for filtration of domestic wastewater by the flat sheet membrane [10]. The cake layer formed by deposition of particles on the membrane surface could be eliminated by providing backwash and aeration. The backwashing was performed with 0.1 (M) solutions of NaOH and HCl successively. The cleaning protocol was found to be cost-effective and energy-efficient. The municipal wastewater, collected after primary biological treatment, was subjected to ultrafiltration using ceramic membranes [62]. The membranes had pore diameter of 20 and 50 nm and were efficient in removal of the BOD, COD, TSS, nitrogen, phosphorous, microorganisms from the wastewater. The permeate flux recovery, obtained by application of backflushing for 0.5 sec at an interval of 1 min, was higher for the 50 nm membrane compared to the 20 nm membrane. However, due to large pore size the separation of the contaminants decreased for the 50 nm membrane. It has been observed that the colloidal substances and organic matter present in domestic sewage are responsible for fouling of the ceramic membrane [63]. Application of the 0.5% NaOH solution and 7.5% of H_2O_2 solution were found effective as membrane-cleaning chemicals. A bench-scale NF membrane system was used to treat secondary wastewater contaminated with PPCPs [64]. The membrane was composed of Al_2O_3 as the support layer and TiO_2 as the active separating layer and possessed MWCO of 200 Da. Size exclusion was the dominant mechanism responsible for separation of PPCPs by the NF membrane, as evident from the increase in their rejection with increase in molecular weight. The difference in hydrophobicity/hydrophilicity of the PPCPs did not affect their separation, thus eliminating the effect of hydrophobic interaction. The negatively charged PPCPs underwent greater removal than the positively charged contaminants due to electrostatic interaction between the negatively charged membrane surface and the contaminants. Membrane fouling was mitigated by flushing with ozonated water (4 mg/L) which resulted in almost complete flux recovery unlike ozone-free water where only 10-20% flux can be recovered. High flux ceramic UF membranes fabricated by dip coating of CuO and ZnO

over clay-Al_2O_3 support tubes were used for removal of the various PPCPs, such as ciprofloxacin, atenolol, and ibuprofen [65, 66]. The process parameters were optimized to maximize the contaminants removal and permeate flux. Negligible toxicity effects of the membrane-treated water upon plants proved the reusability of the membrane-treated water. The concept of hydrophobic interaction among the contaminants and the membrane was implemented for an effective process development to fulfil the target of PPCPs removal [67, 68]. Tubular single-channel clay-Al_2O_3 ceramic membranes were hydrophobically modified by PDMS and applied for ciprofloxacin removal [67]. More than 90% removal was attained within 60 min of filtration. In another study, the authors made an attempt to minimize the membrane preparation cost and utilize the waste-to-resource concept [68]. Waste tannery sludge was immobilized in clay-Al_2O_3 matrix for preparation of ceramic membranes with enhanced permeate flux, mechanical, and chemical stability compared to the pristine clay-Al_2O_3 support. The formation of a needle-like mullite phase in the presence of Cr containing sludge improved the membrane properties compared to the membrane

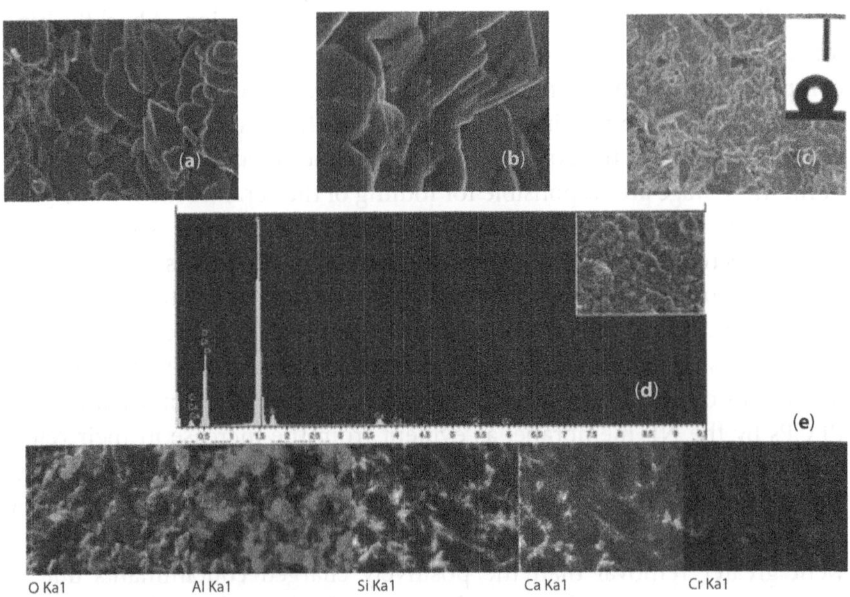

Figure 8.5 FESEM of porous ceramic support (a) without and (b) with tannery sludge; (c) PDMS modified membrane; (d) area analysis and (e) elemental mapping in EDS analysis.

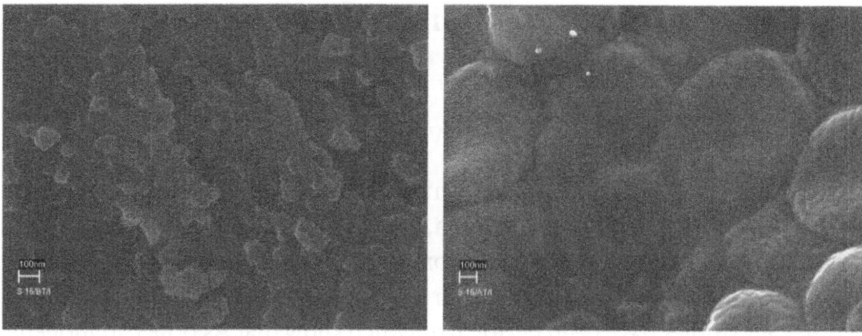

(a) (b)

Figure 8.6 Surface microstructure of γ-alumina coated ceramic ultrafiltration membrane: (a) before and (b) after domestic kitchen-sink wastewater treatment.

formed without sludge addition (Figure 8.5). The membrane was efficient in removing atrazine, a pesticide, at low pressure. Pneumatic backpulsing was applied at optimized intervals to minimize the membrane fouling effects.

In recent years, the trend of using membrane technology for domestic sewage treatment has greatly evolved. However, the main limitations are high membrane preparation cost and membrane fouling which restricts their commercial uses. More research efforts focusing on the aforementioned areas are required for development of robust membrane systems targeting non-specific contaminant removal from wastewater and their subsequent scale-up studies. This can surely replace the existing conventional water treatment systems worldwide in the near future through sustainable, clean and green technology establishment.

Greywater is an important area from wastewater recycling perspectives. Wastewater with high organic loading, oil and grease and detergent content collected from the kitchen-sink of an office canteen during washing of utensils at lunchtime was treated by γ-alumina coated ceramic ultrafiltration membrane [4]. COD removal of 73-90% was achieved in a single-stage UF process which increased to 84-90% by incorporation of the ceramic MF membrane as pretreatment step. The treated water was found suitable for agricultural reuse. Figure 8.6 shows the surface microstructure of the UF membrane before and after greywater treatment as observed in the FESEM analysis. The used membrane surface showed random particle deposition resulting in partial blocking of the active pores. Incorporation of MF as pretreatment substantially contributed to the flux enhancement (~93%).

8.2 Treatment of Contaminated Groundwater and Drinking Water

8.2.1 Arsenic Contaminated Water

Arsenic (As) is the most abundant element found in groundwater and drinking water across several parts of India, Bangladesh and other developed and developing countries. Arsenic can be present in organic as well as in inorganic form. The inorganic form exists as arsenate (As^{+5}) and arsenite (As^{+3}) [11]. Numerous studies have been performed which show that As pose carcinogenic and other serious health hazards to human beings, if the concentration exceeds the maximum allowable limit of 10 ppb [69, 70]. The toxic effect of As^{+3} is more pronounced than As^{+5}. Arsenic contamination of water occurs due to anthropogenic and natural activities. Till date, various processes, such as adsorption, coagulation, biological treatment and oxidation have been applied for arsenic removal. However, a closer look reveals that they are accompanied with some disadvantages, such as low efficiency, sludge production, handling of toxic chemicals, etc., which restrict their wide-scale industrial application. On the other hand, ceramic membranes can overcome these limitations and hence can be successfully applied for treatment of arsenic-contaminated groundwater and drinking water.

Tubular UF Al_2O_3 membranes were coated with iron oxide nanoparticles and were applied for arsenic remediation [71]. Adsorption of arsenic on the membrane surface was the dominant mechanism responsible for removal. The removal was found to be faster and higher for the membrane filtration system compared to batch adsorption using the unsupported Fe_2O_3 nanoparticles. The adsorption capacity was found to vary from 0.021-0.166 mg arsenic/m^2 Fe_2O_3 for initial As concentrations of 145 ppb-10 ppm. The proposed system can find its application as a point of use device due to the cost-effectiveness and simplicity of operation. A low-cost precursor, i.e., local clay of IIT Guwahati campus was used in preparation of ceramic support tubes [72]. The membranes were modified by coating with chitosan to obtain MF and UF membranes with pore diameter ranging from 13-760 nm. Arsenic, upon formation of chelation complex with polyvinyl alcohol resulting in increase in the size, was almost completely separated using the 13 nm UF membrane. The efficiency of the membrane was improved compared to reported NF and RO membranes, in terms of arsenic removal, permeate flux, and operating pressure. A group of researchers fabricated low-cost ceramic membranes using silica containing rice husk ash [73]. The membranes were modified hydrophobically by silane grafting and applied for As^{+5} and As^{+3} removal from water via direct contact membrane distillation. Influence of various parameters, such as feed temperature, solution pH, and As concentration (1-1000 ppm) upon arsenic removal were investigated. The concentration of arsenic was reduced to below 10 ppb, ensuring the water is safe for use. A hybrid process incorporating two tubular ceramic membranes connected in series, was applied for As^{+5} and As^{+3} removal from drinking water [74]. The module 1 consists of Al_2O_3 UF membrane subjected to hydrophobic surface modification by trichloromethylsilane. This membrane acts as ozone distributor to the water stream resulting in oxidation of As^{+3} to As^{+5}. This is followed by a second module consisting of a MF membrane composed of iron oxide nanoparticles coated on Al_2O_3 support for removal of As^{+5}. Almost complete removal of total arsenic from drinking water was obtained. Heat treatment at 110 °C was found to be effective in membrane regeneration. At CSIR-CGCRI Kolkata, ceramic membrane–based process has been developed, patented, and commercially implemented for removal of arsenic from groundwater [75, 76]. The system involves treatment of contaminated water via arsenic adsorbing media followed by filtration through porous multichannel ceramic MF membranes, coated with the aforementioned arsenic specific component. The process was successful in reducing arsenic concentration below 10 ppb from groundwater. The institute has installed several arsenic removal plants for community uses (2500-5000 L/day) at various arsenic contaminated regions in the eastern and northeastern parts of India (Figure 8.7).

Groundwater and drinking water in rural areas is contaminated with arsenic. Use of ceramic membrane instead of typically used sand filters increases

Figure 8.7 Ceramic membrane based arsenic removal plant developed by CSIR-CGCRI technology.

the efficiency, process longevity. The membranes could either be fabricated from arsenic adsorbing component or regulating the pore size. Efforts should be made for development of cost-effective ceramic membrane systems at pilot scale, for installation in various rural areas to provide pure drinking water.

8.2.2 Treatment of Fluoride Contaminated Water

Fluoride (F^-) is an essential element required in human beings at low concentration range of 0.5 mg/L to 1 mg/L, for growth of bones and teeth [77]. However, increase in their concentration above 1.5 mg/L, which is defined as the maximum allowable limit by WHO [78], results in various diseases in humans, such as fluorosis, infertility, arthritis, etc. Human activities and natural phenomenon causes transfer of F^- between rocks, soils, and water. In this regard, efficient, simple defluoridation techniques need to be implemented, such as ceramic membrane–based processes.

Ceramic membrane composed of zirconium metal organic framework (Zr-MOF) deposited over Al_2O_3 support tubes were applied for fluoride removal from drinking water [78]. The membrane having 20 μm thickness removed about 98% fluoride from water collected from a village in China, containing about 2.8 mg/L fluoride. The presence of hydroxyl groups and Zr metal centres in the membrane surface were responsible for fluoride removal from water. The removal was dependent on feed flow rate and initial fluoride concentration. A novel biocompatible membrane comprising hydroxyapatite (HAP) nanowires modified by $Al(OH)_3$ nanoparticles were fabricated and used for fluoride removal [79]. Electrostatic interaction and anion exchange between the membrane surface and fluoride ion played the key role in removal. Systematic studies show that the relative contribution of HAP nanowires and $Al(OH)_3$ nanoparticles towards fluoride removal are 36.7% and 63.3%, respectively. Fluoride was reduced below the maximum allowable limit of 1.5 mg/L. Fluoride containing drinking water collected from Assam, India, was efficiently treated by electrocoagulation followed by ceramic MF process [80]. About 17 states in India, which include Assam, Orissa, Gujarat, Madhya Pradesh, and Andhra Pradesh are immensely affected by high concentrations. The collected water samples had fluoride concentrations of 7.89, 4.79, 1.49 mg/L which were removed by the developed membrane-based process in just 20 min of operation. Incorporation of waste flyash as the membrane-forming precursor, reduced the cost of the process. Additionally, turbidity, dissolved oxygen, conductivity, pH of the membrane treated groundwater was reduced below their acceptable limit, concluding the reusability of the water. Ceramic water filters were developed from clay, sawdust, and bonechar and their efficiency was assessed for

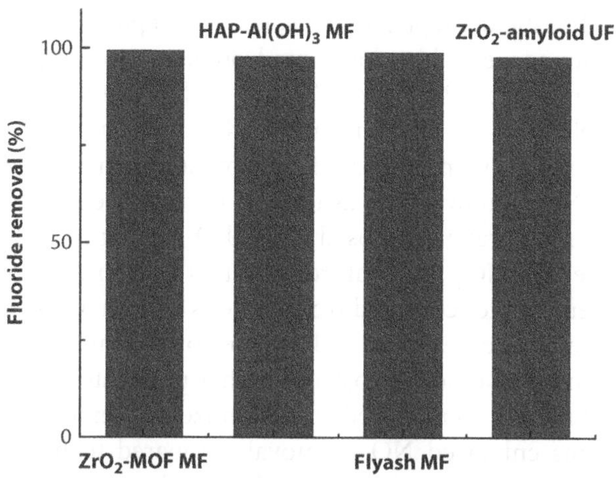

Figure 8.8 Comparative analysis of fluoride removal from contaminated water by ceramic membranes.

fluoride removal from synthetic and contaminated groundwater [81]. The optimized filter possessing pore diameter, surface area, and porosity of 5 nm, 7.30 m2/g, and 35.89%, respectively, resulted in more than 90% removal of fluoride. A novel carbon-based membrane was developed comprising amyloid fibril-ZrO_2 nanoparticles for remediation of fluoride from contaminated water [82]. The amyloid fibrils of β-lactoglobulin acts as template which assists in *in situ* formation of ZrO_2 nanoparticles having particle size less than 10 nm. Carbon matrix enhances the permeability of the membrane. The membrane showed about 99% removal efficiency for low and high initial fluoride concentrations, including for tap water containing 2.8 mg/L of fluoride. The membrane was found to treat 500 L of water/m^2 of membrane area even at high concentrations of 200 mg/L. The membrane could be regenerated by washing with solution of NaOH and NaCl, which results in maximum flux recovery. Figure 8.8 gives a comparative pictorial representation of fluoride removal from contaminated water by various membranes.

Drinking water in India is highly contaminated with fluoride. Ceramic membranes have been efficient in successful removal of fluoride (~100 %) from contaminated water. Application of a pretreatment step such as coagulation and adsorption enhances the removal efficiency of the process.

8.2.3 Treatment of Nitrate Contaminated Water

Nitrate (NO_3^-) is another hazardous contaminant widely found in groundwater and drinking water. It finds its way into the water due to use of fertilizers and pesticide, septic systems, etc. It is reported to cause serious diseases in humans, e.g., blue baby syndrome, stomach cancer, etc. [83]. The US Environmental Protection Agency and WHO have set the maximum allowable limit for NO_3^- in water as 45 mg/L and 50 mg/L, respectively [12]. Ceramic membranes offer viable alternative to conventional processes for NO_3^- removal such as RO, ion exchange, electrodialysis.

A tight UF membrane composed of γ-Al_2O_3 having pore diameter of 5 nm was applied for NO_3^- removal from synthetic water and contaminated drinking water samples collected from various sites in Morocco [12]. The UF membrane overcomes the disadvantages of typically used NF or RO membranes, in terms of lower cost and higher permeates flux. Parameters such as pH, pressure were varied to maximize the removal. Addition of divalent cations enhanced NO_3^- removal compared to monovalent cations due to increased electrostatic repulsion. Change in pH of the feed solution caused changes in the membrane surface charge and maximum NO_3^- removal was obtained at the point of zero charge. NO_3^- removal was found to be directly proportional to increase in pressure. A pilot-scale plant incorporating ceramic membrane filtration combined with ozonation and carbon filtration have been developed for removal of NO_3^- and other contaminants and provide safe drinking water [84]. Flat sheet, multi-channelled membranes of Al_2O_3 having pore diameter of 100 nm were used. Experiments have been conducted for ten months to assess the viability of the process in terms of contaminants removal efficiency and membrane stability. NO_3^- was reduced below the acceptable limit of WHO. COD, turbidity, phosphorous, and organic matter were also removed sufficiently. Ozonation resulted in degradation of contaminants in the water to a certain extent and also reduction in membrane fouling. Backwashing was applied at an interval of 30 days, which shows the low-fouling of the process. TiO_2/clay ceramic membranes were applied for NO_3^- removal by surface adsorption in the range of 0-20 mg/L [85]. The membrane filtration process was compared with adsorption using the aforementioned material as adsorbents.

The scope of NO_3^- removal by ceramic membranes by size exclusion or surface interactions has not been widely explored as compared to other contaminants. Research can be conducted in this area to provide a robust technology for drinking water purification.

8.2.4 Treatment of Water Spiked with Emerging Contaminants

Nations strive to provide safe drinking water to all people. However, population growth leads to increased usage and accumulation of various day-to-day products in groundwater and surface water, both of which act as a drinking water source. Unmonitored release of such pesticides, PPCPs including drugs, cookwares, cosmetics, flame retardants, etc., leads to increasing concentration of their constituting chemicals [86]. These emerging contaminants affect various parts of the human body, including the nervous, reproductive, and endocrine system [87]. Most drinking water treatment processes put minimum focus on these emerging contaminants. In this regard, designating an efficient system to fulfil the aforementioned purpose is the need of the hour.

A microwave-assisted ceramic membrane filtration was developed for removal of a recalcitrant contaminant, 1,4-dioxane from simulated synthetic water [88]. The mechanism involved irradiation and penetration of microwave radiation through the membrane surface, creating hot spots for degradation of the contaminants. Microwave responsive material $BiFeO_3$ had been used to achieve this target. A flat sheet ZrO_2/TiO_2 membrane coated over α-Al_2O_3 support tubes having pore size of 140 nm was used as the ceramic membrane matrix. The target microwave responsive membrane was obtained by dip coating the ceramic membrane with $BiFeO_3$ solution. The membrane degrades the pollutants upon microwave exposure and hence results in minimized fouling effects. Thus it possesses high contaminant separation efficiency and low fouling nature, which are basic criteria for developing a sustainable membrane-based process. Drinking water samples in China contaminated with PPCPs and other emerging contaminants were treated by coagulation, ozonation, and ceramic membrane filtration successively [13]. The 120 m^3/day pilot-scale plant incorporated flat sheet UF membranes having 60 nm pore diameter. The advantage of flat sheet membranes is their easy cleaning and replacement, making them suitable for industrial applications. The system was able to remove 98% of eight EDCs and fourteen PPCPs along with other contaminants from the drinking water. Ozonation resulted in improved removal efficiency as well as reduction of fouling of the membrane. In this integrated process, coagulation could remove only 10% of the emerging contaminants, whereas it increases to 50% and 90% upon UF and UF+ozonation, respectively. Based on the membrane pore size, the probability of size exclusion may be ruled out as the separation mechanism. The physico-chemical property of the membrane and contaminants results in various interactions between them, such as hydrophobic/hydrophobic interactions, etc., which results in their separation. Similar protocal of combining ozonation with membrane filtration have been followed by Stylianou *et al.* for treatment of groundwater [89]. The system acted as gas-liquid membrane contactor, which causes increased mass transfer for membrane with higher length and lower inner diameter. The process efficiently removed atrazine, methyl-tert-butyl-ether, and endosulfan. Endosulfan was reduced below the maximum limit of 0.1 μg/L after 30 min of operation. A pilot plant study has been performed for treatment of drinking water in rural areas in China [90]. The process involving combination of ozonation, ceramic UF membrane followed by activated carbon adsorption had been applied for removal of 37 types of PPCPs, colour, turbidity, organic matter, etc., found in the collected drinking water samples. Twenty flat-sheet membranes, composed of Al_2O_3, each having an area of 0.5 m^2 were used. Contaminants degradation occur in the pores of the membrane, which function as nano-reactors. About 52-62% of PPCPs are removed by ozonation and membrane filtration, and the removal is further enhanced upon activated carbon treatment. Moreover, cost-analysis studies show that the cost of this integrated process is almost similar to that of conventional water treatment systems. The robust, automated, cost-effective pilot-scale system provides pure and safe drinking water, following the stringent criteria of WHO.

A diverse group of emerging contaminants has different properties which results in varied interaction with a particular ceramic membrane. Hence, in order to efficiently remove these contaminants, a thorough study regarding properties of membrane as well as contaminants needs to be performed. Tailoring of pore size, functionality, surface properties of the membrane should be tailored to achieve a desired level of removal. Ceramic membrane–based systems, in amalgamation with coagulation and ozonation, appears to be an effective technology for drinking water treatment.

8.2.5 Treatment of Water Contaminated with Pathogens

The heath of human beings deteriorates due to consumption of drinking water contaminated with bacteria, virus, and other pathogens [91]. Additionally, although virus and bacteria are being used in many applications, their separation from the solution remains inefficient, resulting in accumulation in water [92].

During the last few years, structurally modified ceramic membranes are being fabricated to obtain zero pathogen discharge from drinking water. The size of virus is in nanometer range (~20-100 nm), whereas bacteria falls in micron range (1-4 μm). Thus, bacterial separation usually proceeds via size exclusion unlike virus, where surface interaction is the dominant mechanism. Yttria stabilized ZrO_2 (YSZ) capillary membranes have been fabricated for removal of virus [14]. Variation of microscopic and macroscopic properties of the membrane, such as pore diameters (24 to 146 nm) was achieved by varying particle size of YSZ. Membrane prepared from 40 nm YSZ particles showed log reduction value (LRV)>4 for viruses MS2 and PhiX174, hence satisfying the WHO and US EPA standards. The separation occurs due to size exclusion and surface adsorption, and increases with increase in filtration time due to adsorption on the cake layer. The fouling was mitigated by application of pneumatic backpulsing technique. In another work, a tubular microporous ceramic membrane fabricated from yttria coated diatomaceous earth was applied for virus removal based on electrostatic adsorption [93]. LRV of 5-6 and more than 99% removal of the MS2 virus, having size of 25 nm, were obtained at 3 bar pressure. The membrane surface was positively charged and hence could effectively adsorb oppositely charged viruses from water. However, leaching of the coating material for long-term operations limits further application of this membrane. Low-cost (<10$) water purifiers consisting of ceramic filters are installed in households in Cambodia for purification of drinking water [94]. More than 25% *Escherichia Coli* and MS2 bacteriophage were removed. Slip casting technique was adapted for preparation of low-cost microporous candle filters using lithium alumino-silicates [95]. The efficiency was assessed in removing various bacterial species such as, *E. Coli, B. Cerues, S. Faecalis.* from drinking water samples collected from Africa. Filtration studies show that LRV was greater than 6, thus satisfying the criteria of USEPA. Ceramic membranes were prepared from quartz, kaolin and other precursors via uniaxial pressing and sintering [96]. The optimized membrane is hydrophobic and had pore diameter of 1.4 μm and porosity of 30%. About 85% bacterial removal was obtained by size exclusion. The rejection was found to decrease with increasing pressure and decreasing initial concentration. The membrane cost was calculated to be 61$/m^2 which is much lower compared to typical Al_2O_3 membranes (~500$/m^2). MF membrane prepared from china clay, quartz and calcium carbonate (ore former) via uniaxial compaction successfully removed 90.24% bacterial cultures from water [97]. Ultrasonication was applied for membrane cleaning and high flux recovery was obtained. From cost analysis, the membrane cost was found to be Rs 987/m^2.

Thus, ceramic membranes found wide application in pathogens removal from drinking water via surface interaction and size exclusion. However, desorption and inactivation of the pathogens from the membrane surface needs to be explored further.

Table 8.2 gives an overall glimpse of ceramic membrane application for treatment of contaminated groundwater and drinking water.

Table 8.2 Application of ceramic membranes for contaminated groundwater and drinking water treatment.

Type of membrane	Membrane material	Contamination site	Reference
Arsenic Contaminated Water			
UF	Fe_2O_3	Groundwater	64
UF/MF	Clay	-	65
MF	Rice husk ash modified by silane grafting	-	66
UF-MF	UF- Al_2O_3 subjected to modification by trichloromethylsilane MF- Fe_2O_3	Drinking water	67
Fluoride Contaminated Water			
MF	ZrO_2 MOF	Drinking water	71
MF	Hydroxyapatite nanowires modified by $Al(OH)_3$	-	72
MF	Flyash	Drinking water	73
UF	Amyloid fibril-ZrO_2 nanoparticles	Tapwater, Groundwater	75
Nitrate Contaminated Water			
UF	γ-Al_2O_3	Drinking water	12
MF	TiO_2/clay	-	78
Water spiked with emerging contaminants			
MF	$BiFeO_3$	-	81
UF	Al_2O_3	Drinking water	13
UF	Al_2O_3	Drinking water	83
Water contaminated with pathogens			
UF	Yttria stabilized ZrO_2	Drinking water	14
MF	Yttria coated diatomaceous earth	-	86
MF	Lithium alumino-silicates		88
MF	Quartz, kaolin		89

8.3 Classification of Filtration Based on Configuration

8.3.1 Direct Membrane Filtration

Membrane filtration applies the pressure difference between the feed and permeate side as the driving force for fluid (water in our case) transport across the membrane. Contaminants such as dissolved and suspended solids, organic and inorganic components, etc., are separated from the water due to size exclusion or interactions with the membrane surface. Thus, the permeate produced upon membrane filtration is free from such contaminants.

Size Exclusion

The difference in particle size of contaminants with that of membrane pores results in separation by membrane filtration. MF, UF, NF, and RO are the principle pressure driven one stage membrane separation techniques. The process can be applied for removal of large particles (via MF) to small components (via NF, RO). The process must be selected based on the physical properties of the targeted contaminant. Proper process selection is important to ensure maximum separation efficiency and minimum operational cost. The rejection of a component by membrane process can be expressed according to equation 8.1:

$$\text{Rejection} = \left(1 - \frac{C_p}{C_f}\right) \tag{8.1}$$

C_f and C_p represents the concentration in feed and permeate respectively. The rejection can vary between 0 ($C_p = C_f$), which implies negligible membrane separation, and 1 ($C_p = 0$), indicating 100% removal of the contaminants.

Table 8.3 summarizes the characteristics of these processes [98, 99].

Higher pore size (100 nm to 10 μm) and porosity in MF membrane leads to high permeate flux even at low operating pressure. This makes the system energy-efficient and cost-effective, although its application based on sieving mechanism is limited. The macroporous supports are usually prepared by a one-step method such as extrusion and do not require an additional membrane coating. They can be used in removal of suspended solids, bacteria, and colloidal matter from wastewater stream, as their size is greater than the membrane pores. Hence, they may be used as a pretreatment step for other membrane processes, in order to reduce the

Table 8.3 Classification of types of membranes used in water and wastewater treatment.

Parameters	**Membranes**			
	Microfiltration	**Ultrafiltration**	**Nanofiltration**	**Reverse Osmosis**
Pore size (nm)	100-10000	2-100	0.5-2	<0.5
Operating pressure (bar)	0.1-2	0.1-5	3-20	5-120
Components Rejection:				
Suspended particles	✓	✓	✓	✓
Macromolecules	×	✓	✓	✓
Organic compounds	×	×	✓	✓
Multivalent ions	×	×/✓	✓	✓
Monovalent ions	×	×	×	✓
Applications in water treatment:	Removal of: bacteria suspended solids turbidity	Removal of: virus, bacteria macromolecules organics	Removal of: multivalent ions organics	Removal of: monovalent ions dissolved solids

fouling. Ceramic MF membrane has been applied for removal of microorganisms, suspended solids, and algae from lake water. MF membranes are being used in industries which generate wastewater contaminated with toxic heavy metals and other substances, such as pharmaceutical, food, semiconductor fabrication, etc. [100, 101]. Kaolinite-based ceramic MF membrane was used for treatment of oily wastewater [102]. Oil rejection of 97% was obtained after filtration of 90 min. The MF membranes in most of the studies showed fouling mitigation and considerable recovery of flux upon membrane cleaning and backwashing. UF membranes are applied for removal of smaller-sized contaminants such as organics, viruses, etc., while NF membrane removes ions. Wastewater generated from drilling of crude oil and gas was subjected to treatment using MF (pore size of 100 and 200 nm), UF (50 nm pore size and MWC of 20 kDa), and NF (MWCO of 1 and 0.75 kDa) membrane, respectively [103]. The membrane materials were Al_2O_3 and TiO_2. The pressure was varied from 0.5-2 bar and flux obtained in the range of 3.4 to 330 L/m^{2}hr for the membranes with varying pore size and MWCO. Almost complete removal of oil and about 49% TOC was obtained by the integrated membrane system. The wastewater generated from a paper industry in Italy was treated by tubular ceramic MF membrane followed by RO membrane in pilot scale [104]. This patented process could recover more than 80% of the produced wastewater and was found to be more efficient than polymeric membrane modules. Thus size-based separation using ceramic membranes is widely used in wastewater treatment.

Physico-Chemical Interactions

The membrane-based separation of the contaminants may also occur due to various interactions between the membrane surface and the contaminants, apart from size exclusion. This depends upon their morphological, structural, and chemical composition. The surface hydrophilicity/hydrophobicity, charge, presence of suitable functional groups determines the extent the separation efficiency.

Negatively charged TiO_2 UF membrane (MWCO of 1 and 3 KDA) was used for separation of dissolved phosphate from water [105]. Electrostatic repulsion between negatively charged membrane and negative phosphate ion is the main mechanism responsible for separation. Hence, the 3 KDa membrane, possessing more negative charge, showed higher phosphate rejection (86%) than 1 KDa membrane (75%), although its pores size is more than 1 KDa. pH was found to affect the charge and hence the separation efficiency. The Debye ratios were estimated to 1 and 2 for 3 and 1 KDa membrane, respectively. The proposed system can be applied as a step for RO pretreatment as it can prevent biofouling. Pesticide atrazine and pharmaceutical component ciprofloxacin was removed from surface water by the principle of hydrophobic interaction and π-π interactions [67, 68]. Low-cost tubular ceramic membranes were modified by a hydrophobic polymer PDMS and were successfully applied for removal of more than 90% of the contaminants at low pressure. Application of pneumatic backpulsing resulted in flux recovery and decrease in fouling effects. Al_2O_3 MF membranes were modified with CuO and ZnO in different studies and applied for removal of various PPCPs from water [65, 66, 106]. The effect of process conditions involving the pressure, filtration time and cross flow velocity upon removal and permeate flux were analyzed and optimized. The membranes resulted in more than 95% removal of the contaminants with a low flux decline. Adsorptive interaction between the contaminant and membrane surface are found to be responsible for removal as analyzed from FTIR studies. Two UF membranes were fabricated from locally available precursor kaolinite, zirconia and 3-mercaptopropyl-trimethyoxysilane and used for dye and salt rejection [107]. The negatively charged membrane resulted in about 80% and 100% anionic and cationic dye rejection, respectively, due to electrostatic interactions. Ceramic UF membrane with MWCO of 8800 Da was used for textile wastewater treatment. Electrostatic Donan effect resulted in greater than 98% dye rejection; however, salt rejection was only about 10-30%. The membrane showed improved separation and permeability compared to commercial polymeric membrane. These separations are less energy intensive and cost-effective compared to NF and RO process. Hence, tailoring of surface properties of UF and MF membranes are being performed to obtain a membrane with high flux and high separation efficiency.

8.3.2 Hybrid Approaches

Ceramic membrane filtration is often integrated with other processes, such as biodegradation, adsorption, coagulation, and photocatalysis for water and wastewater treatment. The details of the process are given in the following sections.

Membrane Bioreactor

The membrane bioreactor (MBR) process is the integration of microbial degradation of target contaminants followed by membrane filtration for biomass separation. Hence, the basic components include a bioreactor tank and a membrane module. The membrane module may either be placed outside the bioreactor or linked externally as in side stream mode (Figure 8.9a), or it may be present inside the bioreactor tank as in submerged mode (Figure 8.9b). The advantages of the MBR process over biodegradation or the conventional activated sludge (CAS) process involves reduction of footprint area and compact size. Increased removal of contaminants can be achieved compared to CAS due to variation of hydraulic retention time (HRT) and sludge retention time (SRT). Due to continuous recirculation, sludge production is reduced. Thus, MBR permeate produced is of higher quality, making the water suitable for reuse purpose [108].

In a study conducted by Guerra *et al.*, efficiency of removal of 99 PPCP components by MBR process was studied. While some PPCPs had high removal efficiency greater than 90%, some showed moderate and low removal depending upon their physicochemical properties [109].

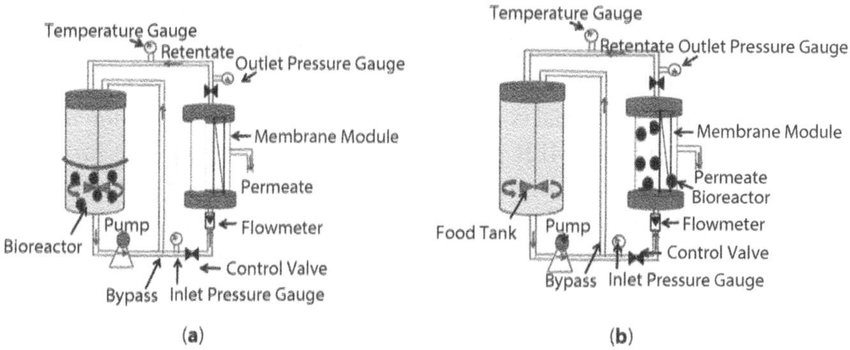

Figure 8.9 Schematic of MBR (a) side-stream mode (b) submerged mode.

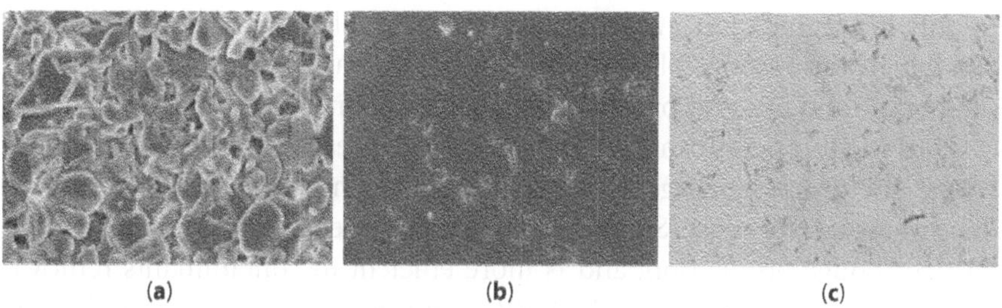

Figure 8.10 FESEM of (a) Al_2O_3 MF support and (b) Fe_2O_3 nanoparticle coated UF membranes used in the MBR process; (c) microscopic image of **E. Coli** retained by the membrane in MBR process.

Acetaminophen, ibuprofene, codein, triclosan, tetracycline, genfibrozil, caffeine, etc., were almost completely removed; components like ofloxacin, carbamazepin, metoprolol, albuterol, etc., were moderately removed whereas almost no removal was found for trimethoprim, oxycodone, meprobamate by the MBR process. Among the components studied, ibuprofen, codeine, caffeine, acetaminophen, gemfibrozil, etc., were mostly removed by biodegradation, whereas tetracycline, ciprofloxacin, norfloxacin, and triclosan were strongly adsorbed on the activated sludge resulting in their remediation by the MBR process. Iron oxide nanoparticle induced UF membrane was prepared by coating it over Al_2O_3 MF membranes (Figure. 8.10) using chitosan as the intermediate matrix [110]. The membrane was integrated with biodegradation in side-stream mode for removal of pesticide atrazine. In this process *E.Coli* was analyzed to be the dominant bacterial strain responsible for the pesticide biodegradation. More than 95% atrazine removal was obtained in the MBR process. Ultrafiltration membrane was prepared using graphene oxide coating on macroporous clay-alumina support and used in

MBR process for >95% removal of atrazine [111]. γ-alumina coated ceramic UF membrane was used in side-stream MBR process for textile wastewater treatment. An in-depth analysis revealed that the biodegradation process was majorly driven by a bacterial consortium of *E.Coli*, *Bacillus* sp., *Pseudomonas* sp. and *Shigella* sp., the MBR process was found highly effective for color (>98%) and COD (>98%) removal from textile wastewater [29]. Toxicity studies using the untreated and MBR treated water were performed on cat-fish, *H. fossilis.* The results showed that raw textile wastewater was toxic to the natural ecosystem, while the treated water could be safely discharged as it did not pose any threat to the aquatic system. Wastewater from biomass gasification, paper, and dairy industries was subjected to MF-MBR treatment for removal of COD and other contaminants [112]. COD removal for all the three systems were increased by about 40% in MBR process compared to only biodegradation. Additionally, the bacterial strain *Rhodococcus opacus,* which was responsible for biodegradation, was recovered from the MF membrane and used for biodiesel production upon extraction of the lipids.

Thus MBR process can be used to efficiently treat various types of wastewater. The process emerges to be cost-effective and highly efficient to produce water safe for reuse. Submerged mode MBR requires less footprint area, has less sludge production, and is more efficient in contaminants removal compared to side-stream mode MBR. The microorganisms are well separated from water by the UF/MF membranes. Ceramic membranes are resistant to bacterial film formation, thereby making them a suitable option for use in MBR system compared to polymeric membranes. Another concern in MBR process is maintaining the permeate flux at desired level. Bentonite clay-coated ceramic UF membrane having average pore diameter of 5 nm was used in the MBR process for treatment of tannery industrial effluent. Figure 8.11 shows permeate flux profile in the MBR process along with a comparison with that obtained in the direct UF process (S. Ghosh *et al.*, unpublished results). It was observed that flux increases with increasing transmembrane pressure for both the processes; however, the flux increment is relatively lower in the case of the MBR process than the UF membrane process since the feed in the MBR process contains activated sludge and biomass with high MLSS values (Figure 8.11a). Figure 8.11(b) show that flux declined with time for both the process due to the fouling and concentration polarisation in the inner surface of the membrane. Significantly, flux declination was higher in the case of the direct UF membrane process compared to the MBR process. This was due to the heavy organic loading in raw tannery effluent which blocks the membrane pores. In the MBR process due to the biodegradation organic loading of raw effluent was adequately reduced, therefore flux declination was less and steady. Figure 8.12 shows the removal efficiency of

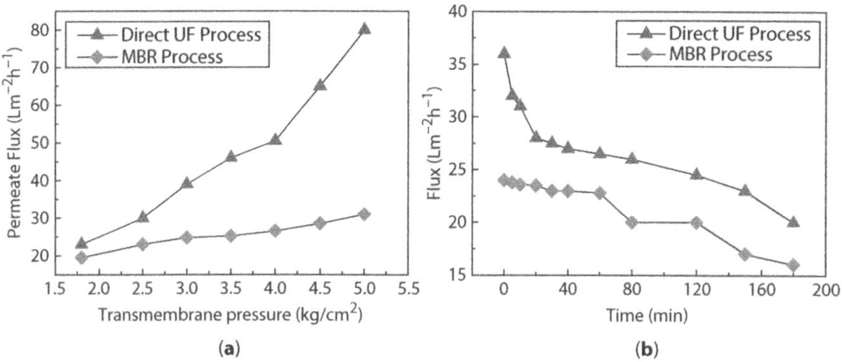

Figure 8.11 Permeate flux profile in bentonite clay-coated UF membrane-assisted MBR process and comparison with the direct UF process during treatment of tannery effluent: (a) variation of transmembrane pressure; (b) variation of operating time.

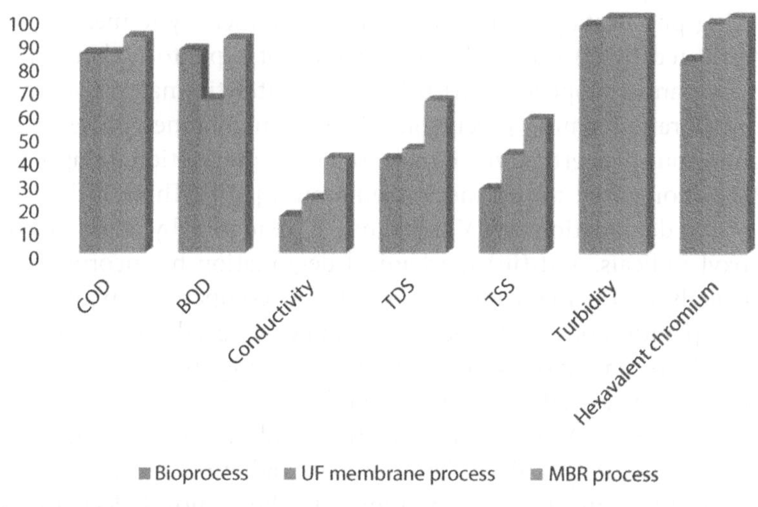

Figure 8.12 Removal efficiency of various parameters observed in the bioprocess, in bentonite clay-coated UF membrane-assisted MBR treatment and in the direct UF process for tannery effluent.

various parameters observed in the bioreactor and in MBR treatment, as well as in the direct UF process for tannery effluent.

Photocatalytic Membrane Reactors

Photocatalytic membrane reactors (PMRs) is a hybrid technology complementing the advantages of photocatalysis and membrane filtration

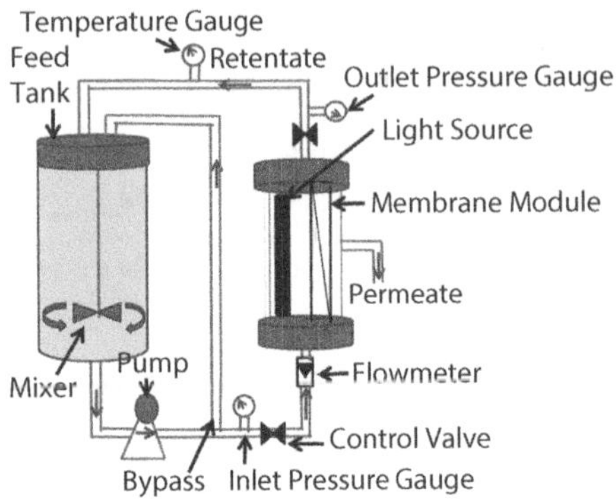

Figure 8.13 Schematic of PMR.

(Figure 8.13). PMRs are found in different configurations, namely (i) photocatalyst suspended in the reactor followed by membrane filtration, and (ii) using photocatalytic membrane [113]. Photocatalytic membrane can be fabricated by (i) immobilization/coating of a photocatalytic material over a ceramic support or (ii) using photocatalytic material as the initial membrane-forming precursor. Ceramic membranes possess certain advantageous properties which make them a better option for application in PMRs compared to polymeric membranes [114]. These include resistance to (i) degradation by UV radiation, (ii) oxidation by photo-generated hydroxyl radicals, and (iii) mechanical degradation by incorporation of photocatalysts. The main advantages of PMRs compared to photocatalysis are (i) separation of the photocatalyst can be achieved, (ii) enhanced efficiency, (iii) regeneration is easier, and (iv) no sludge production.

Humic acid was efficiently removed from wastewater using Ag-TiO_2/hydroxiapiate)/Al_2O_3 MF photocatalytic membrane [115]. Simultaneous effect of photocatalytic degradation by UV radiation and membrane filtration resulted in contaminant removal. The removal decreases with increasing TMP in cross flow filtration mode and was maximized at 1 bar. Mesporous TiO_2 was coated over porous Al_2O_3 support tubes for preparation of UF membrane with pore diameter about 50-70 nm, high pore diameter and porosity [116]. The membrane was applied for removal of Congo red dye (>95%) by the mechanism of photodegradation and adsorption. However, fouling occurred in this membrane, operated in dead-end mode, when filtration time exceeds 6 hr. Cross flow mode PMRs are less prone to fouling compared to dead-end mode due to continuous recirculation in the former case. N-doped TiO_2 was deposited via sol-gel method over α-Al_2O_3 MF membrane [117]. N doping shifted the photo-response towards visible range compared to pristine TiO_2. An emerging contaminant in the pharmaceutical category namely, carbamazepine having 1 mg/L concentration was removed by this membrane with 200 nm and 800 nm pore size. The same group of researchers applied the aforementioned PMR for inactivation of MS2 bacteriophage from surface water [118]. Virus removal of about 4.9 ± 0.1 log (>99.99%) was obtained in this PMR. Studies showed that presence of natural organic matter and increase in pH reduced the removal efficiency while addition of divalent cation Ca^{+2} increased the removal. The coating resulted in flux reduction of about 15%. $TiO_{2,}$ $SiO_{2,}$ silicon carbide (SiC) was deposited over non-oxide SiC substrate for removal of methylene blue dye [118]. The membrane had high hydrophilicity and low MWCO, which enabled high permeate flux while ensuring sufficient membrane separation. About 72% dye removal was obtained via this combined system in dead-end filtration mode.

PMRs offer an efficient method for removal of the catalyst unlike sedimentation. The main challenge lies in immobilization of a photocatalyst upon ceramic membrane matrix. The membrane could either be directly extruded using the photocatalyst as precursor or coated on any inorganic substrate via dip coating, sol-gel, etc. Although very few ceramic based PMRs have been commercialized, it is guaranteed to efficiently remove a plethora of contaminants and hence fulfill the stringent environmental rules.

Membrane Photobioreactor

Membrane photobioreactor (MPBR) emerges as an option to fulfil the energy-environment norms. MPBR consists of a photobioreactor coupled with a membrane filtration unit. Cultivation of microalgae via CO_2 sparging with the aid of membrane and in the presence of suitable nutrients is targeted. Growth of algae in the presence of wastewater, which complements for the nutrients such as nitrogen and phosphorus is performed to make the system environmentally sustainable. The algae can further be processed to produce biofuel. Hence, MPBR is a complete cycle which helps in (i) CO_2 sequestration, (ii) wastewater treatment, and (iii) biofuel production [119, 120].

Hydrophobic surface modified ceramic membranes in capillary configuration (3 mm OD/2 mm ID) were used in a membrane photobioreactor for algal biomass production with the aim of bio-sequestration of CO_2 [120]. Poly dimethylsiloxane (PDMS) was used as hydrophobic agent on bentonite clay-coated UF membrane surface which resulted in contact angle value of 147° (Figure 8.14). The major functions of the membranes were effective dissolution of CO_2 in the form of a nano bubble sparger along with partial

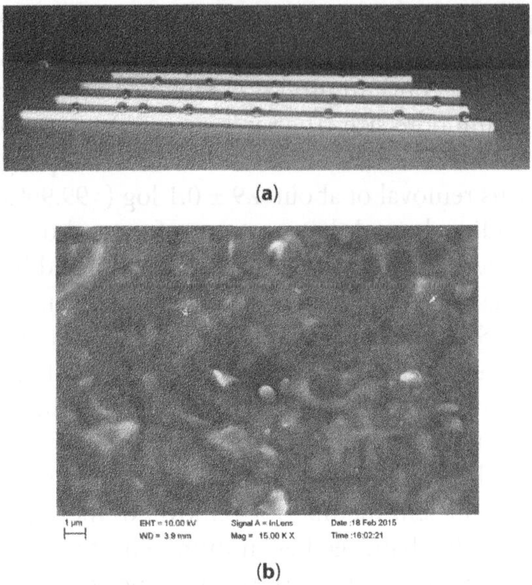

Figure 8.14 (a) Hydrophobic surface modified ceramic capillary membranes (b) FESEM of hydrophobic membrane surface.

elimination of the evolved dissolved oxygen during biomass growth in the culture media. The MPBR system was found efficient with high gas-liquid mass transfer coefficient via membrane module and enhanced biomass growth rate and productivity.

Although the aforementioned process is beneficial in the energy as well as environmental sector, it has not been explored much till date. Scope for improvement in gas-liquid mass transfer rate, stability of the hydrophobic layer, compact density of membrane modules and reduction in the footprint area are some of the critical issues to be addressed in a ceramic membrane-based photobioreactor system for sustainable technology development on algal biomass and biofuel production.

Another major application of ceramic membranes involves algal biomass harvesting. MF and UF membranes are generally used for this purpose. *Chlorella vulgaris* was harvested in a submerged ceramic MPBR and their efficiency was studied for nutrients removal from wastewater and biodiesel production at various hydraulic retention time (HRT) of 72, 24, 6.5 h [119]. The effluent from a submerged MBR was used as the influent for the MPBR. The average concentration of nitrogen in the effluent was about 55.45 mg N/L. The UF membrane was prepared from $ZrO_2/\alpha\text{-}Al_2O_3/TiO_2$ on $\alpha\text{-}Al_2O_3$ support. The membrane had pore diameter of 200 nm. The

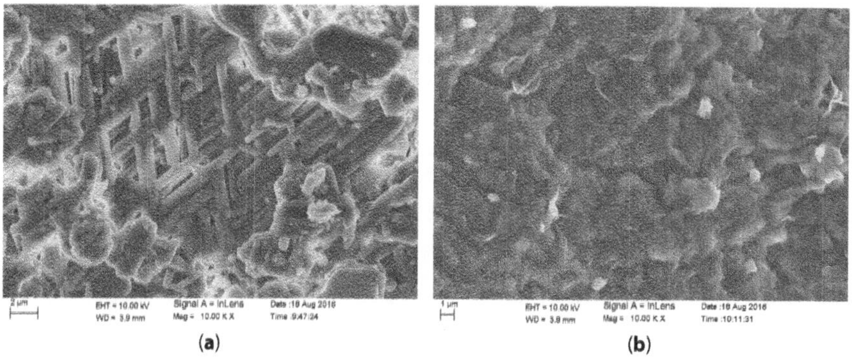

Figure 8.15 Surface microstructure observed in the FESEM of (a) MF support and (b) bentonite-coated UF membrane.

variation of HRT did not produce any significant effect on lipid content of the produced biodiesel, hence HRT was optimized based on optimum nutrient removal from wastewater by *Chlorella vulgaris*. The same group of researchers also investigated the effect varying HRT on membrane fouling [121]. It was concluded that the carbohydrate and proteins generated from the microalgal species were the key foulants. Particle size, as usual, also affected the fouling. Membrane operated at higher HRT of 72 h resulted in reduced fouling compared to HRT of 6.5 h. Higher HRT causes biological reactions at the membrane surface, thus converting high molecular weight components to low molecular weight, and facilitating their passage through the membrane pores. The efficiency of ceramic MF and UF membranes for harvesting of *Arthrospira sp.* was studied in detail [122]. Tubular MF membranes were prepared from clay and Al_2O_3, while the UF membrane had an active separation layer of bentonite clay (Figure 8.15). The coating resulted in reduction of pore size from 1 μm (MF) to around 30 nm (UF). High concentration factor (12.5) and volume reduction factor (11.11) was obtained in the MF process; on the other hand the UF process showed greater flux recovery (93%) and higher energy uptake compared to the MF process. The surface microstructure of the MF and UF membrane as observed in the FESEM images after algal biomass harvesting is shown in Figure 8.16. Due to the larger pore sizes, the MF membranes suffered more fouling with cake layer formation. In case of UF membranes instead of continuous cake layer formation, smaller algal fragments caused partial deposition in the pore mouths which could be eliminated by air back wash/back pulsing thereby, enabling easy regeneration of the membranes and increased harvesting efficiency [122].

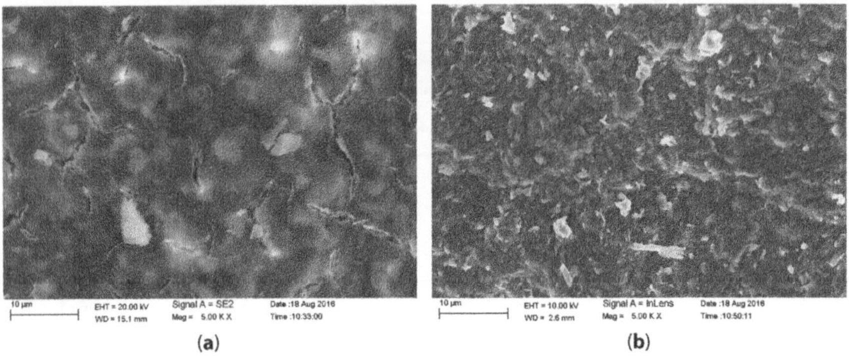

Figure 8.16 Surface microstructure observed in the FESEM of (a) MF support and (b) bentonite coated UF membrane after algal biomass harvesting.

8.4 Pilot-Scale Studies

Several research studies on ceramic membrane based processes have been performed through the years. However, most of the processes are limited to bench-scale application primarily due to lack of (i) funds, (ii) complete study of the system that would be required for scale-up, and (iii) system efficiency for long-term continuous mode operations. In spite of these shortcomings, membrane-based processes are becoming an integral part of the wastewater treatment industries. The demand for the ceramic membranes are increasing exponentially and even reached USD5.183 billion in 2018; an about 9.57% annual hike is expected in the period 2019-2025 [123]. Recent years have witnessed about 80% discharge of wastewater in the surface and landfills. This creates enormous scope for ceramic membranes to be efficiently applied in this end-use industry. Globally some of the producers of ceramic membranes are Pall Corporation, TAMI Industries, Veolia Water Technologies, Atech Innovations GmbH, JIUWU Hi-Tech Company, GEA Group AG, etc. [123].

Ceramic membrane–based pilot plants have been installed in different types of industries, both in India and other countries. A 19-chanelled, Al_2O_3 membrane-based pilot-scale plant has been installed by CSIR-CGCRI, India, for treating the effluent of a common effluent treatment plant which receives wastewater produced from various tannery industries [124]. A number of pilot installations have been performed by CSIR-CGCRI for treatment of arsenic and iron from contaminated groundwater [125]. This technology has been received and implemented by various companies such as, Zeolite (India) Pvt. Ltd., Victor Engineering Co., and Entech Metals Pvt. Ltd. The MBR technology is also being applied worldwide. In North America different manufacturers produce flat sheet and hollow fiber membrane-based MBR system [126]. According to a research study in 2006, about 258 MBR plants were in operation in North America for municipal wastewater treatment, and 2,200 MBR plants have been installed worldwide. While most of such MBR plants in USA have throughput less than 10000 m^3/day, one large-scale plant produces 26900 m^3/day of treated water. MBR technology is being increasingly adapted by Asian countries, specifically South Korea and Japan, for various domestic applications [126]. Drinking water collected from bore wells in Delhi was treated using pilot plant membrane system [127]. Al_2O_3 UF membrane of 50 nm pore diameter and flux of 150 L/m^2hr was used for the aforementioned purpose. Successful removal of suspended particles, microorganisms, turbidity was obtained. UF membranes have witnessed maximum boost in the commercial market in recent years. Among the membrane materials, Al_2O_3 and ZrO_2 finds maximum application.

The ever-increasing demand for producing pure water and the stringent rules of WHO makes it a necessary step to incorporate advances in the existing water treatment technologies. Thus ceramic membrane pilot plants will eventually find their place in academia and commercial industries.

8.5 Challenges of Ceramic Membranes

In spite of the wide application of the ceramic membranes for treatment of various types of contaminants from water and wastewater, some challenges exist which could be the focus of upcoming research studies. Preparation of ceramic membranes requires costly chemicals, as well as equipments/facilities such as extruder, membrane-coating units, high-temperature furnaces, etc. Polymeric membrane, on the other hand, can be prepared by a simple process of solution casting over glass plate. Fabrication of a uniform, homogeneous membrane over a ceramic matrix is comparatively difficult compared to polymeric membranes. Controlling of the pore sizes in the high temperature range is still a challenge in development of ceramic RO membranes. A detailed knowledge about interaction mechanism between the support and membrane material is essential for design and implementation of such processes with respect to the wide range of contaminants present in the water and wastewater systems. Formation of cracks on the membrane surface, as well as obtaining the straightness of the tubes during sintering is another roadblock which results in defective membranes. Tuning of pore diameter and membrane thickness by variation of process parameters, such as coating time and precursor concentrations might not always follow the anticipated path. Hence, the challenge lies in proper optimization of parameters to achieve the desired membrane features. Focus should be given on development of high flux nanofiltration membranes to improve the cost-effectiveness. Various research studies are being performed to mitigate these limitations so that the ceramic membranes of robust and stable nature would become a potential option for water and wastewater treatment in the near future.

8.6 Conclusion and Future Scope of Ceramic Membranes

The chapter aims to provide a glimpse of the demand of ceramic membranes in the commercial market in terms of opportunities, challenges, value and volume. The scope of ceramic membranes for water and wastewater treatment has been categorized according to applications, end-user industry, and composition [128].

Compositional classification: The different composition of the ceramic membranes tailors the morphological and chemical properties and their cost-effectiveness. Al_2O_3, ZrO_2, TiO_2 are typically used as precursors for ceramic membranes. Low-cost waste materials such as fly ash, clay, tannery sludge are used for membrane fabrication due to their negligible cost and scope of waste immobilization.
Classification based on application: More than 55% of share for ceramic membranes has been in the area of wastewater treatment. They are applied in various water and wastewater treatment industries which include tannery, textile, brewery, petrochemical, pharmaceutical and dairy, etc. Treatment of domestic sewage, contaminated groundwater, and drinking water has been studied by ceramic membrane–based processes.
Classification based on separation mechanism: Segmentation of the ceramic membranes market has been done on the basis of configuration of the membrane, i.e., single-stage separation or integrated process. Scope of the MF, UF and NF membranes has been explored in this chapter. MBR also finds a place in the commercial market for water treatment applications. Membrane photobioreactor and algal biomass harvesting are the promising new directions for ceramic membranes.

The primary motive is directed towards designing of ceramic membrane-based water treatment systems which can overcome the existing limitations. The filtration system should possess low fouling and self-cleaning properties, high efficiency for targeted contaminants removal present in the industrial and domestic wastewater samples, high throughput, reduction in energy requirement, and lowering of fabrication and maintenance cost. Upon fulfilling the aforementioned criteria, ceramic membranes may act as suitable replacement of the existing water treatment systems.

References

1. US EPA: United States Environmental Protection Agency. Link: https://www.epa.gov/.
2. Karabelas, A., Plakas, K. (2011). *Membrane Treatment of Potable Water for Pesticides Removal, Herbicides, Theory and Applications*, Marcelo Larramendy (ed.), ISBN: 978-953-307-975-2, InTech, Available from: http://www.intechopen.com/books/herbicides-theory-and-applications/membranetreatment-of-potable-water-for-pesticides-removal.
3. Yoshino, Y., Suzuki, T., Nair, B. N., Taguchi, H., & Itoh, N. (2005). Development of tubular substrates, silica based membranes and membrane modules for hydrogen separation at high temperature. *Journal of Membrane Science, 267*(1-2), 8–17.
4. Bhattacharya, P., Sarkar, S., Ghosh, S., Majumdar, S., Mukhopadhyay, A., & Bandyopadhyay, S. (2013). Potential of ceramic microfiltration and ultra-filtration membranes for the treatment of gray water for an effective reuse. *Desalination and Water Treatment, 51*(22-24), 4323–4332.
5. Sarkar, S., Bandyopadhyay, S., Larbot, A., & Cerneaux, S. (2012). New clay–alumina porous capillary supports for filtration application. *Journal of Membrane Science, 392*, 130–136.
6. Jegatheesan, V., Pramanik, B. K., Chen, J., Navaratna, D., Chang, C. Y., & Shu, L. (2016). Treatment of textile wastewater with membrane bioreactor: a critical review. *Bioresource Technology, 204*, 202–212.
7. Bhattacharya, P., Swarnakar, S., Mukhopadhyay, A., & Ghosh, S. (2016). Exposure of composite tannery effluent on snail, Pila globosa: A comparative assessment of toxic impacts of the untreated and membrane treated effluents. *Ecotoxicology and Environmental Safety, 126*, 45–55.
8. Prodanović, J. M., & Vasić, V. M. (2013). Application of membrane processes for distillery wastewater purification—a review. *Desalination and Water Treatment, 51*(16-18), 3325–3334.
9. Goswami, K. P., & Pugazhenthi, G. (2020). Treatment of poultry slaughter-house wastewater using tubular microfiltration membrane with fly ash as key precursor. *Journal of Water Process Engineering, 37*, 101361.
10. Zhao, Y. X., Li, P., Li, R. H., & Li, X. Y. (2020). Characterization and mitigation of the fouling of flat-sheet ceramic membranes for direct filtration of the coagulated domestic wastewater. *Journal of Hazardous Materials, 385*, 121557.

11. Dhadge, V. L., Medhi, C. R., Changmai, M., & Purkait, M. K. (2018). House hold unit for the treatment of fluoride, iron, arsenic and microorganism contaminated drinking water. *Chemosphere, 199*, 728–736.
12. Breida, M., Younssi, S. A., Bouazizi, A., Achiou, B., Ouammou, M., & El Rhazi, M. (2018). Nitrate removal from aqueous solutions by γ-Al2O3 ultrafiltration membranes. *Heliyon, 4*(1), e00498.
13. Fan, X., Tao, Y., Wang, L., Zhang, X., Lei, Y., Wang, Z., & Noguchi, H. (2014). Performance of an integrated process combining ozonation with ceramic membrane ultra-filtration for advanced treatment of drinking water. *Desalination, 335*(1), 47–54.
14. Werner, J., Besser, B., Brandes, C., Kroll, S., & Rezwan, K. (2014). Production of ceramic membranes with different pore sizes for virus retention. *Journal of Water Process Engineering, 4*, 201–211.
15. Nandi, B.K., Uppaluri, R., Purkait, M.K. (2008). Preparation and characterization of low cost ceramic membranes for micro-filtration applications. *Appl. Clay Sci.*, 42, 102–110.
16. Lee, M., Wu, Z., & Li, K. (2015). Advances in ceramic membranes for water treatment. In *Advances in Membrane Technologies for Water Treatment* (pp. 43-82). Woodhead Publishing.
17. CeraMem_ Full-size membrane modules. 2013 [cited 2013 21/08]. Retrieved from http://www.veoliawaterst.com/ceramem/en/fullsizemodulesceramem.htm.
18. Ceramic membrane filtration system. Drinking water treatment [cited 2014 10/01]. Retrieved from http://www.metawater.co.jp/eng/product/drinking/membrane_clarify/location.html.
19. Likuid-CBR_ filtration systems for MBRS. [cited 2014 10/01]. Retrieved from http://dioxide.com/20130916_Likuid_CBR_EN.pdf.
20. Ceramic membrane unit. 2013 [cited 2014 04/02]. Retrieved from http://water-solution.meidensha.co.jp/download/img/bb60-3163.pdf.
21. Ceramic membrane modules - existing and planned sizes. CeraMem_ Ceramic Membranes. Retrieved from http://www.veoliawaterst.com/ceramem/en/ceramemelementsizes.html.
22. Majouli, A., Tahiri, S., Younssi, S. A., Loukili, H., & Albizane, A. (2012). Elaboration of new tubular ceramic membrane from local Moroccan Perlite for microfiltration process. Application to treatment of industrial wastewaters. *Ceramics International, 38*(5), 4295–4303.
23. Mouratib, R., Achiou, B., El Krati, M., Younssi, S. A., & Tahiri, S. (2020). Low-cost ceramic membrane made from alumina-and silica-rich water treatment sludge and its application to wastewater filtration. *Journal of the European Ceramic Society, 40*(15), 5942–5950.
24. Ma, X., Chen, P., Zhou, M., Zhong, Z., Zhang, F., & Xing, W. (2017). Tight ultrafiltration ceramic membrane for separation of dyes and mixed salts (both NaCl/Na2SO4) in textile wastewater treatment. *Industrial & Engineering Chemistry Research, 56*(24), 7070–7079.

25. Ağtaş, M., Yılmaz, Ö., Dilaver, M., Alp, K., & Koyuncu, İ. (2020). Hot water recovery and reuse in textile sector with pilot scale ceramic ultrafiltration/nanofiltration membrane system. *Journal of Cleaner Production, 256*, 120359.
26. Sahoo, G. C., Halder, R., Jedidi, I., Oun, A., Nasri, H., Roychoudhurry, P., ... & Ben Amar, R. (2016). Preparation and characterization of microfiltration apatite membrane over low cost clay-alumina support for decolorization of dye solution. *Desalination and Water Treatment, 57*(57), 27700–27709.
27. Lafi R., Gzara L., Lajimi R.H., Hafiane A. (2018). Treatment of textile wastewater by a hybrid ultrafiltration/electrodialysis process. *Chemical Engineering and Processing - Process Intensification*, 132, 105–113.
28. Bhattacharya P., Ghosh S., Swarnakar S., Mukhopadhyay A. (2014). Reuse of textile effluent for dyeing using combined technology of ceramic microfiltration and surface treated sugarcane bagasse: toxicity evaluation using *Channa punctatus* as model, *Desalination and Water Treatment*, doi: 10.1080/19443994.2014.887035, 1–21.
29. Banerjee P., Sarkar S., Dey T., Bakshi M., Swarnakar S., Mukhopadhayay A., Ghosh S. (2014). Application of isolated bacterial consortium in UMBR for detoxification of textile effluent: comparative analysis of resultant oxidative stress and genotoxicity in catfish (Heteropneustes fossilis) exposed to raw and treated effluents, *Ecotoxicology*, 23, 1073–1085.
30. Santra B., Ramrakhiani L., Kar S., Ghosh S., Majumdar S. (2020). Ceramic membrane-based ultrafiltration combined with adsorption by waste derived biochar for textile effluent treatment and management of spent biochar, *Journal of Environmental Health Science and Engineering*, 18, 973–992.
31. Bhattacharya P., Dutta S., Ghosh S., Vedajnananda S., Bandyopadhyay S. (2010). Crossflow microfiltration using ceramic membrane for treatment of sulphur black effluent from garment processing industry, *Desalination*, 261, 67–72.
32. Majouli, A., Tahiri, S., Younssi, S. A., Loukili, H., & Albizane, A. (2012). Treatment of effluents coming from beamhouse section of tannery by microfiltration through cordierite/zirconia and alumina tubular ceramic membranes. *Journal of Materials and Environmental Science, 3*, 808–815.
33. Bhattacharya P., Majumdar S., Bandyopadhyay S., Ghosh S. (2016) Recycling of tannery effluent from common effluent treatment plant using ceramic membrane based filtration process: A closed loop approach using pilot scale study, *Environmental Progress & Sustainable Energy*, 35, 60–69.
34. Bhattacharya P., Roy A., Sarkar S., Ghosh S., Majumdar S., Chakraborty S., Mandal A., Mukhopadhyay A., Bandyopadhyay S., (2013) Combination technology of ceramic microfiltration and reverse osmosis for tannery wastewater recovery, *Water Resources and Industry* 3, 48–62.
35. Kaplan-Bekaroglu, S. S., & Gode, S. (2016). Investigation of ceramic membranes performance for tannery wastewater treatment. *Desalination and Water Treatment, 57*(37), 17300–17307.

36. Beqqour, D., Achiou, B., Bouazizi, A., Ouaddari, H., Elomari, H., Ouammou, M., ... & Younssi, S. A. (2019). Enhancement of microfiltration performances of pozzolan membrane by incorporation of micronized phosphate and its application for industrial wastewater treatment. *Journal of Environmental Chemical Engineering, 7*(2), 102981.
37. Lapišová, Kateřina, Vlček, Roman, Klozova, J., Rychtera, M., & Melzoch, K. (2006). Separation techniques for distillery stillage treatment. *Czech J. Food Sci, 24*(6), 261–267.
38. Vasić, V. M., Šćiban, M. B., Jokić, A. I., Prodanović, J. M., & Kukić, D. V. (2012). Microfiltration of distillery stillage: influence of membrane pore size. *Acta Periodica Technologica*, (43), 217–224.
39. Vasić, V. M., Šćiban, M. B., Kukić, D. V., Prodanović, J. M., & Maravić, N. R. (2015). Sequential micro and ultrafiltration of distillery wastewater. *Acta Periodica Technologica*, (46), 177–183.
40. Díaz-Reinoso, B., Moure, A., González, J., & Domínguez, H. (2017). A membrane process for the recovery of a concentrated phenolic product from white vinasses. *Chemical Engineering Journal, 327*, 210–217.
41. Kumar, R. V., Goswami, L., Pakshirajan, K., & Pugazhenthi, G. (2016). Dairy wastewater treatment using a novel low cost tubular ceramic membrane and membrane fouling mechanism using pore blocking models. *Journal of Water Process Engineering, 13*, 168–175.
42. Galvão, D. F. (2018). Membrane Technology and Water Reuse in a Dairy Industry. *Technological Approaches for Novel Applications in Dairy Processing*, 163.
43. Zielińska, M., & Galik, M. (2017). Use of ceramic membranes in a membrane filtration supported by coagulation for the treatment of dairy wastewater. *Water, Air, & Soil Pollution, 228*(5), 173.
44. Khider, K., Akretche, D. E., & Larbot, A. (2004). Purification of water effluent from a milk factory by ultrafiltration using Algerian clay support. *Desalination, 167*, 147–151.
45. Skrzypek, M., & Burger, M. (2010). Isoflux® ceramic membranes—Practical experiences in dairy industry. *Desalination, 250*(3), 1095-1100.
46. Trägårdh, G., & Johansson, D. (1998). Purification of alkaline cleaning solutions from the dairy industry using membrane separation technology. *Desalination, 119*(1–3), 21–29.
47. Almandoz, M. C., Pagliero, C. L., Ochoa, N. A., & Marchese, J. (2015). Composite ceramic membranes from natural aluminosilicates for microfiltration applications. *Ceramics International, 41*(4), 5621–5633.
48. Avula, R. Y., Nelson, H. M., & Singh, R. K. (2009). Recycling of poultry process wastewater by ultrafiltration. *Innovative Food Science & Emerging Technologies, 10*(1), 1–8.
49. Samaei, S. M., Gato-Trinidad, S., & Altaee, A. (2018). The application of pressure-driven ceramic membrane technology for the treatment of

industrial wastewaters–A review. *Separation and Purification Technology, 200*, 198-220.

50. Madaeni, S. S., Monfared, H. A., Vatanpour, V., Shamsabadi, A. A., Salehi, E., Daraei, P., ... & Khatami, S. M. (2012). Coke removal from petrochemical oily wastewater using γ-Al2O3 based ceramic microfiltration membrane. *Desalination, 293*, 87–93.
51. Salehi, E., Madaeni, S. S., Shamsabadi, A. A., & Laki, S. (2014). Applicability of ceramic membrane filters in pretreatment of coke-contaminated petrochemical wastewater: Economic feasibility study. *Ceramics International, 40*(3), 4805–4810.
52. Ghidossi, R., Veyret, D., Scotto, J. L., Jalabert, T., & Moulin, P. (2009). Ferry oily wastewater treatment. *Separation and Purification Technology, 64*(3), 296–303.
53. Abadi, S. R. H., Sebzari, M. R., Hemati, M., Rekabdar, F., & Mohammadi, T. (2011). Ceramic membrane performance in microfiltration of oily wastewater. *Desalination, 265*(1-3), 222–228.
54. Yang, T., Ma, Z. F., & Yang, Q. Y. (2011). Formation and performance of Kaolin/MnO2 bi-layer composite dynamic membrane for oily wastewater treatment: Effect of solution conditions. *Desalination, 270*(1-3), 50–56.
55. Nandi, B. K., Moparthi, A., Uppaluri, R., & Purkait, M. K. (2010). Treatment of oily wastewater using low cost ceramic membrane: Comparative assessment of pore blocking and artificial neural network models. *Chemical Engineering Research and Design, 88*(7), 881–892.
56. Bhattacharya, P., Ghosh, S., Majumdar, S., Dasgupta, S., & Bandyopadhyay, S. (2011). Biosorbent-assisted ceramic microfiltration process for treatment of herbal pharmaceutical wastewater with high organic loading. *International Journal of Environmental Technology and Management, 14*(1-4), 132–146.
57. Systematic evaluation of TiO-GO-modified ceramic membranes for water treatment: Retention properties and fouling mechanisms. Chen Li, Wenjun Sun, Zedong Lu, Xiuwei Ao, Chao Yang, Simiao Li. *Chemical Engineering Journal*, Volume 378, 2019, 122138.
58. Chu, K. H., Fathizadeh, M., Yu, M., Flora, J. R., Jang, A., Jang, M., ... & Yoon, Y. (2017). Evaluation of removal mechanisms in a graphene oxide-coated ceramic ultrafiltration membrane for retention of natural organic matter, pharmaceuticals, and inorganic salts. *ACS Applied Materials & Interfaces, 9*(46), 40369–40377.
59. Mukherjee, D., Banerjee, S., Ghosh, S., & Majumdar, S. (2020). PDMS/ceramic composite membrane synthesis and evaluation of ciprofloxacin removal efficiency. *Korean Journal of Chemical Engineering, 37*(11), 1985–1998.
60. Ghadimkhani, A., Zhang, W., & Marhaba, T. (2016). Ceramic membrane defouling (cleaning) by air Nano Bubbles. *Chemosphere, 146*, 379–384.

61. Zhao, Y. X., Li, P., Li, R. H., & Li, X. Y. (2019). Direct filtration for the treatment of the coagulated domestic sewage using flat-sheet ceramic membranes. *Chemosphere, 223*, 383–390.
62. Salladini, A., Prisciandaro, M., & Barba, D. (2007). Ultrafiltration of biologically treated wastewater by using backflushing. *Desalination, 207*(1-3), 24–34.
63. Liu, J., Zhou, Y. X., Tang, L. H., & Shi, J. J. (2007). Treatment of sewage with ceramic membrane and research for membrane pollution and cleaning. *Journal of Anhui Institute of Architecture & Industry (Natural Science), 3.*
64. Fujioka, T., Hoang, A. T., Okuda, T., Takeuchi, H., Tanaka, H., & Nghiem, L. D. (2018). Water reclamation using a ceramic nanofiltration membrane and surface flushing with ozonated water. *International Journal of Environmental Research and Public Health, 15*(4), 799.
65. Bhattacharya, P., Mukherjee, D., Deb, N., Swarnakar, S., & Banerjee, S. (2020). Application of green synthesized ZnO nanoparticle coated ceramic ultrafiltration membrane for remediation of pharmaceutical components from synthetic water: reusability assay of treated water on seed germination. *Journal of Environmental Chemical Engineering,* 103803.
66. Bhattacharya, P., Mukherjee, D., Dey, S., Ghosh, S., & Banerjee, S. (2019). Development and performance evaluation of a novel CuO/TiO2 ceramic ultrafiltration membrane for ciprofloxacin removal. *Materials Chemistry and Physics, 229*, 106–116.
67. Mukherjee, D., Banerjee, S., Ghosh, S., & Majumdar, S. (2020). PDMS/ceramic composite membrane synthesis and evaluation of ciprofloxacin removal efficiency. *Korean Journal of Chemical Engineering, 37*(11), 1985–1998.
68. Mukherjee, D., Kar, S., Mandal, A., Ghosh, S., & Majumdar, S. (2019). Immobilization of tannery industrial sludge in ceramic membrane preparation and hydrophobic surface modification for application in atrazine remediation from water. *Journal of the European Ceramic Society, 39*(10), 3235–3246.
69. World Health Organization, Guidelines for Drinking Water Quality (3rd ed.) (2011) (Geneva).
70. World Health Organization, Guidelines for Drinking Water Quality (4th ed.) (2011).
71. Sabbatini, P., Yrazu, F., Rossi, F., Thern, G., Marajofsky, A., & de Cortalezzi, M. F. (2010). Fabrication and characterization of iron oxide ceramic membranes for arsenic removal. *Water Research, 44*(19), 5702–5712.
72. Jana, S., Saikia, A., Purkait, M. K., & Mohanty, K. (2011). Chitosan based ceramic ultrafiltration membrane: preparation, characterization and application to remove Hg (II) and As (III) using polymer enhanced ultrafiltration. *Chemical Engineering Journal, 170*(1), 209-219.
73. Hubadillah, S. K., Othman, M. H. D., Kadir, S. H. S. A., Jamalludin, M. R., Harun, Z., Abd Aziz, M. H., ... & Iwamoto, Y. (2019). Removal of As (III) and

As (V) from water using green, silica-based ceramic hollow fibre membranes via direct contact membrane distillation. *RSC Advances*, *9*(6), 3367-3376.

74. Sklari, S., Pagana, A., Nalbandian, L., & Zaspalis, V. (2015). Ceramic membrane materials and process for the removal of As (III)/As (V) ions from water. *Journal of Water Process Engineering*, *5*, 42-47.
75. Bandyopadhyay, S., Kundu, D., Roy, S. N., Ghosh, B., & Maiti, H. S. (2006). U.S. Patent No. 7,014,771. Washington, DC: U.S. Patent and Trademark Office.
76. Bandyopadhy, S., Kundu, D., Roy, S., Ghosh, B., & Maiti, H. (2003). U.S. Patent Application No. 10/109,053.
77. Jia, Y., Zhu, B. S., Zhang, K. S., Jin, Z., Sun, B., Luo, T., . & Liu, J. H. (2015). Porous 2-line ferrihydrite/bayerite composites (LFBC): fluoride removal performance and mechanism. *Chemical Engineering Journal*, *268*, 325–336.
78. He, J., Cai, X., Chen, K., Li, Y., Zhang, K., Jin, Z., ... & Huang, X. (2016). Performance of a novelly-defined zirconium metal-organic frameworks adsorption membrane in fluoride removal. *Journal of Colloid and Interface Science*, *484*, 162–172.
79. He, J., Chen, K., Cai, X., Li, Y., Wang, C., Zhang, K., ... & Liu, J. (2017). A biocompatible and novelly-defined Al-HAP adsorption membrane for highly effective removal of fluoride from drinking water. *Journal of Colloid and Interface Science*, *490*, 97–107.
80. Ghosh, D., Nandi, B. K., Rahaman, M., & Purkait, M. K. (2016). 2 Treatment of Fluoride-Contaminated Water by Electrocoagulation Followed by Microfiltration Technique. In *Advanced Nanomaterials for Wastewater Remediation*, Edited by R.K.Gautam, M.C. Chattopadhyaya, CRC Press, eBook ISBN 9781315368108.
81. Zereffa, E. A., & Desalegn, T. (2019). Preparation and characterization of sintered clay ceramic membranes water filters. *Open Material Sciences*, *5*(1), 24–33.
82. Zhang, Q., Bolisetty, S., Cao, Y., Handschin, S., Adamcik, J., Peng, Q., & Mezzenga, R. (2019). Selective and efficient removal of fluoride from water: in situ engineered amyloid fibril/ZrO2 hybrid membranes. *Angewandte Chemie International Edition*, *58*(18), 6012–6016.
83. Wexler, P., Anderson, B. D., Gad, S. C., Hakkinen, P. B., Kamrin, M., De Peyster, A., ... & Shugart, L. R. (Eds.). (2005). *Encyclopedia of Toxicology* (Vol. 1). Academic Press.
84. Zhang, K., Zhang, Z. H., Wang, H., Wang, X. M., Zhang, X. H., & Xie, Y. F. (2020). Synergistic effects of combining ozonation, ceramic membrane filtration and biologically active carbon filtration for wastewater reclamation. *Journal of Hazardous Materials*, *382*, 121091.
85. Is Fatimah, U. A. Hasanah, Hijrah Purnama Putra. (2014) Preparation of bifunctional ceramic membrane based on TiO2/Kaolinite for Water Desinfection *J. Mater. Environ. Sci.* 5 (6), 1976–1981.

86. Fent, K., Weston, A.A, Carminada, D. (2006). Ecotoxicology of human pharmaceuticals. *Aquat. Toxicol.,* 76,122–159.
87. Pruden, A., Pei, R., Storteboom, H., Carlson, K. H. (2006). Antibiotic Resistance Genes as Emerging Contaminants: Studies in Northern Colorado. *Environ. Sci. Technol., 40,* 7445–7450.
88. Fu, W., & Zhang, W. (2018). Microwave-enhanced membrane filtration for water treatment. *Journal of Membrane Science, 568,* 97–104.
89. Stylianou, S. K., Szymanska, K., Katsoyiannis, I. A., & Zouboulis, A. I. (2015). Novel water treatment processes based on hybrid membrane-ozonation systems: a novel ceramic membrane contactor for bubbleless ozonation of emerging micropollutants. *Journal of Chemistry,* 2015.
91. Chen, L., Li, J., Fu, W., & Zhang, X. (2020). Ceramic membrane based hybrid process for the upgrade of rural water treatment plants: a pilot study. *Water Environment Research.*
92. Shannon, M. A., Bohn, P. W., Elimelech, M., Georgiadis, J. G., Marinas, B. J., & Mayes, A. M. (2010). Science and technology for water purification in the coming decades. In *Nanoscience and Technology: A Collection of Reviews from Nature Journals* (pp. 337–346).
93. Cabral, J. P. (2010). Water microbiology. Bacterial pathogens and water. *International Journal of Environmental Research and Public Health, 7*(10), 3657–3703.
94. Wegmann, M., Michen, B., & Graule, T. (2008). Nanostructured surface modification of microporous ceramics for efficient virus filtration. *Journal of the European Ceramic Society, 28*(8), 1603–1612.
95. Brown, J., & Sobsey, M. D. (2010). Microbiological effectiveness of locally produced ceramic filters for drinking water treatment in Cambodia. *Journal of Water and Health, 8*(1), 1–10.
96. Simonis, J. J., & Basson, A. K. (2011). Evaluation of a low-cost ceramic micro-porous filter for elimination of common disease microorganisms. *Physics and Chemistry of the Earth, Parts A/B/C, 36*(14-15), 1129–1134.
97. Vasanth, D., Pugazhenthi, G., & Uppaluri, R. (2011). Fabrication and properties of low cost ceramic microfiltration membranes for separation of oil and bacteria from its solution. *Journal of Membrane Science, 379*(1-2), 154–163.
98. He, Z., Lyu, Z., Gu, Q., Zhang, L., & Wang, J. (2019). Ceramic-based membranes for water and wastewater treatment. *Colloids and Surfaces A: Physicochemical and Engineering Aspects, 578,* 123513.
99. Padaki, M., Murali, R. S., Abdullah, M. S., Misdan, N., Moslehyani, A., Kassim, M. A., ... & Ismail, A. F. (2015). Membrane technology enhancement in oil–water separation. A review. *Desalination, 357,* 197–207.
100. Ebrahimi, M., Willershausen, D., Ashaghi, K. S., Engel, L., Placido, L., Mund, P., ... & Czermak, P. (2010). Investigations on the use of different ceramic membranes for efficient oil-field produced water treatment. *Desalination, 250*(3), 991-996.

101. Moazzem, S., Wills, J., Fan, L., Roddick, F., & Jegatheesan, V. (2018). Performance of ceramic ultrafiltration and reverse osmosis membranes in treating car wash wastewater for reuse. *Environmental Science and Pollution Research, 25*(9), 8654–8668.
102. Rasouli, Y., Abbasi, M., & Hashemifard, S. A. (2019). Fabrication, characterization, fouling behavior and performance study of ceramic microfiltration membranes for oily wastewater treatment. *Journal of Asian Ceramic Societies, 7*(4), 476–495.
103. Ebrahimi, M., Willershausen, D., Ashaghi, K. S., Engel, L., Placido, L., Mund, P., ... & Czermak, P. (2010). Investigations on the use of different ceramic membranes for efficient oil-field produced water treatment. *Desalination, 250*(3), 991–996.
104. Pizzichini, M., Russo, C., & Di Meo, C. (2005). Purification of pulp and paper wastewater, with membrane technology, for water reuse in a closed loop. *Desalination, 178*(1-3), 351–359.
105. Shang, R., Verliefde, A. R., Hu, J., Zeng, Z., Lu, J., Kemperman, A. J., ... & Rietveld, L. C. (2014). Tight ceramic UF membrane as RO pre-treatment: The role of electrostatic interactions on phosphate rejection. *Water Research, 48*, 498–507.
106. Bhattacharya, P., Mukherjee, D., Deb, N., Swarnakar, S., & Banerjee, S. (2020). Indigenously developed CuO/TiO2 coated ceramic ultrafiltration membrane for removal of emerging contaminants like phthalates and parabens: Toxicity evaluation in PA-1 cell line. *Materials Chemistry and Physics, 258*, 123920.
107. Shang, R., Verliefde, A. R., Hu, J., Zeng, Z., Lu, J., Kemperman, A. J., ... & Rietveld, L. C. (2014). Tight ceramic UF membrane as RO pre-treatment: The role of electrostatic interactions on phosphate rejection. *Water Research, 48*, 498–507.
108. Kim, M., Guerra, P., Shah, A., Parsa, M., Alaee, M., & Smyth, S. A. (2014). Removal of pharmaceuticals and personal care products in a membrane bioreactor wastewater treatment plant. *Water Science and Technology, 69*(11), 2221–2229.
109. De Cazes, M., Abejón, R., Belleville, M. P., & Sanchez-Marcano, J. (2014). Membrane bioprocesses for pharmaceutical micropollutant removal from waters. *Membranes, 4*(4), 692–729.
110. Mukherjee, D., Bhattacharya, P., Jana, A., Bhattacharya, S., Sarkar, S., Ghosh, S., ... & Swarnakar, S. (2018). Synthesis of ceramic ultrafiltration membrane and application in membrane bioreactor process for pesticide remediation from wastewater. *Process Safety and Environmental Protection, 116*, 22–33.
111. Mukherjee, D., Dewanjee, A., Ghosh, S., & Majumdar, S. (2018). Development of graphene oxide/chitosan composite membrane on ceramic support for atrazine remediation by MBR process. *Environmental Science and Pollution Research, 25*(33), 33334–33352.
112. Goswami, L., Kumar, R. V., Pakshirajan, K., & Pugazhenthi, G. (2019). A novel integrated biodegradation—microfiltration system for sustainable

wastewater treatment and energy recovery. *Journal of Hazardous Materials, 365*, 707–715.
113. Molinari, R., Lavorato, C., & Argurio, P. (2017). Recent progress of photocatalytic membrane reactors in water treatment and in synthesis of organic compounds. A review. *Catalysis Today, 281*, 144–164.
114. Chin, S. S., Chiang, K., & Fane, A. G. (2006). The stability of polymeric membranes in a TiO2 photocatalysis process. *Journal of Membrane Science, 275*(1-2), 202–211.
115. Ma, N., Zhang, Y., Quan, X., Fan, X., & Zhao, H. (2010). Performing a microfiltration integrated with photocatalysis using an Ag-TiO2/HAP/Al2O3 composite membrane for water treatment: Evaluating effectiveness for humic acid removal and anti-fouling properties. *Water Research, 44*(20), 6104–6114.
116. Ahmad, R., Kim, J. K., Kim, J. H., & Kim, J. (2017). Well-organized, mesoporous nanocrystalline TiO2 on alumina membranes with hierarchical architecture: Antifouling and photocatalytic activities. *Catalysis Today, 282*, 2–12.
117. Huertas, R. M., Fraga, M. C., Crespo, J. G., & Pereira, V. J. (2017). Sol-gel membrane modification for enhanced photocatalytic activity. *Separation and Purification Technology, 180*, 69–81.
118. Horovitz, I., Gitis, V., Avisar, D., & Mamane, H. (2020). Ceramic-based photocatalytic membrane reactors for water treatment–where to next? *Reviews in Chemical Engineering, 36*(5), 593–622.
119. Low, S. L., Ong, S. L., & Ng, H. Y. (2014). Biodiesel production by microalgae cultivated using permeate from membrane bioreactors in continuous system. *Water Science and Technology, 69*(9), 1813–1819.
120. Jana, A., Bhattacharya, P., Guha, S., Ghosh, S., & Majumdar, S. (2017). Application of a new ceramic hydrophobic membrane for providing CO2 in algal photobioreactor during cultivation of Arthrospira sp. *Algal Research, 27*, 223–234.
121. Low, S. L., Ong, S. L., & Ng, H. Y. (2016). Characterization of membrane fouling in submerged ceramic membrane photobioreactors fed with effluent from membrane bioreactors. *Chemical Engineering Journal, 290*, 91–102.
122. Jana, A., Ghosh, S., & Majumdar, S. (2018). Energy efficient harvesting of Arthrospira sp. using ceramic membranes: analyzing the effect of membrane pore size and incorporation of flocculant as fouling control strategy. *Journal of Chemical Technology & Biotechnology, 93*(4), 1085–1096.
123. Ceramic membrane market gaining traction from the water and waste water industry, https://medium.com/@shruthiragula15/ceramic-membrane-market-gaining-traction-from-the-water-and-waste-water-industry-29460a1f6584
124. Bhattacharya, P., Majumdar, S., Bandyopadhyay, S., & Ghosh, S. (2016). Recycling of tannery effluent from common effluent treatment plant using ceramic membrane based filtration process: a closed loop approach using pilot scale study. *Environmental Progress & Sustainable Energy, 35*(1), 60–69.
125. https://www.cgcri.res.in/research/research-divisions/water-technology/

126. Yang, W., Cicek, N., & Ilg, J. (2006). State-of-the-art of membrane bioreactors: Worldwide research and commercial applications in North America. *Journal of Membrane Science*, *270*(1-2), 201–211.
127. Pillai, S. B., Walder, C., Gupta, P., Koganti, Y. P., & Lahnsteiner, J. (2019). Groundwater treatment by ceramic membranes: pilot tests at the Commonwealth Games Village in Delhi, India. *H2Open Journal*, *2*(1), 83–91.
128. Ceramic Membrane Market by Material (Titania, Alumina, Zirconium Oxide), by Application (Water and Wastewater Technology, Pharmaceuticals, Food and Beverage, Chemical Processing, Biotechnology), by Technology (Ultrafiltration, Microfiltration, Nanofiltration), by Region - Global Trends & Forecasts to 2020, https://www.marketsandmarkets.com/Market-Reports/ceramic-membranes-market 246576534.html.

9

Membrane Distillation for Acidic Wastewater Treatment

Sarita Kalla[1], Rakesh Baghel[2], Sushant Upadhyaya[2]* and Kailash Singh[2]

[1]*Department of Chemical Engineering, Sardar Vallabhbhai National Institute of Technology, Surat, India*
[2]*Department of Chemical Engineering, Malaviya National Institute of Technology, Jaipur, India*

Abstract
Membrane Distillation (MD) is a thermally induced membrane separation process that has been widely used in numerous applications like desalination, oily wastewater treatment, acid concentration, azeotropic mixture separation, separation of volatile organic compounds, etc. Different types of acid/water solutions have been continuously discharged from the different chemical industries and their proper treatment is mandatory for the safety of the aquatic and human life. Membrane distillation is considered a viable method for acidic water treatment and metal or salt recovery from a waste stream coming from metallurgical, mining, Chloralkali, etc., industries. Therefore, in this chapter, about the membrane distillation process, the application of membrane distillation for the acid/water treatment and the safe recovery of the acid from the waste stream has been studied.

***Keywords*:** Membrane distillation, recovery, wastewater treatment, separation, acid/water solution

9.1 Introduction

Due to rapid and continuous population growth, globally, the water requirement for both human beings and industrial applications also increases. Consequently, there is a need for treatment of the wastewater

**Corresponding author*: supadhyay.chem@mnit.ac.in

Siddhartha Moulik, Aditi Mullick and Anirban Roy (eds.) Sustainable Water Treatment: Advances and Technological Interventions, (383–400) © 2022 Scrivener Publishing LLC

coming from all sources including a huge number of chemical industries and domestic. Acidic wastewater is also considered as one of the wastewater types including discharges from the mine sites during leaching process (Kesieme *et al.*, 2012; Kesieme and Aral, 2015), metallurgical industries for steel pickling and electroplating (Liu *et al.*, 2012), nuclear power plants (NPPs) as radioactive wastewater (Jia *et al.*, 2017a), fertilizer industries, explosive industries (Madhumala *et al.*, 2014; Thiruvenkatachari *et al.*, 2006), etc. There are different methods available for the treatment of such acidic effluents for the recovery of pure water and acid constituents like solvent extraction (SX) (Agrawal and Sahu, 2009; Kesieme *et al.*, 2013; Qifeng *et al.*, 2016; Wang *et al.*, 2019), membrane distillation (MD) (Amaya-Vías *et al.*, 2019; Elkina *et al.*, 2013; Kesieme *et al.*, 2014; Kesieme *et al.*, 2012; Thiruvenkatachari *et al.*, 2006), reverse osmosis (RO) (Bártová *et al.*, 2016; Masindi, 2017), membrane distillation bioreactor (MDB) (Yao *et al.*, 2019), nanofiltration (NF) (Ricci *et al.*, 2015; Zacharof *et al.*, 2016), adsorption (Rebecchi *et al.*, 2016), electrodialysis (Zhang *et al.*, 2011), crystallization (Tjus *et al.*, 2007), etc. The above-mentioned techniques have their advantages and disadvantages but the membrane separation processes are comparatively more efficient and cost-effective. MD is also considered as one of the membrane separations processes and is employed for the treatment of the acidic effluent that in turn helps in the recovery of water, acid, and valuable metals (Hu *et al.*, 2020; Hull and Zodrow, 2017; Reis *et al.*, 2018). A combination of MD with other separation techniques has also been used by different researchers (Chen *et al.*, 2019; Kesieme and Aral, 2015; Ryu *et al.*, 2019) to explore the efficiency of the MD hybrid process for the acidic wastewater treatment.

9.2 Membrane Distillation and Its Configurations

MD is a non-isothermal membrane separation process in which only vapors are passed through the hydrophobic polymeric membrane. Initially, the MD process was used by Bodell (Bodell, 1963) in 1963 for producing potable water from polluted water. Subsequently, in 1967, the first paper on MD was published by Findley (Findley, 1967), and become popular in the 1980s. Since then, MD has been applied in different applications such as desalination (Guo *et al.*, 2019; Niknejad *et al.*, 2020), wastewater treatment discharged from pharmaceutical, textile, and food industries (Leaper *et al.*, 2019; Rao *et al.*, 2018; Shirazi *et al.*, 2020; Woldemariam *et al.*, 2017), concentration of aqueous solutions (Julian *et al.*, 2020; Quist-Jensen *et al.*, 2016), volatile organic compounds removal (Kujawska *et al.*, 2016), azeotropic mixture separation (Kalla *et al.*, 2019b), etc. The transmembrane vapor pressure difference worked as the driving force for the MD. The separation in the MD is very much dependent on the membrane used, which can be commercial or laboratory fabricated or modified membranes (Kalla *et al.*, 2019a). The commercial membranes are easily available and are made up of polypropylene (PP), polyethylene (PE), polytetrafluorethylene (PTFE), and polyvinylidenefluoride (PVDF) (El-Bourawi *et al.*, 2006). As the MD process is the emerging membrane separation process, therefore study in the direction of the improvement of membrane morphology is continuously increasing. Fabricated graphene/PVDF membrane (Woo *et al.*, 2016), CF_4 plasma-modified omniphobic PVDF electrospun nanofiber membrane (Woo *et al.*, 2017), superhydrophobic PVDF membrane (Lin *et al.*, 2020), etc., are some examples of the recently developed MD membranes.

According to the way in which driving force across the membrane is maintained, the MD process has been sorted into four configurations: direct contact membrane distillation (DCMD), air gap membrane distillation (AGMD), vacuum membrane distillation (VMD) and sweeping gas membrane distillation (SGMD). All these configurations are displayed in Figure 9.1. In DCMD, the membrane, integrated in the membrane module, acts as the partition wall between the feed section and permeate section. The vapors coming from the feed section pass through the membrane and condensate is collected in the permeate section. There are mainly three sections: feed section, permeate section, and cooling section in AGMD configuration. The vapors coming from the feed section condense when it comes in contact with the cooling plate after passing through the membrane and air gap or permeate section. In VMD, the vapors condense the exterior of the membrane module, and a vacuum pump is used to create the vacuum downstream. In SGMD, the sweeping gas which is generally an inert gas is used as condensation medium and transports the vapor coming from the feed side to outside the module. In DCMD and AGMD configurations the condensation of feed vapors occurs inside the membrane module while in VMD and SGMD, the same occurs outside. Among all these configurations, the DCMD is the most popular and vastly used MD configuration because of its simple module geometry and relatively high flux (Kalla *et al.*, 2019a).

9.3 Sources of Acidic Effluent

Amidst all the acidic wastewater sources, the acid mine drainage (AMD) discharges during hydrometallurgical processes from mining industries,

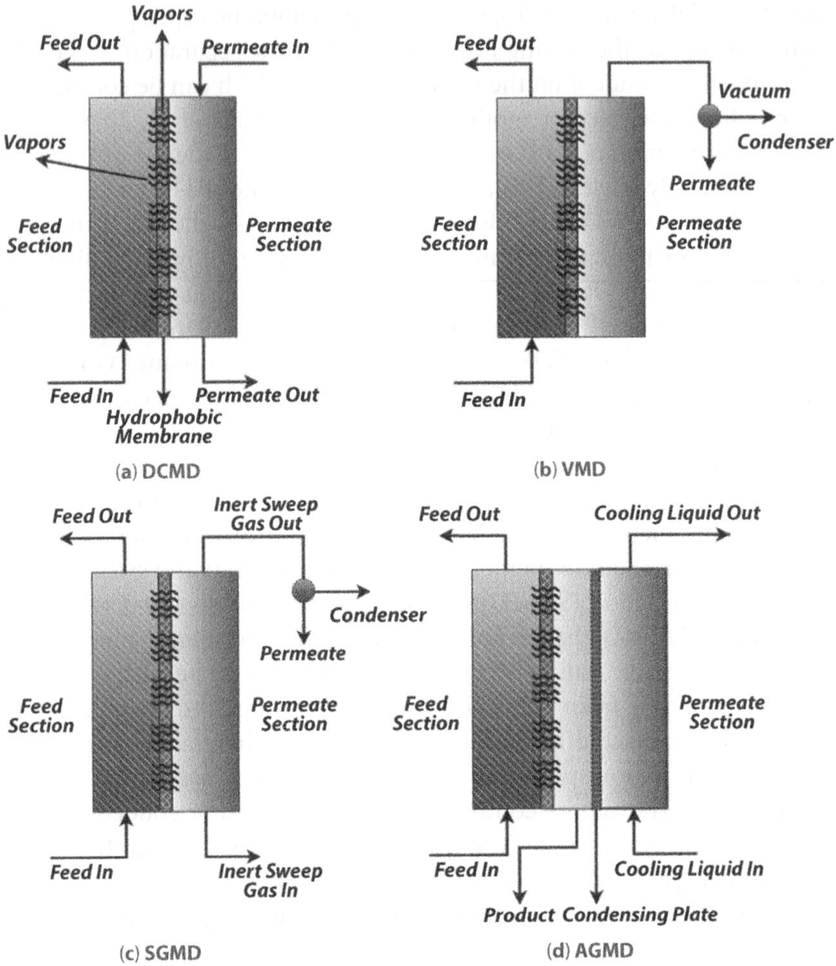

Figure 9.1 Schematic presentation of MD configurations.

and acidic pickling solutions generated during pyrometallurgical processes in the metal industries are the prominent ones (Agrawal and Sahu, 2009; RoyChowdhury *et al.*, 2015). AMD mainly consists of sulphates (SO_4^{2-}) and toxic metals (like As, Cd, Zn, Pb, Mn, etc.); it also has low pH (up to 3, highly acidic) and these contents cause contamination of both water and soils, therefore it is of utmost important to treat such effluent before discharging it into the waterbodies (Liu *et al.*, 2018; Luo *et al.*, 2020). So far, AMD has adversely affected more than 100,000 km of river and stream structure, causing poor water quality (Dettrick *et al.*, 2019). Stainless steel

(SS) is the most versatile material and is used in a large number of applications (Wu *et al.*, 2019). In CY – 2019, India was the second-largest crude steel producer country (111 million tons) just after China (largest producer) (Ministry of Steel, 2019). During the production of SS, strong acids such as nitric acid, hydrofluoric acid, hydrochloric acids, etc., are used for the removal of oxide layer from the surface (also known as pickling) (Su *et al.*, 2020). Besides pickling, a large amount of water is also used during washing and rinsing purposes. Likewise, in other metallurgical industries there are some processes like electroplating, galvanizing, etc., generated bleed streams that contain corrosive and toxic contents (Agrawal and Sahu, 2009). Thus, the wastewater generated from the metallurgical industries is the complex mixture of different mixed acids and heavy metal ions such as cyanide, chromium (Cr), nickel (Ni), iron (Fe), etc., and known as spent pickle solution (SPS). The presence of these toxic, corrosive, hazardous, and polluting contents in the wastewater creates the necessity of waste stream treatment. The other benefits of AMD and SPS treatments are the recovery of valuable metals and acidic contents that can be further reused for some other applications and simultaneously pure permeate can be used for some other purpose.

9.4 Applications of MD for Acidic Wastewater Treatment

MD is considered a viable technique for acidic water treatment because of its certain benefits; availability of the developed membranes that are suitable for working at extremely low pH (Lin *et al.*, 2020), because hydrophobic membrane only vapor molecules can pass through easy retention of the valuable metals and both pure water and metal recovery are possible (Hull and Zodrow, 2017; Moradi *et al.*, 2016); concentration of the acidic solution can be increased either in feed or permeate depending on the nature of the feed components (Madhumala *et al.*, 2014); MD is operated at low temperature and low pressure, therefore, waste heat or low-grade thermal energy like solar and geothermal energy can be applied (Jaafar and Sarbatly, 2012; Kumar and Martin, 2014). However, the membrane wetting, lack of fully developed and commercial MD membrane, complex membrane module, and high energy consumption as compared to reverse osmosis process are some downsides of the MD process (Drioli *et al.*, 2015). But extensive research in the MD field has been carried out by researchers worldwide to make it commercially successful (Duong *et al.*, 2016; Hitsov *et al.*, 2017; Woldemariam *et al.*, 2017).

A lot of work has been carried out by various researchers around the world, as depicted in Table 9.1, for acidic wastewater treatment using different MD configurations. But it has been observed, as the table shows, that most of the workers use DCMD for their work because of previously mentioned advantages of DCMD configuration. Amaya-Vías *et al.*, 2019, explored the feasibility of both AGMD and water gap membrane distillation (WGMD) for the treatment of acid mine drainage coming from Tinto river situated in SW Spain. The flat-sheet PTFE membrane of 0.45 μm pore size is used for both the MD modes. The schematic of the co-current AGMD/WGMD module is shown in Figure 9.2. The feed sample is characterized by pH ≈2.12 (due to humic acids) and different metals and other ions such as sulphate (SO_4^{-2}), iron (Fe), copper (Cu), cobalt (Co), zinc (Zn), Ni (Nickel), magnesium (Mg^{+2}), calcium (Ca^{+2}), etc. For both MD modes the performance was measured in terms of permeate flux, salt rejection, heat recovery (in terms of GOR), and specific thermal energy consumption (STEC). The rejection factor of above 99.9% for both MD mode and no membrane fouling proves the success of the MD process for AMD treatment. However, WGMD showed higher permeate flux (16.80 $L/h \cdot m^2$) as compared to AGMD (10.16 $L/h \cdot m^2$), SWRO (Seawater reverse osmosis) and FO fluxes, and hence it (WGMD) works better for the similar feed water characteristics. Also, WGMD showed lower STEC and higher GOR as compared to the AGMD. Couto *et al.*, 2019 studied the effect of humic acid (HA, organic matter) concentration on the pharmaceutically active compounds (PhACs) pollutant removal from the water using the DCMD process. Hydrophobic PTFE porous membrane DCMD module was used by the authors for the removal of selected 25 types PhACs and found that at all HA concentrations the removal of all 25 PhACs is ≥ 99% and recovery rate is 60%. This rejection of PhACs by the MD membrane is mainly contributed by the PhACs compounds volatility and membrane hydrophobia. Membrane fouling was also observed at high HA concentration due to its deposition on the surface. However, the PhACs rejection by the membrane is not affected by the membrane fouling, therefore DCMD was found more efficient then NF and RO, in terms of the effect of feed nature on the PhACs rejection.

9.5 Hybrid MD Process

Sometimes, the hybrid of MD with other traditional or recently surfaced separation methods leads to more efficient separation and cost-effective separation because in the hybrid process the benefit of individual technique can be utilized for the separation and if required separation is not possible

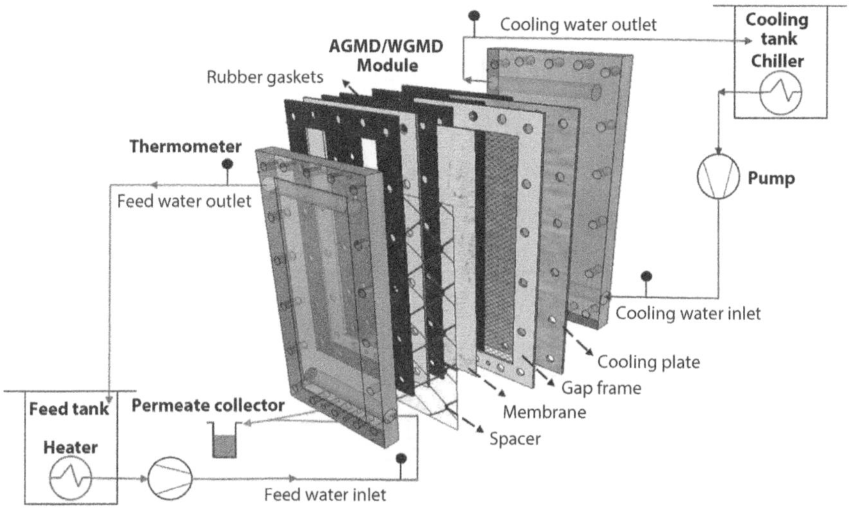

Figure 9.2 Schematic of AGMD/WGMD co-current module used by Amaya-Vías *et al.*, 2019, with permission from Elsevier.

with the single method; at that time integrated methods are also very useful. Similarly, in the case of acidic effluent treatment, different researchers (Chen *et al.*, 2019; Jia *et al.*, 2017a; Kesieme *et al.*, 2014; Ren *et al.*, 2018; Ryu *et al.*, 2019) worked on the integrated MD processes for increasing the performance criteria, i.e., acid or water recovery, flux, and salt rejection. Table 9.2 shows some studies related to the hybrid MD processes for acidic wastewater treatment. Chen *et al.*, 2019, explored the performance of hybrid nano-filtration and VMD process to purify the radioactive wastewater from the nuclear power plant by removing ionic radionuclides and subsequently concentrate the boric acid to a nuclear-grade. This hybrid membrane separation process turns out to be a promising method with a high degree of radioactive wastewater purification. A laboratory fabricated ceramic NF membrane and hydrophobic ceramic membrane were used for the NF and VMD processes, respectively. The simulated feed solution consists of divalent Co^{2+} and monovalent Ag^{+} as radionuclides ions and boric acid. Firstly, Co^{2+} and Ag^{+} ions are removed by using double-stage NF processes. The effect of different operating parameters like solution pH, boric acid concentration, transmembrane pressure (TMP) and Co^{2+} and Ag^{+} ion concentration on NF process was studied and authors observed the ≈99.2% Co^{2+} ions removal and ≈75.8% Ag^{+} ions removal at 3 pH, 10 g/L boric acid concentration, 1 mg/L Co^{2+}, and Ag^{+} ion concentration and 0.9 TMP. Similarly, during MD processes the effect of feed temperature, vacuum

Table 9.1 Some studies carried out by researchers for acidic wastewater treatment using the MD process.

Reference	MD type	Membrane type	Feed solution	Feed characteristics	Operating conditions	Fouling observed	Permeate flux	Recovery	Salt or metal rejection
Kesieme *et al.*, 2012	DCMD	Flat sheet PTFE membrane	Synthetic spent acid solution from mining industry	0.5 M H_2SO_4, 0.2 M Na_2SO_4, 0.2 M NaCl	Feed Temperature: 60 °C Distillate Temperature: 20°C Feed and Distillate Flow Rate: 900 ml/min		20-30 kg/ $m^2 \cdot h$	Water recovery > 80%	> 99.9%
Elkina *et al.*, 2013	DCMD	Flat sheet commercial MF membrane made up of tetrafluoroethylene/ vinylidenefluoride copolymer	Volatile inorganic acid solutions from metal pickling	Hydrochloric acid and nitric acid in the concentration range of 0.01 to 7.0 N	Feed Temperature: 40 to 60 °C Distillate Temperature: 20°C Feed and Distillate Flow Rate: 40 L/h	Pore wetting due to high concentrated feed solution	8-20 kg/$m^2 \cdot h$	Concentration of HCl and HNO3 in feed; 1.02 N and 1.00 N, respectively Concentration of HCl and HNO3 in permeate; 0.0296 N and 0.0240 N, respectively	-

(*Continued*)

Table 9.1 Some studies carried out by researchers for acidic wastewater treatment using the MD process. (*Continued*)

Reference	MD type	Membrane type	Feed solution	Feed characteristics	Operating conditions	Fouling observed	Permeate flux	Recovery	Salt or metal rejection
Kesieme *et al.*, 2014	DCMD	Flat sheet PTFE supported on the PP scrim	Real Leach Solution from of HCl and H_2SO_4	HCl and H_2SO_4 leach solutions consist of Free acid, Co, Cu, Fe, Mg, Mn, Ni, Zn, etc.	Feed Temperature: 60 °C Distillate Temperature: 20°C Feed and Distillate Flow Rate: 900 ml/min	No notable scale formation	15-35 kg/$m^2 \cdot h$	Water recovery > 80%	> 99.9%
Madhumala *et al.*, 2014	VMD	Flat sheet PTFE membrane	Chloralkali industrial effluent	Hydrochloric acid (32.8 wt%), a minute amount of iron salt and hydrocarbons of order C_9-C_{14}	Feed Temperature: 25-60 °C Permeate pressure: 5 mmHg	Non-wetting of membrane	□8.57 kg/$m^2 \cdot h$	Wt% HCl in permeate and feed = 33.6 and 32.8, respectively	-
Moradi *et al.*, 2016	VMD	Modifies PVDF/ TiO_2 Electrospun Nanofibrous Hybrid Membrane	Synthetic heavy metal contaminated wastewater	Metals like Cobalt, Zinc, Copper, Nickel, cadmium, lead	Feed Temperature: 65-70 °C Permeate Pressure: 2000-2500 Pa Feed flow rate: 0.1 L/min	Pore wetting at vacuum < 500 pa	10 kg/$m^2 \cdot h$	-	100 % (Complete Rejection)

(*Continued*)

Table 9.1 Some studies carried out by researchers for acidic wastewater treatment using the MD process. (*Continued*)

Reference	MD type	Membrane type	Feed solution	Feed characteristics	Operating conditions	Fouling observed	Permeate flux	Recovery	Salt or metal rejection
Hull and Zodrow, 2017	DCMD	Flat sheet PVDF membrane	Synthetic ARD (Acid Rock Drainage) Solution	Acidic (pH=2.47), metals like Iron, Sulfate, Aluminum, Copper, Zinc	Feed Temperature: 90 °C Distillate Temperature: 6°C Feed and Distillate Flow Rate: 3.5 L/min	Scaling by iron oxyhydroxide and scale induced pore wetting	38.4 ± 1.1 $L/m^2 \cdot h$	Water Recovery = 80 %	Average salt rejection 99.96% for 2 runs
Reis *et al.*, 2018	DCMD	Flat Sheet PTFE	Acidic effluent from gold mining industry of Minas Gerais (Brazil)	Acidic (pH= 1.80 ± 0.23), metals like sulfate, arsenic, calcium, magnesium	Feed Temperature: 60 °C Distillate Temperature: 20°C Feed Velocity: 0.1 m/s (Re = 581)	Fouling due to $CaSO_4$ precipitation	16 $kg/m^2 \cdot h$	Permeate (water) recovery = 40%	> 90 %
Hu *et al.*, 2020	DCMD	PVDF hollow fiber membrane	Titanium white waste acid from titania production	Sulphuric Acid (~20 wt%), Metals such as iron, magnesium, small amount of polyacrylamide (PAM) (0.0002 wt%)	Feed Temperature: 65 °C Distillate Temperature: 16 °C Feed Flow Rate: 0.15 L/min Distillate Flow Rate: 0.04 L/min	Fouling due to PAM molecule	~10 $kg/m^2 \cdot h$	-	~ 90 %

Table 9.2 Hybrid MD processes for acidic effluent treatment carried out by some researchers.

References	Hybrid process	Feed solution	MD membrane	Feed characteristics	Operating conditions: Process 1	Operating conditions: Process 2	Major findings	Recovery	Salt or metal rejection
Kesieme and Aral, 2015	Hybrid DCMD - SX system	Synthetic acid mining wastewater	Flat sheet PTFE supported on PP scrim	Sulphuric acid, and metal ions such as Al, Co, Cu, Fe, Ni, etc.	Feed Temp = 60 °C Permeate Temp = 20 °C Flow Rate: Feed and Permeate = 900 mL/min	Organic Solvent TEHA (tris-2-ethylhexylamine) for extraction of H_2SO_4 and metals from concentrate acidic MD output	Concentration factor for all metals was found > 4 except calcium because of its precipitation and nearly all (≈ 99%) H_2SO_4 acid extracted in three consecutive extraction stages	Water recovery > 80%	Sulphate and metal rejection > 99.99%
Jia *et al.*, 2017b	Vacuum membrane distillation crystallization (VMDC)	Simulated radioactive wastewater	Hollow fiber PP membrane	Nuclides like Co, Sr, Cs, Boric acid (about 500 mg/L)	Concentration factor (CF) = 200 Feed temperature = 70 °C Vacuum degree = 0.97 atm Feed flow rate = 41.8 L/h	Crystallization Temperature = 20 °C	Feed temperature found as most significant parameter and observed that VMDC is quite independent of the feed concentration, therefore more decent as compared to RO for getting high CF	Boric Acid Recovery ≈ 50%	Nuclides Rejection ≈ 96.0% Boric Acid Rejection ≈ 99.5%

(*Continued*)

Table 9.2 Hybrid MD processes for acidic effluent treatment carried out by some researchers. (*Continued*)

References	Hybrid process	Feed solution	MD membrane	Feed characteristics	Operating conditions		Major findings	Recovery	Salt or metal rejection
					Process 1	Process 2			
Ryu *et al.*, 2019	Integrated submerged DCMD-zeolite sorption system	Synthetic acidic mine drainage	Hollow fiber PVDF membrane	pH = 2.0±0.2, Heavy metals like Fe, Al, Zn, Cu, Ni	Feed Temp = 55.0±0.5 °C Permeate Temp = 22.0±0.5 °C Flow Rate: Feed and Permeate = 0.8 L/min	Finely powdered and heat-treated (500 °C) zeolite as absorber at pH 4 and with 30 hr contact time	High rejection of the Fe and Al metal ions, less the membrane fouling, selective recovery of Sulphuric acid as the concentration of sulfate ions in permeate reduces up to 0.13-0.15 mg/L	Water Recovery -50%	High rejection of all ions with permeate TDS 0.01 g/L

degree, crossflow rate, and boric acid concentration were studied and 1 g/L of feed boric acid concentrated up to 107 g/L with permeation flux 20 L/ $m^2 \cdot h$. Ren *et al.*, 2018, worked on the hybrid electrodialysis-VMD process for acid separation. The synthetic acidic rare earth element wastewater which consists of only ceric ammonium nitrate and nitric acid was used for the treatment. The types of anion exchange membranes (AEMs) were used in ED and hollow fiber PVDF membranes for the VMD process. In the ED, the Ce^{4+} ions were resisted by the anionic membrane and ions were recovered that can be reused. The nitrate ions recovery depends on the voltage and pH of the ED system, current density, flow rate and feed nitrate concentration. The Ce^{4+} ions were concentrated further using VMD process and the rejection rate observed was more than 99.9%. Due to the high VMD rejection rate, the clean permeate can be further used as cooling water. Thus, the authors found an effective performance of ED+VMD system for rare earth metal wastewater treatment. However, the fouling of the membrane due to vapor condensation and salt precipitation was overserved in the VMD and ED, respectively, and drying was considered as the prominent solution.

9.6 Implications

The MD process alone or in combination with other separation techniques has emerged as a promising and efficient technology for the acidic wastewater treatment and can serve as the replacement of other traditional methods of RO and NF. MD, which is a transmembrane vapor pressure-driven process, is capable of achieving high salt rejection factor, i.e., $\geq$ 99.9%, and records good pure water recovery of greater than 80%. Thus, in some sense, MD is capable of reducing the pure water requirement for the ever-increasing population and generated high-quality distillate can be further reused for some industrial applications. Among all the MD operating modes, DCMD is most commonly used because of its simple module design and consequent low operating cost and high flux rate. The acid recovery and metal or salt rejection characteristic of the MD makes it practicable and can be implemented at a pilot scale for further study.

References

Agrawal A, Sahu KK. An overview of the recovery of acid from spent acidic solutions from steel and electroplating industries. *J. Hazard. Mater.* 2009; 171(1–3): 61–75.

Amaya-Vías D, Tataru L, Herce-Sesa B, López-López JA, López-Ramírez JA. Metals removal from acid mine drainage (Tinto River, SW Spain) by water gap and air gap membrane distillation. *J. Memb. Sci.*; 2019; 582 (February): 20–9. Available from: https://doi.org/10.1016/j.memsci.2019.03.081

Bártová Š, Kůs P, Skala M, Vonková K. Reverse osmosis for the recovery of boric acid from the primary coolant at nuclear power plants. *Nucl. Eng. Des.* 2016; 300: 107–16.

Bodell BR. Silicone rubber vapor diffusion in saline water distillation United States Patent. 1963. p. 2–4.

Chen X, Chen T, Li J, Qiu M, Fu K, Cui Z, *et al.*, Ceramic nanofiltration and membrane distillation hybrid membrane processes for the purification and recycling of boric acid from simulative radioactive waste water. *J. Memb. Sci.*; 2019; 579 (February): 294–301. Available from: https://doi.org/10.1016/j.memsci.2019.02.044

Couto CF, Amaral MCS, Lange LC, Santos LV de S. Effect of humic acid concentration on pharmaceutically active compounds (PhACs) rejection by direct contact membrane distillation (DCMD). *Sep. Purif. Technol* 2019; 212: 920–8. Available from: https://doi.org/10.1016/j.seppur.2018.12.012

Dettrick D, Costelloe J, Arora M, Yuen S. A comparison of measured and predicted diffusion coefficients applied to sand and silt sized acid mine drainage materials. *J. Environ. Manage.* 2019; 231: 1106–16. Available from: https://doi.org/10.1016/j.jenvman.2018.08.011

Drioli E, Ali A, Macedonio F. Membrane distillation: Recent developments and perspectives. *Desalination;* 2015; 356: 56–84.

Duong HC, Cooper P, Nelemans B, Cath TY, Nghiem LD. Evaluating energy consumption of air gap membrane distillation for seawater desalination at pilot scale level. *Sep. Purif. Technol.* 2016;166:55–62. Available from: http://dx.doi.org/10.1016/j.seppur.2016.04.014

El-Bourawi MS, Ding Z, Ma R, Khayet M. A framework for better understanding membrane distillation separation process, *J. Memb. Sci.* 2006; 285: 4–29.

Elkina IB, Gilman AB, Ugrozov V V., Volkov V V. Separation of mineral acid solutions by membrane distillation and thermopervaporation through porous and nonporous membranes. *Ind. Eng. Chem. Res.* 2013; 52(26): 8856–63.

Findley ME. Vaporization through Porous Membranes. *Ind. Eng. Chem. Process Des. Dev.* 1967; 6(2): 226–30.

Guo J, Deka BJ, Kim KJ, An AK. Regeneration of superhydrophobic TiO2 electrospun membranes in seawater desalination by water flushing in membrane distillation. *Desalination* 2019;468:114054. Available from: https://doi.org/10.1016/j.desal.2019.06.020

Hitsov I, De Sitter K, Dotremont C, Cauwenberg P, Nopens I. Full-scale validated Air Gap Membrane Distillation (AGMD) model without calibration parameters. *J. Memb. Sci.;* 2017; 533: 309–20. Available from: http://dx.doi.org/10.1016/j.memsci.2017.04.002

Hu B, Ouyang J, Jiang LY. Influence of Flocculant polyacrylamide on concentration of titanium white waste acid by direct contact membrane distillation. *Chinese J. Chem. Eng.*. 2020. Available from: https://doi.org/10.1016/j.cjche.2020.03.022

Hull EJ, Zodrow KR. Acid Rock Drainage Treatment Using Membrane Distillation: Impacts of Chemical-Free Pretreatment on Scale Formation, Pore Wetting, and Product Water Quality. *Environ. Sci. Technol.* 2017; 51(20): 11928–34.

Jaafar S, Sarbatly R. Geothermal Water Desalination by using Nanofiber Membrane. *Int. Conf. Chem. Environ. Biol. Sci. Penang, Malaysia.* 2012; 46–50.

Jia F, Li J, Wang J. Recovery of boric acid from the simulated radioactive wastewater by vacuum membrane distillation crystallization. *Ann. Nucl. Energy;* 2017a; 110: 1148–55. Available from: http://dx.doi.org/10.1016/j.anucene.2017.07.024

Jia F, Li J, Wang J. Recovery of boric acid from the simulated radioactive wastewater by vacuum membrane distillation crystallization. *Ann. Nucl. Energy;* 2017b; 110: 1148–55.

Julian H, Yaohanny F, Devina A, Purwadi R, Wenten IG. Apple juice concentration using submerged direct contact membrane distillation (SDCMD). *J. Food Eng.*; 2020; 272: 109807. Available from: https://doi.org/10.1016/j.jfoodeng.2019.109807

Kalla S, Upadhyaya S, Singh K. Principles and advancements of air gap membrane distillation. *Rev. Chem. Eng.* 2019a; 35(7).

Kalla S, Upadhyaya S, Singh K, Baghel R. Experimental and mathematical study of air gap membrane distillation for aqueous HCl azeotropic separation. *J. Chem. Technol. Biotechnol.* 2019b ;94(1).

Kesieme UK, Aral H. Application of membrane distillation and solvent extraction for water and acid recovery from acidic mining waste and process solutions. *J. Environ. Chem. Eng.*; 2015; 3(3): 2050–6. Available from: http://dx.doi.org/10.1016/j.jece.2015.07.008

Kesieme UK, Aral H, Duke M, Milne N, Cheng CY. Recovery of sulphuric acid from waste and process solutions using solvent extraction. *Hydrometallurgy*; 2013; 138: 14–20. Available from: http://dx.doi.org/10.1016/j.hydromet.2013.06.005

Kesieme UK, Milne N, Cheng CY, Aral H, Duke M. Novel application of membrane distillation for acid and water recovery from mining waste waters. *Int. Mine Water Assoc. Annu. Conf. 2012.* 2012; 623–30.

Kesieme UK, Milne N, Cheng CY, Aral H, Duke M. Recovery of water and acid from leach solutions using direct contact membrane distillation. *Water Sci. Technol.* 2014; 69(4): 868–75.

Kujawska A, Kujawski J, Bryjak M, Kujawski W. Removal of volatile organic compounds from aqueous solutions applying thermally driven membrane processes. 2. Air gap membrane distillation. *J. Memb. Sci.* Elsevier; 2016; 499: 245–56.

Kumar NTU, Martin A. Co-generation of Drinking Water and Domestic Hot Water Using Solar Thermal Integrated Membrane Distillation System. *Energy Procedia*. 2014; 61: 2666–9.

Leaper S, Abdel-Karim A, Gad-Allah TA, Gorgojo P. Air-gap membrane distillation as a one-step process for textile wastewater treatment. *Chem. Eng. J.*; 2019; 360: 1330–40. Available from: https://doi.org/10.1016/j.cej.2018.10.209

Lin J, Lin F, Liu R, Li P, Fang S, Ye W, *et al.*, Scalable fabrication of robust superhydrophobic membranes by one-step spray-coating for gravitational water-in-oil emulsion separation. *Sep. Purif. Technol.*; 231:115898, 2020. Available from: https://doi.org/10.1016/j.seppur.2019.115898

Liu Q, Chen B, Haderlein S, Gopalakrishnan G, Zhou Y. Characteristics and environmental response of secondary minerals in AMD from Dabaoshan Mine, South China. *Ecotoxicol. Environ. Saf.*; 2018; 155: 50–8. Available from: https://doi.org/10.1016/j.ecoenv.2018.02.017

Liu R, Qin Y, Li X, Liu L. Concentrating aqueous hydrochloric acid by multiple-effect membrane distillation. *Front. Chem. Sci. Eng.* 2012; 6(3): 311–21.

Luo C, Routh J, Dario M, Sarkar S, Wei L, Luo D, *et al.*, Distribution and mobilization of heavy metals at an acid mine drainage affected region in South China, a post-remediation study. *Sci. Total Environ.*; 2020; 724:138122.

Madhumala M, Madhavi D, Sankarshana T, Sridhar S. Recovery of hydrochloric acid and glycerol from aqueous solutions in chloralkali and chemical process industries by membrane distillation technique. *J. Taiwan Inst. Chem. Eng.* Taiwan Institute of Chemical Engineers; 2014; 45(4): 1249–59.

Masindi V. Recovery of drinking water and valuable minerals from acid mine drainage using an integration of magnesite, lime, soda ash, CO2 and reverse osmosis treatment processes. *J. Environ. Chem. Eng.*; 2017; 5(4): 3136–42. Available from: http://dx.doi.org/10.1016/j.jece.2017.06.025

Ministry of steel G of I. An overview of steel sector [Internet]. 2019 [cited 2020 Jun 4]. Available from: https://steel.gov.in/overview-steel-sector

Moradi R, Monfared SM, Amini Y, Dastbaz A. Vacuum enhanced membrane distillation for trace contaminant removal of heavy metals from water by electrospun PVDF/TiO2 hybrid membranes. *Korean J. Chem. Eng.* 2016; 33(7): 2160–8.

Niknejad AS, Bazgir S, Sadeghzadeh A, Shirazi MMA. Styrene-acrylonitrile (SAN) nanofibrous membranes with unique properties for desalination by direct contact membrane distillation (DCMD) process. *Desalination*; 2020; 488:114502. Available from: https://doi.org/10.1016/j.desal.2020.114502

Qifeng W, Xiulian R, jingjing G, Yongxing C. Recovery and separation of sulfuric acid and iron from dilute acidic sulfate effluent and waste sulfuric acid by solvent extraction and stripping. *J. Hazard. Mater.*; 2016; 304: 1–9.

Quist-Jensen CA, Macedonio F, Conidi C, Cassano A, Aljlil S, Alharbi OA, *et al.*, Direct contact membrane distillation for the concentration of clarified orange juice. *J. Food Eng.* 2016; 187: 37–43. Available from: http://dx.doi.org/10.1016/j.jfoodeng.2016.04.021

Rao U, Posmanik R, Hatch LE, Tester JW, Walker SL, Barsanti KC, *et al.*, Coupling hydrothermal liquefaction and membrane distillation to treat anaerobic digestate from food and dairy farm waste. *Bioresour. Technol.* 2018; 267(July): 408–15.

Rebecchi S, Pinelli D, Bertin L, Zama F, Fava F, Frascari D. Volatile fatty acids recovery from the effluent of an acidogenic digestion process fed with grape pomace by adsorption on ion exchange resins. *Chem. Eng. J.*; 2016; 306: 629–39. Available from: http://dx.doi.org/10.1016/j.cej.2016.07.101

Reis BG, Araújo ALB, Amaral MCS, Ferraz HC. Comparison of Nanofiltration and Direct Contact Membrane Distillation as an alternative for gold mining effluent reclamation. *Chem. Eng. Process. - Process Intensif.*; 2018; 133: 24–33. Available from: https://doi.org/10.1016/j.cep.2018.08.020

Ren M, Ning P, Xu J, Qu G, Xie R. Concentration and treatment of ceric ammonium nitrate wastewater by integrated electrodialysis-vacuum membrane distillation process. *Chem. Eng.* J; 2018; 351: 721–31. Available from: https://doi.org/10.1016/j.cej.2018.06.155

Ricci BC, Ferreira CD, Aguiar AO, Amaral MCS. Integration of nanofiltration and reverse osmosis for metal separation and sulfuric acid recovery from gold mining effluent. *Sep. Purif. Technol.* 2015; 154: 11–21.

RoyChowdhury A, Sarkar D, Datta R. Remediation of Acid Mine Drainage-Impacted Water. *Curr. Pollut. Reports*. 2015. p. 131–41.

Ryu S, Naidu G, Hasan Johir MA, Choi Y, Jeong S, Vigneswaran S. Acid mine drainage treatment by integrated submerged membrane distillation–sorption system. *Chemosphere* 2019; 218: 955–65. Available from: https://doi.org/10.1016/j.chemosphere.2018.11.153

Shirazi MMA, Bazgir S, Meshkani F. A novel dual-layer, gas-assisted electrospun, nanofibrous SAN4-HIPS membrane for industrial textile wastewater treatment by direct contact membrane distillation (DCMD). *J. Water Process Eng.*; 2020; 36:101315. Available from: https://doi.org/10.1016/j.jwpe.2020.101315

Su P, Zhang J, Li Y. Chemical fixation of toxic metals in stainless steel pickling residue by Na2S·xH2O, FeSO4·6H2O and phosphoric acid for beneficial uses. *J. Environ. Sci.* (China). 2020; 90: 364–74.

Thiruvenkatachari R, Manickam M, Ouk Kwon T, Shik Moon I, Woo Kim J. Separation of water and nitric acid with porous hydrophobic membrane by air gap membrane distillation (AGMD). *Sep. Sci. Technol.* 2006; 41(14): 3187–99.

Tjus K, Bergström R, Fortkamp U, Forsberg K, Rasmuson Å. Development of a Recovery System for Metals and Acids from Pickling Baths using Nanofiltration and Crystallisation. 2007.

Wang Y, Chen F, Ma X, Zhang G. Recovery of nitric and acetic acids from etching waste solutions using a synergistic system consisting of N235 and TRPO in cyclohexane. *Hydrometallurgy*; 2019; 185: 23–9.

Woldemariam DM, Kullab A, Martin AR. District heat-driven water purification via membrane distillation: New possibilities for applications in Pharmaceutical Industries. *Ind. Eng. Chem. Res.* [Internet]. 2017; Available from: http://pubs.acs.org/doi/abs/10.1021/acs.iecr.6b04740

Woo YC, Chen Y, Tijing LD, Phuntsho S, He T, Choi J-S, *et al.*, CF4 plasma-modified omniphobic electrospun nanofiber membrane for produced water brine treatment by membrane distillation. *J. Memb. Sci.* Elsevier; 2017; 529: 234–42.

Woo YC, Kim Y, Shim WG, Tijing LD, Yao M, Nghiem LD, *et al.*, Graphene/PVDF flat-sheet membrane for the treatment of RO brine from coal seam gas produced water by air gap membrane distillation. *J. Memb. Sci.* Elsevier; 2016; 513: 74–84.

Wu MT, Li YL, Guo Q, Shao DW, He MM, Qi T. Harmless treatment and resource utilization of stainless steel pickling sludge via direct reduction and magnetic separation. *J. Clean. Prod.;* 2019; 240: 118187. Available from: https://doi.org/10.1016/j.jclepro.2019.118187

Yao M, Woo YC, Ren J, Tijing LD, Choi JS, Kim SH, *et al.*, Volatile fatty acids and biogas recovery using thermophilic anaerobic membrane distillation bioreactor for wastewater reclamation. *J. Environ. Manage.*; 231: 833–42, 2019. Available from: https://doi.org/10.1016/j.jenvman.2018.11.009

Zacharof MP, Mandale SJ, Williams PM, Lovitt RW. Nanofiltration of treated digested agricultural wastewater for recovery of carboxylic acids. *J. Clean. Prod.* [Internet]. Elsevier Ltd; 2016; 112: 4749–61. Available from: http://dx.doi.org/10.1016/j.jclepro.2015.07.004

Zhang X, Li C, Wang Y, Luo J, Xu T. Recovery of acetic acid from simulated acetaldehyde wastewaters: Bipolar membrane electrodialysis processes and membrane selection. *J. Memb. Sci.* 2011; 379(1–2): 184–90. Available from: http://dx.doi.org/10.1016/j.memsci.2011.05.059

10

Demonstration of Long-Term Assessment on Performance of VMD for Textile Wastewater Treatment

Rakesh Baghel[1], Sarita Kalla[2], Sushant Upadhyaya[1]* and S. P. Chaurasia[1]

[1]*Department of Chemical Engineering, Malaviya National Institute of Technology Jaipur, JLN Marg, Jaipu, India*
[2]*Department of Chemical Engineering, Sardar Vallabhbhai National Institute of Technology, Surat, India*

Abstract
VMD is a prominent technique for separation of non-volatile solutes from wastewater using porous hydrophobic membrane. In VMD, the difference in vapor pressure results in the transport of vapor molecules. The feed side is in contact with feed and the applied vacuum on permeate helps to maintain the pressure difference across membrane which is known as driving force. VMD has several applications in different areas such as desalination for removal of various salts, food industry to remove aroma compounds, and wastewater treatment to treat textile wastewater. In this chapter, information regarding the heat and mass transfer, transport mechanism such as Knudsen and Poiseuille flow, impact of process variables like flow rate, temperature, and vacuum degree, long-term performance analysis of VMD, and the fouling in long-term assessment is elaborated in detail.

Keywords: Heat transfer, mass transfer, transport mechanism, fouling

10.1 Introduction

Dyes are well-defined aromatic organic structures which consist of delocalized electron systems in aryl rings which have binding affinity with surfaces. Dye molecule is to have within auxochrome, chromogen and

**Corresponding author*: supadhyay.chem@mnit.ac.in

Siddhartha Moulik, Aditi Mullick and Anirban Roy (eds.) Sustainable Water Treatment: Advances and Technological Interventions, (401–414) © 2022 Scrivener Publishing LLC

chromophore. The unsaturated groups present in chromophores absorb light and reflect at a definite angle to provide different shades, e.g., nitroso, thio, keto, etc; the affinity and the final shade is determined by the chromogen present in dyes [1, 2]. The intensity and the depth of the hue is determined by the cationic and anionic groups substitute of auxochromes present in dye molecules. Nowadays, a large amount of dyes are used in different industrial applications such as food coloring, paper production, tanning of leather, coloring of textile, hair coloring, etc. Synthetic dyes are also used to find specific areas of sewage and wastewater treatment, activated sludge, etc.

At the end of 2015, the overall material market was worth an estimated USD1557.1 billion with the generation of an overabundance of 88.5 million tons. The dedication of different countries in the overall material market has come to 58.6% to the extent cost inside 10 years prior [3, 4]. By far most of the hues used in material dealing with are lacking in the all-out fixation of hues to surfaces, and along these lines, 20% of hues enter the earth which produces coloring containing toned effluents. The overall age of dyestuffs is around 10 million kg consistently; about 2 million kg each year enters into the atmosphere and water.

For safe drinking water, the presence of dye in a small quantity is not acceptable due to chromophores which show clear visibility in water. Due to dye in water, the aquatic life is disturbed because the coloring compound stops the sunlight penetration into water body which decreases the photosynthesis process and the level of dissolved oxygen is decreased, which may cause harmful effects to aquatic animals [2, 4]. The dyes are toxic and carcinogenic in nature so their presence in water may lead to damaging of aquatic life. Due to this, the dyes separation from wastewater is essential before being disposed of into a water body.

Nowadays, an advanced membrane-based separation process is used for separation of dyes found in industrial wastewater. VMD is a prominent technique for removal of non-volatile dye molecules due to its simple nature, ease of operation and high removal efficiency. In the process, vapor molecules are passed via hydrophobic porous membrane due to vapor pressure difference (driving force) across membrane. Figure 10.1 represents the schematic diagram. In VMD the vacuum is maintained on permeate side via physical means and the hot working fluid is in contact on the membrane's other side. The driving force enhanced the convective mass transfer via membrane pores which results in an increase in total mass transfer and due to this reason VMD provides more permeate flux as compared to other membrane techniques. However, pore wetting is considered as a major problem due to higher vacuum.

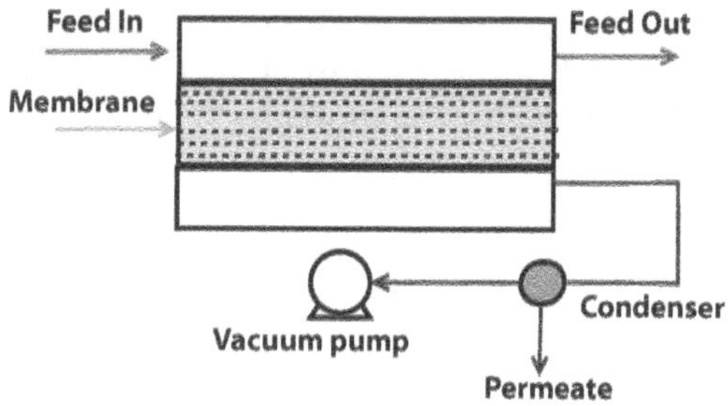

Figure 10.1 Schematic diagram of VMD.

10.2 Transport Mechanism

Heat Transfer

Heat transfer via membrane is associated with mass transfer. There are three ways to analyze heat transfer in VMD: (i) feed bulk to membrane's surface; (ii) via solid membrane matrix and porous structure; (iii) membrane surface to permeate side [5–7]. The resistances on permeate side is negligible due to applied vacuum. So, the series model of resistances represents the overall heat transfer coefficient for VMD process as given in equation (10.1).

$$\frac{1}{H}=\frac{1}{h_{bf}}+\frac{1}{h_m+(N\Delta H_v/\Delta T_m)} \tag{10.1}$$

Heat transfer due to convection on the feed side is determined using the following equation (10.2).

$$q_{bf}=h_{bf}(T_{bf}-T_{fm}) \tag{10.2}$$

In feed section, the heat flux is increased with temperature difference in feed and membrane. If this difference is higher then the heat flow is high but the boundary layer on membrane surface applied resistances from feed to membrane surface which reduced temperature difference at interface. So, the coupled mass transfer due to vapor transport is decreased and called temperature polarization. The h_{bf} is calculated using dimensionless empirical correlations.

Mass Transfer

Due to coupling, mass transfer is also studied in three ways: (i) feed solution to membrane surface; (ii) via membrane pores; (iii) pores to permeate side.

The applied vacuum on permeate reduced the resistances on permeate side. So, the series model can be used to predict the overall mass transfer behaviour in VMD as shown in equation (10.3).

$$\frac{1}{k}=\frac{1}{k_{bf}}+\frac{1}{k_m} \tag{10.3}$$

The mass transfer in feed section is determined using the following equation (10.4):

$$N=k_{bf}C\left[\frac{x_{fm}-x_p}{x_{bf}-x_p}\right] \tag{10.4}$$

The mass transfer coefficient is determined using mass transfer analogy $Sh = a\, \mathrm{Re}^b\, Sc^c$ in the form of dimensionless number. The MS excel solver can be used for fitting of empirical correlation by reducing the error which provide a, b and c values.

Inside the membrane pores, the dusty gas and kinetic theory model is used to predict the mass transfer mechanism (diffusive and convective both). According to this model, the vapor flux is proportional to the difference in vapor pressures which is created due to temperature difference [8–10].

$$N = B_m M\,[P_{fm} - P_{pm}] \tag{10.5}$$

The feed side membrane surface vapor pressure is evaluated using Antoine equation:

$$\log(P_{fm})=\left(A-\frac{B}{T_{fm}+C}\right) \tag{10.6}$$

In equation 10.5, B_m is the proportionality constant comprises the combined impact of the constant parameters of all three mechanisms like Knudsen, Molecular and Poiseuille flow. In most cases authors have only considered Knudsen and Poiseuille flow as dominating mass transfer mechanism. The selection of governing mechanism (Knudsen and Poiseuille flow) is determined using Knudsen number (Kn) which is the ratio of mean free path of gases to the membrane pore size. If the collisions between membranes pore with water molecules are dominant (Kn > 10) than only Knudsen diffusion is considered to describe the mass transfer.

$$B_m = \frac{2}{3RT}\frac{\varepsilon \bar{r}}{\tau \delta}\left(\frac{8RT}{\pi M_i}\right)^{\frac{1}{2}} \tag{10.7}$$

If the collisions between the molecules are dominant (Kn < 0.01), then Poiseuille flow is considered as mass transfer governing mechanism.

$$B_m = \frac{1}{RT\delta}\frac{\varepsilon \bar{r}^2}{8\tau \eta_i}\bar{p} \tag{10.8}$$

If the value of Knudsen number lies between 0.01 and 10 then both molecule-molecule and molecule-wall are dominant.

$$B_m = \frac{2}{3RT}\frac{\varepsilon \bar{r}}{\tau \delta}\left(\frac{8RT}{\pi M_i}\right)^{\frac{1}{2}} + \frac{1}{RT\delta}\frac{\varepsilon \bar{r}^2}{8\tau \eta_i}\bar{p} \tag{10.9}$$

Various authors have considered different combinations to define the mass transfer mechanism depending upon the physical properties of membrane and module for treatment of wastewater using vacuum membrane distillation.

10.3 Impact of Process Variables on Permeate Flux

The impact of different process variables such as feed temperature varies from 25 to 85 °C, vacuum degree varies from 670 to 750 mmHg, and dye concentration varies from 10 to 130 ppm was analyzed on permeate flux to evaluate the performance and removal efficiency of VMD process.

Impact of feed temperature

Various studies have been carried out to analyse the feed temperature's impact of on permeate flux for removal of dyes from wastewater. Figure 10.2 shows the variation in flux with change in feed temperature and their associated energy consumption of the VMD process for removal of dyes using PTFE membrane of 0.22μm pore size. A significant increment in flux was observed varying from 1.950 to 53.510 kg/m^2.h with changing the temperature from 25 to 85 °C at constant condition of 5 lpm, 750mmHg, and 30ppm for feed rate, vacuum degree and dye concentration, respectively. This exponential increment is due to increased in trans-membrane vapor pressure difference which increased mass transfer across membrane. The experimental results were found to be best fitted with CFD and CCD model developed by Baghel *et al.* (2018) with R^2 value of 0.989 and 0.995 and the MAPE value of 10.16 and 7.63 [11]. From this statistical analysis, the CFD model was seen as best fitted as compared to the CCD model with experimental results. In the CFD model author considered the impact of Knudsen and Poiseuille flow during estimating of flux, and due to this the CFD model developed by Baghel *et al.* (2020) is more suitable as compared to the CCD model developed by Baghel *et al.* (2018). The sudden reduction was also observed in consumed specific energy from 7.120 to 0.960 kWh/kg with change in temperature from 25 to 85 °C. This reduction in specific energy is due to increase in permeate flux because it has inverse relation with the amount of permeate collected. Similarly, Criscuoli *et al.* (2008) also analysed the impact of feed temperature on permeate flux. The authors found the linear increment in permeate flux at constant permeate flux of 10mbar using 0.20 μm polypropylene membrane during the treatment of wastewater [12, 13].

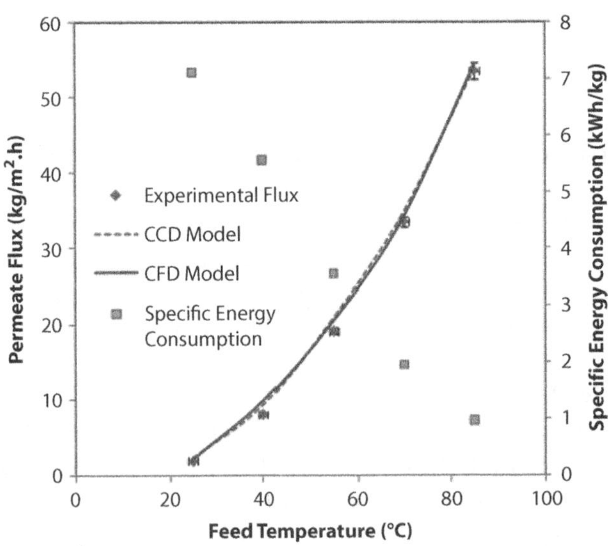

Figure 10.2 Impact of feed temperature on permeate flux (reprint with permission from Elsevier).

Impact of vacuum degree on permeate flux

The change in flux with increment in vacuum degree is given in Figure 10.3. It was observed that the flux is increased linearly from 46.81 to 53.50 kg/m^2h with increase in vacuum degree from 670 to 750mmHg at constant temperature, concentration and flow rate as 85 °C, 30 ppm 2 lpm for PTFE membrane. The applied vacuum on permeate side decreased the diffusion resistances arising due to presence of air molecules are reduced which enhanced the vapor transport across pores and resulted in higher permeate flux. The decrement in consumed energy was seen from 2.53 to 0.95 kWh/kg with increased in vacuum degree. This reduction in energy consumption is due to reduction in temperature polarization coefficient which increased the permeate flux and results in linear increment in permeate collected. In model analysis, it was found that both CFD and CCD models are best fitted with experimental data. The R-squared for CFD and CCD model was found to be 0.989 and 0.995 with laboratory results. Moreover, MAPE value was seen to be 1.8 and 0.95 for CFD and CCD model with operating conditions for PTFE membrane. Various studies have been carried out to analyze the impact of vacuum degree flux and a similar trend was obtained. However, the moderate range of vacuum degree from 650 to 720mmHg was used by various researchers to save the membrane from getting punctured/pore wetting [14, 15].

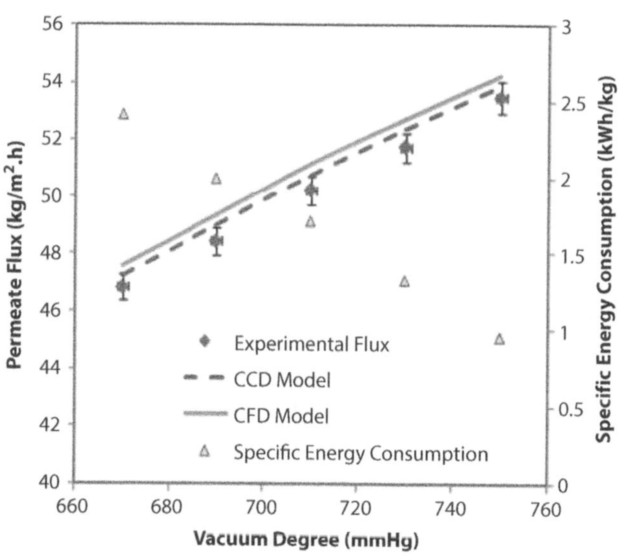

Figure 10.3 Impact of vacuum degree on permeate flux (reprint with permission from Elsevier).

Impact of flow rate on permeate flux

The increment in flow rate of working fluid at feed side also influenced the permeate flux. The impact on flux is shown in Figure 10.4 on varying feed rate from 2 to 10 lpm. The linear increment in permeate flux was observed from 48.30 to 53.50 kg/m^2.h operating at 85 °C, 750mmHg of temperature and vacuum degree with increased flow rate. On increasing flow rate, the boundary layer created on membrane surface due to accumulated dye particles is reduced and results in reduction of mass transfer resistances. The higher flow rate influences the hydrodynamic conditions which reduced the accumulation rate on membrane which directly impact permeate flux. The reduction in SEC was also observed from 1.210 to 0.610 kWh/kg due to increased in permeate volume. The models CCD and CFD are best fitted with experimental conditions with R^2 as 0.980 and 0.990 and MAPE as 0.908 and 1.39, respectively.

10.4 Long-Term Performance Analysis of VMD

The removal efficiency and the percentage reduction in flux were depicted for continuous performance analysis for vacuum membrane distillation in continuous operation. The performance was analyzed continuously for the period of 60 hrs for dye removal using PTFE membrane. The decrement in flux was observed from 45.50 to 43.46 kg/

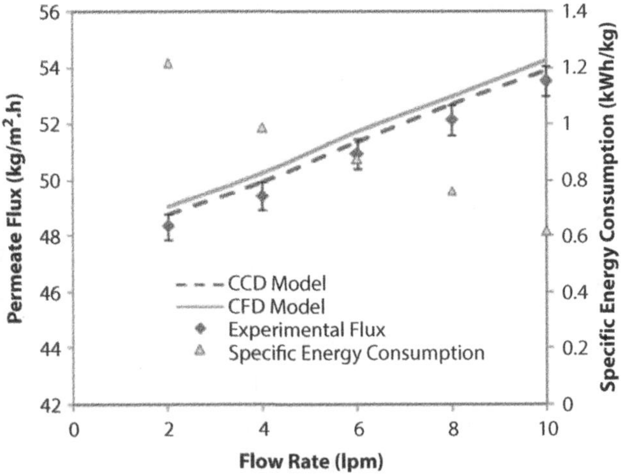

Figure 10.4 Impact of flow rate on permeate flux (reprint with permission from Elsevier).

m^2h in 60 hrs at 85 °C, 5lpm, 750 mmHg and 30 ppm of temperature, vacuum, flow rate and dye concentration. In PTFE membrane, about 4.5% decrement was estimated in permeate flux. This decrement is due to accumulation of larger colored particles on membrane surface which can be removed by various physical means such as water washing, acid washing, etc., depending upon the type of fouling on membrane surface. This accumulation is determined using the SEM analysis and the

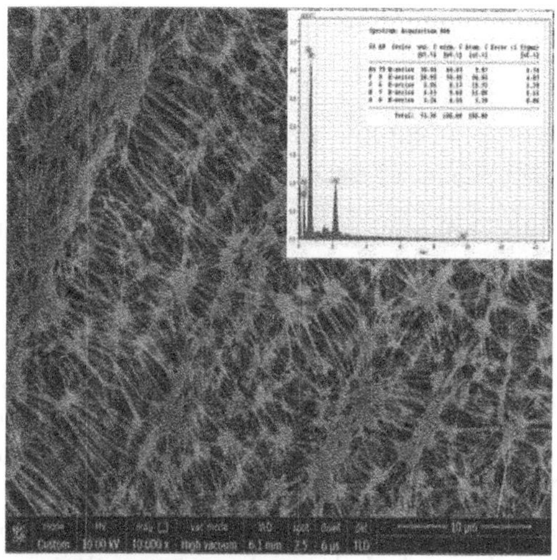

(a) Fresh Membrane

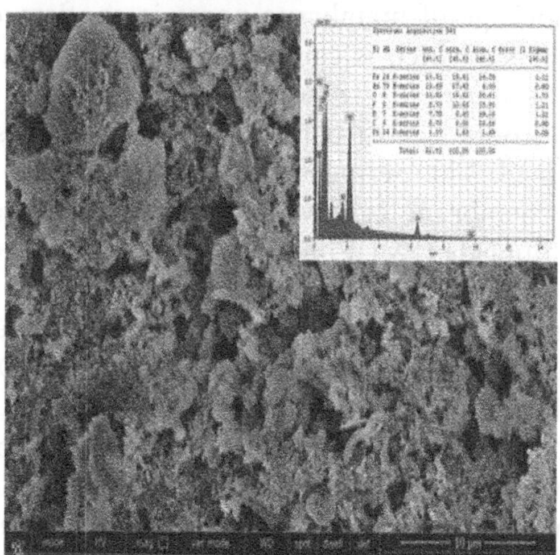

(b) After use membrane

Figure 10.5 SEM with EDS images of before and after use of PTFE membrane.

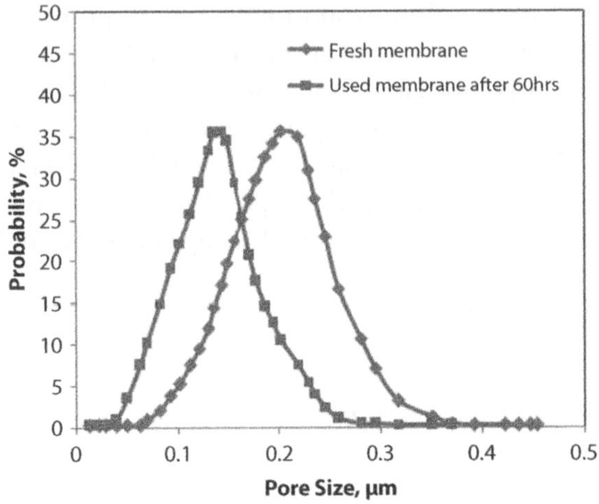

Figure 10.6 PSD analysis before and after use PTFE Membrane (reprint with permission from Elsevier).

SEM images of PTFE membrane fresh and used are given in Figure 10.5. SEM and corresponding EDS analysis confirmed the accumulated of dye molecules and reduction in pore size. After washing with water, the permeate flux was recovered nearly up to 44.94 kg/m^{2}h for PTFE. However, the original flux was not recovered because some pores were getting permanently occluded by dye accumulation on membrane surface. The blockage of membrane pores is confirmed by pore size distribution analysis which is carried out before and after use of the membrane and is shown in Figure 10.6. In PSD shown in Figure 10.6, the reduction in pore size distribution of fresh and used membrane pores is confirmed. The permeate flux obtained from the polytetrafluoroethylene membrane is lower compared to polypropylene and polyvinylidene fluoride (PVDF) membrane but the thermal stability of PP membrane is lower (less than 95 °C) and due to this polypropylene membranes are operated at lower temperature. In case of PTFE and PVDF membrane, the permeate flux obtained from PVDF is higher but the mechanical stability is good enough to resist scale formation and pore wetting. In PVDF membranes, nearly 8% reduction in permeate flux was observed when working at the same operating conditions for 60 hrs operating time in a continuous manner. Moreover, the mechanical stability of PVDF membrane is lower, which is why the chances of pore wetting increases with increase in membrane pores size when the system is operating at constant operating conditions.

10.5 Scale Formation in Long-Term Assessment

The deposition of inorganic salts, organic compounds and other biological particles at surface is known as fouling which decreased the performance of VMD processes. The study of rate of scale formation is required during the continuous operation because it provides information regarding the washing of membrane and/or permanent replacement of membrane. In VMD, three types of fouling are possible on membrane surface: (i) inorganic; (ii) organic, and (iii) biological fouling [16–18]. All types of fouling can be possible in dye contaminated wastewater depending upon the source. However, the biological fouling is negligible in VMD process because it is a thermal-driven process which operates normally higher than 50 °C which decompose the biological microorganism present in water.

Inorganic fouling is due to accumulation of colloidal and salts on the surface of membrane from the liquid feed. Depending upon the nature, the inorganic fouling is categorized as alkaline, non-alkaline and neutral material fouling. This fouling occurred at saturation conditions arising due to evaporation and temperature change at the membrane surface. These saturation conditions start the nucleation (crystallization) of foulant on membrane surface and growth of this crystallization depend upon the rate of vapor transfer and the change in temperature. The nucleation is known as heterogeneous crystallization and after some time a thick layer of scale is formed which is known as bulk (homogeneous) crystallization.

The adherence of organic materials such as acids, proteins, etc., at membrane surface is known as organic fouling which form the gel-like layer at membrane. These compounds can damage or alter the membrane structure permanently by reacting with the membrane. Generally, the fouling due to accumulation of natural organic matter was observed in VMD processes when used to treat the wastewater which had the source as ponds, rivers and oceans. The organic fouling affects the membrane systems in three ways: (i) acts as inorganic fouling and damages the membrane by blocking the pores; (ii) formed the gel-like structure which reduced the flow passage; (iii) bind with membrane surface and permanently alter the membrane morphology.

In dye-containing wastewater systems, both inorganic and organic fouling is present due to chromophores and chromogens along with inorganic salts. Dye-contaminated wastewater has higher fouling as compared to other desalination processes and due to this pressure-driven processes like reverse osmosis, ultra filtration, etc., are not preferred for the treatment of wastewater. The chromophores, chromogens and inorganic salts present in dye molecules, exert vander-waal forces and electrical double layer forces during interaction of dye molecules with membrane surface. Due to availability of opposite charges on dye and membrane surface the fouling is more due to attraction force. So, the proper washing is required at constant time interval difference to protect the membrane from damage during long-term operation.

Conclusion

Compared with other membrane processes, the VMD process is considered as economic and efficient for separation of dyes from wastewater. In VMD, heat and mass transfer are associated in couple, and transport mechanism is explained using molecular, Knudsen, and Poiseuille flow. The consideration of transfer mechanism depends on the physical properties of membrane and its module. The impact of process variables like feed temperature, vacuum degree and flow rate was analyzed and it was found that feed temperature increased permeate flux exponentially. The long-term performance assessment is carried out and the dye precipitation on surface is confirmed by SEM and EDS analysis during the continuous operation of 60 hrs. It was also noticed that the pore size was also reduced due to dye accumulation on membrane surface which was also confirmed by pore size distribution analysis. At the end, it was seen that VMD process can be efficiently used to produce fresh water from dye-contaminated wastewater. The VMD process also has the capability to combine with other renewable surfaces which can significantly reduce the water treatment problems and energy crisis.

Nomenclature

A	Antoine Coefficient
B	Antoine Coefficient
C	Antoine Coefficient
H	Overall Heat transfer Coefficient
h_{bf}	Bulk feed side heat transfer coefficient
h_m	Membrane heat transfer coefficient
k	Overall Mass Transfer Coefficient
k_{bf}	Bulk feed side Mass transfer coefficient
k_m	Membrane side mass transfer coefficient
M	Molecular weight of working fluid
N	Permeate flux

P_{fm}	Feed side membrane surface vapor pressure
P_p	Permeate side vapor pressure
P_{pm}	Permeate side membrane surface vapor pressure
q_{bf}	Bulk feed side heat flux
R	Gas Constant
T_{bf}	Bulk feed side Temperature
T_{fm}	Feed side membrane surface temperature
x_{bf}	Bulk feed side solute fraction
x_{fm}	Feed side membrane surface solute fraction
x_p	Permeate side solute fraction

Greek Symbols

ΔH_v	Latent heat of vaporization
ΔT_m	Temperature difference across membrane
ε	Membrane Porosity
τ	Membrane tortuosity
δ	Membrane Thickness
η	liquid Viscosity

References

1. Tan, I. A. W., B. H. Hameed, and A. L. Ahmad, *Chemical Engineering Journal*, Vol. 127, p. 111–119, 2007.
2. Gürses, A., Açıkyıldız, M., Güneş, K., & Gürses, M. S., *Dyes and Pigments*, p. 13–29, 2016.
3. Christie, R. M., *Environmental aspects of textile dyeing*, Elsevier, 2007.
4. Mengual, J. I., Khayet, M., & Godino, M. P., *International Journal of Heat and Mass Transfer*, Vol. 47(4), p. 865–875, 2004.
5. Khayet, M., & Matsuura, T., *Membrane distillation: principles and applications*, Elsevier, 2011.
6. Criscuoli, A., Bafaro, P., & Drioli, E., *Desalination*, Vol. 323, p. 17–21, 2013.
7. Abu-Zeid, M. A. E. R., Zhang, Y., Dong, H., Zhang, L., Chen, H. L., & Hou, L., *Desalination*, Vol. 356, p. 1–14, 2015.
8. Baghel, R., Upadhyaya, S., Singh, K., Chaurasia, S. P., Gupta, A. B., & Dohare, R. K., *Reviews in Chemical Engineering*, Vol. 34(1), p. 73–106, 2017.
9. Kalla, S., Upadhyaya, S., Singh, K., & Baghel, R., *Journal of Chemical Technology & Biotechnology*, Vol. 94(1), p. 63–78, 2019.
10. Upadhyaya, S., Singh, K., Chaurasia, S. P., Dohare, R. K., & Agarwal, M., *Desalination and Water Treatment*, Vol. 57(26), p. 11956–11971, 2016.

11. Kalla, S., Upadhyaya, S., & Singh, K., *Reviews in Chemical Engineering*, Vol. 35(7), p. 817–859, 2019.
12. Criscuoli, A., Zhong, J., Figoli, A., Carnevale, M. C., Huang, R., & Drioli, E., *Water Research*, Vol. 42(20), p. 5031–5037, 2008.
13. Baghel, R., Upadhyaya, S., Chaurasia, S. P., Singh, K., & Kalla, S., *Journal of Cleaner Production*, Vol. 199, p. 900–915, 2018.
14. Upadhyaya, S., Singh, K., Chaurasia, S. P., Dohare, R. K., & Agarwal, M., *Desalination and Water Treatment*, Vol. 57(55), p. 26886–26898, 2016.
15. Baghel, R., Kalla, S., Upadhyaya, S., Chaurasia, S. P., & Singh, K., *Chemical Engineering Research and Design*, 2020.
16. Curcio, E., Ji, X., Di Profio, G., Fontananova, E., & Drioli, E., *Journal of Membrane Science*, Vol. 346(2), p. 263–269, 2010.
17. Gryta, M., *Desalination*, Vol. 216(1-3), p. 88–102, 2007.
18. Krivorot, M., Kushmaro, A., Oren, Y., & Gilron, J., *Journal of Membrane Science*, Vol. 376(1-2), p. 15–24, 2011.

Section IV

EMERGING TECHNOLOGIES & PROCESSES

11

Application of Zero Valent Iron to Removal Chromium and Other Heavy Metals in Metallurgical Wastewater

Khac-Uan Do[1]*, Thi-Lien Le[1] and Thuy-Lan Nguyen[2]

[1]*School of Environmental Science and Technology, Hanoi University of Science and Technology, Hanoi, Vietnam*
[2]*Center of Industrial Environment, National Institute of Mining - Metallurgy Science and Technology, Hanoi, Vietnam*

Abstract

The metallurgical wastewater contains a high content of heavy metals. Currently, the management and treatment of wastewater sources at the metallurgical factories have not met the requirements for discharge conditions. Therefore, metallurgical wastewater must be treated properly before being discharged into the receiving waters. In this work Fe^o nanoparticles were used in the batch experiments to investigate the efficiency of treating heavy metals in metallurgical wastewater. In particular, the effects of pH, Fe^o dosage, reaction time, PAC dosage and PAM dosage on the reduction efficiency of Cr(VI) and other heavy metal ions were investigated. The obtained results showed that the proper amount of Fe^o was found to be 10 mg/L, at the condition pH of 1.5, and the reaction time of 60 minutes could maintain high treatment efficiency. The other operational conditions including pH of 8, PAC of 250 mg/L, and PAM of 3,5 mg/L could maintain the effective removal up to 98% for other metals (Pb, Cd, Zn, Ni, Fe, ...). In conclusion, using Fe^o nanomaterials would be a suitable solution for treating metallurgical wastewater.

***Keywords*:** Heavy metals, hexavalent chromium, metallurgical wastewater, polyacrylamide, polyaluminium chloride, zero valent iron

**Corresponding author*: uan.dokhac@hust.edu.vn

Siddhartha Moulik, Aditi Mullick and Anirban Roy (eds.) Sustainable Water Treatment: Advances and Technological Interventions, (417–440) © 2022 Scrivener Publishing LLC

11.1 Introduction

11.1.1 Wastewater Sources from Metallurgical Factories

Metallurgical factories could generate wastewater from several activities such as electrolysis, metallurgy, and mineral sorting workshops [1]. In addition, wastewater could be discharged from the analytical laboratory or from the electrolytic sludge treatment areas. A schematic production process with waste stream from a general small metallurgical factory in Vietnam is presented in Figure 11.1. Water used for different activities of the factory is described in Table 11.1.

Wastewater characteristic in the ore sorting stage usually contains high suspended solids. Wastewater is also generated from washing activity [4]. It could be acidic and could contain heavy metals from dissolved substances of the crude ore. In the smelter workshop, water was only used in the cooling process [5]. Water was also used for gas treatment [6]. Therefore, wastewater often contains impurities of ore and heavy metal [7]. In addition, wastewater discharged from gas treatment could be acidic and could contain sulfide compounds [8]. Wastewater from electrolysis sludge treatment was only about 5 m^3/day. In this step, electrolytic sludge was collected and put into a tank. It was then filtered and burned in a coal furnace. Sludge after incineration was dissolved in a tank containing hydrochloric acid to recover metals. Therefore, wastewater from this activity contains

Table 11.1 Water used and wastewater generated from a general small metallurgical factory in Vietnam [2].

No.	Water used and wastewater generated	Average flowrate (m^3/day)	Maximun flowrate (m^3/day)
1	Water used at the mineral sorting activity	10	20
2	Water used at the smelting worshop	10	15
3	Wastewater from electrolysis sludge treatment	3	5
4	Wastewater from physical and chemical analysis activities	1,5	2
Total		**24,5**	**42**

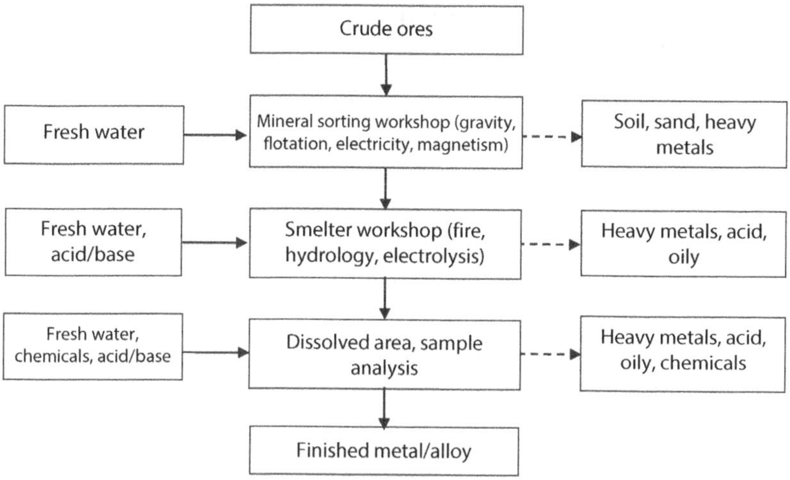

Figure 11.1 A schematic production process with waste stream [3].

a relatively low pH. A metallurgical factory normally has a physical and chemical analysis center. This center uses chemicals to analyze the quality of input ore and products from the refining process. Wastewater from this area is not large but is highly polluted. Wastewater flow is about 2 m^3/day, but contains heavy metals, acids and some other chemicals.

11.1.2 Characteristics of Wastewater in Metallurgical Factories

A metallurgical factory includes many different mineral sorting and metallurgical technologies [9]. The amount of wastewater that was generated depended on the production scale. Its composition is very complicated [10]. For example, in ore sorting workshop, it often contains inorganic impurities with high suspended solids content. In fire-treatment process, wastewater often contains impurities of ore and metallurgical metals. In hydrolysis process, it could contain dissolving solution of HCl, H_2SO_4, NaOH, other salts, suspended solids or heavy metals. In metal processing, water was used in different processes, such as cleaning, polishing, coating [11]. Therefore, wastewater could contain high pollutions of rust, heavy metals, grease, caustic soda, acids, detergents [12]. Currently, the management and treatment of wastewater sources at the metallurgical factories have not met the requirements for discharge conditions. Most of the facilities have only had preliminary treatment due to limited technology. It could be seen that wastewater in metallurgical factories must be treated properly before being discharged into the receiving waters.

11.1.3 Conventional Technologies for Treating Wastewater in Metallurgical Factories

So far, there have been many methods of reating metallurgical wastewater such as chemical precipitation, adsorption, redox, electrochemical or biological methods [13–15]. Chemical precipitation method is commonly based on precipitation of metals in the form of hydroxide [16]. The chemicals used in the precipitation process include NaOH, $Ca(OH)_2$, Na_2CO_3, Na_2S [17]. The metal precipitation reaction is generally described as follows [18].

$$Me^{+n} + X^{-m} \rightarrow Me_mX_n$$

In which: Me^{+n} are metal cations in water such as Cu^{2+}, Ni^{2+}, Zn^{2+}, Cr^{3+}.

X^{-m}: The chemical anions introduced such as OH^-, S^{2-}, SO_4^{2-}, PO_4^{3-}.

The adsorption method used suitable materials which could adsorb heavy metals such as activated carbon, peat, iron oxide, manganese oxide, ash and slag, bentonite, chemical polymer materials or biopolymers [19, 20]. The adsorption method could treat heavy metals at low concentrations effectively [21]. This method could use industrial wastes such as Fe_2O_3. This absorption material could be reused after adsorption releasing and regeneration steps [22].

Redox method is the process of converting harmful substances into less toxic substances [23]. Several reducing agents such as Na_2SO_3, $NaHSO_3$, $FeSO_4$ could be used to reduce Cr(VI) to Cr^{3+} in metallurgical wastewater [24].

$$2CrO_3 + 6FeSO_4 + 6H_2SO_4 \rightarrow 3Fe_2(SO_4)_3 + 6H_2O + Cr_2(SO_4)_3$$

$$4CrO_3 + 6NaHSO_3 + H_2SO_4 \rightarrow 2\ Cr_2(SO_4)_3 + 3Na_2SO_4 + 6H_2O$$

$$2H_2Cr_2O_7 + 3\ Na_2S + 6H_2SO_4 \rightarrow 2Cr_2(SO_4)_3 + 3Na_2SO_4 + 7H_2O + 3S$$

Among those reducing agents, using $FeSO_4$ could generate more Fe^{3+} ions in the effluent. Sludge wasted could contain iron hydroxide precipitation. Using $NaHSO_3$ or Na_2SO_3 to reduce Cr(VI) could reduce the sludge wasting [13, 25].

The ion exchange method was based on chemical interactions between ions in the liquid phase and ions in the solid phase [26]. General reaction mechanism of this method is as follows.

$$A_mB + {}_mC \rightarrow mAC + B.$$

For example, using cationite and anionite could remove heavy metals in wastewater [27], such as.

$$\text{Cationit: } nR\text{-}H + B^{n+} \rightarrow RnB + nH^{+}$$

In which: B^{n+} could be Cu^{2+}, Ni^{2+}, Mn^{2+}, Pb^{2+}

$$\text{Anionite: } nR\text{-}OH + B^{n-} \rightarrow R_nB + nOH$$

In which B^{n-} could be $Cr_2O_7^{2-}$, MnO_4^{2-}

Wastewater treatment by ion exchange method could achieve high treatment efficiency (about 99%) [28]. This method could regenerate ion exchangers and could recover precious metals [29]. In addition, it could be operated simply and suitably with wastewater containing high metal concentrations. Moreover, it does not generate secondary wastes [30]. However, this method has a high treatment cost which would not be suitable for large amounts of wastewater.

The electrochemical method was based on a direct current flowing through the electrode's two poles [31]. In this method, oxidation and reduction processes occurred on the anode and the cathode. The process of cation reduction occurred at the metal anodes in which metals were reduced to less toxic ions or were formed on the electrodes as in the following equation.

$$M^{n+} + me \rightarrow M^{n-m} \ (n>m)$$

m: The valence of metals

n: Number of electrons that changes valence (n > m)

The electrochemical method could be used to treat wastewater containing high concentrations of metals [32]. This method does not use chemicals, and could recover valuable metals [33]. However, this method could consume high power which could increase the treatment costs.

The biological method is based on the principle that some species of organisms and plants in water could use metals as trace substances during development of biomass [34, 35]. The common microorganisms could be algae, fungi, bacteria. In addition, some kind of plants, such as vertiver grass, watercress, and fern living in the environment with heavy metal pollution, could be able to absorb the toxic heavy metals [36]. Moreover, the *Saccharomyces cerevisiae* yeast could absorb some heavy metals (Cu^{2+}, Pb^{2+}, Zn^{2+}) in water [37, 38]. The *Saccharomyces cerevisiae* yeast has a good ability to absorb heavy metals at pH of 5 [39].

Using biological methods to treat heavy metals in wastewater could save chemical treatment costs [40, 41]. However, this method requires a very large surface. In addition, it could generate biomass containing heavy metals.

11.1.4 Zero Valent Iron for Removing Heavy Metals

As mentioned above, metallurgical wastewater contains a lot of suspended solids and dissolved metals. There have been several studies on treatment of heavy metals from metallurgical wastewater [42]. The chemical precipitation and redox methods were most commonly used. The redox method using $FeSO_4$ could reduce Cr(VI) to Cr^{3+} at pH less than 1.5. It could remove other heavy metals at pH value from 7 to 9 [18]. Waste scrap iron could be also used to remove Cr(VI) in wastewater. In Vietnam, research works on metallurgical wastewater treatment were still limited. The physical method and chemical precipitation method have been applied. The electrochemical flocculation method combined with electrochemical method was tested for treatment of heavy metal ions and fluoride in wastewater [32]. In recent years, Fe^{o} nanomaterials were developed and applied in the environmental field [43, 44]. Fe^{o} nanomaterials were used to remove heavy metals such as Pb, Cr, Mn [45]. At pH of 4.5 - 7.5, the removal efficiency of Pb^{2+} could be 325 g Pb^{2+} /1 kg Fe^{o}, in which the initial concentration of Pb^{2+} was less than 50 mg/L and Fe^{o} nanomaterials amount was of 0.1 g/L [46]. In addition, Fe^{o} nanomaterials could be used to treat nitrate. The treatment efficiency reached 98.9% at initial nitrate concentration of 30 mg NO_3^-N/L, reaction time of 40 minutes, at pH of 2 and the Fe^{o} nanomaterials of 1 g/L [47]. Application of Fe^{o} nanoparticles in wastewater treatment could save treatment time and reduce sludge waste [48]. Fe^{o} nanoparticles were applied in domestic wastewater or general industrial wastewater [49]. It could be also applied for the metallurgical wastewater. This could be an alternative way to ensure high treatment efficiency and to save costs for treating metallurgical wastewater.

11.1.5 Objectives of the Study

Wastewater generated from metallurgical factories containing high concentration of heavy metals could pollute the receiving source and could affect the aquatic plants and animals. Currently, the management and treatment of metallurgical wastewater were not really effective. Facing this situation, it is necessary to select a suitable method to treat heavy metals in metallurgical wastewater. This could contribute to establish a wastewater treatment technology for treating wastewater from a metallurgical factory or for enhancing the current wastewater treatment system in other metallurgical factories. In this study, Fe^{o} nanomaterials was used to reduce Cr(VI) in the metallurgical wastewater. In addition, the work was aimed to determine the optimal treatment conditions to remove heavy metals (Cr, Pb, Cd, Cr, Zn, Mn, Ni, Fe) in metallurgical wastewater.

11.2 Materials and Methods

11.2.1 Metallurgical Wastewater

In this work, the samples were selected and taken from the experimental metallurgical factory (which belongs to the National Institute of Mining-Metallurgy Science and Technology), and is located in Thai Nguyen Mining and Metallurgy Co, Ltd. (Thai Nguyen, Vietnam). The wastewater sample was taken from production processes. Sampling location was selected at the storage tank standing before the current treatment system of the factory (Figure 11.2). The sampling device is a wide-mouthed flask with a volume of 2 L. The sampling device was rinsed by wastewater twice before taking the wastewater samples.

It should be noted that wastewater samples were taken at different times of the day (morning, noon, afternoon) under the normal operation of the

Figure 11.2 Selection of location for wastewater sampling.

factory. Wastewater samples were then poured into 1.5 L plastic bottles. They were put into the dark foam containers to store at 4°C. Total sample volume were 10 L and then transported to the laboratory at the Center of Industrial Environment. Samples were stored in a dedicated sample cabinet at 4°C to analyze heavy metals.

11.2.2 Preparation of Zero Valent Iron

The chemicals used in the study include the standard stock solution of HCl 35%, NaOH, distilled water, $NaBH_4$ salt. The chemical solutions were prepared as follows. Preparation of 10% NaOH solution: 10 g of pure NaOH was dissolved in 90 mL of distilled water in a 250 mL triangle. A magnetic stirrer was used at a speed of 200 rpm for 10 minutes. Zero valence Fe is a strong reducing agent. There are many methods to make nanoscale iron materials such as crushing method, micro-emulsion method, co-precipitation, and chemical reduction. A method of liquid phase reduction through $FeSO_4.7H_2O$ desalination by $NaBH_4$ in the presence of PAM (polyacrylamide) is most commonly used to fabricate materials nano-size iron. Preparation of nano iron is carried out by the following steps. 4 g of $FeSO_4.7H_2O$ (using the GT310-6 technical balance) was dissolved in 50 mL of distilled water in a 150 mL triangle. Solution was shaken by a shaker for 20 minutes at 150 rpm to obtain solution A. Then 0.1 g of PAM was dissolved in 1000 mL of distilled water to achieve 0.01% PAM solution. This solution was put overnight. Then 2 g of $NaBH_4$ was dissolved in 18 mL of distilled water in a 150 mL flask. 2 mL of 0.01% PAM dispersant solution was added and stirred for about 5 min at high speed of 300 rpm to obtain solution B. Finally, solution B was gradually added to solution A on a magnetic stirrer at a low speed of 5 mL/min. A black precipitate of nano Fe^o was formed. The magnets were used to separate Fe^o nanoparticles from solution. It was then washed with alcohol. The obtained Fe^o nanomaterials were put into the desiccator to preserve and to be used in the work. In addition, the PAC (polyaluminium chloride) flocculation chemicals were mixed into a 10% solution. PAM deposition chemical was mixed into a 0.1% solution to ensure a good mixing in the wastewater.

11.2.3 Batch Experiments

The experiments (Exp.) were performed in batch tests to determine the optimum conditions. The experiments were summarized in Table 11.2. Samples before and after treatment at different conditions are illustrated in Figure 11.2.

Table 11.2 The bath tests used in this study.

Objectives	Notes	Experimental description
Reduction of Cr(VI) to Cr^{3+}	Exp. 1	Investigate the effect of pH on the treatment efficiency of Cr(VI) by nano Fe^0.
	Exp. 2	Investigate the effect of Fe^0 nanoscale content on Cr(VI) treatment efficiency.
	Exp. 3	Investigate the effect of time on the treatment efficiency of Cr (VI) by nano Fe^0.
Treating heavy metals	Exp. 4	Investigate the effects of pH on the efficiency of treating heavy metals.
	Exp. 5	Investigate the effect of PAC dosage on the efficiency of heavy metal removal.
	Exp. 6	Investigate the effect of the PAM dosage on the efficiency of heavy metal removal.

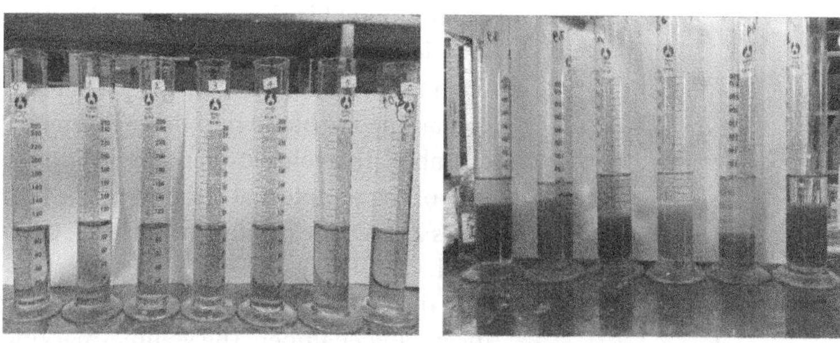

Samples before treatment   Samples after treatment

Figure 11.3 Samples before and after treatment at different conditions.

11.2.4 Analysis Methods

The pollutant parameters were selected for analysis Pb, Cd, Cr, Zn, Mn, Ni, Fe and pH. The analysis procedures are described in Table 11.3.

Other heavy metals such as Pb, Cd, Zn, Ni, Fe, ... are analyzed in the laboratory of the Center for Industrial Environment - Institute of Mining Science and Technology. The analysis procedures followed the standard methods [50].

Table 11.3 The analysis procedures used in this work.

Parameters	Method	Analysis procedures
pH	pH meter	- Pour the water sample in a clean glass. - Adjust the pH meter with standard buffer solutions. Rinse the electrode with distilled water, dry the electrode and immerse in the test water and read the result on the pH meter. The measured pH value was calculated based on the average of 3 time measurements of the same sample.
Cr(VI)	8023 - Hach DR2800	- Press the machine to select the test at a wavelength of 540 nm. - Pour 10 mL of water sample into a clean sample bottle; - Put 1 pack of ChromaVer 3 Chromium reagent in the sample vial and shake well (if the test contains Cr(VI), the solution will appear purple). - Press TIMER> OK. The reaction time of the reagent and test sample is 5 minutes. - Preparation of blank sample: Pour 10 mL of sample into the second sample vial. - When the reaction time is over, place the vial containing the blank into the measurement chamber. The volume-marking line on the wall of the sample vial is facing to the right. Press ZERO, the display will display: 0.00 mg/L Cr(VI) - Place the vial containing the test sample in the measuring chamber. The volume-marking line on the wall of the sample vial is facing to the right. Press READ. The result is mg/L Cr(VI). The range of concentration limits is from 0.010 to 0.700 mg/L Cr(VI).

(*Continued*)

Table 11.3 The analysis procedures used in this work. (*Continued*)

Parameters	Method	Analysis procedures
Cu^{2+}	8506 - Hach Dr2800	- Press machine to select the test at a wavelength of 560 nm. - Pour 10 mL of water sample into a clean sample bottle; - Add 1 pack of CuVer® 1 Copper reagent to the sample vial and mix. Use a 25 mL sample vial and CuVer® 2 Copper reagent for high Al, Fe, and high hardness samples. - Press TIMER> OK. The reaction time of the reagent and test sample is 2 minutes. - Sample preparation: Pour 10 mL of sample into the second sample vial. - Place the vial containing the blank into the measurement chamber. The volume-marking line on the wall of the sample vial is facing to the right. Press ZERO, the screen will display 0.00 mg/L Cu. - Within 30 minutes after the reaction time has elapsed, place the vial containing the test sample in the measuring chamber. The volume-marking line on the wall of the sample vial is facing to the right. Press READ. The results are mg/L Cu. The range of concentration limits is from 0.04 to 5.00 mg/L.
Mn^{2+}	8304 - Hach Dr2800	- Press the machine to select the test at a wavelength of 525 nm. - Pour 10 mL of water sample into sample bottle No. 1. - Add 1 pack of Citrate Type reagent to the sample vial. Cap tightly and shake upside down to dissolve the reagent. Add another pack of Sodium Periodate reagent to the sample vial. Cap tightly and shake upside down to dissolve the reagent. The solution will appear purple if it contains Mn.

(*Continued*)

Table 11.3 The analysis procedures used in this work. (*Continued*)

Parameters	Method	Analysis procedures
Mn^{2+}	8304 - Hach Dr2800	- Press TIMER > OK. The reaction time of the reagent and test sample is 2 minutes. - Sample preparation: Pour 10 mL of sample into sample bottle 2. - Place the vial containing the blank into the measurement chamber. The volume-marking line on the wall of the sample vial is facing to the right. Press ZERO, the screen will display 0.0 mg/L Mn. - Within 8 minutes after the reaction time has elapsed, place the vial containing the test sample in the measuring chamber. The volume-marking line on the wall of the sample vial is facing to the right. Press READ. Results showed mg/L Mn. Limit concentration range from 0.1 to 20 mg/L.

11.3 Results and Discussion

11.3.1 Effects of pH on Hexavalent Chromium Removal

Experiment 1 was performed to find the optimal pH value for the treatment proccesses. The obtained results of Cr(VI) in the treated wastewater samples using Nano Fe^{o} at different pH values were presented in Figure 11.4.

The results of Figure 11.3 show that when the pH was changed from 1.5 to 4, the concentration of Cr (VI) was also changed in the increasing direction and the treatment efficiency was decreased. In this test, the initial Cr (VI) concentration was 3.4 mg/L. In the environment, the pH of wastewater was pH = 1.5, the concentration of Cr (VI) after treatment was only 0.05 mg/L. The treatment efficiency was about 98%. When pH was raised to 2, the concentration of Cr (VI) was decreased to 0.04 mg/L and the treatment efficiency was increased to about 99%. When pH was continuously raised to 2.5, there was a significant change in treatment efficiency. The treatment efficiency was decreased to about 87%. The concentration of Cr (VI) was 0.45 mg/L which exceeded the permitted standard by about 4.5 times. In case of increasing pH to 3, the treatment efficiency was decreased significantly. The concentration of Cr (VI) after treatment was increased to 0.98 mg/L. The treatment efficiency was only about 70%. The treatment

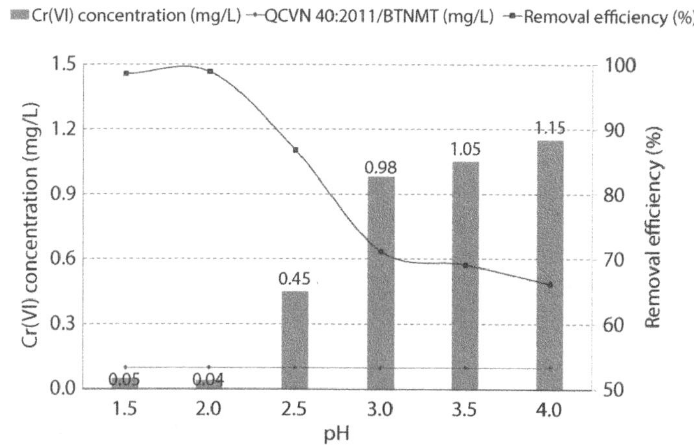

Figure 11.4 Effects of pH on the treatment efficiency of Cr (VI).

efficiency was reduced to only about 65% at pH of 4. Therefore, the pH value was selected as the initial pH of 1.5 to ensure the high treatment efficiency.

It should be noted that in water Cr (VI) usually exists in the form CrO_4^{2-} and $Cr_2O_7^{2-}$. When Cr (VI) is in contact with Fe^0 solution, the reduction reaction of Cr (VI) to Cr^{3+} and Fe^0 oxidizes to Fe^{2+} according to the following equation [51].

$$2CrO_4^{-} + 2H^+ \rightarrow Cr_2O_7^{2-} + H_2O$$

$$Cr_2O_7^{2-} + 3Fe^0 + 14H^+ \rightarrow 3Fe^{2+} + 2Cr^{3+} + 7H_2O$$

Fe^{2+} is formed and continues to reduce Cr (VI)

$$Cr_2O_7^{2-} + 6Fe^{2+} + 14H^+ \rightarrow 6Fe^{3+} + 2Cr^{3+} + 7H_2O$$

The reaction equation shows that the amount of H^+ was higher than 14 times the amount of Cr(VI) in wastewater in the reaction occurs, so the most favorable environmental conditions are low pH (pH $\leq$ 2). The Cr^{3+} and Fe^{2+} ions were generated when Fe^0 nanoparticles were used to reduce $Cr_2O_7^{2}$ [52]. The produced Fe^{2+} would reduce Cr (VI) continuosly to increase the treatment efficiency to 98% [53]. This would also save chemicals.

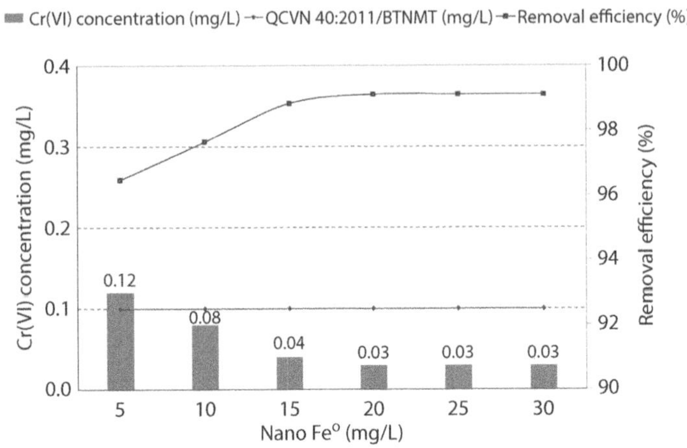

Figure 11.5 Effects of nano Fe^0 on the treatment efficiency of Cr(VI).

11.3.2 Effects of Fe° on Hexavalent Chromium Removal

After performing the experiment 1, the appropriate pH value was determined for the treatment process in the experiment 2. In this test, the original pH of wastewater was kept at pH = 1.5 to determine the amount of nano Fe° used in the treatment process. The experimental results are presented in Figure 11.5.

The results in Figure 11.5 show that when the Fe° nano dosage was increased from 5 mg/L to 30 mg/L, the concentration of Cr (VI) was also changed and decreased while the treatment efficiency was increasing. 5 mg/L nano Fe° was added into 100 mL of wastewater. It was then stirred well for 60 minutes by a magnetic stirrer at 100 rpm. The treatment efficiency reached 96% while the concentration of Cr (VI) after treatment was about 0.12 mg/L which exceeded the permissible standard limit of 1.2 times. When the Fe° nano dosage was increased to 10 mg/L, the treatment efficiency was increased about 98%. The concentration of Cr (VI) after treatment was 0.08 mg/L which met the discharge standards. In case of increasing the Fe° nano dosage from 15 mg/L to 30 mg/L, the treatment efficiency was almost stable at 99%. The concentration of Cr (VI) after treatment was maintained at 0.03 mg/L. Thus, the amount of Fe° nanoparticles of 10 mg/L would be suitable for the treatment process.

The composition of Fe^0 nanoparticles consists of 56% FeO and 44% Fe^0 (in which the shell is FeO, the core layer is nano Fe^0). When Fe^0 nanoparticles reacted with $Cr_2O_7^{2-}$ could create Cr^{3+} ions and Fe^{2+} ions. The generated Fe^{2+} could again reduce Cr(VI). When the amount of Fe° is large enough,

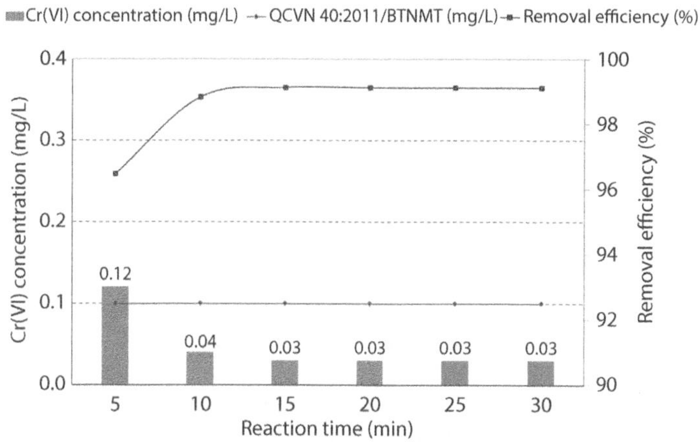

Figure 11.6 Effect of time on Cr (VI) treatment efficiency.

the reduction reaction would be complete with high treatment efficiency [53]. However, the treatment efficiency was remained as described in Figure 11.5 at a stable level of Fe^{o}, even it was added more.

11.3.3 Effects of Contact Time on Hexavalent Chromium Removal

After experiments 1 and 2, the pH value of 1.5 and the amount of Fe^{o} nanoparticles were determined to be 10 mg/L. Those conditions were applied to perform the experiment 3 which investigated the effect of time on treatment efficiency of Cr(VI). The experimental results were shown in Figure 11.6.

The results in Figure 11.6 show that when the reaction time was increased from 30 to 180 minutes, the concentration of Cr (VI) was also changed and decreased while the treatment efficiency was increased. In the first 30 minutes, the treatment efficiency was about 96%. The concentration of Cr (VI) after treatment was 0.12 mg/L which was higher by about 1.2 times than the permitted standard. The reaction time was maintained of 60 minutes and stirred at a speed of 200 rpm, the treatment efficiency was increased 98%. The concentration of Cr (VI) after treatment decreased significantly was 0.04 mg/L which was below the discharged standard. When the reaction time was increased continuously from 90 to 180 minutes, the treatment performance is almost unchanged, reaching about 99%. The concentration of Cr(VI) after treatment was 0.03 mg/L. Thus, the appropriate time for effective treatment was selected at 60 minutes.

It should be noted that the size of Fe^0 nanoparticles was very small, about 18 - 70 nm. Therefore, the Fe^0 nanoparticles could be in the dispersed form which could make a larger surface area with the surface contact area of about 60 m^2/g [54, 55]. This could affect strongly the treatment time and treatment efficiency [56, 57]. This could be due to the fact that a small size could generate higher contact capacity. Therefore, nano Fe^0 could react easily with Cr(VI) at a short time, and resulted in a higher treatment efficiency.

11.3.4 Effects of pH on Heavy Metals Removal

The removal process of reducing other metals was carried out after the reduction phase of Cr(VI). Experimental results of investigating the effects of pH on the treatment efficiency of heavy metals are shown in Figure 11.7.

The results in Figure 11.7 show that when pH was changed from 5.5 to 10, the treatment efficiency was increased gradually. The color of wastewater was decreased gradually as well. In an acidic environment, metals existed in the ionic state. Their concentrations were shown based on the color of wastewater. Therefore, to investigate the metal treatment efficiency, it would be only necessary to examine the color of the wastewater at different pH values. At pH of 5.5, the color of the wastewater was 150 Pt-Co. When pH was increased to 6, the color was 95 Pt-Co. At pH of 7, the color of wastewater was decreased to 60 Pt-Co. However, at pH of 8, the color of the wastewater was decreased significantly to 15 Pt-Co. It was stabilized

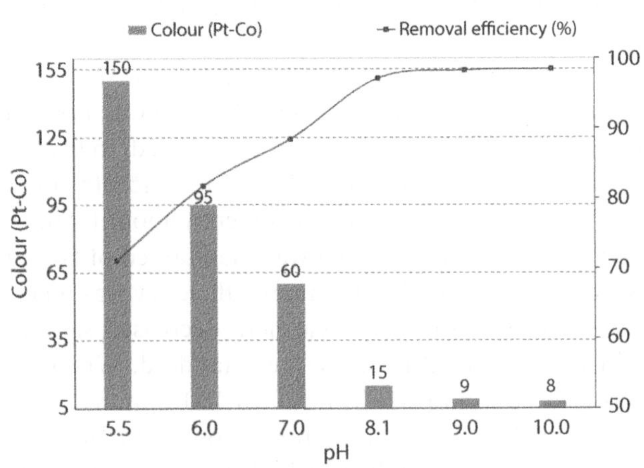

Figure 11.7 Effect of pH on the efficiency of heavy metal treatment.

gradually when pH was increased at 9 and 10. Therefore, the pH value of 8 was selected to ensure high treatment efficiency of heavy metals and also to save chemicals for pH raising. It is clearly that when the pH is raised, most metals could be joined in the hydroxide precipitation reaction. This would reduce the concentration of ions in the solution, resulting in reducing the color of the wastewater as the following reaction.

$$Me^{n+} + nOH^{-} \text{ à } Me(OH)_n \downarrow$$

In this experiment, PAC chemical was used at the rate of 250 mg/L. This is a flocculation chemical that helps to increase the binding capacity of the colloidal $Me(OH)_n$ produced during the pH raising process [58].

11.3.5 Effects of PAC on Heavy Metals Removal

From the above experiment, the pH value of 8 was determined as the most appropriate pH for treatment process. The next experiment was carried out to determine the amount of PAC addition for the best sedimentation efficiency. In this work, the PAC coagulant was mixed into a solution with a concentration of 10%. The obtained results after adding PAC at different values are presented in Figure 11.8.

Figure 11.8 shows that when the amount of PAC was added from 150 to 650 mg/L, the treatment efficiency was increased but it was not increased linearly with increasing PAC dosage. When adding a PAC quantity of 150 mg/L, the color of wastewater was 18 Pt-Co. The amount of PAC

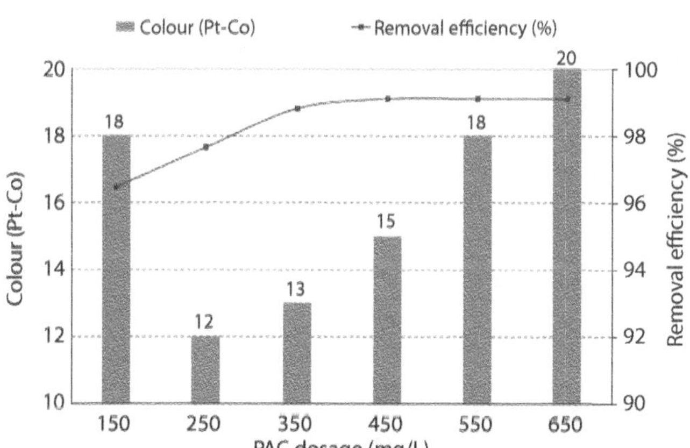

Figure 11.8 Effect of PAC on the efficiency of heavy metal handling.

was increased continuously to 250 mg/L, the color of the wastewater was decreased sharply to 12 Pt-Co. When the amount of PAC was increased to 350 mg/L, the treatment efficiency was changed and slightly reduced due to the increased color of 13 Pt-Co. When the PAC was increased to high concentration of 550 - 650 mg/L, the color of wastewater was increased markedly 20 Pt-Co. Therefore, the appropriate PAC dosage for the treatment process was selected as 250 mg/L. This could be explained by the fact that PAC ($[Al_2(OH)_nCl_6\text{-}nXH_2O]_m$), which was based on the chemical composition of Polyaluminium Chloride, could be a good water solubility agent. This chemical is very effective at low doses. So over range of PAC addition could create the colloidal seeds, resulting in increasing the color of the water. In this experiment, when the amount of PAC was higher than 450 mg/L, the PAC would be redundant. Some of them were dissolved again in the wastewater, increasing the turbidity and color of wastewater [59, 60].

11.3.6 Effects of PAM on Heavy Metals Removal

After determining the appropriate pH and suitable dosage of PAC flocculation chemicals, the next experiment was focused on addition of PAM solution with 0.1% concentration to increase the settling rate of colloidal particles. The experimental results of the sedimentation time of wastewater samples are shown in Figure 11.9.

Figure 11.9 shows that the amount of PAM could be determined based on the settling time in the sediment column (this should be estimated from the end of stirring to the level of sludge to the ¼ level of of the column height) at different times. The amount of PAM was adjusted from 0.5 to 5.5 mg/L

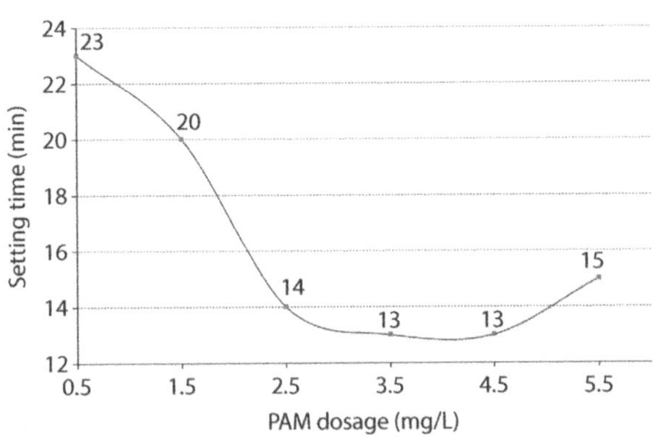

Figure 11.9 Effect of PAM on heavy metal removal efficiency.

in the wastewater samples. In this work, the PAM solution was dissolved with 0.1% concentration. As seen from Figure 3.6, at 0.5 mgL of PAM and after 23 minutes, the height of the sediment column in the test tube reached ¼ column height. When the amount of PAM was increased to 1.5 mg/L, the time of sediment column to reach ¼ of sediment column was just 20 minutes. The amount of PAM was increased continuously to 2.5 mg/L, the settling time was decreased to 14 minutes. When the amount of PAM was fluctuated from 3.5 to 4.5 mg/L, the settling time was relatively stable and fluctuated in about 13 minutes. Therefore, the PAM dosage was selected to be 3.5 mg/L to ensure the best settling time. It should be noted that the flocculation process of with polymers is based on the bridging mechanism which could be carried out by the following steps: (i) polymer solution was dispersed into the suspension; (ii) polymer in the system was transported to the particle surface; (iii) polymer was adsorpted on the particle surface; (iv) the polymer adsorbed particles were linked together. Therefore, the efficiency of the coagulation process with the polymer based on the bridging mechanism was strongly depended on the polymer molecular weight. If the PAM was increased too high, its solubility was decreased. In that case the solution would have a higher viscosity, resulting in reduction of settled residues. Conversely, if the amount of PAM was too little, it would not be enough to adhere the residue particles, thus it would reduce the settle residues as well [61, 62]. Therefore, the optimal dosage could help to form a larger residue, and help the sedimentation process faster.

11.4 Conclusion

Based on the obtained results, the following conclusions could be reached. The metallurgical wastewater contains a high content of heavy metals. In this work, Fe^o nanoparticles, flocculation chemicals, and deposition aids were used in the batch experiments to investigate the efficiency of treating heavy metals in wastewater. In the batch experiments, the effects of pH, Fe^o dosage, reaction time, PAC dosage and PAM dosage on the reduction efficiency of Cr(VI) and other heavy metal ions were investigated. In the reduction phase of Cr (VI), the proper amount of Fe^o was found to be 10 mg/L, at the condition pH of 1.5, and the reaction time of 60 minutes. In the experiments of the precipitation of other metals (Pb, Cd, Zn, Ni, Fe,...), the proper conditions could be pH of 8, the content of PAC could be used at 250 mg/L, and the PAM dosage should be 3,5 mg/L to maintain the effective removal of heavy metal ions up to 98%. In conclusion, based on the observed results, using Fe^o nanomaterials would be a suitable solution for treating metallurgical wastewater. At proper operational conditions, Fe^{o} nanomaterials could remove heavy metals effectively.

Acknowledgements

The authors greatly appreciate the support from Center of Industrial Environment, National Institute of Mining - Metallurgy Science and Technology. The support from Hanoi University of Science and Technology, Vietnam (No. 3961/QD-DHBK-SDH for master project of 2014BQLMT-KT05) is also acknowledged.

References

1. Kruzhanov, V.S., Modern Manufacturing of Powder-Metallurgical Products with High Density and Performance by Press–Sinter Technology. *Powder Metallurgy and Metal Ceramics*, 2018. **57**(7): p. 431-446.
2. Thai Nguyen Mining and Metallurgy Co, L., Report on results of implementation of environmental protection works and measures for the operation phase, 2014, Thai Nguyen Mining and Metallurgy Co, Ltd.
3. Ba, M.T., Research on technology to treat residual chemicals and heavy metals in wastewater at metallurgical experimental research facilities. 2015, National Institute of Mining - Metallurgy Science and Technology.
4. Kalderis, D., *et al.*, Characterization and treatment of wastewater produced during the hydro-metallurgical extraction of germanium from fly ash. *Desalination*, 2008. **230**(1): p. 162-174.
5. Wu, P., *et al.*, Treatment of metallurgical industry wastewater for organic contaminant removal in China: status, challenges, and perspectives. *Environmental Science: Water Research & Technology*, 2017. **3**(6): p. 1015-1031.
6. Gingerich, D.B. and M.S. Mauter, Flue Gas Desulfurization Wastewater Composition and Implications for Regulatory and Treatment Train Design. *Environmental Science & Technology*, 2020. **54**(7): p. 3783-3792.
7. Akpor, O.B., G.O. Ohiobor, and T.D. Olaolu, Heavy Metal Pollutants in Wastewater Effluents: Sources, Effects and Remediation. *Advances in Bioscience and Bioengineering*, 2014. **2**(4): p. 37-43.
8. Agoro, M.A., *et al.*, Heavy Metals in Wastewater and Sewage Sludge from Selected Municipal Treatment Plants in Eastern Cape Province, South Africa. *Water*, 2020. **12**(10): p. 2746.
9. Maiti, D., *et al.*, Comprehensive review on wastewater discharged from the coal-related industries – characteristics and treatment strategies. *Water Science and Technology*, 2019. **79**(11): p. 2023-2035.

10. Radovanović, D.Đ., *et al.*, Solidified structure and leaching properties of metallurgical wastewater treatment sludge after solidification/stabilization process. *Journal of Environmental Science and Health*, Part A, 2016. **51**(1): p. 34-43.
11. Beh, C.L., *et al.*,, Removal of Heavy Metals from Steel Making Waste Water by Using Electric Arc Furnace Slag. *E-Journal of Chemistry*, 2012. **9**: p. 128275.
12. Radovanović, D., *et al.*,, Solidified structure and leaching properties of metallurgical wastewater treatment sludge after solidification/stabilization process. *J Environ Sci Health A Tox Hazard Subst Environ Eng*, 2016. **51**(1): p. 34-43.
13. Mitra, S., A. Sarkar, and S. Sen, Removal of chromium from industrial effluents using nanotechnology: a review. *Nanotechnology for Environmental Engineering*, 2017. **2**(1): p. 11.
14. Stegemann, J.A. and Q. Zhou, Screening tests for assessing treatability of inorganic industrial wastes by stabilisation/solidification with cement. *Journal of Hazardous Materials*, 2009. **161**(1): p. 300-306.
15. Ye, L., *et al.*, Chemical precipitation granular sludge (CPGS) formation for copper removal from wastewater. *RSC Advances*, 2016. **6**(115): p. 114405-114411.
16. Houari, B., *et al.*, New synthetic material removing heavy metals from aqueous solutions and wastewater. *Arabian Journal of Chemistry*, 2019. **12**(8): p. 5040-5048.
17. Barakat, M.A., New trends in removing heavy metals from industrial wastewater. *Arabian Journal of Chemistry*, 2011. **4**(4): p. 361-377.
18. Zhang, Y. and X. Duan, Chemical precipitation of heavy metals from wastewater by using the synthetical magnesium hydroxy carbonate. *Water Science and Technology*, 2020. **81**(6): p. 1130-1136.
19. Hegazi, H.A., Removal of heavy metals from wastewater using agricultural and industrial wastes as adsorbents. *HBRC Journal*, 2013. **9**(3): p. 276-282.
20. Zhou, X.Y. and X.X. Xue, Study on Adsorption of Heavy Metalion in Metallurgical Wastewater by Sepiolite. *Advanced Materials Research*, 2013. **726–731**: p. 726–731, 2585–2588.
21. Yan, L., *et al.*, Evaluating adsorption media for simultaneous removal of arsenate and cadmium from metallurgical wastewater. *Journal of Environmental Chemical Engineering*, 2016. **4**(3): p. 2795-2801.
22. Coz, A., *et al.*, Influence of commercial and residual sorbents and silicates as additives on the stabilisation/solidification of organic and inorganic industrial waste. *Journal of Hazardous* Materials, 2009. **164 2-3**: p. 755-61.
23. Ksepko, E., J. Klimontko, and A. Kwiecinska, Industrial wastewater treatment wastes used as oxygen carriers in energy generation processes. *Journal of Thermal Analysis and Calorimetry*, 2019. **138**(6): p. 4247-4260.
24. Tandon, P.K. and S.B. Singh, Redox processes in water remediation. *Environmental Chemistry Letters*, 2016. **14**(1): p. 15-25.

25. Jiang, X., *et al.*, Removal of Cr(vi) from wastewater by a two-step method of oxalic acid reduction-modified fly ash adsorption. *RSC Advances*, 2019. **9**(58): p. 33949-33956.
26. Padmavathi, R., M. Minnoli, and D. Sangeetha, Removal of heavy metal ions from waste water using anion exchange polymer membranes. *International Journal of Plastics Technology*, 2014. **18**(1): p. 88-99.
27. Kononova, O.N., *et al.*, Ion exchange recovery of chromium (VI) and manganese (II) from aqueous solutions. *Arabian Journal of Chemistry*, 2019. **12**(8): p. 2713-2720.
28. Verbych, S., *et al.*, Ion Exchange Extraction of Heavy Metal Ions from Wastewater. *Separation Science and Technology*, 2005. **39**(9): p. 2031-2040.
29. Cetin, G., S. Kocaoba, and G. Akcin, Removal and Recovery of Chromium from Solutions Simulating Tannery Wastewater by Strong Acid Cation Exchanger. *Journal of Chemistry*, 2013. **2013**: p. 158167.
30. Zewail, T.M. and N.S. Yousef, Kinetic study of heavy metal ions removal by ion exchange in batch conical air spouted bed. *Alexandria Engineering Journal*, 2015. **54**(1): p. 83-90.
31. Rodriguez, J., *et al.*, Feasibility assessment of electrocoagulation towards a new sustainable wastewater treatment. *Environmental Science and Pollution Research - International*, 2007. **14**(7): p. 477-482.
32. Jin, W. and Y. Zhang, Sustainable Electrochemical Extraction of Metal Resources from Waste Streams: From Removal to Recovery. *ACS Sustainable Chemistry & Engineering*, 2020. **8**(12): p. 4693-4707.
33. Sillanpää, M. and M. Shestakova, Chapter 2 - Electrochemical Water Treatment Methods, in *Electrochemical Water Treatment Methods*, M. Sillanpää and M. Shestakova, Editors. 2017, Butterworth-Heinemann. p. 47-130.
34. Dominguez-Benetton, X., *et al.*, Metal recovery by microbial electro-metallurgy. *Progress in Materials Science*, 2018. **94**: p. 435-461.
35. Puyol, D., *et al.*, Resource Recovery from Wastewater by Biological Technologies: Opportunities, Challenges, and Prospects. *Frontiers in Microbiology*, 2017. **7**: p. 2106-2106.
36. Civilini, M., L. Ceccon, and M. De Bertoldi, Biological wastewater treatments for metallurgical industries. *Annals of Microbiology*, 2006. **56**(1): p. 7.
37. Massoud, R., *et al.*, Bioremediation of heavy metals in food industry: Application of Saccharomyces cerevisiae. *Electronic Journal of Biotechnology*, 2019. **37**: p. 56-60.
38. Machado, M.D., *et al.*, Removal of heavy metals using a brewer's yeast strain of Saccharomyces cerevisiae: advantages of using dead biomass. *Journal of Applied Microbiology*, 2009. **106**(6): p. 1792-1804.
39. Soares, E.V. and H.M.V.M. Soares, Bioremediation of industrial effluents containing heavy metals using brewing cells of Saccharomyces cerevisiae as a green technology: a review. *Environmental Science and Pollution Research*, 2012. **19**(4): p. 1066-1083.

40. Căpriță, F.C. and A. Ene, Biosorption of heavy metals from the metallurgical industry wastewater by macroalgae. *AIP Conference Proceedings*, 2020. **2218**(1): p. 030011.
41. Ferdous, A., *et al.*, Removal of heavy metal from industrial effluents using Baker's yeast. *AIP Conference Proceedings*, 2016. **1754**(1): p. 060011.
42. Kang, S., *et al.*, Highly efficient removal of hexavalent chromium in aqueous solutions via chemical reduction of plate-like micro/nanostructured zero valent iron. *RSC Advances*, 2017. **7**(88): p. 55905-55911.
43. Yang, J., *et al.*, Nanomaterials for the Removal of Heavy Metals from Wastewater. *Nanomaterials* (Basel, Switzerland), 2019. **9**(3): p. 424.
44. Ramos, P.B., *et al.*, Treatment of real non-biodegradable wastewater: Feasibility analysis of a zero-valent iron/H2O2 process. *Journal of Environmental Chemical Engineering*, 2020. **8**(4): p. 103954.
45. Li, S., *et al.*, Recovery of gold from wastewater using nanoscale zero-valent iron. *Environmental Science: Nano*, 2019. **6**(2): p. 519-527.
46. Zhang, B., *et al.*, Nanoscale Zero Valent Iron Supported by Biomass-Activated Carbon for Highly Efficient Total Chromium Removal from Electroplating Wastewater. *Water*, 2020. **12**(1): p. 89.
47. Chaung, S.-H., *et al.*, Nanoscale Zero-Valent Iron for Sulfide Removal from Digested Piggery Wastewater. *Journal of Nanomaterials*, 2014. **2014**: p. 518242.
48. Chen, S.Y., W.H. Chen, and C.J. Shih, Heavy metal removal from wastewater using zero-valent iron nanoparticles. *Water Sci Technol*, 2008. **58**(10): p. 1947-54.
49. Brasili, E., *et al.*, Remediation of hexavalent chromium contaminated water through zero-valent iron nanoparticles and effects on tomato plant growth performance. *Scientific Reports*, 2020. **10**(1): p. 1920.
50. Rice, E.W., R.B. Baird, and A.D. Eaton, *Standard methods for the examination of water and wastewater*. 23rd ed. Co-publishers: American Public Health Association (APHA) and Water Environment Federation (WEF), Washington, pp 422-648., 2017.
51. Chen, X., *et al.*, Application of nanoscale zero-valent iron in hexavalent chromium-contaminated soil: A review. *Nanotechnology Reviews*, 2020. **9**(1): p. 736-750.
52. Du, Y., *et al.*, Fabrication of a low-cost adsorbent supported zero-valent iron by using red mud for removing Pb(ii) and Cr(vi) from aqueous solutions. *RSC Advances*, 2019. **9**(57): p. 33486-33496.
53. Yin, Y., *et al.*, Removal of hexavalent chromium from aqueous solution by fabricating novel heteroaggregates of montmorillonite microparticles with nanoscale zero-valent iron. *Scientific Reports*, 2020. **10**(1): p. 12137.
54. Elwakeel, K.Z., *et al.*, Perspectives regarding metal/mineral-incorporating materials for water purification: with special focus on Cr(vi) removal. *Materials Advances*, 2020. **1**(6): p. 1546-1574.

55. Soni, R., *et al.*, An overview of nanoscale materials on the removal of wastewater contaminants. *Applied Water Science*, 2020. **10**(8): p. 189.
56. Kakavandi, B., *et al.*, Enhanced chromium (VI) removal using activated carbon modified by zero valent iron and silver bimetallic nanoparticles. *Journal of Environmental Health Science and Engineering*, 2014. **12**(1): p. 115.
57. Carvalho, F.P., Mining industry and sustainable development: time for change. *Food and Energy Security*, 2017. **6**(2): p. 61-77.
58. Zhou, F.-s., *et al.*, Preparation and Characteristics of Polyaluminium Chloride by Utilizing Fluorine-Containing Waste Acidic Mother Liquid from Clay-Brine Synthetic Cryolite Process. *Journal of Chemistry*, 2014. **2014**: p. 274126.
59. Niu, X., *et al.*, Preparation and coagulation efficiency of polyaluminium ferric silicate chloride composite coagulant from wastewater of high-purity graphite production. *Journal of Environmental Sciences*, 2011. **23**(7): p. 1122-1128.
60. Das, P., *et al.*, Effluent Treatment Technologies in the Iron and Steel Industry - A State of the Art Review. *Water Environment Research*, 2018. **90**(5): p. 395-408.
61. Jiang, W.-l. and H.-f. Xu, Treatment and Recycling of the Process Water in Iron Ore Flotation of Yuanjiacun Iron Mine. *Journal of Chemistry*, 2017. **2017**: p. 9187436.
62. Oraeki, T., G. Skouteris, and S. Ouki, Optimization of coagulation-flocculation process in the treatment of wastewater from the brick-manufacturing industry. *Water Practice and Technology*, 2018. **13**(4): p. 780-793.

12

Removal of Arsenic and Fluoride from Water Using Novel Technologies

Ishita Sarkar[1], Sankha Chakrabortty[2*], Jayato Nayak[3] and Parimal Pal[4]

[1]CSIR-Central Mechanical Engineering Research Institute, Durgapur, West Bengal, India
[2]School of Chemical Technology, Kalinga Institute of Industrial Technology, Bhubaneswar, Odisha, India
[3]Department of Chemical Engineering, Kalasalingam Academy of Research and Education, Tamil Nadu, India
[4]Department of Chemical Engineering, National Institute of Technology Durgapur, M.G. Avenue, Durgapur, India

Abstract

In the backdrop of ever-increasing demand for safe potable water, contamination of groundwater by leached out arsenic and fluoride in several parts across the world has thrown a challenge to the scientific community in finding efficient, low-cost and flexible treatment systems for removal of such contaminants from groundwater. Over the decades, extensive research has established adsorption, ion-exchange, chemical coagulation–precipitation, photo-catalytic oxidation, electro-dialysis, pressure-driven membrane separation and membrane distillation as the broad technological options for water purification. However, till now, over 1.5 billion people across the world do not have access to safe drinking water. There is still a need for simple, low-cost and easily implementable technologies which can stand as a sustainable solution for the production of safe drinking water in the vast arsenic-fluoride affected areas of the world.

Although there has been widespread research on groundwater contamination by arsenic and fluoride and their removal from drinking water by laboratory techniques, millions of people continue to suffer, particularly in the developing countries like India and Bangladesh. Among all the methods available in literature, membrane-based separation techniques have proved to be very promising in purifying arsenic and fluoride-contaminated groundwater to a WHO-prescribed

**Corresponding author*: sankha.chakrabortty@kiitbiotech.ac.in

Siddhartha Moulik, Aditi Mullick and Anirban Roy (eds.) Sustainable Water Treatment: Advances and Technological Interventions, (441–486) © 2022 Scrivener Publishing LLC

level. Through a comprehensive review of the existing literature on the occurrence, causes, impacts and remedial measures, the current review focuses on the shortcomings and tries to emphasize the remedial measures.

Keywords: Arsenic, fluoride, groundwater, contamination, membrane-based technologies

12.1 Background Study of Arsenic

12.1.1 Source and Existence of Arsenic

Arsenic naturally occurs in soil, rocks, sedimentary, deposits, water, air, etc., and is the major constituent of at least 245 different minerals at different levels, like elemental arsenic, arsenides, sulphides, oxides, arsenites and arsenates. High concentrations of arsenic are associated with sulfide minerals and metal oxides, especially iron oxides such as Orpiment (As_2S_3), realgar (AsS), mispickel (FeAsS), loellingite ($FeAs_2$), which are the most significant arsenic-bearing minerals (Bissen *et al.*, 2003).

Arsenic present in water streams or under groundwater has been considered highly toxic and harmful to the environment and human health; it is known as a Class I Human Cancer and is a public concern due to its extensive presence in both industry and agriculture. As arsenic can only be transformed into different organic and inorganic forms or converted into insoluble compounds in binding with other metal elements such as iron, it cannot be easily destroyed. Moreover, many impurities, such as lead, iron, and selenium, can be mixed with arsenic, which can be very economical to reduce. The toxicology and carcinogenicity of arsenic depends not only on its oxidation states but also on its different chemical forms. The trivalent form of arsenic is more hazardous than the pentavalent form, whereas the inorganic form of arsenic is more poisonous than the organic form (except MMA (III) and DMA (III)) (Mascher *et al.*, 2002). As (V) (arsenate) can substitute the function and position of phosphate molecule in the human body (in citric acid cycleIt) as it has very similar structure and properties with phosphate. During this cycle, arsenate disrupts the synthesis process of ATP, the primary energy storage form in the cell. This disruption results in loss of energy. On the other hand, the activities of various enzymes, co-enzymes and receptors containing thiol groups (-SH) can be destroyed by As (III) (arsenite) because of the binding of arsenite with the thiol group in the enzymes as a catalyst. For example, As (III) can react with lithic acid, which contains dithiol groups. This reaction prevents the formation of pyruvate dehydrogenase, which requires lipic acid for enzymatic activity. This inhibition disrupts the Krebs cycle by destroying the function of PDH.

Arsenate is considered less toxic than Arsenite and it has been found that Arsenite is 4 to 10 times more soluble in water and 10 times more toxic than Arsenate (Squibb and Fowler 1983; Mascher *et al.*, 2002) and 70 times more toxic than trivalent methylated arsenic species like MMA(III) and DMA(III). The toxicity scale of arsenic can be presented as per the following decreasing order: arsine > inorganic As(III) > organic As(III) > inorganic As(V) > organic As(V) > arsenium compounds and elemental arsenic.

In inorganic compounds, As(III) species exits as arsenious acid (H_3AsO_3) and arsenite ions ($H_2AsO_3^-$, $HAsO_3^{2-}$, and AsO_3^{3-}) while the arsenic As(V) species exists as arsenic acid (H_3AsO_4) and arsenate ions ($H_2AsO_4^-$, $HAsO_4^{2-}$ and AsO_4^{3-}).

12.1.2 Effects of Arsenic

Approximately, 137 million people in 70 countries around the world are at a risk of arsenic poisoning. Currently, populations in the Bengal-Delta Basin of Bangladesh and India are the largest part which are at high risk in the world (Bhattacharyya *et al.*, 2003; Wickramasinghe *et al.*, 2004a; Violante *et al.*, 2006; Pal *et al.*, 2009). As per the largest poisoning in world's history, about 70 million people are suffering from an arsenic problem alone in these regions (Nigam *et al.*, 2013). Presently, 1 million people from 111 blocks of 12 districts in West Bengal (Basu *et al.*, 2014) are mostly affected by arsenic contamination in groundwater which is present in the range from 50 to 3600 μg/L in (De, 2008; Mondal *et al.*, 2011). Out of 27 districts, 19 districts of Assam state are seriously affected by the problem of arsenic contamination in groundwater and the detection level in the range of 50 ppb to 1000 ppb was found. Different states of India affected by the arsenic contaminated groundwater are shown in Figure 12.1.

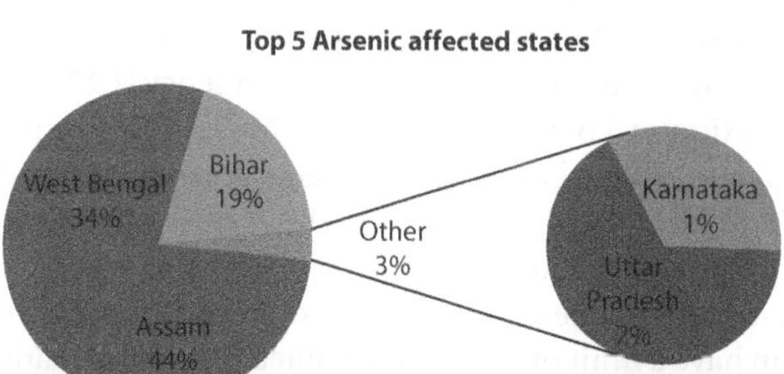

State	Maximum As conc.
Assam	2.4 ppm
West Bengal	1.9 ppm
Bihar	1.1 ppm
Uttar Pradesh	0.127 ppm
Karnataka	0.07 ppm

Figure 12.1 States and concentration of arsenic present.

Millions of people from West Bengal (India) and Bangladesh are seriously suffering from the problem of arsenic contaminated groundwater and claim the biggest misfortune in the world. Above 60 mg/L of arsenic concentration is dangerous for human consumption (Wickramasinghe *et al.*, 2004). Different countries have set the maximum contaminant level (MCL) of arsenic in drinking water to stop arsenic-related illnesses. Health ailments caused by the presence of inorganic arsenic in drinking water include stomach and intestinal irritation, decreased red and white blood cell production, skin changes and lung irritation. Some medical research has shown that ingestion of a significant amount of inorganic arsenic exacerbates the chances of developing cancer, especially skin cancer, lung cancer, liver cancer, and lymphoma. Two thousand villages in Bangladesh have detected arsenic in excess of 50 mg/L in groundwater, and more than 50 million have been exposed to per capita arsenic poisoning, while six million people from 3,000 villages in India have been adversely affected. The presence and toxicity of arsenic has been widely reported (Jain *et al.*, 2000). Acute and sub-acute are the main classification of arsenic toxicology, which is basically a very complex phenomenon. Severe arsenic poisoning, usually caused by consuming arsenic-contaminated food or beverages, requires immediate medical attention.

12.1.3 Regulation and Permissible Limit of Arsenic in Drinking Water

Severe and chronic arsenic diseases by ingestion of arsenic contaminated drinking water have been reported in many countries, mainly in Bangladesh, Argentina, India, Mongolia, Mexico, Thailand and Taiwan, where 100 to 30 contaminated groundwater with arsenic. More than 2,000 mg/L, the toxic effect of arsenic on the human body is well recognized in small doses. In 1942, the United States Public Health Service first set the maximum limit of arsenic of 0.05 mg/L drinking water. But in early 1968, researchers concluded that arsenic as a ration could cause skin cancer at 0.05 mg/L. From 1980 to 1990, the United States Environmental Protection Agency (USEPA) attempted to set the maximum contamination level (MCL) for arsenic. Supported by strong epidemiological evidence of arsenic genotoxicity and carcinogenicity, the World Health Organization (WHO, 1993) revised the MCL in drinking water from 50 ppb in 1993 to 10 ppb, and this was later accepted by the USEPA in 2001. While the developing countries Bangladesh, India, China and Taiwan have a limit of 50 ppb, the United States, Germany and Japan maintain a limit of 10 ppb of their drinking water. Canada and Australia maintain a limit of 25 ppb and 7 ppb, respectively.

12.2 Background Study of Fluoride

12.2.1 Source and Existence of Fluoride

Fluoride is the ion of fluorine derived from the 13th most abundant element and is found in the atmosphere, soil and water, due to its natural presence in the earth's crusts and excessively agricultural activities and different industrial effluents especially from steel, aluminium, glass, phosphate, electroplating, welding, and the ceramic industries in the form of brackish water sometimes. Four super most fluoride-rich rocks, such as (i) fluorspar (CaF_2) (ii) cryolite (Na_3AlF_6) (iii) fluorapatite ($Ca_5(PO_4)_3F$) and (iv) sellaite (MgF_2) are the cause of fluoride contamination in groundwater and this contamination depends not only on the availability and solubility of parent fluoride minerals but also on contact time, rock porosity, velocity of flowing water, temperature of the interactions between the Ration of stones and water, water pH and calcium ions present in water.

When the fluoride compound is contaminated with groundwater, the fluorine element is mainly fluoride ions. In addition, fluorides in the solution HF_2 and dissolved HF depending on the ionic concentrations and pH of the solvent. Suppose the calcium fluoride and sodium fluoride are the representative of the natural fluorides and artificial fluoride, respectively, dissolved in water under the equivalent conditions where the ionization takes place as shown in the following equations:

Calcium Flouride	Sodium Flouride
CaF_2 ---- Ca_{++} + $2F^-$	NaF ---- Na^+ + F^-
F^- + H^+ ----HF	F^- +H^+ ---- HF
HF + - ----HF_{z^-}	HF + F- ----HF_{z^-}

12.2.2 Effects of Fluoride

Like arsenic, the problem of occurrence of fluoride in high concentration in groundwater is now one of the most serious toxicological and geo-environmental issues in several countries. Approximately 200 million people in 29 countries are suffering from fluoride contamination of groundwater. Fluoride contamination of groundwater in India, Argentina and Ethiopia has been well documented as the major fluoride-affected regions in the world.

In India, some states like West Bengal, Jharkhand, Uttar Pradesh, Assam, and Chhattisgarh suffer from both arsenic and fluoride contamination of groundwater, which makes groundwater resource management more challenging. Over 19 states are seriously affected by fluoride-contaminated groundwater and by a recent study, it is estimated that in the Indian population, 18,197,000 are affected with dental fluorosis and 7,889,000 with skeletal fluorosis (Saxena and Sewak, 2015). Thirty-two districts of Gujrat, 32 districts of Rajasthan, 17 districts of Andhra Paradesh, 9 districts of Bihar and 4 districts (Birbhum, Bankura, Bardhman and Purulia) of West Bengal (Figure 12.2) are seriously affected by the problem of contaminated fluoride in groundwater.

12.2.3 Regulation and Permissible Limit of Fluoride in Drinking Water

A certain level of fluoride, i.e., 0.5-0.7 ppm, provides a healthy protection against tooth decay. But a long-term consumption of excess concentration (1.5-4 ppm) of fluoride through drinking water may cause dental fluorises (decay of teeth), skeletal fluorosis (decay of bone), osteoporosis, arthritis, memory loss, kidney damage, cancer, genetic damage and gastrointestinal problems (Murray, 1986).

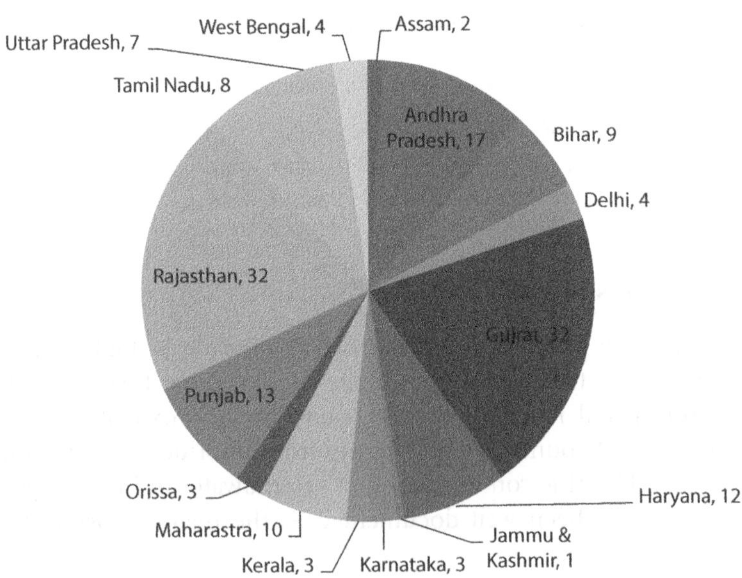

Figure 12.2 Districts affected by fluoride-contaminated groundwater in different states.

Above 4 ppm of fluoride concentration levels in drinking water cause stiffness in bone-joints (chronic skeletal fluorosis) that causes osteoporosis and growth in bone mass in old age. For extreme cases, premature aging or paralysis may happen. Additionally, fluorides get stored in the space of the neck, knee bone, girdle and shoulder bones, making it difficult to move or walk. Moreover, excessive consumption of fluoride might result in degeneration of muscle fiber, low levels of haemoglobin, malformations in RBCs, high tendency toward headache and nausea, skin rashes, abdominal pain, etc. The halides alter the useful mechanisms of the liver, kidney, neuro system, and genital system as well as inhibit the functions of 60 enzymes.

According to World Health Organization (International Standards), the maximum acceptable concentration of fluorides is 1.5 mg/L and drinking water must not exceed 1.5 mg/L level of fluoride. In addition, the US Environmental Protection Agency has established a secondary MCL of 2 mg/L.

12.3 Technologies Used for Arsenic Removal from Contaminated Groundwater

Various technologies available for removal of arsenic from contaminated water are based mainly on six principles:

- Oxidation and filtration
- Biological Oxidation: The oxidation of As (V) to As (V) by microorganisms and subsequent removal of As (V) by iron and manganese oxides.
- Co-precipitation: Adds an appropriate oxidizing agent and oxidizes Ys (III) to Ys (V) by freezing, precipitation, and filtration.
- Absorption: Activated Alumina, Activated Carbon, Iron-Based Sorbents, Zero Valent Iron and Hydrated Iron Oxide, etc.
- Ion exchange through appropriate cation and ion exchange resins.
- Membrane Technology: Reverse Osmosis, Nanofiltration and Electrodialysis.

12.3.1 Oxidation Method

Arsenite [As(III)] is less removed than Arsenate [As (V)] from water as it exists as a neutral form (H_3AsO_3) whereas As (V) is predominantly present

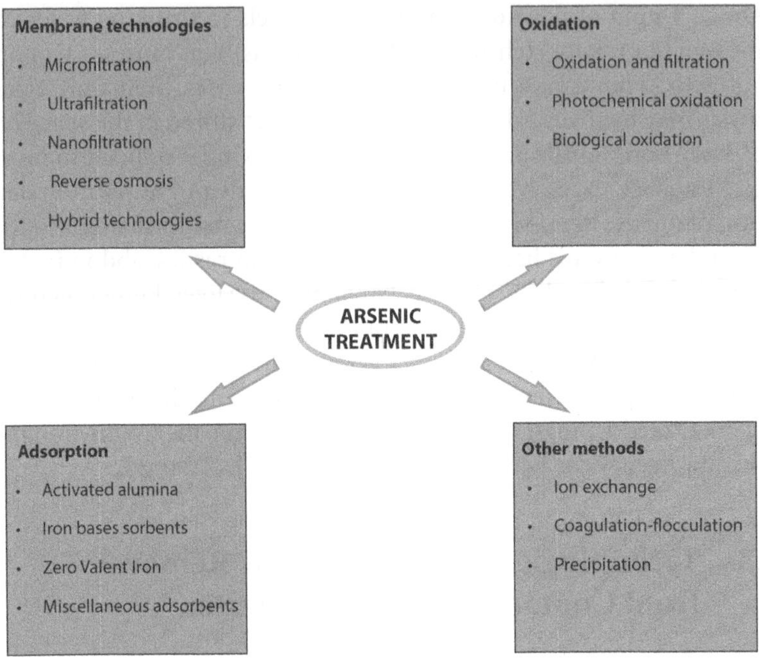

Figure 12.3 Different remediation technology of arsenic.

as monovalent or divalent anions ($H_2AsO_4^-$, $HAsO_4^{2-}$) in natural sources of water (Ferguson and Gavis, 1972). Unlike arsenate anions, the neutral arsenite does not adsorb strongly to mineral surfaces (Hug and Leupin, 2003). Hence, a peroxidation step often includes converting arsenite to arsenate for the treatment of groundwaters in which arsenite is the dominant species. Several types of oxidant have been used for oxidation, such as chlorine, hypochlorite, ozone, permanganate and hydrogen peroxide. Permanganate and solid manganese dioxide are the most promising oxidants used in the peroxidation step as they produce no harmful by-products as well as the fact that permanganate is an effective agent for oxidation of As(III) in a wide range of pH (Na *et al.*, 2007). The reaction of arsenic oxidation by permanganate is defined as follows:

$$3As^{3+} + 2MnO_4^- + 4H_2O \rightarrow 2MnO_2 + 3As^{5+} + 8OH^-$$

An oxidation and coagulation study with potassium permanganate and chlorine was performed by Shen *et al.* (2006) where both oxidants significantly increased arsenic removal from well-water, unlike the coagulation process alone. Three different types of combinations were used during the investigation, like aeration, aeration combined with coagulation, and aeration combined with oxidation/coagulation and results showed that aeration had no apparent effect on arsenic removal.

Recently, Iron and Manganese were subsequently used as the oxidant for the removal of arsenic. In this process, the following directions of reactions are followed:

(i) Oxidation of Mn^{II} to Mn^{IV} and Fe^{II} to Fe^{III};
(ii) Oxidation of As^{III} to As^{V},
(iii) Precipitation of manganese oxides,
(iv) Similarly, the abiotic oxidation of As^{III} by manganese oxides and finally
(v) As^{V} sorption by manganese oxides.

Here, the first two steps are biotic and the next three steps are once abiotic.

García *et al.* (2004) implemented solar oxidation technology in the individual units (SORAS) (Solar Oxidation and Removal of Arsenic) of their developed system for the removal of arsenic from water. In this process, As (III) gets oxidised to As (V) by photochemical oxidation process and removed by precipitation-filtration method of As(V) adsorbed on Fe(III) (Figure 12.4). At pH 0.5-2.5, the rate of oxidation becomes faster.

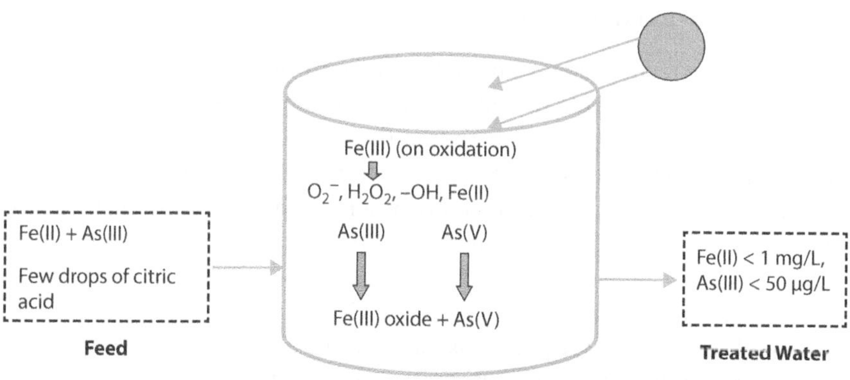

Figure 12.4 Treatment of arsenic-contaminated water by photo-oxidation method.

The photons coming from the radiation of sunlight get adsorbed on the Fe(III)-hydroxide and Fe(III)-chloride complexes and produce highly oxidizing agents.

The following two steps are adopted in the SORAS method for the removal of arsenic:

- First step: As(III) oxidized to As(V) which strongly gets adsorbed onto the surface of iron hydroxides.
- Second step: As(V) adsorbed Fe(III) hydroxide gets precipitated and settles at the bottom of the container where clear water is decanted.

12.3.2 Coagulation-Precipitation Method

The precipitation-coagulation technique is the most traditional technology employed for heavy metal removal from contaminated water. Iron, calcium and aluminium salts have been used extensively worldwide as coagulant for arsenic separation (Jekel, 1994) as these two salts get easily hydrolysed to their respective hydroxides which act as sink for arsenic present in water. Furthermore, Ferric hydroxide [$Fe(OH)_3$] has a strong affinity for dissolved arsenic. Arsenic can be removed most commonly by precipitation as ferric arsenate, aluminium arsenate, or calcium arsenate. For calcium generated precipitation, operating pH must be maintained at 10.5. It is necessary to oxidize As(III) to As(V) using different oxidizers, like $KMnO_4$, Cl_2, NaOCl, and H_2O_2 as it gets less precipitated than As(V). The conventional coagulation-based technology is shown in Figure 12.5.

12.3.3 Ion-Exchange Method

Ion-Exchange technology is one of the most promising technologies nowadays. Many undesirable dissolved solids can be removed from water by synthetic ion exchange resins based on the cross-linked polymer skeleton, which is the matrix of the time and is composed of polystyrene, cross-linked with divinylbenzene. The following four charged functional groups are added to the matrix by covalent bonding:

- Strongly acidic (e.g. sulfonate, $-SO_3^-$)
- Weakly acidic (e.g. carboxylate, $-COO^-$)
- Strongly basic [e.g. quaternary amine, $-N+(CH_3)_3$]
- Weakly basic [e.g. tertiary amine, $-N(CH_3)_2$]

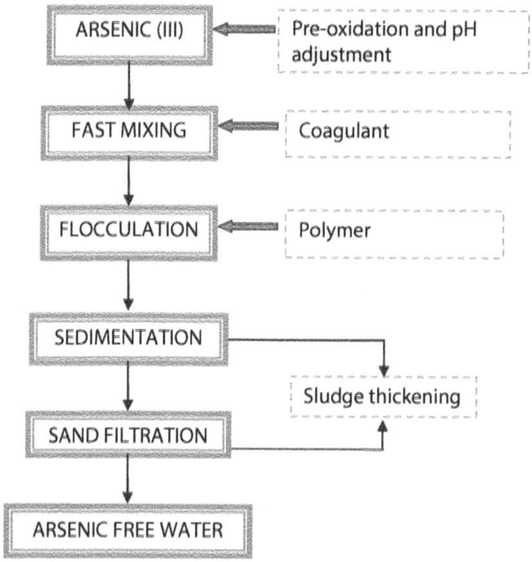

Figure 12.5 Conventional coagulation-flocculation-based arsenic mitigation technology.

In this method, arsenic ions are stabilized electrically on the surface of strong base ion exchange resins and exchanged from the resin for ions of similar charge in solution. Sulfate, phosphate and nitrate are common ions that play an important role in arsenic removal by ion exchange. Strong-base ion resins are more effective than weak-base resins over wide-range pH. Due to the neutral orthography of As (III) in natural water, oxidation of As (III) to As (V) is an important pre-treatment step for ion exchange processes (Kartinen and Martin, 1995). The traditional ion-exchange method for arsenic removal is shown in Figure 12.6.

12.3.4 Adsorption Method

This process is effective for the removal of arsenic from groundwater or arsenic contaminated water. In this method, two forces, such as physical and chemical forces, help to attach arsenic species on the outward surface of the adsorbent. The active surface area, surface energy of adsorbent, and the solution pH greatly influence the arsenic removal efficiency. Activated alumina, greensand, activated charcoal (carbon), granular form of ferric hydroxide, iron oxide coated sand; and granules of copper-zinc mixture are the well-known, widely used adsorbent. Nevertheless, these adsorbents still remain expensive materials.

Table 12.1 Chemicals used in different arsenic precipitation processes (Jing *et al.*, 2005).

Parameters	Initial As conc. (μg/l)	Chemicals used for pre-treatment	Chemicals used in the process	pH	Arsenic removal (%) after chlorination	Arsenic removal (%) without chlorination
Alum	300	Cl_2	Alum, Acid, NaOH	<6.5	90	20
Iron	300	Cl_2	$Fe_2(SO_4)_3$	6-8	90	60
Lime softening	400	Cl_2	Lime, Acid	>10.5	90	80
Iron and manganese	<100	Cl_2	$FeCl_3$, SO_2, $KMnO_4$, Polymeric aluminum silicate sulfate, organic polymer	>7	40-90	-

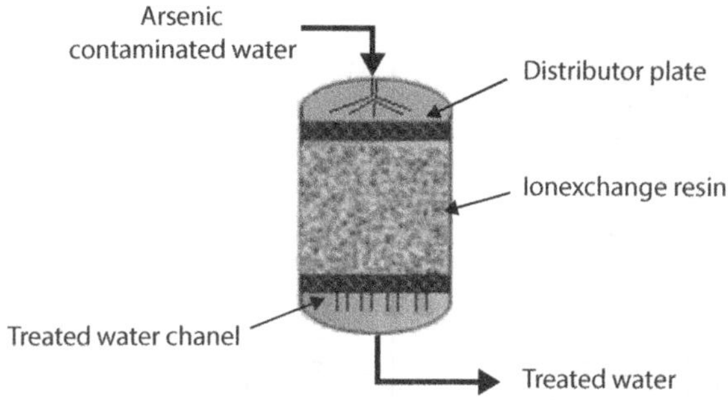

Figure 12.6 Arsenic removal from contaminated water by ion exchange resin.

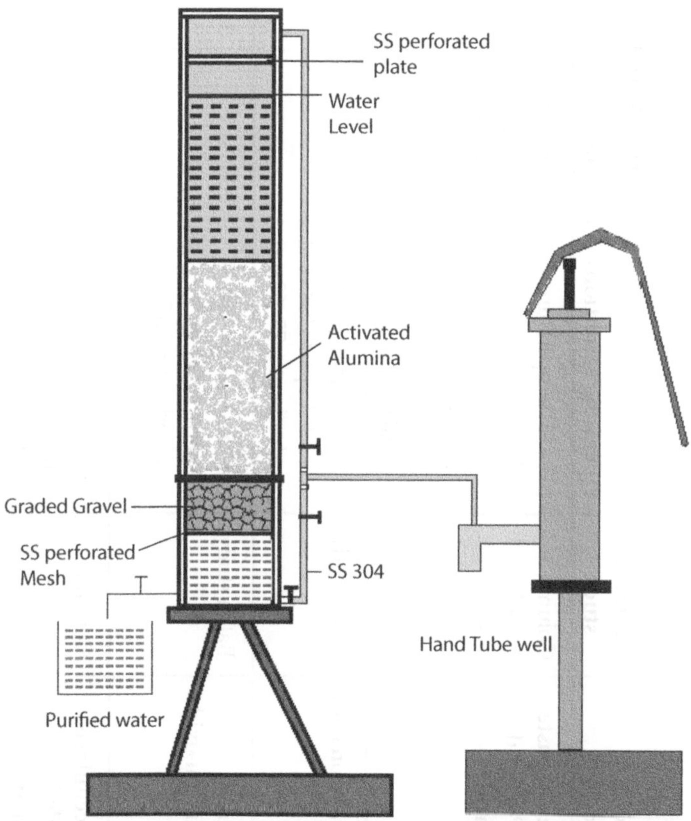

Figure 12.7 Conventional activated alumina-based arsenic treatment unit attached to a tube well. Adopted from Pal, 2015.

Table 12.2 Arsenic removals by recently developed or surface modified adsorbent at different operating conditions.

Name of modified adsorbent	Nature of study	Operating conditions	As concentration (μg/l)		Results in terms of removal efficiency	References
			Initial	Final		
Hardened paste of Portland cement	Column and batch study	pH 4–5, temperature = 24 °C, As_0 = μg/l (batch), As_0 = 400–500 μg/l (column), particle size = 4.88–4.92mm	200 (batch study), 500 (column study)	20 μg/l (batch), 50 μg/l (column study)	Removal efficiency =90%	[Kundu *et al.*, 2004]
Natural iron ores	Batch study	pH 4.5–6.5, temperature = 24 °C, particle size of adsorbent = 75–63 μm	100	5	Removal efficiency =95%	[Zhang *et al.*, 2004]
Granular titanium oxide	Batch study	pH 7±0.1, temp = 24 °C, particle size of adsorbent = 0.15–0.6mm	300	15	Removal efficiency 95%,	[Bang *et al.*, 2005a]
Bead cellulose loaded with iron oxy hydroxide	Column study	pH 7, temp = 25±0.5 °C, particle size of adsorbent = 20–60 mesh	500	As(3)= 25, As(5) = 100	Removal efficiency As(3) = 95%, As(5) = 80%	[Guo *et al.*, 2005]

(Continued)

Table 12.2 Arsenic removals by recently developed or surface modified adsorbent at different operating conditions. (*Continued*)

Name of modified adsorbent	Nature of study	Operating conditions	As concentration (μg/l)		Results in terms of removal efficiency	References
			Initial	Final		
Zero-valent iron and filter sand	Batch study	pH 6, temperature = 24 °C, Fe = 200 μg/l, DO= 8 mg/l, SiO_2 = 560 mg/l, particle size of Fe(o) = 100 and 40 mesh, for filter sand = 0.35–0.45mm	8500	85	Removal efficiency =99%	[Bang *et al.*, 2005b]
Iron oxide coated cement	Column study	pH 4, temp = 27±2 °C, particle size of adsorbent = 0.212mm	2000	10	Removal efficiency 99.5%.	[Kundu and Gupta, 2005]
Polymetallic sea nodule	Batch study	pH < 6, temp = 20 °C, particle size 75 μm, shaking time 30 min	240	24	Removal efficiency =90%	[Maity *et al.*, 2005]

12.4 Technologies for Fluoride Removal from Contaminated Groundwater

Defluoridation of fluoride-rich waters includes:

- Chemical coagulation-precipitation method
- Adsorption method
- Flocculation
- Softening with lime
- Sedimentation
- Ion-exchange

12.4.1 Coagulation-Precipitation Method

Alum (Aluminum sulfate, $Al_2(SO4)_3$) is one of the most commonly used coagulants for the removal of dissolved species through chemical coagulation-precipitation in fluoride-contaminated water treatment where Aluminum hydroxide $Al(OH)_3$ is formed after addition of alum into water. Addition of hydrated lime ($Ca(OH)_2$), calcium oxide (CaO) or other calcium salts, such as $CaSO_4$, $CaCl_2$ will precipitate with fluoride by the formation of insoluble calcium fluoride compound (CaF_2).

Recently, Battula *et al.* (2014) studied an experiment on the removal of fluoride from fluoride-contaminated groundwater using the electrocoagulation method where the treatment was performed without addition of any chemical coagulant or flocculants. The effective surface area for the reaction was found to be 40 cm^2 while distance of inter-electrode was maintained 1 cm. Results show that above 85% of fluoride concentration was mitigated at 30V electrical potential and at 20 minutes of reaction time. Results also exhibited that the rate of removal got increased with the increase of reaction time from 10 to 30 minutes.

12.4.2 Nalgonda Technique

The Nalgonda process is a very well-known and most effective technique for fluoride removal. It is based on combined use of lime and alum in a two-step process (Figure 12.8). This technology was developed by National

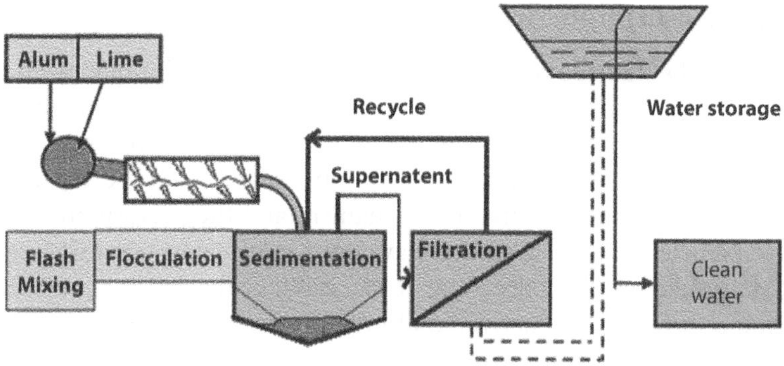

Figure 12.8 Nalgonda technique.

Environmental Engineering Research Institute (NEERI), Nagpur, under the Rajiv Gandhi Drinking Water Mission project.

Limitations of Nalgonda Technique

Several limitations of this process have been pointed out as follows:

- Less effective as it removes only a small amount of fluoride (around 30%) in the form of precipitates.
- Formed toxic compound through the conversion of ionic fluoride (67–82%) to soluble aluminium fluoride complex ion.
- Because aluminum sulfate is used as a coagulant, the rate of sulfate ion is increased and the maximum permissible limit in "treated" water exceeds 400 mg/L.
- Due to presence of excess amount of aluminium (>0.2 mg/L) in the residual in "treated" water causes dangerous dementia disease.
- The operation and maintenance cost of this process is very high.
- The process is not self-controlled.
- High volume of sludge is produced and a large space is required for drying of sludge.

12.4.3 Adsorption Method

Nowadays, the fluoride removal through adsorption technique by an activated alumina column is one of the most favorable techniques as activated alumina has a good capacity and selectivity. Fluoride is removed by the activated alumina through electrosorption technique where the electric field of the adsorbent affects the mechanism and the performance of the removal efficiency. Several studies have been reported by several researchers (Maliyekkal *et al.*, 2006; Tang *et al.*, 2009; Camacho *et al.*, 2010) for fluoride removal by the activated alumina processes which were carried out in sorption filters.

Manganese dioxide is another adsorbent which is insoluble and nonhazardous in nature as well as being able to efficiently remove fluoride from contaminated water. Several researchers (Gadhari *et al.*, 2010; Sanghavi and Srivastava, 2010; de Oliveira *et al.*, 2007; Mobin *et al.*, 2010; Kotkar *et al.*, 2007; Shahrokhian *et al.*, 2009) have shown interest in the use of manganese oxide for defluoridation. MnO_2 coated alumina, bentonite clay with incorporated MnO_2, and MnO_2 coated activated carbon are very well-known manganese-based adsorbents.

12.4.4 Ion-Exchange Method

Ion exchange is the process of exchanging carbonate radicals with fluoride at active sites bound with resin. It is nothing but an absorption process

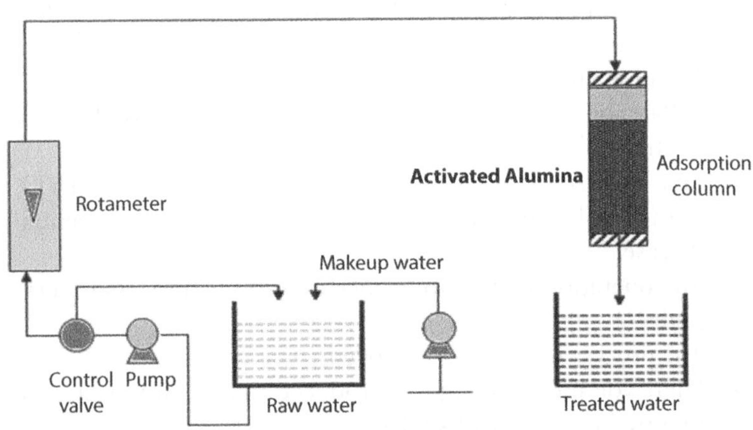

Figure 12.9 Conventional activated alumina-based adsorption treatment plant. Adopted from Pal, 2015.

similar to that of activated alumina. The ion exchange resin attached to carbonate ions at the exchange sites is placed in a vessel.

Different ion-exchange materials, such as bone, bone char, activated alumina, snake, ion and cation exchange resins, are used for the defluoridation process. The various ion-exchange processes for defluoridation are described as follows:

- **Bone Char:** Bone coal is a very promising ion-exchange material, in which the cellular structure is composed of a large number of minute tubes and channels of calcium phosphate and carbonates. The granularity of bone coal (28 to 48 meshes) is in the form of its induced carbon content. The removal mechanism of fluoride is in place of the carbonate radical with fluoride. Caustic soda is commonly used for the regeneration of an exhausted bed.
- **Activated alumina:** Activated alumina is an excellent adsorbent used as an ion-exchange material to remove fluoride from contaminated water and reduce the ration level of fluoride to <1.5 mg/l, which is the maximum allowable level for drinking water (WHO). The absorption capacity of activated alumina is very effective for the treatment of fluoride-contaminated water, which requires only 20 mg of activated alumina to treat 1 mg of fluoride.
- **Anion exchange resins:** Strongly basic anion-exchange resin effectively remove fluoride from water as it contains quaternary ammonium functional groups. The removal of fluoride takes place according to the following reaction:

$$\text{Matrix-NR}^{3+}\text{Cl}^- + \text{F}^- \rightarrow \text{Matrix-NR}^{3+}\text{F}^- + \text{Cl}^-$$

The chloride ions of the resin are replaced by fluoride ions, and this will continue until all the active sites of the resin's surface are occupied by fluoride. Strong electronegativity is the driving force of the ion-exchange process.

In another study, Rao *et al.* (1998) performed a study for defluoridation in water using a variety of materials, including aluminum salts, calcium salts, alumina salts, magnesium salts, lime, activated carbon, sulfonated carbonaceous materials and ion-exchange resins. The results showed that sulfonated carbonaceous materials and ion-exchange resins effectively reduced the fluoride concentration from raw water (from 5 mg/l to 1.5 mg/l).

12.5 Membrane Technology Used for Arsenic and Fluoride Mitigations

12.5.1 Introduction of Membrane Technology

Scientists have developed several membranes-based technologies, such as forward osmosis, reverse osmosis, microfiltration, ultrafiltration, nanofiltration, etc., as an alternative arsenic or fluoride remediation technique. Membrane separation has been found to be attractive as it does not generate any sludge though the rejected contains high concentration of the pollutants. Synthetic membranes are available that are selectively permeable: the structure of the membrane passes through certain molecules, some are excluded or rejected. Membrane filtration has the primary purpose of removing many contaminants from water, including bacteria, salts and various heavy metals.

In terms of operating pressure arrays, membrane filtration can be classified as: low-pressure operated membranes (microfiltration and ultrafiltration), moderate-pressure operated membrane (nanofiltration) and high-pressure operated membranes (reverse osmosis). Low-pressure membranes have large nominal pore sizes and operate at pressures below 4 bar. The moderately tighter and moderate-pressure operated membranes are typically operated at pressures from 5-20 bar where the high-pressure membranes operated at pressures higher than 20 bar.

From Figure 12.10, it is evident that reverse osmosis (RO) and nanofiltration (NF) membranes have sufficient pore sizes to remove dissolved arsenic or fluoride, which is within the range of the metal ion count. Both the RO and NF membranes work in the most frequent lateral configurations in which only a small amount of feed water (10–15%) is permeable through the membrane. In household systems that require a small amount of purified water for cooking and drinking, this low recovery rate is acceptable. Municipal systems achieve high recovery rates (over 80 to 90%) by using multiple layer units in series.

Membrane filtration has the benefit of reducing concentrations of arsenic and many other components. Ultrafiltration (UF) membranes can also eliminate over 99.9% of bacteria, giardia and viruses. Also, the layer itself does not accumulate arsenic, so disposal of used layers is very easy. Operation and Maintenance Requirements Very Low: Chemicals do not

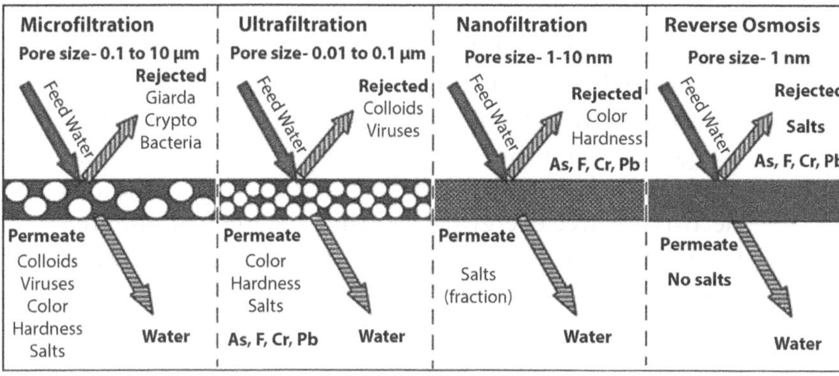

Figure 12.10 Pore size of various membranes and size of materials subject to filtration.

need to be added, and maintenance involves ensuring reasonable consistent pressure and periodically wiping the membrane clean.

The main disadvantages are lower water recovery rates (typically only 10-20% of raw water passes through the membrane), the need to operate at high pressure, relatively high capital and operating costs, and the risk of membrane fouling. Also, especially with RO, purified water contains very low levels of dissolved solids and can be important micronutrients for humans due to its corrosive and mineral deficiency.

In nanofiltration, the distribution of the non-charged solution at the boundary layer/layer interface is determined by the steric exclusion procedure. The steric exclusion is not typical for nanofiltration but is also applicable to ultrafiltration and microfiltration. Because of its size, a solvent has access to only a fraction of the total surface area of a hole. This causes the solvent to be geometrically excluded from the membrane. Differentiation between solvents occurs only when there is a difference in the size of the solution. Two additional distribution methods can be identified for charged solutions:

- The Donnan exception, which, in comparison to other pressure-driven membrane processes, has a clear effect on the separation in NF. Due to the slightly charged nature of the membrane, solvents are attracted with the opposite charge compared to the membrane (counter-ions), and the solutions with similar charge (co-ions) are repulsed. The distribution of co- and counter-ions occurs at the membrane surface, resulting in additional separation
- Dielectric exception, which usually does not play a role in ultrafiltration and microfiltration, but is very important in

electrodialysis. Due to the charge of the membrane and the dipole momentum of the water, the water molecules show polarity in the hole. This polarity reduces the dielectric constant inside the hole, making it less favorable to enter the charged solvent. However, even when the dielectric constant inside the hole is similar to that of water, a change in the ion's electrostatic free energy occurs when the ion is transferred from the bulk to the hole. The relative importance of the two mechanisms in the NF is still under debate in the scientific community. Much of the literature on NF uses the Donnan exception as a distribution mechanism.

12.5.2 Arsenic Removal by Membrane Filtration

12.5.2.1 Arsenic Removal by Microfiltration System

Microfiltration is similar to a low-pressure membrane filter or conventional coarse filtration process that separates colloidal and suspended particles in the range of 0.1 to 10 μm and does not remove arsenate and arsenite strains from arsenic contaminated water. Therefore, arsenic species can be removed by microfiltration system only by increasing the particle size of arsenic-bearing species by adding some binding agent.

Chwirka *et al.* (2000) prepared a microfiltration layer through a mixture of esters cellulose acetate and cellulose nitrate with pore size of 0.22 to 1.2 μm and showed the removal of arsenic by binding to ferric ions and cationic polymeric flocculants. The combination of flocculation and MF technique provided higher arsenic removal efficiency than single-stage use of microfiltration. Due to low energy requirements and high fluxes, compared to other membrane processes such as reverse osmosis and nanofiltration, this system may be a cost-effective method for arsenic removal from drinking water.

Han *et al.* (2002) showed that flocculation followed by MF leads to high flow along with significant arsenic removal in the permeate. They used MF membranes with pore sizes of 0.22 micrometer and 1.2 micrometer, made using cellulose acetate and cellulose nitrate. The residual turbidity, after flocculation and MF, may be used as a guide to the level of arsenic removal. MF is a cost-effective method due to the low energy requirement.

A coagulation-microfiltration-based hybrid treatment scheme (Figure 12.11) was developed for arsenic removal by Wickramasinghe *et al.* (2004) for the treatment of groundwater obtained from a city in southern Colorado in the United States and from Sonargaon in Bangladesh. Results of a bench-scale experiment illustrate that the coagulation with ferric ions followed by filtration effectively removed arsenic from the contaminated groundwater and maintained the residual arsenic concentration below the maximum permissible limit of 10 μg/l (as per WHO). Moreover, the raw water quality and presence of polyelectrolyte coagulant affected the actual removal efficiency of the arsenic though extra addition of polyelectrolyte coagulant may lead to improved permeate fluxes during cross-flow filtration.

A few years back, Ghosh *et al.* (2011) achieved quite effective removal performance of arsenic by the combination of elctrocoagulation-microfiltration system. The microfiltration membrane was made by ceramic materials. Around 96% arsenic was removed from the contaminated water though fluoride and iron was present in feed solution.

A comparative study between electrocoagulation-microfiltration and chemical coagulation-microfiltration processes was conducted by Malgora *et al.* (2013) for the removal of arsenic from contaminated groundwater at an initial concentration of 100 μg/l. Several operating parameters, such as iron concentration, reaction time, and solution pH, were used during the research. However, 97% of arsenic was removed during the chemical freeze-microfiltration process, which required 26 hours for the system to maintain a pH of 7.0 and a ferric ion of ration as low as 4.0 mg/L, as well as a residual arsenic concentration below the limit value (10 μg)./l) recommended by the World Health Organization (WHO). This study also showed that the chemical coagulation-microfiltration system performed better than the electrocoagulation-microfiltration system in terms of arsenic removal. Moreover, due to the extra chlorine addition (to carry out

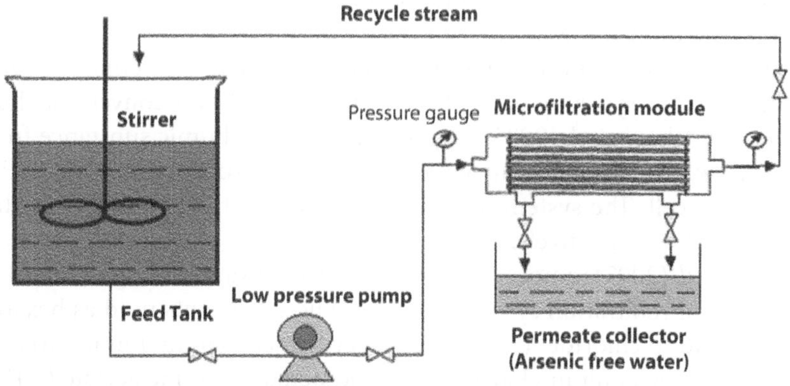

Figure 12.11 Coagulation-microfiltration-based arsenic treatment unit (Wickramasinghe *et al.*, 2004).

the oxidation of Fe^{2+} into Fe^{3+}) in the electrocoagulation-microfiltration system, the chemical coagulation-microfiltration system was chosen as a cost-effective system.

12.5.2.2 Arsenic Removal by Ultrafiltration System

A size exclusion-based pressure-driven membrane separation process is ultrafiltration which has pore size in the range from 10 to 10000 A^0 and capable of separating sugars, bio-molecules, polymers, and colloidal particles in the molecular weight range of 300 to 500,000 Dalton. Due to its high pore size than arsenic radius, it cannot separate arsenic species from the aqueous solution. But charged ultrafiltration can separate arsenic through the repulsion mechanism.

Brandhuber and Amy (2001) introduced a series of bench-level experiments for arsenic removal from natural and synthetic water using a negatively charged ultrafiltration (UF) membrane under cross-flow regime. Several operating conditions like membrane recovery, pH, bulk arsenic concentration in the feed, presence of mono- and di-valent ions and feed stream temperature have been investigated during the experiments. This study illustrates that the removal of As(V) was considerably influenced by the presence of co-occurring divalent ions in the feed. Moreover, due to small pore size of the membrane, the membrane gets fouled due to concentration polarization. As(V) rejection by the UF membrane was also shown to be sensitive to membrane operating conditions. After investigation, the authors suggested that relatively "loose" charged membranes are capable of partial rejection of As(V) from feed waters through its Donnan Exclusion mechanism and to meet regulatory requirements, the application of "loose" charged membranes could be an effective treatment method over the investigated system.

Electro-ultrafiltration is another system which could remove arsenic from water; the system is a significant improvement over conventional UF treatment. Weng *et al.* (2005) removed arsenic and humic substance from aqueous solution by applying of voltage to the electro-ultrafiltration (EUF) membrane cell. The system removed 71% As(III) at pH 10 and over 90% AS-V at pH 6, respectively.

Iqbal *et al.* (2007) removed arsenic from groundwater through a micellar-enhanced ultrafiltration system. Four cationic surfactants used as hexadecylpyridinium chloride (CPC), hexadecyltrimethylammonium bromide (CTAB), benzalkonium chloride (BC) and octadecylamine acetate (ODA) micellar. Out of four used surfactants CPC showed the highest 96% removal efficiency of arsenic followed by CDAB (94%), ODA (80%), BC (57%).

The authors explained the pH effects on arsenic removal. As the solution pH decreases, the valance of arsenate decreases from trivalent to monovalent, thus resulting in less removal efficiency at lower pH. Likewise, the authors showed that 0.45 mM of nitrate and 0.01 mM of phosphate addition in the feed solution reduced the removal efficiency by 5–8% due to the competition between the arsenate, nitrate and phosphate for the binding sites of the surfactant micelle. Furthermore, before being discharged to the environment of the effluent, powdered activated carbon (1 g/l) was used to regenerate 98% of used surfactant to lower the surfactants concentration in the effluent. Lastly they concluded that the micellar-enhanced ultrafiltration system feasible process using CPC or CTAB to remove the arsenate from groundwater compared with the other solid-based adsorbent processes.

Another electro-ultrafiltration-based experiment was performed by Hsieh *et al.* (2008) to remove arsenic from groundwater which was brought from two different wells of the northeastern part of Taiwan. They reported that very less amount (1-14%) of arsenic was removed from the investigated samples by the ultrafiltration membrane (100-kDa) in the absence of electrical voltage but around 80% of overall arsenic was removed by ultrafiltration membrane when it was under an electrical voltage of 25 V. It happened due to association between As(III) species and dissolved organic matter over the membrane surface. Higher water flux also been achieved by the authors through the electro-osmosis phenomena of the system. The authors also showed that the system was more competitive than other conventional arsenic-removing technologies due to its higher water flux, higher removal efficiency and less electrical cost.

Lohokare *et al.* (2008) removed arsenic from synthetic solution by a modified polyacrylonitrile based negative charged ultrafiltration (UF) membrane. Results showed that removal efficiency of As (V) was drastically decreased from 100% to 65% with increase in As(V) concentration from 50 ppb to 1000 ppb where transmembrane pressure and cross-flow velocity were maintained 200 kPa and 0.39m/s, respectively. Due to formation and attachment of carboxylate (COO−) groups over the membrane surface, high percentage of arsenic got removed by the membrane.

Agarwal *et al.* (2013) performed an experiment on arsenic rejection via modified polyacrylonitrile ultrafiltration (UF) membrane and showed the affection of removal performance on different foulants, such as ovalbumin, humic acid and egg white and also exhibited that removal performance was affected by the total quantity of proteins and not the variety of proteins present in a solution. The removal efficiency was found to decrease in nature with increase in proteins and humic acid concentration; a strong positive correlation with pH of the solution (90% at pH 9) was also found.

12.5.2.3 *Arsenic Removal by Nanofiltration System*

Nanofiltration (NF) is a pressure-driven membrane process that runs between 3-20 bar pressure range, less than the RO process. The nominal pore size is 0.4-2 nm and is capable of separating solute particles of less than 1000Da of molecular weight. NF layers behave asymmetrically and are negatively charged at neutral and alkaline media. The charge interaction between the membrane surface and the charged particles plays a major role in the removal efficiency of solvents through the membrane. The elimination of both dissolved arsenate and arsenite through nanofiltration membranes can be explained by the quantity exclusion and Donnan exclusion mechanisms.

Brandhuber and Amy (1998) performed an arsenic removal study using three commercial NF membranes: NF1 (NF70 4040-B), NF2 (HL-4040F1550) and NF3 (4040-UHA-ESNA). All layers show that As (V) rejects more than 95% due to its ionic nature in the range of pH 5 to 8, while (III) is rejected in the range of 20-53% due to neutral properties at the same pH. Range. Due to the electrostatic repulsion between the arsenate molecules and the negatively charged NF layers in the neutral and alkaline pH regimes, arsenate ions are rejected much higher than the arsenite.

Urase *et al.* (1998) conducted an investigation regarding arsenic removal from groundwater in which arsenic-bearing species, like arsenate, arsenite and dimethyl arsinic acid (DMMA) were artificially added. A low-pressure aromatic polyamide NF membrane namely ES-10 was employed in the investigation. Results show that As(V) got more rejected (90% in average) than As(III) species (70% in average) in the pH range of 5-8 and the authors justified that As(V) predominant as $H_2AsO_4^-$ and $HAsO_4^{2-}$ at neutral pH which behave as anionic molecule whereas at the same pH range, As(III) is mainly present as uncharged species (H_3AsO_3).

A laboratory-scale plate and frame cross-flow membrane filtration test unit was developed by Vrijenhoek and Waypa (2000) for the separation of arsenic from drinking water using an aromatic, polyamide, thin-film composite NF membrane, namely NF-45 (FilmTec., Minnetonka, MN). The membrane was characterized through pore size determination and molecular weight cut-off by the retention measurements of saccharide whereas membrane electrokinetic properties was determined through streaming potential measurements. Moreover, salt (single and mixed salt) transportation by the membrane was predicted through the model development. Results show that As(III) was not removed efficiently at different pH or concentrations due to its neutral charge characterstics and relatively low molecular weight compared to the MWCO of the membrane. But with increase in pH 8.1 to 8.2 and in the presence of 10^{-2} M NaCI, As(V) was removed 60-90% from synthetic feed waters containing 10, 32, 100, and 316 g/L due to its anionic nature in natural water.

Sato *et al.* (2002) compared the conventional fast sand filtration inter-chlorination system and the nanofiltration (NF) system; the removal performance of the arsenic and nanofiltration system has been shown to be more effective in terms of removal performance than the rapid sand filtration. ES-10 (aromatic polyamide), NTR-729HF (polyvinyl alcohol) and NTR-7250 (polyvinyl alcohol) were used in nanofiltration research and 95% As (V) was found to be rejected. Due to the large pore sizes of these membranes (III) less than 22% were removed by NTR-729HF and NTR-7250. The removal efficiency of As (III) through the compact ES-10 membrane was found to be above 75% without prior addition of chemical compounds. It cannot be removed by rapid sand filtration without pre-oxidation from trivalent arsenic to pentavalent arsenic.

Saitúa *et al.* (2005) removed arsenic from synthetic and surface water by a spiral wound thin film composite polyamide (192-NF 300) nanofiltration membrane module (Figure 12.12) with an active membrane surface of 1.5 m^2. Results showed that arsenic rejection increased from 86% to 99% with increase in arsenic retentate concentration from 100 to 382 μg/L whereas rejection of arsenic didn't get much affected with increase in transmembrane pressure, cross-flow rate and temperature. However, permeate flux increased with the increase in transmembrane pressure and temperature. In surface water, the mean rejection of As V was found to be 95%.

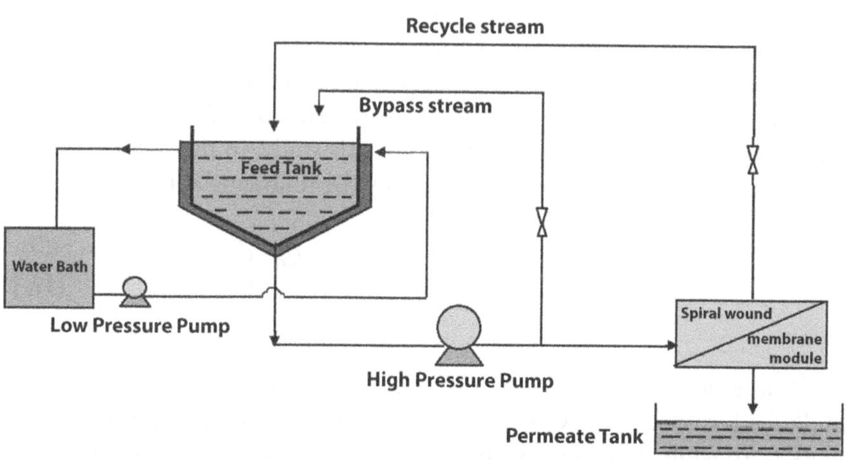

Figure 12.12 Spiral wound–based nanofiltration process (Saitúa *et al.*, 2005).

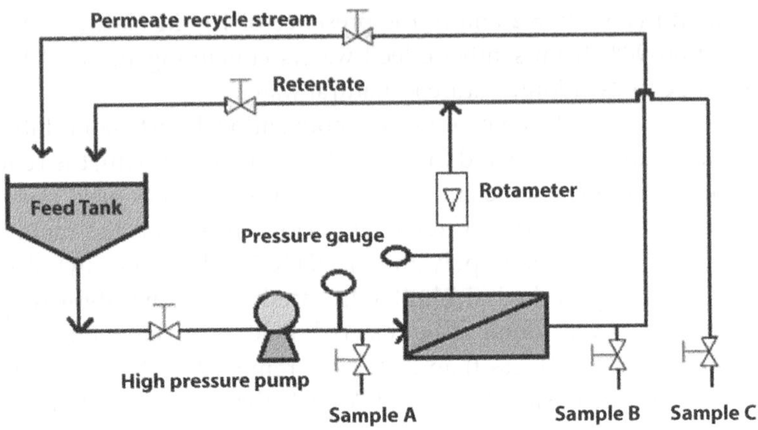

Figure 12.13 Schematic of the experimental unit for NF (Xia *et al.*, 2007).

The co-occurrence of dissolved inorganics does not significantly influence arsenic rejection.

In another study, a pilot scale spiral nanofiltration module (Figure 12.13) was developed by Xia *et al.* (2007) for the removal of arsenic from the synthetic water. The module contained polyamide-based nanofiltration membrane. The authors showed that the arsenic rejection by the membrane increased with increasing pH and found 100% removal efficiency of As (V) at pH 7.5, while As (III) got removed only by 5% at the same pH. As (III) within a pH range from 3 to 7, shows neutral characterstics. The presence of additional salts have been shown to have an impact on the rejection of As (V).

Nguyen *et al.* (2009) evaluated the theoretical elucidation of arsenic separation by a thin film composite of commercially charged polyamide nanofiltration membrane (NE 90, Wuongjin Chemical, South Korea), as well as the bench-scale experiment. In addition, several membrane properties, such as molecular weight cut-off (MWCO), zeta potential and rejection of individual salts, were also determined in this study. In the theoretical section, they calculate Donnan's efficiency by the following equation:

$$\psi_D = \psi_M - \psi = \frac{RT}{z_i F} \ln\left(\frac{C_i}{c_i^{m}}\right) \tag{12.1}$$

Where, ψ_D and ψ_M are the electric potentials of the membrane and bulk feed, respectively; C_i is the concentration of charged species i in the bulk feed, while c_i^m is the concentration at the pore inlet of the membrane of ion i, and z_i is the valence of the species i. Evaluation of counter-ion concentration in terms of effective membrane charge density (X_m) at the membrane pore inlet was also shown in this study. The authors expressed the electroneutrality conditions as:

$$c_{\text{counter}}{}^{\text{m}} = C_{\text{counter}} \left(\frac{|z_{counter}| C_{\text{counter}}}{|z_{counter}| c_{\text{counter}}{}^{\text{m}} + X} \right)^{\frac{|z_{counter}|}{|z_{co}|}} \tag{12.2}$$

Arsenic elimination results in the initial arsenic concentration, arsenic specification and the presence of various electrolytes in the feed tank. When (V) was removed at approximately 95% pH 7 and the applied pressure of 5.5 bar, 44% As (III) was removed from the feed solutions containing 50 μg/L As (III) and As (V). This is due to the superiority of the Donnan exclusion (charge repulsion) over the steric exclusion (size exclusion). The mutual effects of monovalent and divalent ions were also shown in this study where the rejection of As(V) was enhanced by the presence of Cl^-, but reduced by the presence of SO_4^{2-} due to mutual interactions between anions.

A commercial thin film composite NF-300 (Permionics, Vadodara, India) was used by Harisha *et al.* (2010) for arsenic removal from contaminated groundwater, collected from one of the local sites of Manvi, India. A pilot-scale skid-mounted system (Figure 12.14) was developed for this study. Results showed high rejection (99.80%) of arsenic and total dissolved solids (TDS) and other contaminates. The flux declination was found very insignificant at the end of each experiment with operation time of 3 hours

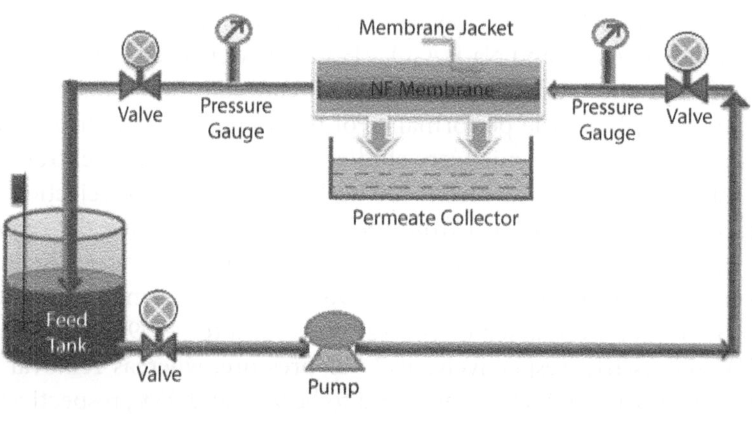

Figure 12.14 Schematics of nanofiltration pilot plant (Harisha *et al.*, 2010).

which indicated that fouling phenomenon of the membrane was reduced by using appropriate cleaning procedures.

In another study, influence of operating parameters on the arsenic removal by nanofiltration was investigated by Figoli *et al.* (2010) using synthetic arsenic water which contained pentavalent arsenic. The investigation was performed in a laboratory scale set-up employing spiral-wound nanofiltration modules (N30F by Microdyn-Nadir and NF90 by Dow Chemical). Arsenic rejection was found to decrease with increase in operating temperature and initial arsenic concentration whereas arsenic rejection got increased from 74% to 88%, and from 94% to 98%, with increase in pH 3.4 to 10, for N30F and NF90, respectively. In the other side, the permeate flux increased with temperature and pressure, and was found to be maximum at the pH of 8.

Sen *et al.* (2010) carried out an experiment with a novel flat-sheet cross-flow nanofiltration set-up integrated with a reactor for the separation of arsenic from contaminated groundwater, collected from arsenic-affected areas of West Bengal, India. The oxidation step was used as the pre-treatment technology in this hybrid treatment process for the conversion of trivalent arsenic to pentavalent form where $KMnO_4$ was used as the oxidising agent. Several effects, such as oxidant ($KMnO_4$) dose, applied pressure, iron concentration in groundwater and pH on arsenic removal were studied during the investigation to get reasonable industrial acceptance flux as well as high removal efficiency. Results showed that applied pressure, oxidant dose and pH has a strong impact on flux and retention of arsenic. All three types of membranes removed about 90 to 98% of As(V) at an applied pressure of 12 bar.

Akbari *et al.* (2010) carried out an experiment to determine the arsenic removal by a thin film composite and aromatic polyamide nanofiltration membrane (model NF90, Dow-Film TecCo) using spiral wound membrane module. As_2O_3 and $Na_2HAsO_4.7H_2O$ salts were used to make As(III) and As(V) solution. Effects of initial arsenic concentration transmembrane pressure and pH on the performance of the membranes in terms of percentage rejection of arsenic were studied. As the initial concentration in feed water increased As (V) from 120 μg/L to 1026 μg/L, the rejection percentage of As (V) decreased from 98.35% to 96.59%, while the removal efficiency of As (III) was reduced from 118 μg/L in feed water. The initial concentration of As (III) increased to 985 μg/L, decreasing from 94.07% to 87.51%. Moreover, arsenic removal was found to be 99% and 95% for As(V) and As(III), respectively, at 7 bar pressure, whereas removal efficiency of As(V) and As(III) was found to be 99% and 95%, respectively, at pH 11.

A Poly (m–phenylene isophthalamide) based nanofiltration membrane was synthesized by Zhao *et al.* (2012) to remove arsenic from simulated arsenic solution that contained $Na_2HAsO_4.7H_2O$ salt and the authors found over 90% rejection of arsenate. They also found increase in rejection of As(V) from 83% to 99%, with increase in pH 3 to 9. Moreover, As(V) removal efficiency increased 92% to 98% with increase in pressure in the range of 0.4 MPa to 0.8 MPa. A significant change on arsenic removal was found by the authors with the presence of various inorganic substances in the feed. For example, the presence of NaCl enhanced As(V) rejection which reached upto 99% at a feed As concentration of 100 μg/L.

A few years back, Xie *et al.* (2013) showed the effect of dissolved organic matter on arsenic removal by nanofiltration. They performed the investigation in a laboratory bench-level plate-and-frame cross-flow nanofiltration membrane module. Results of this study exhibited that the presence of humic acid as a dissolved organic matter increased removal efficiency of arsenic because of the reaction between arsenic and humic acid produced humic/arsenic complexes. With increase of humic acid from 2 to 10 mg TOC/L arsenic removal efficiency gets increased from 80 to 100%. The concentration of the membrane flux gets decreased 20% at various concentration of dissolved organic matter after 400 min the same cross-flow velocity.

A similar type of effect, such as natural organic matter (humic acid) and ion concentration on As(V) removal from synthetic aqueous solution by nanofiltration membrane under different transmembrane pressures was studied by Yu *et al.* (2013). They performed the investigation using four different nanofiltration membranes: ESNA-1-K1, NF270, ESNA-1-LF, and HODRA-CORE. Highest 94% arsenic was removed by the ESNA-1-LF membrane whereas HODRA-CORE membrane got removed only 47% (lowest). Increase in ion concentration and addition of humic acid in the feed solution decreased the arsenic removal efficiency by the HODRA-CORE membrane whereas increase of the ion concentration and of humic acid in the feed solution made the rejection increase for the other three membranes. Due to fouling phenomena of the NF membranes, the permeate fluxes got decreased to some extent after addition of humic acid to the solutions after 6 hours of operation.

Recently, Xi *et al.* (2014) employed spiral wound nanofiltration membrane by which they simultaneously removed fluoride and arsenic from groundwater obtained from a 300 m deep well in a rural area of Beijing, China. They optimized a novel membrane configuration by some bench-scale contrast experiments using commercial NF membranes and the optimized configuration played a significant role on operation performances and membrane fouling. The optimized arrangement provided a reasonable removal performance of 73% and 94% for fluoride and arsenic, respectively, as well as producing 73.37 L/m^{2}h of permeate flux and acceptable water flux recovery rate of 65% at 16°C. By lowering the inter stage Reynolds number, arrangement could further restrict the potential of membrane fouling. Furthermore, the water permeation cost was lowest by the NF system with this optimal arrangement.

Arsenic (III) and Arsenic (V) rejection along with fouling study by various NF membranes is presented in Table 12.3.

12.5.2.4 Arsenic Removal by Other Membrane-Based Process

Waypa *et al.* (1997) performed a research for arsenic removal from synthetic water and surface water sources using a thin-film composite type layer. The results showed that As (V) and As (III) both reject 99% of the water through the RO layer. The novel thin-film composite type layer shows high selectivity and operates at a lower operating pressure than other layers and produces a higher rate of permeability. Four types of RO membranes, such as TFC 4921 (Fluid Systems), TFC 4820-ULPT (Fluid Systems), AG 4040 (DESAL) and 4040LSA-CPA2 (Hydranautics) were used by Brandhuber and Amy (1998) to remove the arsenic from arsenic-contaminated feed. They found 95% and 85% removal efficiency for As(V) and As(III), respectively, by all RO membranes.

In another RO experiment, Amy *et al.* (1998) found 96% removal efficiency of arsenate while 5% removal efficiency for Arsenite using a bench scale single element RO unit. In the other side, the removal efficiency of arsenate was found around 88% using flat sheet RO testing by a DK2540F (DESAL) RO membrane.

Recently, Teychene *et al.* (2013) performed a comparative study of the rejection of boron and As (III) from brine by the osmonic cell unit (Cepa CF II, General Electric -USA) using the "brine" and "seawater" layers using flat sheet reverse osmosis layers. Transmembrane pressure, membrane type and pH showed a strong influence on the removal of the metalloid percentage. In addition, the results showed that the "brine" layers had a higher water flow and the lower metalloid was removed under the same operating conditions (pH, transmembrane pressure). "Saltwater" layers remove 99% As (III) at pH = 9.6 and 40 bar, but for seawater it is more than 99%, low pH and low pressure (pH = 7.6 and 24 bar). The same resolutions were made for boron as well.

A novel solar-driven direct contact membrane distillation (DCMD) treatment method was applied by Manna *et al.* (2010) for the removal

Table 12.3 Removal efficiency of As(III) and As(V) by different NF membranes.

Nanofiltration module used	Membrane used (code)	Characterstics of feed	Total As conc. (µg/L)	Rejection As(III)/As(V) (%)	References
Spiral wound	ES-10	Groundwater spiked with arsenic compounds	600	89/93	Urase *et al.* (1998)
Spiral wound	NF70 4040-B HL- 4040F1550 4040-UHA-ESNA	Groundwater spiked with arsenic compounds	15-30	53/99 21/99 30/97	Brandhuber and Amy (1998)
Plate and Frame	NF-45	Synthetic water	100	<10/90	Vrijenhoek and Waypa (2000)
Spiral wound	ES-10 NTR-729HF NTR-7250	Groundwater spiked with arsenic compounds	50	82/97 24/91 10/88	Sato *et al.* (2002)
Spiral wound	NF 300	River water spiked with arsenic compounds	100	93 for As(V)	Saitúa *et al.* (2005)

(*Continued*)

Table 12.3 Removal efficiency of As(III) and As(V) by different NF membranes. (*Continued*)

Nanofiltration module used	Membrane used (code)	Characterstics of feed	Total As conc. (µg/L)	Rejection As(III)/As(V) (%)	References
Plate and Frame	NE 90	Synthetic water	50	63/97	Nguyen *et al.* (2009)
Spiral wound	NF-300	Groundwater	51	99 for As(V)	Harisha *et al.* (2010)
Spiral wound	N30F NF90	Synthetic water	500	90 for As(V) 98 for As(V)	Figoli *et al.* (2010)
Spiral wound	NF90	Synthetic water	500	93/97	Akbari *et al.* (2010)
Spiral wound	Synthesized NF membrane	Synthetic water	100	>96 for As(V)	Zhao *et al.* (2012)
Spiral wound	ESNA-1-K1 NF270 ESNA-1-LF HODRA-CORE	Synthetic water	200	94 for As(V) 90 86 46	Yu *et al.* (2013)

of arsenic from contaminated groundwater. During the investigation, a hydrophobic polyvinylidenefluoride (PVDF) based microfiltration membrane was used in the membrane module. The unique design of the solar-driven direct contact membrane distillation (DCMD) system successfully removed all the arsenic present in the contaminated groundwater in a largely fouling-free operation as well as yielding high fluxes under reduced temperature polarization. For a feed flow rate of 0.120 m^3/h, the 0.13 mm PVDF membrane yielded a high flux of 74 kg/m^2 h and 95 kg/m^2h at a feed water temperature of 40°C and 60°C, respectively.

12.5.3 Fluoride Removal by Different Membrane Filtration System

Coagulation-precipitation along with microfiltration and ultrafiltration process, nanofiltration, electrodialysis, and reverse osmosis are the major five membrane-based processes used for defluoridation. Though RO process provides 100% pollutant-free water due to selective demineralization it should be avoided for drinking water purification whereas nanofiltration (NF) is much more appropriate for defluoridation process as in this process the remineralization step is absent.

Hu and Dickson (2006) showed performance of three thin-film composite negatively charged commercial nanofiltration membranes (DS-5-DL, DS-51-HL and SR-1) for fluoride removal from high fluoride content water. The performance of the membrane was investigated in terms of three importants membrane parameters: water permeability, membrane pore radius and electrical potential of the membrane surface. Extended Nernst–Planck and Grahame equation were used for the model development whereas model predicted parameters were found by the validation of experimental and model values of fluoride rejection and flux for two different levels of sodium fluoride concentrations (100 and 1000 ppm). Both experimental and model results showed that the removal efficiency of sodium fluoride increased with the pressure while the solution flux increased with decreases of feed concentration.

In the same year, Tahaikt *et al.* (2006) also studied a performance comparison of three commercial nanofiltration membranes (NF90, NF270 and TR60) for fluoride removal from a fluoride-contaminated groundwater. The experiments were carried out on an NF pilot plant which was

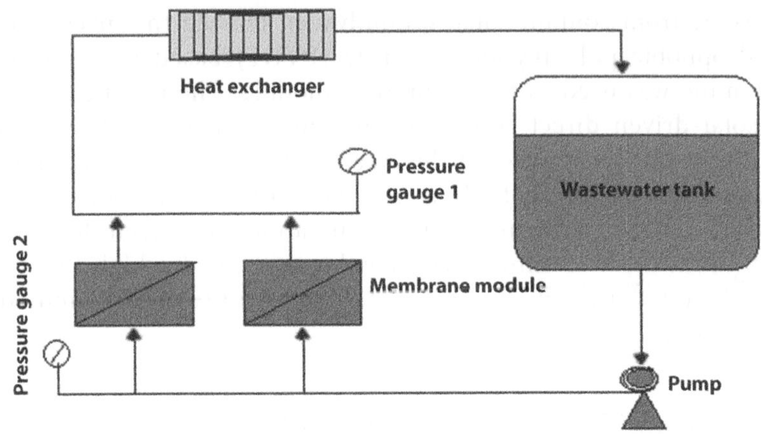

Figure 12.15 Diagram of the nanofiltration pilot plant (Tahaikt *et al.*, 2006).

equipped with two identical spiral wound membrane modules working in series mode. Figure 12.15 shows the scheme of the pilot plant used in the investigation. With the help of a similar type of experimental scheme, Tahaikt *et al.* (2006), also performed an experiment for fluoride removal from contaminated water by another polyamide-based nanofiltration membrane (NF400).

Elazhar *et al.* (2009) performed an economic evaluation study for the removal of fluoride by spiral wound membranes (NF90 8040) having a capacity of 2400 m^3/d in a rural area of Morocco. The plant was designed with pre-set optimized conditions where the recovery rate and removal efficiency of fluoride were found to be 84% and 97.8%, respectively, at applied pump pressure of 10 bar. The capital cost and operational cost were found to be 748,003€ and 0.212€/m^3.

Wei-fang *et al.* (2009) showed the influence of different factors for removal of fluoride from groundwater by spiral wound nanofiltration membrane module. The effect of co-existing ions, like chloride, sulfate, feed water characterstics, like pH, temperature were studied to remove fluoride from contaminated groundwater using nanofiltration membrane. The authors also studied membrane fouling in their study. Results showed that retention of fluoride decreases with the increase of initial concentration of fluoride in feed, concentration of co-existing ions and pH of the groundwater. With the increase in 1 mol Cl^- ion concentration and 0.5 mol of SO_4^{2-} ion concentration, the removal percentage rate of decreasing was the same. Insignificant removal efficiency was observed above 220mg/L of feed chloride ion concentration. The removal efficiency of fluoride increased with the increase in pH, but above pH 4.4, no significant changes in rejection was observed. Membrane fouling was analysed by environmental scanning electron microscope (ESEM) and its energetic chart analysis apparatus, X-ray diffraction (XRD) analysis.

Influence of some physical (pressure applied and recovery rate) and chemical (feed concentration and ionic strength) constraints on fluoride separation by nanofiltration was studied by Mnif *et al.* (2009). The experiment was carried out with a spiral wound module which contained a thin film composite polyamide based nanofiltration membrane (HL 2514 T, Osmonics Inc.). The transport mechanism of the membrane also investigated in this study where the Kedem–Katchelsky model was applied in order to define reflexion coefficient of the membrane and solute permeability of ions. Monovalent and bivalent salts get rejected by the membrane as it carries negative charge over the membrane surface. Above 80% removal efficiency of fluoride by HL membrane was found in their study. Theoretical model corroborated very well with the experimental results of various salts.

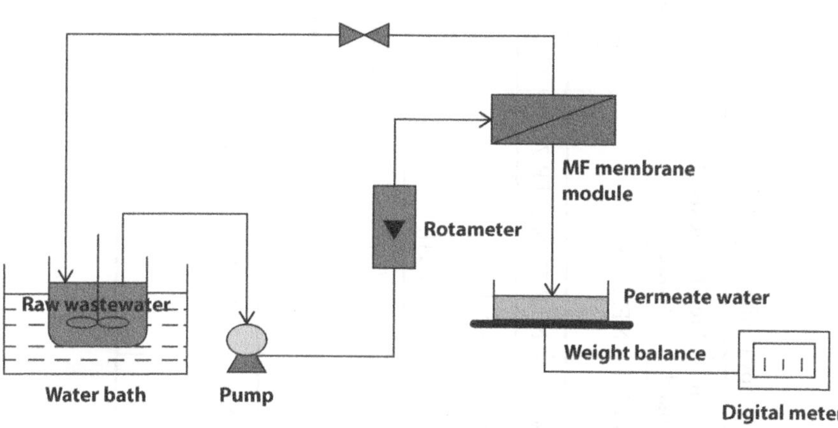

Figure 12.16 Schematic diagram for fluoride and phosphate separation by a microfiltration-based hybrid system (Lu *et al.*, 2010).

Table 12.4 Removal efficiency of fluoride by different NF membranes.

Nanofiltration module used	Membrane used (code)	Characteristics of feed	Total F conc. (mg/L)	Rejection (%)	References
Spiral wound	NF70-2540	Tap water spiked with fluoride compounds	238	94	
Hollow fiber	NTR 7250/ NTR 7450	Groundwater	5	72/70.4	Choi *et al.* (2001)
Hollow fiber	SR-1 DS-5-DL DS-51-HL	Synthetic water	100/1000	92/73 92/72 98/90	Hu and Dickson (2006)
Spiral wound	NF90/NF400	Groundwater spiked with fluoride compounds	1.8 5 10 20	95/50 96/65 96.5/75 97/86	Tahaikt *et al.* (2007)
Spiral wound	NF90/NF270/TR60	Groundwater spiked with fluoride compounds	2.32 3.32 6.32 22.32	97/82/74 98/84/78.9 98.5/88/85.7 99/82.6/82	

(Continued)

Table 12.4 Removal efficiency of fluoride by different NF membranes. (*Continued*)

Nanofiltration module used	Membrane used (Code)	Characteristics of feed	Total F conc. (mg/L)	Rejection (%)	References
Spiral wound	NF90 8040	Groundwater	2.32	97.8	Elazhar *et al.* (2009)
Spiral wound	HL1812T	Groundwater spiked with fluoride compounds	0.40	80	Wei-fang *et al.* (2009)
Spiral wound	HL 2514 T	Synthetic water	123 410 820 2050 4100	98 96 95 94 92	Mnif *et al.* (2010)
Spiral wound	Synthesized NF membrane	Groundwater spiked with fluoride compounds	4	73	Xi *et al.* (2014)
Flat sheet	NF90 NF270 TFC-SR2 TFC-SR3	Surface water Groundwater River water River water	240 22 2.2 2.2	97 80 68 96	

Fluoride removal by various NF membranes with different initial concentration is presented in Table 12.4.

References

Agarwal G P., Karan R, Bharti S, Kumar H, Jhunjhunwala S, Sreekrishnan, T.R, Kharul U, Effect of foulants on arsenic rejection via polyacrylonitrile ultrafiltration (UF) membrane, *Desalination* 309 (2013) 243–246.

Akbari H. R., Mehrabadi A. R., Torabian A. determination of nanofiltration efficency in arsenic removal from drinking water. *Iran. J. Environ. Health. Sci. Eng.*, 2010, Vol. 7, No. 3, pp. 273–278.

Amy, G.L., Edwards, M., Benjamin, M., Carlson, K., Chwirka, J.,Brandhuber, P., McNeill, L. and Vagliasindi, F., Draft Report, AWWARF, 1998.

Bang S., Patel M., Lippincott L., Meng X., Removal of arsenic from groundwater by granular titanium dioxide adsorbent, *Chemosphere* 60(2005a) 389–397.

Bang S., George P., Korfiatis, Meng X, Removal of arsenic from water by zerovalent iron, *J. Haz. Mater.* 121 (2005b) 61–67.

Basu A, Saha D, Saha R, Ghosh T and Saha B (2014): A review on sources, toxicity and remediation technologies for removing arsenic from drinking water, *Research on Chemical Intermediates*, 40(2), 447–485.

Battula S.K, Cheukuri J, N. V. V. S. S Raman, Himabindu V, Bhagawan D, Effective Removal of Fluoride from Ground Water Using Electro-Coagulation, *Int. Journal of Engineering Research and Applications*, Vol. 4, Issue 2 (Version 1), February 2014.

Bissen M., Frimmel F. H, Arsenic - a review. Part I: Occurrence, toxicity, speciation, mobility, *Acta hydrochim. hydrobiol.*, 31, 1, 9–18, 2003.

Bhattacharyya R, Chatterjee D, Nath B, Jana J, Jacks G, Vahter M. High arsenic groundwater: mobilization, metabolism and migitation — an overview in the Bengal Delta Plant, *Molecular Cell Biochemistry* 2003; 253:347–355.

Brandhuber, P., Amy, G., 1998, Alternative Methods for Membrane Filtration of Arsenic from Drinking Water, *Desalination*, 117, 1–10.

Brandhuber P, Amy G. Arsenic removal by a charged ultrafiltration membrane influences of membrane operating conditions and water quality on arsenic rejection. *Desalination* 140 (2001) 1–14.

Camacho J, Wee HY, Kramer TA, Autenrieth R. Arsenic stabilization on water treatment residuals by calcium addition. *Journal of Hazardous Materials* 2000;165:599-603.

Camacho, L.M., Torres, A., Saha, D., Deng, S., 2010. Adsorption equilibrium and kinetics of fluoride ion sol–gel-derived activated alumina adsorbents. *J. Colloid Interface Sci.* 349, 307–313.

Choi S, Yun A, Hong S, Ahn K (2001) The effect of co-existing ions and surface characteristics of nanomembrane on the removal of nitrate and fluoride. *Desalination* 133: 53–64.

Chwirka, J., Thomsom, B., Stomp, J., 2000. Removing Arsenic from Groundwater. Jour. *AWWA*, 92 (3), 79–88.

De AK (2008): *Environmental chemistry*, New Age International, New Delhi.

de Oliveira, I.R.W.Z.R.E., de Barros Osorio, H.M., Neves, A., Vieira, I.C., 2007. *Sens. Actuators, B* 122, 89–94.

Elazhar F, Tahaikt M, Achatei A, Elmidaoui F, Taky M, El Hannouni F, Laaziz I, Jariri S, Amrani El M., Elmidaoui A, Economical evaluation of the fluoride removal by nanofiltration, *Desalination* 249 (2009) 154–157.

Figoli A, Cassano A, Criscuol A, Mozumder M S I, Uddin M T, Islam M A, Drioli E, 2010. Influence of operating parameters on the arsenic removal by nanofiltration. *Water Research*, 44(1): 97–104.

Ferguson JE Gavis J. 1972. A review of the arsenic cycle in natural waters. *Water Res* 6:1259–1274.

García MG, d'Hiriart J, Giullitti J, Lin H, Custo G, *et al.* (2004) Solar light induced removal of arsenic from contaminated groundwater: the interplay of solar energy and chemical variables. *Solar Energy* 77: 601–613.

Gadhari, N.S., Sanghavi, B.J., Karna, S.P., Srivastava, A.K., 2010. Electrochim. *Acta* 56, 627–635.

Ghosh D, Medhi C.R, Purakait M.R, Treatment of fluoride, iron and arsenic containing drinking waterby electrocoagulation followed by microfiltration, *Proceeding of 12th International conference on environment science and technology, Rhodes, Greece, 8-10 September, 2011.*

Guo X., Chen A., Removal of arsenic by bead cellulose loaded with iron oxyhydroxide from groundwater, *Environ. Sci. Technol.* 39 (2005) 6808–6818.

Han B, Runnells T, Zimbron J, Wickramasinghe R, Arsenic removal from drinking water Microfiltration by flocculation and microfiltration, *Desalination* 145 (2002) 293-298

Harisha R.S., Hosamani K.M., Keri R.S., Nataraj S.K., Aminabhavi T.M. Arsenic removal from drinking water using thin film composite nanofiltration membrane. *Desalination* 252 (2010) 75–80.

Hsieh L, Weng Y, Huang C, Li K. Removal of arsenic from groundwater by electro-ultrafiltration. *Desalination* 2008;234:402–408.

Hu K, Dickson JM (2006) Nanofiltration membrane performance on fluoride removal from water. *J Membr Sci* 279:528–529.

Hug, S.J., Leupin, O., 2003. Iron-catalyzed oxidation of arsenic (III) by oxygen and by hydrogen peroxide: pH-dependent formation of oxidants in the Fenton reaction. *Environ. Sci. Technol.* 37, 2734–2742

Iqbal, J., Kim, H.J., Yang, J.S., Baek, K. and Yang, J.W. (2007). Removal of arsenic from ground water by micellar-enhanced ultrafiltration (MEUF). *Chemosphere*, 66(5), 970–976.

Jain C.K, Ali I. Arsenic: occurrence, toxicity and speciation techniques. *Water Res* 2000;34:4304–12.

Jekel M.R (1994) Removal of arsenic in drinking water treatment. In: Nriagu JO (ed.) *Arsenic in the environment. Part I: cycling and characterization*. Wiley, New York p 119.

Jing C, Liu S, Meng M. Arsenic leachability and speciation in cement immobilized water treatment sludge. *Chemosphere* 2005; 59:1241–7.

Jing C, Meng X, Liu S, Baidas S, Patraju R, Christodoulatus C, Korfiatis G (2005) Surface komlexation of organic arsenic on nanocrystalline titanium dioxide. *J Colloid Interf Sci* 290:14–21.

Kartinen EO, Martin CJ (1995) An overview of arsenic removal process. *Desalination* 103:78–88.

Kotkar, R.M., Desai, P.B., Srivastava, A.K., 2007. *Sens. Actuators, B* 124, 90–98.

Kundu S., Gupta A.K., Analysis and modeling of fixed bed column operations on As(V) removal by adsorption onto iron oxide-coated cement (IOCC), *J. Colloid Interf. Sci.* 290 (2005) 52–60.

Kundu S., Paul S., Kavalakatta S., Pal A., Ghosh S.K., Mandal M., Pal T., Removal of arsenic using hardened paste of Portland cement batch adsorption and column study, *Water Res.* 38 (2004) 3780–3790.

Lohokare H.R., Muthu M.R. Agarwal G.P., Kharul U.K. (2008) Effective arsenic removal using polyacrylonitrile-based ultrafiltration (UF) membrane. *J. Membrane Sci.*, 320: 159–166.

Ma, Y., Wang, S.G., Fan, M., Gong, W.X., Gao, B.Y., 2009. Characteristics and defluoridation performance of granular acti, A study to remove fluoride ions from ground water 823 vated carbon coated with manganese oxides. *J. Hazard. Mater.* 168, 1140–1146.

Maity S., Chakravarty S., Bhattacharjee S., Roy B.C., A study on arsenic adsorption on polymetallic sea nodule in aqueous medium, *Water Res.* 39 (2005) 2579–2590.

Maliyekkal, S.M., Sharma, A.K., Philip, L., 2006. Manganese oxide coated alumina: a promising sorbent for defluoridation of water. *Water Res.* 40, 3497–3506.

Manna, A.K., Sen, M., Martin A.R., Pal, P., Removal of arsenicfrom contaminated ground water by solar-driven membrane distillation, *Envirinmental Pollution* 158, p. 805–811 (2010).

Mascher R, Lippmann B, Bergmann H. Arsenic toxicity: effects on oxidative stress response molecules and enzymes in red clover plants. *Plant Sci* 2002; 163(5):961–9.

Mnif A, Ali M.B.S, Hamrouni B, Effect of some physical and chemical parameters on fluoride removal by nanofiltration, *Ionics* 16 (2010) 245–253.

Mnif A, Ben Sik Ali M, Hamrouni B (2010) Effect of some physical and chemical parameters on fluoride removal by nanofiltration. Ionics 16, 245–253.

Mobin, S.M., Sanghavi, B.J., Srivastava, A.K., Mathur, P., Lahiri, G.K., 2010. *Anal. Chem.* 82, 5983.

Mólgora, C.C., Domínguez, A.M., Avila, E.M., Drogui, P. and Buelna, G. (2013) Removal of arsenic from drinking water: A comparative study between electrocoagulation-microfiltration and chemical coagulation-microfiltration processes. *Separation and Purification Technology* 118(0), 645–651.

Mondal NK, Roy P, Das B and Datta JK (2011): Chronic arsenic toxicity and it's relation with nutritional status: a case study in Purabasthali–II, Burdwan, West Bengal, India, *International Journal of Environmental Sciences*, 2(2), 1103–1118.

Murray J., Appropriate Use of Fluorides for Human Health, World Health Organization, Geneva, 1986, p. 66.

Na L., Maohong F., Johannes V.L Basudeb S., Hongqun Y., Huang C. P, Oxidation of As(III) by potassium permanganate, *Journal of Environmental Sciences* 19 (2007) 783–786.

Nigam S, Vankar PS and Gopal K (2013b): Biosorption of arsenic from aqueous solution using dye waste, *Environmental Science and Pollution Research*, 20(2), 1161–1172. membrane. *Desalination*, 130, p. 262–277 (2000).

Nguyen C. M., Bang S., Cho J., Kim K, Performance and mechanism of arsenic removal from water by a nanofiltration membrane. *Desalination* 245 (2009) 82–94.

Pal P, Sen M, Manna AK, Pal J, .Roy SK, Roy P. Contamination of Groundwater by Arsenic: a review of occurrence, causes, impacts, remedies and membrane-based purifications, *Journal of Integrative Environmental Sciences*, 2009; 6:1–22.

Pal P., Adsorption Method of Arsenic Separation from Water,Book name "Groundwater Arsenic Remediation", CHAPTER 3, (2015), http://dx.doi.org/10.1016/B978-0-12-801281-9.00003-5

Rao, N.V., Mohan, R., Bhaskaran, C.S., 1998. Studies on defluoridation of water. *J. Fluorine Chem.* 41, 17–24.

M., Cerutti, S. and Padilla, A.P., Effect of operating conditions in removal of arsenic from water by nanofiltration membrane. *Desalination*, 172, p. 173–180 (2005).

Sato, Y., Kang, M., Kamei, T., Magara, Y., Performance of nanofiltration for arsenic removal. *Water Research*, 36 (13), p. 3371–3377 (2002).

Saxena K. L., Sewak R. Fluoride Consumption in Endemic Villages of India and Its Remedial Measures, *International Journal of Engineering Science Invention*, 2015; 4(1): 58-73.

Sen M, Manna A, Pal P. Removal of arsenic from contaminated groundwater by membrane-integrated hybrid treatment system. *Journal of Membrane Science* 2010; 354:108–113.

Shahrokhian, S., Ghalkhani, M., Amini, M.K., 2009. *Sens. Actuators, B* 137, 669.

Shen, F. Zhang, Hideaki. (2006). Photocatalytic oxidation and removal of arsenite from water using slag-iron oxide-TiO2 adsorbent, *Chemosphere*, 65: 125

Squibb KS, Fowler BA. The toxicity of arsenic and its compounds. In: Fowler BA, ed. *Biological and environmental effects of arsenic.* Amsterdam: Elsevier; 1983. p. 233–69.

Tahaikt M, Habbani RE, AitHaddou A, Achary I, Amor Z, Taky M, Alami A, Boughriba A, Hafsi M, Elmidaoui A (2007) Fluoride removal from groundwater by nanofiltration. *Desalination* 212: 46–53.

Tahaikt M, Habbani R.E, Haddou A.Ait Achary I. Amor Z. Taky M. Alami A. Boughriba A.Hafsi M. Elmidaoui A (2006) Desalination Volume 212, Issues 1–3, 25 June 2007, Pages 46-53, https://doi.org/10.1016/j.desal.2006.10.003.

Tang, Y., Guan, X., Su, T., Gao, N., Wang, J., 2009. Fluoride adsorption onto activated alumina: modeling the effects of pH and some competing ions. *Colloids Surf.* 337, 33–38.

Teychene, B., Collet, G., Gallard, H., Croue, J.P., 2013. A comparative study of boron and arsenic (III) rejection from brackish water by reverse osmosis membranes. *Desalination* 310, 109e114.

Urase, T., Oh, J. and Yamamoto, K., Effect of pH on rejection of different species of arsenic by nanofiltration. *Desalination*, p. 11–18 (1998).

Violante A, Ricciardella M, Gaudio SD, Pigna M. Coprecipitation of arsenate with metal oxides: nature, mineralogy, and reactivity of aluminum precipitates. *Environmental Science and Technology* 2006; 40: 4961–4967.

Vrijenhoek E M, Waypa J J, 2000. Arsenic removal from drinking water by a "loose" nanofiltration membrane. *Desalination*, 130(3): 265–277.

Waste Classification Guidelines Part 1: Classifying Waste, Department of Environment and Climate Change NSW, Sydney (2008).

Waypa, J.J., Elimelech, M. & Hering, J.G., Arsenic removal by RO and NF membranes. J. Am. Water Works Assoc.(AWWA) 89(10) : p. 102-114 (1997) Fluoride removal from groundwater by nanofiltration

Weng Y-H, Hsieh LHC, Lee HH, Li KC, Huang CP (2005). Removal of arsenic and humic substances (HSs) by electro-ultrafiltration (EUF). *J. Hazard. Mater.*, 122: 171–176.

Wei-fang M.A, WEN-jun L, Guo-wei C, Factors influencing the removal of fluoride from groundwater by nanofiltration, *Proc. ICBBE'09. International Conference on Bioinformatics and Biomedical Engineering, Beijing, 2009.*

Wickramasinghe SR, Han B, Zimbron J, Shen Z, Karim MN (2004) Arsenic removal by coagulation and filtration: comparison of groundwaters from the United States and Bangladesh. *Desalination* 169(3):231–244.

WHO. Guidelines for drinking-water quality. World Health Organization. Geneva. 1993

Xia S, Dong B, Zhang Q, Xu B, Gao N, Causseranda C (2007) Study of arsenic removal by nanofiltration and its application in China. *J Membr Sci* 204:374–379.

Xi, B., Wang, X., Liu, W., Xia, X., Li, D., He, L., Wang, H., Sun, W., Yang, T. & Tao, W. (2014). Fluoride and Arsenic Removal by Nanofiltration Technology from Groundwater in Rural Areas of China: Performances with Membrane

Optimization. Separation Science and Technology (Philadelphia), 49, pp. 2642–2649. DOI: 10.1080/01496395.2014.939761

Xie Y, Guo C, Ma Rui, Xu B, Gao N, Dong B, Xia S, Effect of dissolved organic matter on arsenic removal by nanofiltration, *Journal Desalination and Water Treatment* Volume 51, 2013 - Issue 10-12.

Yu Y., Zhao C., Wang Y., Fan W., Luan Z. Effects of ion concentration and natural organic matter on arsenic(V) removal by nanofiltration under different transmembrane pressures. *Journal of Environmental Sciences* 2013, 25(2) 302–307.

Zhao C., Du S., Wang T., Zhang J., Luan z., Arsenic removal from drinking water by self-made PMIA nanoiltration membrane. *Advances in Chemical Engineering and Science*, 2, p. 366–371 (2012).

Zhang W., Singh P., Paling E., Delides S., Arsenic removal from contaminated water by natural iron ores, *Min. Eng.* 17 (2004) 517–524.

Zouboulis A, Katsoyiannis I. Removal of arsenates from contaminated water by coagulation-direct filtration. *Separation Science and Technology* 2002; 37:2859–2873.

13

A Zero Liquid Discharge Strategy with MSF Coupled with Crystallizer

Jasneet Kaur Pala[1], Siddhartha Moulik[2], Asim K. Ghosh[3], Reddi Kamesh[2*] and Anirban Roy[1]

[1]*Water Energy Nexus Laboratory, Department of Chemical Engineering Goa, BITS Pilani, Goa, India*
[2]*Department of Process Engineering and Technology Transfer, CSIR-Indian Institute of Chemical Technology, Hyderabad, India*
[3]*Desalination and Membrane Technology Division, Bhabha Atomic Research Centre, Trombay, Mumbai, India*

Abstract

The current state-of-art desalination technologies leave little scope for improvement in terms of technology or energetics in obtaining potable water. However, the challenge lies in the discharge of the high saline brine water back to the sea that would adversely affect marine life and environment. The current study is focused on the MLD (Minimum Liquid Discharge) and ZLD (Zero Liquid Discharge) options by integrating Crystallizer with MSF-OT (Multi-Stage Flash Once-Through) consisting of 15 stages. The authors have considered different combinations of crystallizer with MSF-OT to achieve the maximum recovery of potable water and to study the thermal energy required by the system. Additionally, the thermal energy required for desalination of saline water was calculated for different salinities with MSF-OT. Simulations were carried out using Aspen Plus, ELECNRTL property method to recover potable water and precious salts (NaCl, KCl, $MgCl_2$). The thermal demand for desalination of seawater (35000 ppm) turned out to be 69.763 kWh/m^3 with 47.92% of water recovery while for 2,15000 ppm, the water recovered and thermal energy required by the plant was 38.984% and 72.355 kWh/m^3 respectively. These results were further compared with the minimum energy demand for desalination irrespective of the technologies and it was observed that there is an enormous gap

Corresponding author: kamesh@iict.res.in

Siddhartha Moulik, Aditi Mullick and Anirban Roy (eds.) Sustainable Water Treatment: Advances and Technological Interventions, (487–516) © 2022 Scrivener Publishing LLC

between the numbers. Therefore, reducing the thermal energy is the utmost next challenging task which when overcome could be a better alternative to reverse osmosis membranes. Also, when the crystallizer was integrated with MSF-OT, the best possible case revealed the solid fraction of precious minerals to be 1 with 57.25% of the recovery of water. Yet, the thermal energy requirement turned out to be extremely high (349.264 kWh/m^3).

Keywords: Desalination, multi-stage flash (MSF-OT), crystallizer, zero liquid discharge (ZLD), minimum liquid discharge (MLD)

13.1 Introduction

The importance of water can never be overemphasized under the current circumstances [1, 2]. The human race has traversed a journey from one century to the next and experienced the search for energy through petroleum to make way to search for water through desalination. Interestingly, what was thought to be a utility has now started being considered as a resource and soon water is going to be traded on Wall Street. Through advanced research and manufacturing solar power has become extremely affordable and together with wind is well placed to replace the non-renewables in certain sectors [3–6]. However, water has no alternative, and hence energy at the cost of water and water at the cost of energy, i.e., Water-Energy Nexus (WEN) is the approach adopted by researchers and engineers worldwide to provide holistic solutions [7–9].

It is well established that desalination is one alternative that can support the surging water demands [10, 11]. This is also well documented in the publication trends (Figure 13.1) [12]. It is reported that since 1980, around 16,000 journals have been published exploring the technological, economic, environmental, and socio-political angle of the technology. In terms of technology, it can be seen that ever since Loeb and Sourirajan developed the first phase inversion membrane, reverse osmosis (RO) has always been the most popular choice for the desalination of seawater [13–15]. Thermal alternatives like multi-effect distillation (MED) and multi-stage flash (MSF) also have been extremely promising along with other emerging technologies like forward osmosis (FO) [16–18]. However, thermodynamics plays an important part in the popularity of technology. The theoretical minimum energy of desalination of seawater is 1.04 kWh/m^3, whereas the state-of-the-art RO desalination units work at 2-3 kWh/m^3 (excluding the pre and post-treatment energies). The same values for

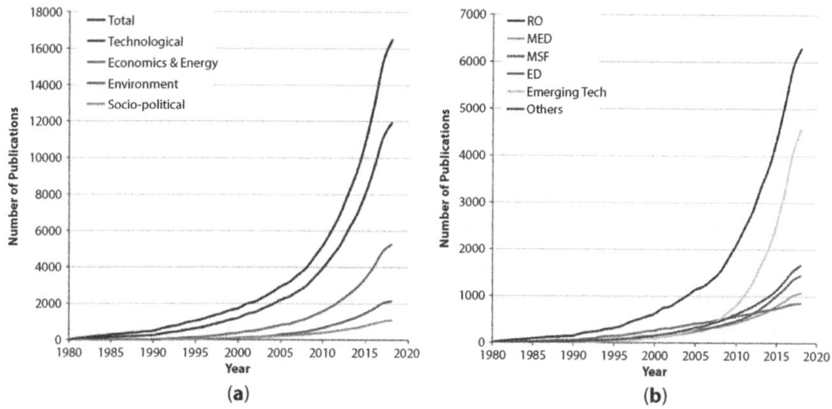

Figure 13.1 (a) Number of desalination publications by categorisation (total, technical, social, environment, energy & economic), (b) number of publications by type of desalination technology (reverse osmosis [RO], multi-effect distillation [MED], multi-stage flash [MSF], electrodialysis [ED]), emerging technologies (nanofiltration [NF], forward osmosis [FO] and membrane distillation [MD]) and other (humidification-dehumidification [HDH], solar stills [SS] and vapour compression [VC]) (adapted from ref. 9 with permission).

MSF and MED are higher (20-30 kWh/m^3). However, MSF and MED have stood the test of time due to the robust performance and ability to handle high salinity streams. Currently, distillation processes produce 50% of the worldwide desalination capacity of which 84% is produced by the MSF technology [19]. Hence, it is not surprising that MSF publications have also shown a steady growth (Figure 13.1b).

The authors opine that as seawater desalination has reached maturity worldwide [20], it is time that the next big challenge is addressed and that is handling brine. This is one issue that can raise serious doubts over the sustainability of desalination. The authors delved into the literature and found that a thorough and detailed Aspen simulation of brine handling flowsheet is not reported, which becomes a challenge when quick decisions on process parameters have to be made. This current effort is undertaken to develop an understanding of this challenge and answer the following objectives: (i) developing a Zero Liquid Discharge flowsheet using seawater as feed which finally ends with crystallizer, (ii) detailed description on modeling the unit operations, and (iii) analyzing the energetics of such a process along with the water recovery. The novelty of the current work lies in the simulation of the ZLD plant using Aspen with detailed performance characteristics and energy analysis. The authors feel that the paper will help engineers and researchers develop related strategies for various streams and understand the feasibility of the same. To the best of the authors' knowledge, this is a unique report on crystallizer integration with MSF using Aspen and validation of same using reported literature. The paper is organized as follows: (i) derivation of minimum energy demand (in terms of work and heat) irrespective of desalination technology are described in Section 13.2, (ii) process description and detailed model simulation of MSF and MSF coupled with crystallizers in Aspen Plus and input parameters are reported in Section 13.3, (iii) results and discussion (Section 13.4), and (iv) conclusion (Section 13.5).

13.2 Minimum Energy Required for Desalination Process

Thermodynamics helps in the determination of the minimum requirement of energy to carry out the desalination process irrespective of the technology used [21]. The minimum work is calculated by assuming the separation as a reversible process. Theoretical calculations of minimum energy demand give us the practical vision of how much energy has to be reduced to improve the economic efficiencies along with the thermodynamics of the desalination plant.

Minimum energy can be calculated by using Gibbs free energy and it can be transferred to the system in two ways, heat or work as shown in Figure 13.2. Heat is the transfer of thermal energy while work is the transfer of mechanical energy. Equation (13.1) represents the equation of Gibbs free energy which would be equal to the chemical potential at constant temperature and pressure. Also because the separation process is non-spontaneous,

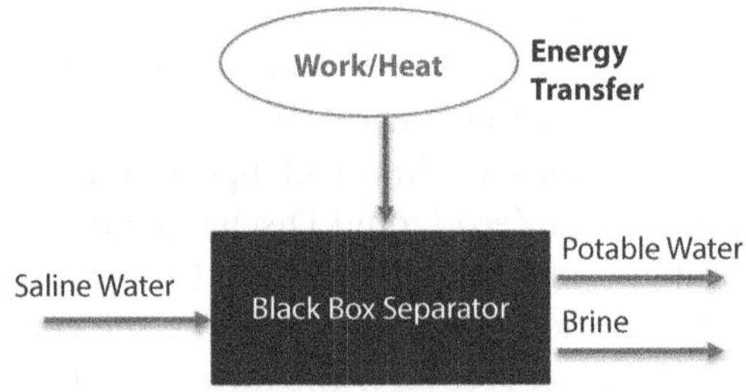

Figure 13.2 Heat or work input for separation process.

Gibbs free energy will be positive in nature. Gibbs free energy of separation can be calculated by Equation (13.2). Additionally, due to the reason that Gibbs free energy gives us minimum work for separation, saline water cannot be separated with energy less than Gibbs free energy of separation [22–25]. As discussed earlier, the least energy of separation can be calculated by two-mode, (i) least work of separation, (ii) least heat of separation.

$$\Delta G = -S\Delta T + V\Delta P + \Sigma\mu\Delta n_i \quad (13.1)$$

$$\Delta G_{sep} = \Delta G_{feed} - \Delta G_{brine} - \Delta G_{product} \quad (13.2)$$

Least Work of Separation: Figure 13.3a represents the separation process via work (W_{sep}) where to satisfy the first and second law of thermodynamics, heat equivalent to ambient temperature ($\dot{Q}_0$) is considered as an input as shown in equation (13.3) and equation (13.4). Multiplying equation (13.4) by temperature (T_a) and subtracting from equation (13.3), we get the equation (13.5). Entropy generated (S_{gen}) in equation (13.5) may be caused by mixing, fast reaction, friction, etc., which is an irreversible process. To obtain the least work of separation, entropy generated will be equal to zero (separation is assumed to be a reversible process), and the minimum work required for separation is represented in equation (13.6).

$$\dot{W}sep + \dot{Q} + \dot{H}_f = \dot{H}_b + \dot{H}_p \quad (13.3)$$

$$\frac{\dot{Q}}{T_a} + \dot{S}_f + \dot{S}_{gen} = \dot{S}_b + \dot{S}_p \quad (13.4)$$

$$\dot{W}sep = \dot{G}_b + \dot{G}_p - \dot{G}_f + T_a\dot{S}_{gen} \quad (13.5)$$

Case: Work of Separation = reversible (least) work of separation

$$\dot{S}_{gen} = 0, \qquad \dot{W}sep = \dot{G}_b + \dot{G}_p - \dot{G}_f \quad (13.6)$$

Least Heat of Separation: Figure 13.3b represents the same black box separator in which instead of work, heat is supplied from an external source, and the minimum heat required for the desalination process is shown by equation (13.7). The equation signifies that the minimum heat required for separation is always greater than the minimum work required for the same external source of energy.

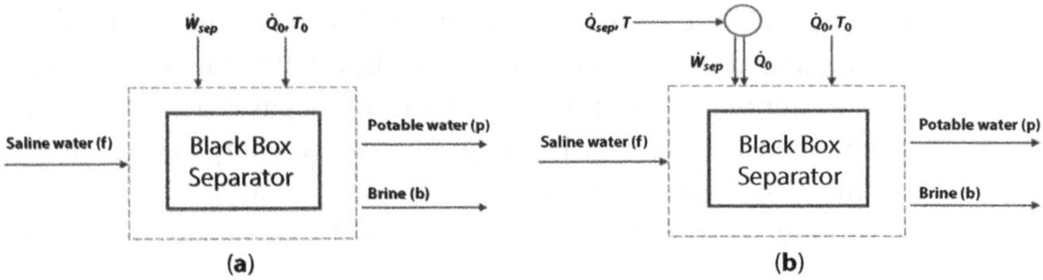

Figure 13.3 (a) Least work of separation; (b) least heat of separation.

$$q_{\min} = w_{\min}\left(1-\frac{T_0}{T}\right)^{-1} \tag{13.7}$$

13.2.1 Minimum Work Requirement

The minimum work input for the separation process is represented by equation (13.8) where R is the universal gas constant, T_0 is the temperature of saline feed water, N is the molar flow rate of component i, and x_i is the mole fraction of component i in the mixture. Since our feed is composed of saline water, the equation of the two-component mixture is represented by equation (13.9). Furthermore, separation of two outgoing impure streams (for instance, brine and product) requires an additional work that is mentioned in form of equation (13.10) and (13.11). For partial separation, the minimum work required to be done is shown by equation (13.12)

$$\dot{W}_{min} = -RT_0 \sum \dot{N}_l \ln x_i \tag{13.8}$$

$$\dot{W}_{min} = -RT_0(\dot{N}_S \ln x_S + \dot{N}_W \ln x_W) \tag{13.9}$$

$$\dot{W}_{min,BRINE} = -RT_0(\dot{N}_{S,BRINE} \ln x_{S,BRINE} + \dot{N}_W \ln x_{W,BRINE}) \tag{13.10}$$

$$\dot{W}_{min,PRODUCT} = -RT_0(\dot{N}_{S,PRODUCT} \ln x_{S,PRODUCT} + \dot{N}_W \ln x_{W,PRODUCT} \tag{13.11}$$

$$\dot{W}_{min} = \dot{W}_{complete} - (\dot{W}_{min,BRINE} + \dot{W}_{min,PRODUCT}) \tag{13.12}$$

By substituting equation (13.9), (13.10), (13.11) in (13.12) we get equation (13.13)

$$\dot{W}_{min} = RT_0\left[\dot{N}_{BRINE}\left(x_{S,BRINE} \ln \frac{x_{S,BRINE}}{x_S} + x_{W,BRINE} \ln \frac{x_{W,BRINE}}{x_W}\right) + \dot{N}_{PRODUCT}\left(x_{S,PRODUCT} \ln \frac{x_{S,PRODUCT}}{x_S} + x_{W,PRODUCT} \ln \frac{x_{W,PRODUCT}}{x_W}\right)\right] \tag{13.13}$$

To calculate the minimum work input per unit mole of freshwater produced, equation (13.13) is divided by $\dot{N}_{PRODUCT}$. Also, the molar flow rate of brine and product are shown by equations (13.14) and equation (13.15).

$$x_s\dot{N}_m = x_{s,BRINE}\dot{N}_{BRINE} + x_{s,PRODUCT}\dot{N}_{PRODUCT} \tag{13.14}$$

$$x_w\dot{N}_m = x_{w,BRINE}\dot{N}_{BRINE} + x_{w,PRODUCT}\dot{N}_{PRODUCT} \tag{13.15}$$

Solving for $\dot{N}_{BRINE}$ and $\dot{N}_{PRODUCT}$ from equation (13.14) and (13.15) and substituting it in equation (13.13), the resultant equation is divided by $M_{PRODUCT}$ to get minimum work input per unit mass of produced water and the final equation is represented as equation (13.16).

$$\dot{W}_{min} = \frac{RT_0}{M_{PRODUCT}}\left[\frac{x_s x_{w,PRODUCT} - x_w x_{s,PRODUCT}}{x_w x_{s,BRINE} - x_s x_{w,BRINE}}\left(x_{S,BRINE} \ln \frac{x_{S,BRINE}}{x_S} + x_{W,BRINE} \ln \frac{x_{W,BRINE}}{x_W}\right) + x_{S,PRODUCT} \ln \frac{x_{S,PRODUCT}}{x_S} + x_{W,PRODUCT} \ln \frac{x_{W,PRODUCT}}{x_W}\right] \tag{13.16}$$

13.2.2 Recovery Ratio

The recovery ratio plays an important role in estimating the overall performance of the plant, the more potable water produced with less consumption of energy, the more efficient is the plant. It is the ratio of the mass flow rate of outgoing product water to the mass flow rate of incoming saline water (in terms of mass fraction) as mentioned in equation (13.17). Substituting equation (13.18) in (13.19), we get the required equation depicted in equation (13.19) [26, 27].

$$r_r = \frac{\dot{m}_{PRODUCT}}{\dot{m}_M} \tag{13.17}$$

$$mf_s \dot{m}_M = mf_{s,BRINE} \dot{m}_{BRINE} + mf_{s,PRODUCT} \dot{m}_{PRODUCT} \tag{13.18}$$

$$mf_{s,BRINE} = \frac{mf_s - r_r mf_{s,PRODUCT}}{1 - r_r} \tag{13.19}$$

13.3 Methodology and Simulation

13.3.1 MSF Process Description

Multi-Stage Flash Once-Through (MSF-OT) consist of two sections, heat recovery section and brine heater, as depicted in Figure 13.4 [28] and works

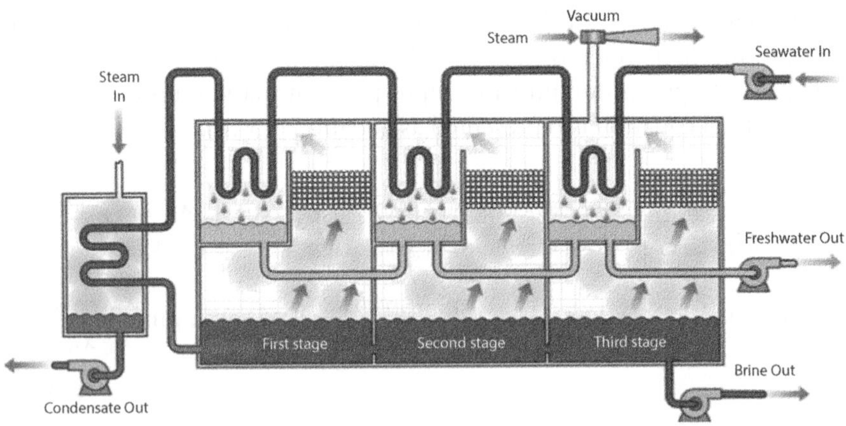

Figure 13.4 MSF-OT desalination system (adapted from ref. [18] with permission).

on the principle of evaporation at low pressure and condensation by the release of heat via latent heat of condensation at distinct temperatures [29]. It is formed by the attachment of Single-Stage Flash (SSF) in series while their physical properties remain the same. The seawater enters the tube via tube bundles located at the top of the MSF system and flows through the N-Stages towards the brine heater. Meanwhile, the water inside the tubes gets heated due to the heat exchange with the vapor outside the tubes that get condensed because of the release of the latent heat of condensation. After the seawater reaches the end of the tube (1st stage of MSF) as shown in Figure 13.4, the water flows through the brine heater and gets heated to about 90-120°C called Top Brine Temperature (TBT) by the external source of heat, i.e., steam. Once the seawater attains TBT, the brine water enters the first stage of the MSF where it is flashed at low pressure, and after passing through the demister, the vapor gets condensed and form distilled water which is collected in a distillate tray [30].

As the water flows towards the last stage of MSF, the temperature and the pressure of the stage decreases. Every stage in the system consists of a demister that wouldn't allow the entrainment of the brine droplets with the flashed vapor. Once the vapor is flashed, the brine water enters the next stage through orifice and baffles that are installed in between the stages. The motive of their installment in between the stages is to reduce the brine's pressure to the pressure required for boiling and flashing at the next flash system. The remaining concentrated brine from the last stage of MSF is generally discharged to the sea.

There are often non-condensable gases present in raw seawater like O_2 and CO_2 that would increase the resistance of the heat transfer, decrease the flow rates, and could also lead to corrosion. To solve this purpose, the deaerator is designed that would eject the non-condensable gases before it enters the brine heater. Also, antiscalants are added in the raw water during pretreatment to avoid corrosion before it enters the MSF system [31–34].

13.3.2 Crystallizer Process Description

The crystallization process is an old technique that was originally used for the extraction of salt from the seawater. The main principle behind the crystallization process is based on the limited solubility of the solute in the solvent at a specific temperature and pressure. According to the literature review, the solubility of most of the solute increases with the increase in temperature; however, the solubility curve of NaCl (in grams) as a solute in 100 grams of water is nearly flat with an increase in temperature. Therefore, with the increase in temperature, the solubility of salt wouldn't increase which would assist us to recover maximum water from the crystallizer at high temperature leaving solid residue behind [35].

13.3.3 Modeling and Simulation

The simulations were performed on a Dell laptop with Intel(R) Core (TM) i5-5200U processor, using ASPENONE ® for Universities r6, Media version 11.1 – ESD kit (Agreement number: 77856, Customer number: 37586).

The simulation of Multi-Stage Flash (MSF-OT (Once-Through)) 15 Stages with crystallizer was carried out using Aspen Plus software. The flowsheet consists of the Heat Recovery Section, Brine Heater, Crystallizer, Vapor Compressor, and Heat Exchanger models. The thermodynamic property of the mixture was represented by the ELECNRTL model. A lot of literature that the authors have considered while writing this article suffered from the following drawbacks: (i) a detailed description of MSF modeling is not available, (ii) component-wise analysis of section in MSF is not reported, and (iii) similar data for crystallizer is also not available. Keeping these considerations in mind the authors have delved into deeper details and have explained each component in the following sections. The assumptions while creating the model are (i) the distillate produced are salt-free, (ii) the heat produced while mixing is considered negligible, and (iii) no heat losses [29].

Single-Stage Flash (SSF): As already mentioned, MSF is formed by increasing the number of stages of SSF. In the SSF system shown in Figure 13.5a, brine (Flashing Brine In) enters the flash system and flashes at a particular temperature and pressure. The flashed vapor gets condensed and the heat released due to the latent heat of condensation is gained by the

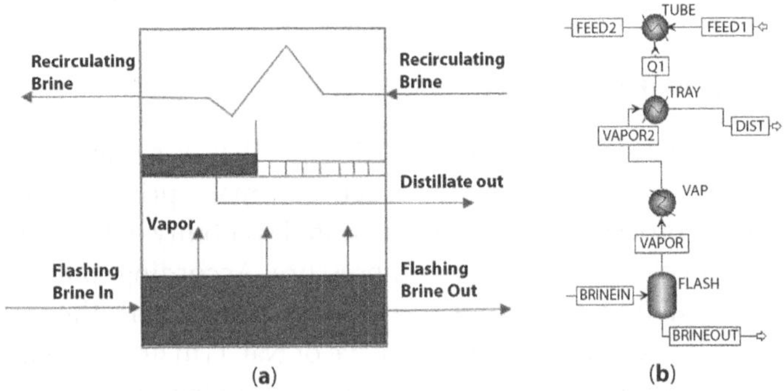

Figure 13.5 (a) Single-stage flash (b) the aspen plus model of single-stage flash.

feed seawater (Recirculating Brine) that passes through the tubes inside the system. The remaining brine water is rejected to the outlet stream (Flashing Brine Out) towards the next stage of MSF whereas the feed water (Recirculating Brine) moves towards the brine heater. Single-Stage Flash was simulated in Aspen Plus software using the FLASH 2 model for flash drum, HEATER for VAPOR, TRAY, and TUBE as shown in Figure 13.5b. Likewise, SSF were connected in continuity to build 15 Stages.

According to Figure 13.5b, the brine (BRINEIN) enters the Flash system (FLASH) where the vapor gets flashed. In the actual flash system, the salt-free vapor passes through demister, but due to the lack of features in Aspen, the demister is not considered during simulations. The reason for designing HEATER for vapor (VAP) is that the temperature of the vapor is 1 or 2 °C less than the temperature of brine in the flash model. The vapor formed is passed to the tray (TRAY), where due to latent heat of condensation, the vapor gets condensed to liquid form, and the condensed water is separated via DIST stream. Meanwhile, the heat released during condensation (Q1) gets exchanged with the water inside the tube (TUBE) which gets heated before it flows to the tube of the next stage or towards the brine heater (FEED2). During the process, brine that is not flashed is rejected via stream (BRINEOUT) which would enter the next flash system.

Deaerator and Brine Heater: As discussed in previous section, a deaerator is used to remove non-condensable gases like O_2, CO_2, etc. The deaerator (DA) as shown in Figure 13.6 is modeled using FLASH 2 and it is located before the raw water enters the brine heater (BHEATER). The top outlet stream of the DA consists of NC (Non-Condensable gases) and a fraction of water that is lost with non-condensable gases during evaporation while the bottom stream leads to BHEATER.

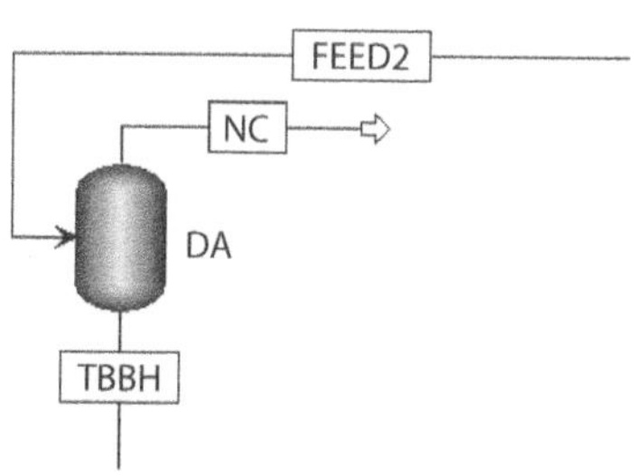

Figure 13.6 Deaerator.

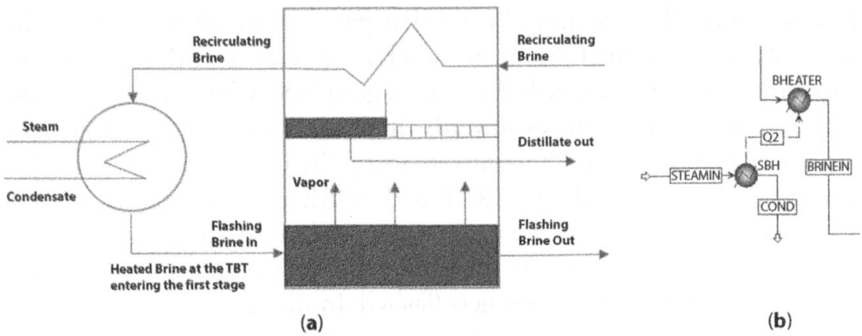

Figure 13.7 (a) Actual plant flowsheet (b) aspen plus flowsheet.

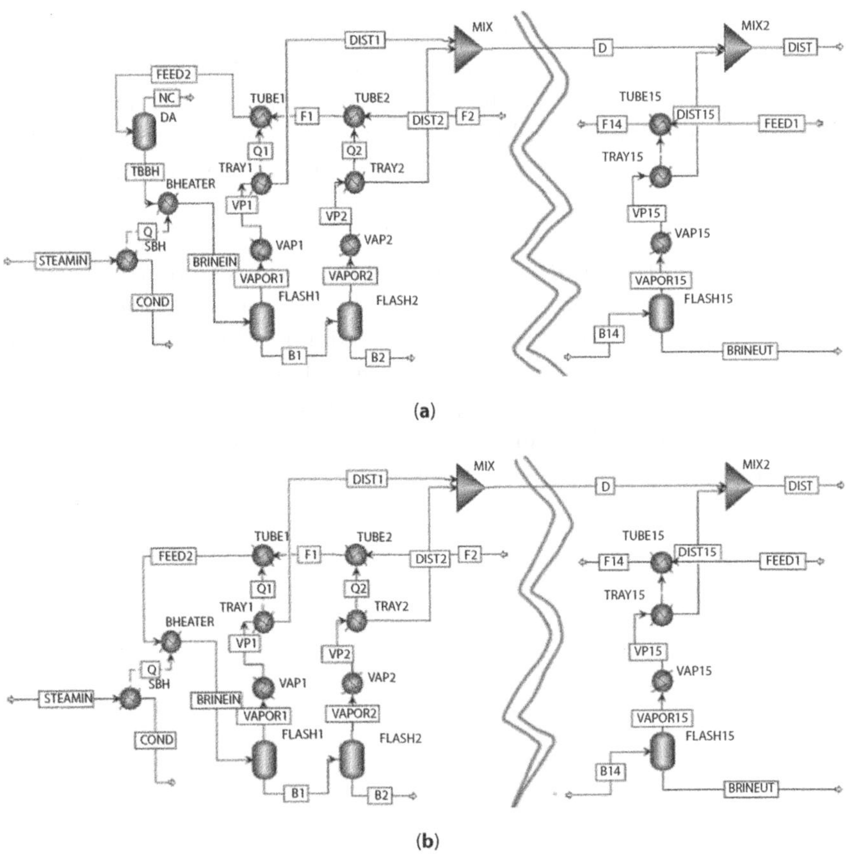

Figure 13.8 (a) MSF-OT with deaerator (b) MSF-OT without deaerator.

A BHEATER is used to heat the seawater to TBT that enters after its exit from the tube bundles of the first stage of MSF (FEED2) (in absence of deaerator) or the bottom stream of the deaerator (TBBH) (in presence of deaerator). The TBT is attained in the brine heater by the external source, i.e., steam. The brine heater model was designed by using the HEATER model. The comparison of the actual plant flowsheet and Aspen Plus model is shown in Figure 13.7a and 13.7b, respectively. As mentioned earlier, the role of BHEATER is to heat the seawater extensively to 90-120°C called TBT, and the heat supplied to BHEATER is from the heat exchanger SBH (Q2) which condenses the steam and heat released due to latent heat of condensation is passed to the BHEATER. Figure 13.8a and 13.8b depicts the Aspen Plus MSF-OT model with and without deaerator, respectively.

Crystallizer: The crystallizer is modeled with a vapor compressor, heat exchanger system, mixer, and splitter. During the simulation, the crystallizer was integrated with MSF in four different techniques shown in Figure 13.9a, 13.9b, and 13.9c.

Case I includes the crystallizer that is used to crystallize the dissolved salts with no recycle stream in the system as shown in Figure 13.9a, while Case II and Case III (Figure 13.9b and 13.9c) involve the recycling streams in order to get maximum solid fraction and potable water. The above three cases include deaerator. Moreover, the components used for crystallization in the second and third case consists of crystallizer, vapor compressor, mixer, splitter, and two heat exchangers. The purpose of using a vapor compressor after crystallizer is to saturate vapor by increasing the pressure which would help in exchanging heat with the recycling stream in the heat exchanger. At the same time, Case IV is designed with no recycling stream and deaerator.

Case I as shown in Figure 13.9a is modelled using FLASH 2 (CRYST) and HEATER (HX). The HX is used to condense the vapor (WATER) to liquid state (C) before it is mixed with potable water that is pumped from the MSF distillate trays from all stages. The concentrated brine (BRINEUT) discharged from the last stage of MSF (15th Stage) enters the CRYST, where the maximum amount of salt-free water is recovered through the C stream and the salt in the solid-state is taken out from the SOLID stream.

Figure 13.9b and 13.9c consists of recycling streams with a slight difference in the design. In both the simulations, the FLASH 2 model is used to represent crystallizer, while VC represents Vapor Compressor, HX1 and HX are heat exchangers and MIX is MIXER while SPLIT is the splitter. The concentrated brine in Figure 13.9b that is discharged from the BRINEUT

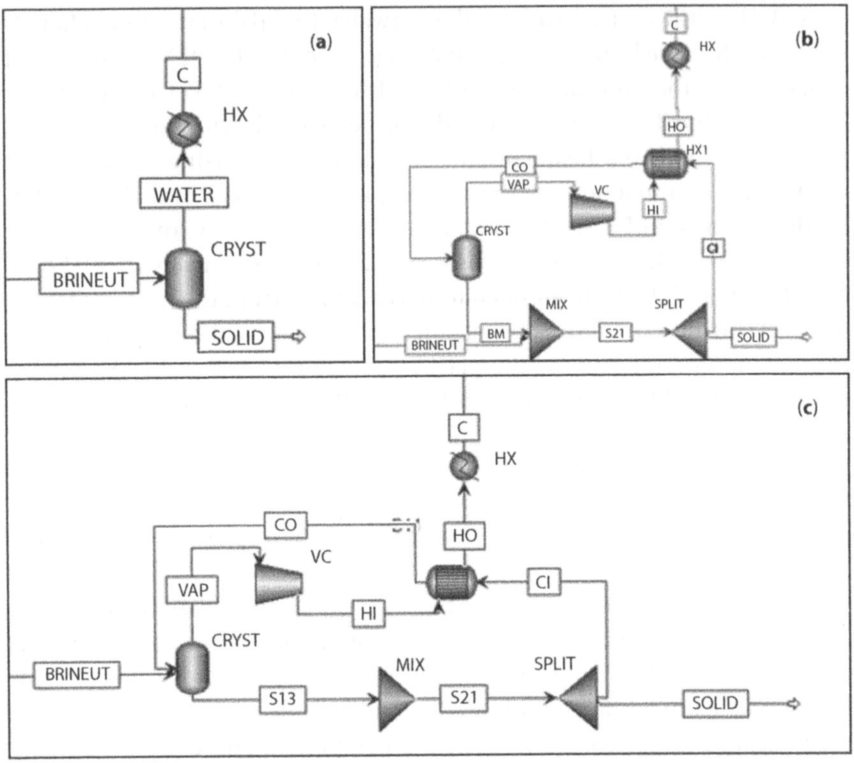

Figure 13.9 (a) Case I (b) case II (c) case III.

stream is mixed with (BM) stream that is rejected from the bottom outlet of the crystallizer (CRYST) in Mixer (MIX) which then proceeds towards the splitter (SPLIT) that split the fraction of concentrated brine back to crystallizer through the heat exchanger (HX1). Meanwhile, the vapor (VAP) flashed from the CRYST enters in the VC which compresses the vapor that eventually increases the temperature and pressure of the saturated vapor. After that, the saturated vapor passes through the HX where it gets condensed to the liquid phase and the stream is merged with the DIST stream. HI, HO, CI, CO represents hot inlet, hot outlet, cold inlet, and cold outlet, respectively.

The only difference between Case II and Case III is that in the third case the BRINEUT stream is not mixed with BM in the mixer, instead the BRINEUT stream directly enters CRYST, and stream S13 from the bottom outlet stream proceeds towards the MIX followed by SPLIT.

13.3.4 Input Parameters

Increase in salinity of seawater: Firstly, MSF was simulated in order to determine the thermal energy demand with an increase in the saline concentration of feed water. The flow rate of feed was kept at 31,25,000 kg/hr at 35 °C. The mass fraction of NaCl was varied from 0.035 (35000 ppm) to 0.215 (215000 ppm) while water consisted of the remaining mass fraction. Besides that, the GOR (Gain Output ratio) (mass of water produced/ mass of steam used) was kept 9 throughout the simulation. The other input parameters of MSF are mentioned in Table 13.1, Table 13.2 and the steam input value is shown in Table 13.3 [29]. In this case, the system didn't consist of a crystallizer and deaerator. The reason for not considering the deaerator in the flowsheet is that we only considered NaCl and water as the components during simulation.

MSF integrated with Crystallizer: When MSF-15 stages were integrated with crystallizer, the seawater composition was taken into account as shown in Table 13.4. The temperature of the seawater was set at 35°C while the flow rate of the feed was 246772 kg/hr [36]. The input variables of the flash system remained the same as shown in Table 13.1 and Table 13.2. Also, Deaerator (DA) was set at 86°C and 0.25 vapor fraction.

The overall reaction taking place inside the system were introduced in the Aspen Plus as follows:

Table 13.1 Input variables of MSF-15 stages.

Flash Stages	1	2	3	4	5	6	7	8	9
Temperature (°C)	88.9	86.7	84.4	82	79.6	77.3	75	72.7	70.5
Vapor Faction	0.02	0.02	0.03	0.05	0.05	0.05	0.05	0.05	0.05
Vapor									
Temperature (°C)	87.9	85.7	83.4	81	78.6	76.3	74	71.7	69.5
Vapor Fraction	1	1	1	1	1	1	1	1	1
Tray									
Temperature (°C)	87.7	85.5	83.2	80.8	78.4	76.1	73.8	71.5	69.3
Vapor Fraction	0	0	0	0	0	0	0	0	0
Tube									
Pressure (bar)	0.55	0.53	0.51	0.49	0.47	0.45	0.43	0.41	0.39

Table 13.2 Input variables of MSF-15 stages.

Flash	10	11	12	13	14	15
Temperature (°C)	68.3	66.2	64.1	62	59.9	57.8
Vapor Fraction	0.05	0.05	0.05	0.05	0.05	0.05
Vapor						
Temperature (°C)	67.3	65.2	63.1	60.9	58.8	56.6
Vapor Fraction	1	1	1	1	1	1
Tray						
Temperature (°C)	67.1	65	62.9	60.7	58.6	56.4
Vapor Fraction	0	0	0	0	0	0
Tube						
Pressure (bar)	0.37	0.35	0.33	0.31	0.29	0.27

Table 13.3 Input variables of steam.

Block	**Input value**
1. Steam Inlet	Temperature- 101.5°C Vapour Fraction- 0
2. Steam Heater	Temperature- 100°C Vapour Fraction- 0
3. Brine Heater	Pressure- 0.702 bar

$$Na^{+} + Cl^{-} \rightarrow NaCl(s) \tag{13.20}$$

$$Mg^{2+} + Cl^{-} \rightarrow MgCl_2(s) \tag{13.21}$$

$$K^{+} + Cl^{-} \rightarrow KCl(s) \tag{13.22}$$

The ionic reactions mentioned in equations (13.20), (13.21), and (13.22) were generated by using the ELEC WIZARD function in the properties section. The input parameters of the crystallizer for four cases are given in below Table 13.5.

Table 13.4 Seawater composition.

Mass flow rate	Kg/hr
H_2O	197870
NA+	14602
Cl-	30362
CO_2	232
O_2	213
K+	1167
MG++	2326

Table 13.5 Input parameters of crystallizer.

Crystallizer	Case I/ IV	Case II	Case III
Temperature (°C)	92	92	92
Vapor Compressor (VC) Isentropic		2 bar	2 bar
Splitter (SOLID stream)		0.2	0.2
Heat Exchanger (HX1)		HO Temperature decreases 85 °C	HO Temperature decreases 85°C
Temperature (°C)		25	25
Vapor Fraction		0	0

<u>Calculation of Thermal Energy and water recovery:</u> Aspen Plus does not directly calculate and display the thermal energy of the simulated model. Therefore, thermal energy for all the simulations were calculated using equation (13.23), and the units were set as kWh/m^3. According to the equation, the Heat duty (cal/sec) of all the components were added and converted in kW which was later divided with the potable water produced (in m^3/hr). While the percentage of water recovered was calculated with the use of equation (13.24).

$$Thermal\ Energy = \frac{Heat\ Duty}{Distilled\ water} \tag{13.23}$$

$$Water\ recovered(\%) = \frac{Distilled\ water\ produced}{Feed\ Water} \tag{13.24}$$

13.4 Results and Discussion

13.4.1 Comparison of Energy Demand Between Simulated Model and Theoretical Model

As previously discussed the derivation for minimum energy required to attain desalination irrespective of desalination technology, Figure 13.10a represents the minimum demand of energy in terms of work and heat for saline concentration 35000 ppm (seawater) and 215000 ppm (hypersaline water) with different recovery ratio. Although only the estimation of thermal consumption is required while simulating MSF, because minimum heat is proportional to minimum work as expressed in equation (13.7), we would be focusing on both work and heat demand to compare the deviation between the two modes of energy in order to look for room of improvement to increase the efficiency of the plant.

According to the graph, the minimum work required to treat seawater at 0% recovery is 0.437 kWh/m^3 while minimum thermal energy is 2.508 kWh/m^3 which is five times more than the minimum work required. Since in our simulation, water recovered while treating seawater turned out to be 47.92%, theoretical work, and heat for the same parameter were between 0.564 (40% recovery) - 0.681 (60% recovery) kWh/m^3 and 3.238 (40% recovery) -3.912 (60% recovery) kWh/m^3. While for hypersaline water (2,15000 ppm), with the use of equation (13.21), (13.24), and (13.12), the minimum work and heat for 0% recovery were 3.197 kWh/m^3 and 18.347 kWh/m^3 respectively. However, the potable water recovered was 38.98% according to Table 13.6, so the minimum work and heat must lie between 3.66 (20% recovery) - 4.39 (40% recovery) kWh/m^3 and 21.042 (20% recovery) -25.203 (40% recovery) kWh/m^3 respectively. Thermal consumption was seven times more than the work required in treating hypersaline water; this is due to the presence of extensively high saline concentration because of which boiling point significantly rises, eventually demanding

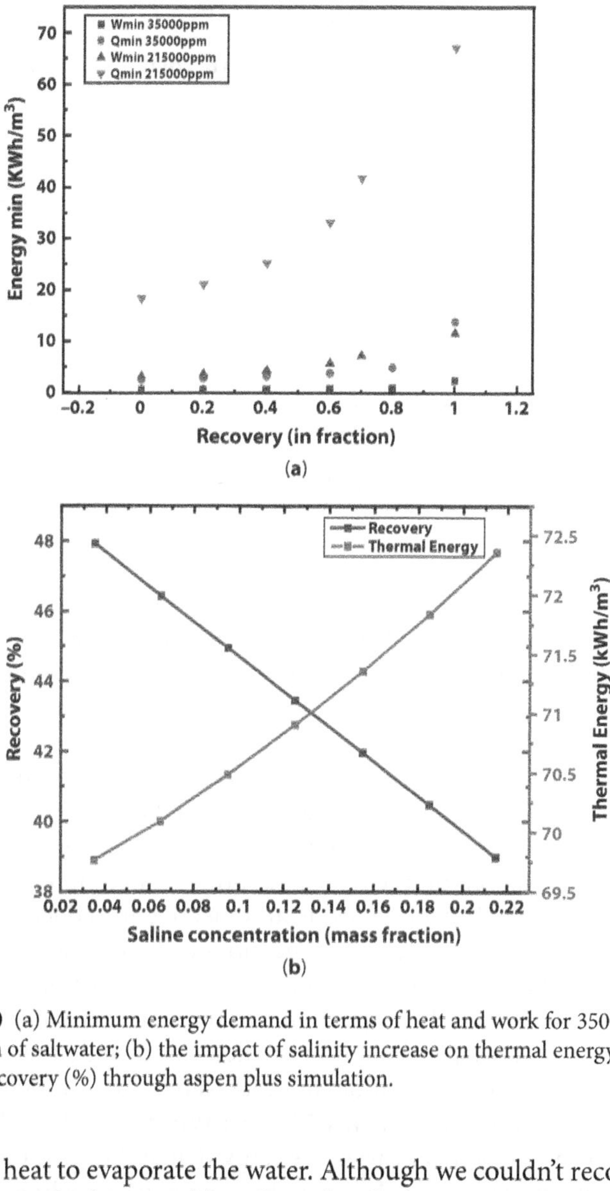

Figure 13.10 (a) Minimum energy demand in terms of heat and work for 35000 ppm and 2,15000 ppm of saltwater; (b) the impact of salinity increase on thermal energy (kWh/m^3) and water recovery (%) through aspen plus simulation.

additional heat to evaporate the water. Although we couldn't recover 100% of feed water, which is possible only in the ideal case, in such cases the minimum work and heat required for seawater and hypersaline came out to be 2.419 kWh/m^3, 13.882 kWh/m^3, 11.704 kWh/m^3, and 67.163 kWh/m^3 respectively. Moreover, these cases are commonly termed as Zero Liquid Discharge (ZLD) where feed water is completely recovered. We would discuss the energy consumption for ZLD in the next section.

Table 13.6 Thermal energy & water recovery (%).

S. no	Salinity (mass fraction)	Distilled water (m^3/hr)	Thermal energy (kWh/m^3)	Water recovery (%)
1	0.035	1497.58	69.7637	47.92
2	0.065	1451.02	70.0902	46.43
3	0.095	1404.47	70.4829	44.94
4	0.125	1357.91	70.9031	43.45
5	0.155	1311.35	71.3532	41.96
6	0.185	1264.8	71.8358	40.47
7	0.215	1218.24	72.3559	38.98

For the simulated MSF-OT model, the saline concentration was increased from 35000 ppm to 2,15000 ppm. The inlet stream only consisted of NaCl and water and steam input value varied for different compositions of the feed stream to maintain the GOR 9. As seen from Figure 13.10b, for treating seawater, 69.76 kWh/m^3 of thermal energy was required and 47.92% of the water was recovered. While comparing it with the results obtained through theoretical calculation, we could observe that practically, thermal requirement is 17 (approx.) times more than theoretical obtained value. This provides us an objective to reduce the thermal energy demand to make the system economically feasible. According to the Figure 13.10b, saline concentration is directly proportional to the thermal energy while inversely proportional to the water recovery. Moreover, as the salinity increased, water recovery decreased while thermal energy rose. When saline concentration of feed water was 2,15000 ppm, the thermal energy consumption was 72.356 kWh/m^3 which is three times higher than the theoretical thermal energy demand with water recovery of 38.98%. It can be observed that with an increase in saline concentration, difference between simulated thermal energy and calculated value decline which bring us to the light that MSF can be suitable for treating hypersaline water in prospect of thermal energy. Otherwise, our motive could be to increase the efficiency of the plant by reducing thermal energy to minimum heat consumption.

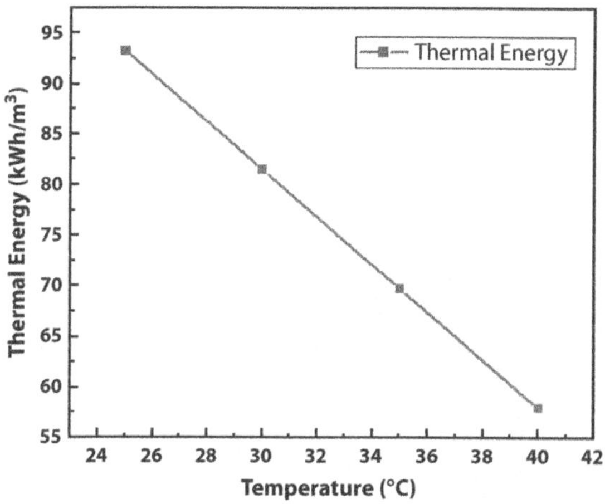

Figure 13.11 Effect of temperature on thermal energy.

13.4.2 Impact of Temperature and Flowrate on Thermal Energy

The temperature of feedwater is one of the factors that influences the thermal demand of the whole system. Figure 13.11 depicts the relation between the temperature of feed water and the thermal energy demand (kWh/m^3). According to the figure, with an increase in the temperature of feed water, the demand for thermal energy decreases. For instance, if the feed water temperature was set at 25°C, 93.229 kWh/m^3 of thermal energy was required which is more than the thermal energy demand simulated for 35 °C (69.7637 kWh/m^3) while for 45°C, thermal energy turned out to be 57.977 kWh/m^3. It can also be concluded that as the difference between the temperature of feed water and brine water get narrower, thermal demand decreases. As for flowrate, the rise of flowrate didn't affect the thermal energy because the recovery ratio was fixed (temperature and vapor fraction of each column were fixed). Therefore, water that was not recovered ultimately proceeded toward the brine discharge.

13.4.3 Impact on Thermal Energy During MLD and ZLD

As discussed in Section 13.3.3, four cases were considered while simulating the crystallizer with MSF. The simulation in all four cases were run when the vapor fraction of the crystallizer was set at 0.5 and 1 in order to have a clear vision on the requirement of thermal energy when the system was

Table 13.7 Requirement of thermal energy on integration of crystallizer with MSF.

Case	Vapor fraction	Solid fraction	Distilled water (m^3/hr)	Heat duty (mcal/sec)	Thermal energy (kWh/m^3)	Water recovery (%)
I	0.5	0.2395	114.574	10.635	388.399	46.429
	1	1	141.281	11.221	332.314	57.251
II	0.5	0.345	124.425	8.016	365.241	50.421
	1	0.407	129.088	10.402	374.219	52.310
III	0.5	0.520	135.452	7.382	337.882	54.889
	1	1	141.281	10.365	349.264	57.251
IV	0.5	0.167	159.777	1.615	42.311	64.746
	1	1	198.315	2.003	42.245	80.363

operated at MLD (vapor fraction 0.5) and ZLD (vapor fraction 1). The literature related to MLD and ZLD along with the energy demand were studied and validated with the existing published work [37, 38].

Table 13.7 shows the demand for thermal energy in four cases when the crystallizer is integrated with MSF. In case I, where no recycling stream was involved, thermal energy demand to carry out MLD was 388.399 kWh/ m^3 along with the 46.42% of water recovery. But when vapor faction was turned 1, heat duty rose from 10.635 Mcal/sec to 11.22 Mcal/sec in order to completely evaporate feed water with an increase in water recovery to 57.25% (23% more than in MLD case). The thermal energy came out to be 332.314 kWh/m^3 which was less than the thermal demand in MLD because according to equation (13.23), the potable water recovered (in the denominator) dominates which in turn decreases the output value. Additionally, the solid fraction (salts recovered) in MLD was 0.239 while all the salts were recovered (solid fraction 1) during ZLD.

Case II and Case III consist of a recycling stream but in Case II the stream coming out from the last stage of the MSF connects to the mixer while in Case III, the BRINEUT stream connects directly to the crystallizer. Unlike in Case I, thermal energy in case II (vapor fraction of crystallizer set at 0.5 &1) was highest among all the cases with consumption 365.241 kWh/ m^3 and 374.219 kWh/m^3 respectively. Besides, water recovery during ZLD rose to 3.7% as compared to MLD, and even though salts recovered during MLD were more than in case I, all salts were not recovered during ZLD. Hence, Case II was the most unfavorable among all cases, while the MLD process in Case III was the most favorable case among all with salt recovery and water recovery to 0.520 and 54.889%, respectively. Also, thermal energy consumption was less as compared to Case I by 13.17%. However, for ZLD purposes, Case I was still better than Case III.

Lastly, in Case IV where no deaerator and recycling stream were involved, maximum water recovery was observed during MLD and ZLD. Additional to the potable water, it had all the non-condensable gases in it and because the distilled water produced was extremely high, thermal energy being inversely proportional, heat consumption for MLD and ZLD were 42.31 kWh/m^3 and 42.245 kWh/m^3 respectively. Even though thermal energy consumption is extremely low and water recovery percentage is high, (64.74% at vapor fraction 0.5 and 80.363% at vapor faction 1) the presence of non-condensable gases in the system could lead to the decrease in efficiency of the plant. Hence, it is not advisable to consider case IV even when the results obtained are best among all.

Table 13.8 Crystallization of salts.

Case	**Case I**		**Case II**		**Case III**		**Case IV**	
Vapor Fraction	0.5	1	0.5	1	0.5	1	0.5	1
Salt (kmol/hr))								
NaCl	599.079	635.156	629.808	634.542	634.801	635.156	538.667	635.156
$MgCl_2$	0	95.703	0	0	0	95.703	0	95.703
KCl	0	29.847	26.090	29.275	29.486	29.847	0	29.847
Total solid fraction	0.239	1	0.345	0.407	0.520	1	0.167	1

13.4.4 Crystallization of Salts

From the previous section, it was discovered that Case III was most favorable during MLD while all the salt were recovered during ZLD in Cases I, III, IV. The salts considered during simulation were sodium chloride (NaCl), Magnesium Chloride ($MgCl_2$), and potassium chloride (KCl). Table 13.8 illustrates the amount of salts recovered (in kmol/hr) in all cases during MLD and ZLD. From the table we could observe that NaCl was the easiest to be recovered among all cases but the maximum recovery was seen in Case III (634.801 kmol/hr) for MLD and all the NaCL was recovered in Cases I, III and IV during ZLD (635.156 kmol/hr). While $MgCl_2$ was not recovered in any case during MLD but in ZLD except for Case II, all the $MgCl_2$ was recovered. Interestingly, KCl seemed to be recovered in MLD only when recycling streams were involved, though Case III achieved the highest recovery (0.520 kmol/hr), at the same time similar to previous cases, KCl was totally recovered during ZLD in Cases I, III, and IV. Overall, in term of salt recovery, Case III turned out to provide maximum salts among all cases whereas Case II was the most unfavorable case.

13.5 Conclusion

It is of utmost importance that for sustainability of an important process like desalination, the brine handling strategy and process should be developed and understood comprehensively. An energy-efficient technology capable of handling RO brine, recover maximum water and value-added minerals is the "holy grail" of desalination. This work is a small step towards understanding the same. For the simulated MSF-OT, it was observed that for ~ 48% recovery the thermal energy demand was around 70 kWh/m^3. This is in close agreement with reported literature and hence proves the validity of the simulation. Next, the feed salinity was increased to 0.215 mass fraction (215,000 ppm) the recovery drastically decreased to around 39% with thermal energy of around 72 kWh/m^3. The MSF-OT flowsheet was then integrated with a crystallizer to cater to the ZLD strategy. It was observed that for successful ZLD, the optimum case was Case III, with maximum water recovery of 57.25% with thermal energy 349.26 kWh/m^3, while the least favorable was Case II when the vapor fraction of crystallizer was set as 1 (100% recovery), maximum salt recovery was not met and thermal energy requirement was the highest. In Case IV, the thermal energy requirement was least with 42.24 kWh/m^3 with maximum recovery of 80.36%, but due to the presence of non-condensable gases, it was not considered as a desirable alternative. Thus, this paper serves as a facile method to model the ZLD approach using Aspen and understand the performance for any feed stream for quick decision making. The future direction of this work would delve into simulating the same using open-source software like DWSIM and comparing the same with Aspen.

13.6 Acknowledgment

Anirban would like to thank the OPERA award (FR/SCM/230117/CHE, dt. 19/08/2017) by BITS Pilani Goa for carrying out the work. "Dr. Siddhartha Moulik would like to thank the Director, CSIR-IICT for the support [Ref. IICT/Pubs./2022/118]".

References

1. S.D. Alexandratos, N. Barak, D. Bauer, F.T. Davidson, B.R. Gibney, S.S. Hubbard, H.L. Taft, P. Westerhof, Sustaining Water Resources: Environmental and Economic Impact, ACS *Sustainable Chemistry and Engineering.* 7 (2019) 2879–2888. https://doi.org/10.1021/acssuschemeng.8b05859.
2. S. Chakraborty, A. Mukherjee, S. Das, N.R. Maddela, S. Iram, P. Das, Study on isotherm, kinetics, and thermodynamics of adsorption of crystal violet dye by calcium oxide modified fly ash, *Environmental Engineering Research.* 26 (2021) 1–9. https://doi.org/10.4491/eer.2019.372.
3. A. Al-Karaghouli, L.L. Kazmerski, Energy consumption and water production cost of conventional and renewable-energy-powered desalination processes, *Renewable and Sustainable Energy Reviews.* 24 (2013) 343–356. https://doi.org/10.1016/j.rser.2012.12.064.
4. N. Ghaffour, J. Bundschuh, H. Mahmoudi, M.F.A. Goosen, Renewable energy-driven desalination technologies: A comprehensive review on challenges and potential applications of integrated systems, *Desalination.* 356 (2015) 94–114. https://doi.org/10.1016/j.desal.2014.10.024.
5. C. Koroneos, A. Dompros, G. Roumbas, Renewable energy driven desalination systems modelling, *Journal of Cleaner Production.* 15 (2007) 449–464. https://doi.org/10.1016/j.jclepro.2005.07.017.
6. S.A.K.-P. in energy, combustion science, undefined 2005, Seawater desalination using renewable energy sources, Elsevier. (n.d.). https://www.sciencedirect.com/science/article/pii/S0360128505000146.
7. J.L. Schnoor, Water-energy nexus, *Environmental Science and Technology.* 45 (2011) 5065. https://doi.org/10.1021/es2016632.
8. C. Olkis, H. Dong, S. Brandani, G. Santori, Ionogels at the Water-Energy Nexus for Desalination Powered by Ultralow-Grade Heat, *Environmental*

Science and Technology. 54 (2020) 3591–3598. https://doi.org/10.1021/acs.est.9b06037.

9. L.M. Rehman, R. Dey, Z. Lai, A.K. Ghosh, A. ROY, A Reliable & Novel approach based on Thermodynamic Property estimation of Low to High Salinities Aqueous Sodium Chloride Solutions for Water-Energy Nexus (WEN) Applications, *Industrial & Engineering Chemistry Research*. (2020). https://doi.org/10.1021/acs.iecr.0c02575.
10. K. Roy, A. Mukherjee, N.R. Maddela, S. Chakraborty, B. Shen, M. Li, D. Du, Y. Peng, F. Lu, L.C. Garciá Cruzatty, Outlook on the bottleneck of carbon nanotube in desalination and membrane-based water treatment-A review, *Journal of Environmental Chemical Engineering*. 8 (2020) 103572. https://doi.org/10.1016/j.jece.2019.103572.
11. S. Lanjewar, A. Mukherjee, P. Khandewal, A.K. Ghosh, A. Mullick, S. Moulik, A. Roy, Thermodynamics of synthesis and separation performance of Interfacially polymerized "loose" reverse osmosis membrane: Benchmarking for greywater treatment, *Chemical Engineering Journal*. (2020) 127929. https://doi.org/10.1016/j.cej.2020.127929.
12. E. Jones, M. Qadir, M.T.H. van Vliet, V. Smakhtin, S. mu Kang, The state of desalination and brine production: A global outlook, *Science of the Total Environment*. 657 (2019) 1343–1356. https://doi.org/10.1016/j.scitotenv.2018.12.076.
13. S. Loeb, LOEB-SOURIRAJAN MEMBRANE: HOW IT CAME ABOUT, in: *ACS Symposium Series, ACS*, 1981: pp. 1–9. https://doi.org/10.1021/bk-1981-0153.ch001.
14. S. Loeb, L. Titelman, E. Korngold, J.F.-J. of membrane science, undefined 1997, Effect of porous support fabric on osmosis through a Loeb-Sourirajan type asymmetric membrane, Elsevier. (n.d.). https://www.sciencedirect.com/science/article/pii/S0376738896003547.
15. M.T. So, F.R. Eirich, H. Strathmann, R.W. Baker, Preparation of asymmetric loeb-sourirajan membranes, *Journal of Polymer Science: Polymer Letters Edition*. 11 (1973) 201–205. https://doi.org/10.1002/pol.1973.130110311.
16. O.J.M.- Desalination, undefined 1993, Design and operating comparison of MSF and MED systems, Elsevier. (n.d.). https://www.sciencedirect.com/science/article/pii/0011916493800977.
17. S. Ihm, O.Y. Al-Najdi, O.A. Hamed, G. Jun, H.C.- Desalination, undefined 2016, Energy cost comparison between MSF, MED and SWRO: Case studies for dual purpose plants, Elsevier. (n.d.). https://www.sciencedirect.com/science/article/pii/S0011916416307263.
18. R.G. Raluy, L.M. Serra, J. Uche, LCA of Different Commercial Desali-nation Technologies (MSF, MED, RO), *Int J LCA*. 10 (2005) 285–293. https://doi.org/10.1065/lca2004.09.179.2.
19. A.Y. Hoekstra, M.M. Mekonnen, A.K. Chapagain, R.E. Mathews, B.D. Richter, Global monthly water scarcity: Blue water footprints versus blue

water availability, *PLoS ONE*. 7 (2012). https://doi.org/10.1371/journal.pone.0032688.

20. L.M. Rehman, A. Mukherjee, Z. Lai, A. Roy, Membrane Technology, in: *Modeling in Membranes and Membrane-Based Processes*, Wiley, 2020: pp. 327–373. https://doi.org/10.1002/9781119536260.ch10.
21. A. Mukherjee, S. Lanjewar, R. Kumar, A. Chakraborty, A. Abdelrasoul, A. Roy, Role of Thermodynamics and Membrane Separations in Water-Energy Nexus, in: *Modeling in Membranes and Membrane-Based Processes*, Wiley, 2020: pp. 145–199. https://doi.org/10.1002/9781119536260.ch4.
22. L. Wang, C. Violet, R.M. Duchanois, M. Elimelech, Derivation of the Theoretical Minimum Energy of Separation of Desalination Processes (2020). https://doi.org/10.1021/acs.jchemed.0c01194.
23. E.S. Karakas, Y. Cengel, Y. Çerçi, Improving the Thermodynamic and Economic Efficiencies of Desalination Plants Minimum Work Required for Desalination and Case Studies of Four Working Plants Study of Taylor-Couette-Poiseuille Flow at Submerged Cryogenic Induction Motor Air Gap View project Design and Development of a Jet Exducer for Two-Phase Cryogenic Expander View project, n.d. https://www.researchgate.net/publication/326891832.
24. K.H. Mistry, R.K. McGovern, G.P. Thiel, E.K. Summers, S.M. Zubair, J.H. Lienhard, Entropy generation analysis of desalination technologies, *Entropy*. 13 (2011) 1829–1864. https://doi.org/10.3390/e13101829.
25. K.H. Mistry, J.H. Lienhard, Generalized least energy of separation for desalination and other chemical separation processes, *Entropy*. 15 (2013) 2046–2080. https://doi.org/10.3390/e15062046.
26. E.S. Karakas, Y. Cengel, Y. Çerçi, Improving the Thermodynamic and Economic Efficiencies of Desalination Plants Minimum Work Required for Desalination and Case Studies of Four Working Plants Study of Taylor-Couette-Poiseuille Flow at Submerged Cryogenic Induction Motor Air Gap View project Design and Development of a Jet Exducer for Two-Phase Cryogenic Expander View project, n.d. https://www.researchgate.net/publication/326891832.
27. L. Wang, C. Violet, R.M. Duchanois, M. Elimelech, Derivation of the Theoretical Minimum Energy of Separation of Desalination Processes, *Journal of Chemical Education*. 97 (2020) 4361–4369. https://doi.org/10.1021/acs.jchemed.0c01194.
28. A. Al-Karaghouli, L.L. Kazmerski, Energy consumption and water production cost of conventional and renewable-energy-powered desalination processes, *Renewable and Sustainable Energy Reviews*. 24 (2013) 343–356. https://doi.org/10.1016/j.rser.2012.12.064.
29. K.A. Al-Shayji, Modeling, simulation, and optimization of large-scale commercial desalination plants (1998). https://vtechworks.lib.vt.edu/handle/10919/30462.

30. W. Lian-ying, X. Sheng-nan, G. Cong-jie, Simulation of Multi-stage Flash (MSF) Desalination Process, *Advances in Materials Physics and Chemistry*. 02 (2012) 200–205. https://doi.org/10.4236/ampc.2012.24b052.
31. H. Al-Fulaij, A. Cipollina, H. Ettouney, D. Bogle, Simulation of stability and dynamics of multistage flash desalination, *Desalination*. 281 (2011) 404–412. https://doi.org/10.1016/j.desal.2011.08.012.
32. M. Al-Sahali, H. Ettouney, Developments in thermal desalination processes: Design, energy, and costing aspects, *Desalination*. 214 (2007) 227–240. https://doi.org/10.1016/j.desal.2006.08.020.
33. N.M. Abdel-Jabbar, H.M. Qiblawey, F.S. Mjalli, H. Ettouney, Simulation of large capacity MSF brine circulation plants, *Desalination*. 204 (2007) 501–514. https://doi.org/10.1016/j.desal.2006.02.047.
34. W. Lian-ying, X. Sheng-nan, G. Cong-jie, Simulation of Multi-stage Flash (MSF) Desalination Process, *Advances in Materials Physics and Chemistry*. 02 (2012) 200–205. https://doi.org/10.4236/ampc.2012.24b052.
35. P. Bharmoria, H. Gupta, V.P. Mohandas, P.K. Ghosh, A. Kumar, Temperature invariance of NaCl solubility in water: Inferences from salt-water cluster behavior of NaCl, KCl, and NH4Cl, *Journal of Physical Chemistry B*. 116 (2012) 11712–11719. https://doi.org/10.1021/jp307261g.
36. A.M. Rempel, K.G. Berger, E.A. Gensib, A.N. Walker, MgCl2 and KCl Recovery from Brine Wastewater, (2016). http://repository.upenn.edu/cbe_sdrhttp://repository.upenn.edu/cbe_sdr/82.
37. M. Ahmad, P. Williams, Assessment of desalination technologies for high saline brine applications - Discussion paper, *Desalination and Water Treatment*. 30 (2011) 22–36. https://doi.org/10.5004/dwt.2011.1374.
38. V.C. Onishi, J.A. Reyes-Labarta, J.A. Caballero, Membrane desalination in shale gas industry: Applications and perspectives, in: *Current Trends and Future Developments on (Bio-) Membranes: Membrane Desalination Systems: The Next Generation*, Elsevier, 2018: pp. 243–267. https://doi.org/10.1016/B978-0-12-813551-8.00010-3.

14

A Critical Review on Prospects and Challenges in "Conceptualization to Technology Transfer" for Nutrient Recovery from Municipal Wastewater

Shubham Lanjewar[1], Birupakshya Mishra[2], Anupam Mukherjee[1,2], Aditi Mullick[2], Siddhartha Moulik[2]* and Anirban Roy[1]

[1]*Water-Energy Nexus Lab, Department of Chemical Engineering Goa, BITS Pilani, Goa, India*
[2]*Cavitation and Dynamics Lab, Department of Process Engineering and Technology Transfer, CSIR-Indian Institute of Chemical Technology, Hyderabad, India*

Abstract

Municipal wastewater is being viewed as a source of nutrients than a waste stream. A variety of nutrients like nitrogen and phosphorus are available aplenty, and large volumes of discharge ensure the same high availability. The challenge lies in developing processes that can recover the nutrients from such streams since untreated discharge of wastewater results in eutrophication and accelerated eutrophication, baneful for the aquatic ecosystem. This paper reviews the possible technologies which can be utilized for recovering water and nutrients from wastewater streams, particularly municipal wastewater. The recovery of nutrients and recovery of water is constrained by laws of thermodynamics and involves energy. Hence, it is of utmost importance that such recovery is carried out at minimum energy penalties. Keeping this in perspective, the work also reviews technologies that can harvest energy from waste streams. Summarizing the approach, the current work develops a framework through available and developed technologies which can be used to (i) recover nutrients, (ii) recover water, and (iii) generate power from municipal wastewater streams. A closed-loop operation can result in a self-consistent approach where the generation of wastewater and recovery of water and nutrients can be achieved, minimizing wastage and carbon footprint.

**Corresponding author*: smoulik@iict.res.in

Siddhartha Moulik, Aditi Mullick and Anirban Roy (eds.) Sustainable Water Treatment: Advances and Technological Interventions, (517–566) © 2022 Scrivener Publishing LLC

Keywords: Resource recovery, wastewater treatment, eutrophication, circular economy

14.1 Introduction

With the growth in population worldwide, there is a sharp increase in the wastewater generated from various sources [1]. This wastewater is rich in various natural resources such as nitrogen, phosphorous,etc. When deposited in stagnant water bodies such as lakes or ponds cause a natural phenomenon called eutrophication [2, 3]. It occurs as simple plants such as planktons or algae growth in water bodies take up dissolved oxygen from water leading to less oxygen in water bodies, leading to asphyxiation in aquatic organisms. Also, the growth spurt leads to less or sometimes no sunlight inside the water body leading to the dissolved plants not able to perform photosynthesis. The surface blooms occasionally release some toxins in water bodies leading to the death of the aquatic life. Hence it becomes crucial to check the concentration of various minerals in treated water before leaving them out into the open [2]. In the previous decade, there has been an upsurge in the number of sewage treatment plants worldwide, out of which the majority are not meant for nutrient recovery. As a result, the extent of eutrophication has increased, and the rate has escalated, leading to accelerated eutrophication. The best method to curb the situation would be to extract the minerals and resources from the water before discharging it [4, 5].

The demand for freshwater can be met by various means, such as desalination of seawater or treatment of existing wastewater [6]. If energy demands are taken into consideration, desalination of seawater would prove futile. Moreover, the release of highly concentrated salt solution, required for desalination would again need to be treated before the release into environment. Hence for a sustainable growth in future it must be observed that the wastewater which was previously regarded as unusable shall be used to economically extract various minerals from it by using methodologies based on the laws of thermodynamics [7]. This ensures the successful extraction along with the expenditure of minimum energy making its reusability favorable. There exists a plethora of technologies that may recover the same [8].

There are variety of methods chemically to extract the resources economically such as, crystallization or adsorption [9]. Compounds such as a mixture of calcium and magnesium can render the resources to be precipitated, which can further be treated simply to remove phosphorous and even re-usage of magnesium ions which tend to be adding more on the cost factor of this method. In this perspective of sustainable extraction, the reusability of the magnesium ions aid in maintaining the cost whilst the extracted phosphorous can be reused Some studies reveal salt water to be used for the recovery of materials too [10]. With the exception of space and time requirements biological treatment of resource extraction can also be a decent process for extraction of resources from wastewater. It employs certain plants and microorganisms for the uptake of the resources which on short span of time incorporate it into biomass which can be utilized in agricultural fields as source of manure [5]. While there is no such high requirement of energy in both chemical or biological treatment, there would be faint requirement for energy for irrigation in case of biological treatment and stirring in case of chemical methods. Citing both the above to be slow and can only be operated in batch formation only, hence the moderate requirement of energy for the usage in membrane method of removal of resources is seen to be a suitable method for extraction of materials from the water. Various types of membranes having different opening or pore sizes can be utilized for the removal of minerals having different concentrations in effluents from various sources [11]. The reject from the membranes would be concentrated in the minerals. The fate of the reject marks the sustainability of the process. However good the method can be, the usage of combination method or hybrid method employs the usage of membranes along with some combined chemical or biological method is also used in many places for optimum recycling of minerals. The time and cost requirement of this method is high pertaining to the demand to be combines in multiple processes.

Here in this review article, the methods highlighted above are discussed vividly and establishing clear differentiation between the methods thereby providing a clear image of the processes. Moreover, the process efficiency and other factors are considered to find an approach to maximize resource removal. The approach adopted in this review is (i) a comprehensive review of state-of-the-art technologies, including chemical, biological, membrane, and membrane-based hybrid technologies. (ii) Understanding the inherent challenges in each of the technologies, (iii) understanding and developing the links between individual technologies, and (iv) percentage recovery of resources and related economics.

14.2 Chemical Processes for Resources Recovery

One of the commercial processes for utilizing phosphorus from wastewater effluents as fertilizer is chemical precipitation with metal salts such as magnesium and calcium salts (Figure 14.1). Crystallization can be achieved via chemical precipitation which is essentially a phase change process that converts dissolved components into a particulate, inorganic compound for easy separation from the liquid bulk.

Some more reagents can also be poured into the wastewater during this process. The pH of the solution is maintained to control the precipitation. The ions form bonds with PO_4^{3-}-P and leave a product that settles out as a precipitate from wastewater. Then the removed P can be collected in sludge. On further investigation, more methods have been explored by the researchers, such as the iron reactive filtration system [12]. The chemical processes can be conducted in comparatively compact, stable, and easily operated reactors than the biological processes.

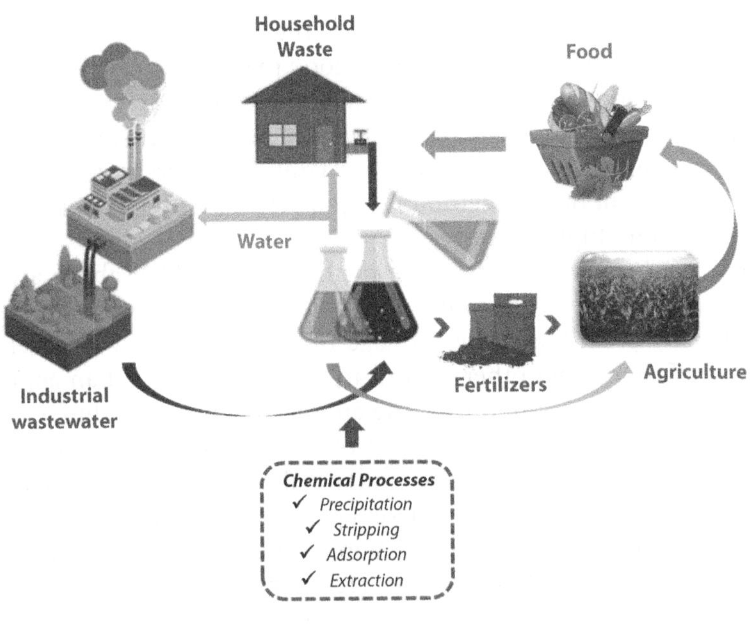

Figure 14.1 Resources recovery by chemical processes.

14.2.1 Chemical Precipitation

14.2.1.1 Magnesium and Calcium – Phosphorous Precipitation

Struvite ($MgNH_4PO_4{\cdot}6H_2O$) is one of the good fertilizers in agriculture, which contains 12 % P and 5 % N with a trace amount of heavy metal [13–15]. It can be produced from nutrient-rich urban/municipal wastewater by adding the required amount of Mg^{2+} salts ($MgCl_2$, $MgSO_4Mg(OH)_2$,etc.) [16]. The addition of magnesium salt increases the pH level of the solution, which reducesthe solubility of PO_4^{3-}, induces supersaturation, and results in precipitation of struvite [17]. The reaction can be described as follows:

$$Mg^{2+} + PO_4^{3-} + NH_4^{+} + 6H_2O \rightarrow MgNH_4PO_4.6H_2O \downarrow$$

The molar ratio of $1(Mg^{2+}):1(NH_4^{+}):1(PO_4^{3-})$ is the most favorable condition for struvite precipitation in water, and other conditions include a lower concentration of suspended solids and pH should be more than 7.5. The addition of Mg^{2+} salt in this process is the most expensive cost component (75 % of the overall production costs) in struvite production. To reduce the processing cost, researchers are looking for an alternative and cost-effective magnesium sources such as seawater, $MgCO_3$–magnesite, bittern solution, etc. [18]. The more economical way of production of struvite could be satisfying magnesium salt demand using seawater. However, the presence of impurity may affect the quality of the final product. For example,the presence of excess calcium may impure struvite with calcium phosphate precipitate.

Calcium-phosphorus precipitation is another widely used method for producing hydroxyapatite ($Ca_5(PO_4)_3OH$), which is a feedstock of the phosphate industry. This is cost-effective precipitationas calcium cost is low. The efficiency of hydroxyapatite production using this process ranged from 75 to 85% [19]. The reaction can be described as follows:

$$5Ca^{2+} + 3PO_4^{3-} + OH^{-} \rightarrow Ca_5(PO_4)_3OH \downarrow$$

TheMg: P: N / Ca: P ratios and favorable pH in the chemical precipitation mainly depend on wastewater composition.Good knowledge of the effect of pH and Mg/Ca selections are required for significant improvement and effectiveness of recovery of nutrients, $MgNH_4PO_3{\bullet}6H_2O$,

$Ca_3(PO_4)_2$, especially $MgNH_4PO_3 \cdot 6H_2O$. This method can be utilized to remove NH_4^+-N and PO_4^3-P.

14.2.1.2 Aluminum – Phosphorous Precipitation

Especially for orthophosphate and condensed phosphate precipitation the aluminum hydroxide, $Al(OH)_3$ is a good adsorption agent. The organic phosphate can only be precipitated at pH 3.6 using this compound. A theoretical study proved that phosphorous should be in excess in the wastewater to precipitate using Aluminum hydroxide. When conventional alum is applied to the wastewater, the orthophosphate removal can be achieved via aluminum hydroxide phosphate precipitation rather than precipitated $AlPO_4$. The reaction can be described as follows.

$$Al^{3+} + H_2PO_4^- \rightarrow AlPO_4 \downarrow + 2H^+$$

Precipitation of phosphate and aluminum was governed more by the integrated particles than $AlPO_4$ and $Al(OH)_3$. A packed bed column of limestone, silica, and activated aluminum oxide after two years also can remove phosphorous over 99% [20]. Another study investigated a 55% reduction of phosphorous in 20 mins of exposure using spent alum sludge [21]. The activated alumina can reduce the lower concentrations (1 mg/L) of phosphorous [22]. The enhancement of adsorption utilizing aluminum sulfate can be done by adding polyelectrolytes, such as clay, tannin, synthetic anionic polyelectrolyte [23]. The massive amount of phosphate can be adsorbed from wastewater by using Hydrotalcite [24]. The aluminum-based precipitates depend on the amount of organic load exist and the type of wastewater. The solid aluminum of variety in surface properties forms on the effect of components present in the water. One such example was the precipitation of aluminum hydroxy phosphate by adding aluminum in the presence of tannic acid into the wastewater. The tannic acid and other organic substances impede the removal of phosphorous.

The impedance was found to increase with increasing concentrations of organic matter. When aluminum is exposed to tannic acid, the organic matter was found to form a coat around the surface of the inorganic solid, preventing adsorption of phosphorus. It should be advantageous to add alum treatment at the end of the cleaning process to harvest more phosphorus when concentrations of organic matter were found to be relatively lower [23].

14.2.1.3 Ferric – Phosphorous Precipitation

The development of better phosphorous precipitation has been reported using ferric cations from industrial byproducts such as blast furnace slag. The compound has a very high phosphorous adsorption capacity [25]. In the literature, bi-layered Iron-based hydroxides were prepared, and the compounds release metal cations (Mg, Ca, and Fe). The released cations and their hydroxides can effectively serve as coagulants for phosphorus removal [26]. The reaction can be described as

$$Fe^{3+} + H_2PO_4^- \rightarrow FePO_4 \downarrow + 2H^+$$

The gradual increase in phosphorous content in sludge is the major drawback of this process [27]. The addition of $FeCl_2$ into the aeration tank can monitor the filamentous accumulation with the removal of phosphate. The ferric salt significantly reduces the phosphate release and production of sulfide through waste by enriching the iron-reducing bacteria selectively over the sulfate-reducing one on leaving behind the other bacterial species [28].

14.2.1.3.1 Reactor Configuration

The reactor configuration is an essential aspect for phosphorous and ammonium recovery through struvite crystallization. The most common types of reactors are stirred batch and continuous reactors. The processes crystallize struvite by adding $MgCl_2$ to obtain the molar ratio Mg:P 1:1. The NaOH adjusts the pH of the solution required for nucleation. The propeller mixes the solutions and enhances struvite crystallization. The settling zone is required in addition to the reactor for particle accumulation. In the batch operation, the crystallization occurs at the lower portion. Another configuration is a continuously stirred tank reactor (CSTR). It was tested at a laboratory level by integrating CSTR and a tank that recycles the fine particles back to the CSTR [29]. Another reactor tank ensures the effect of recycling fine particles on crystal growth. The advantage of a CSTR reactor is its simplicity in operation [29]. The continuous mode causes the fouling over the impeller, which is a significant drawback for an industrial application of CSTR. Mixing speed is kept at 600 rpm as it is required for the growth limitation and vast quantities of fines production and also helps the solution to homogenize and achieve particles in suspension [30].

The most commonly used reactors are fluidized bed reactors in order to crystallize struvite from wastewater. The particles spontaneously precipitate from supernatants on adding chemicals to reach the molar ratio Mg: P: N - 1:1:1. As soon as the nucleation initiates, the small struvite particles interact together (agglomerates) in air-agitated reactors. The liquid flow rates control the particle suspension in the reactor [31, 32]. Considering particle size struvite can be further used as a fertilizer. To let the struvite particles the larger struvite particles can be produced by using seed material so that struvite particles agglomerate with seeds. This method has proved to be efficient for particle sizes ranging from 0.1 mm up to 1.0 mm [33]. High flow rates and high mixing energies to ensure the seeds of the bed are fluidizing continuously are required for successful operation of these processes. The wastewater treatment plants limit the FBR application due to high energy and raw material costs [34]. The data of nutrient recovery achievable by chemical precipitation is as given in Table 14.1.

14.2.2 Adsorption and Ion-Exchange

The N, P and K from the wastewater can be accumulated by adsorption and ion exchange. Used sorbents can be recycled using inexpensive, highly concentrated aqueous cationic or anionic solutions such as chloride, sodium, or sulfate. The main design parameter is the volume of the bed/the amount of liter the given sorbent can handle. For adsorption to occur ions are transferred from solvents to the active sites present on the surface of insoluble and stiff sorbents present in a vessel or column. The sorbent is a material with a porous substructure interlinked with small cavities and a large surface area. This technology is best suitable for the streams containing nutrients between the ranges (1-2000 mg/L). The waste streams mostly from sewage treatment plants and forged lakes having a concentration of nutrients below 5 mg/L have been treated using state of the art polymer-based sorbents and reduced P concentration in the effluent to 50-100 μg/L [62]. For the waste streams above 2000 mg/L concentration, the metal oxide or hydroxide, red mud, and zirconium-based sorbents are reduced P concentration in the effluent to 50-100 μg/L [62]. For the waste streams above 2000 mg/L concentration, the metal oxide or hydroxide, red mud, and zirconium-based sorbents are preferred for the recovery of P, cliptotinolite, and modified zeolite to recover K and N, respectively. The 57 g/Kg P is the highest loading capacities reported for zirconium filled orange waste gel [63]

Table 14.1 Literature data on nutrients recovery through chemical precipitation.

Type of waste water	Source of Mg	Source of P	(Struvite) MAP ratio	pH controlling agent	Optimum pH	Reaction time (Min)	Hydraulic retention time (Min)	Nutrient removal		Reactor configuration	References
Landfill Leachate	MgO Seawater Bittern $MgCl_2.6H_2O$ $MgCl_2.6H_2O$ $MgCl_2.6H_2O$	H_3PO_4 Bone Meal Na_2HPO_4 KH_2PO_4 H_3PO_4	2: 1: 1 1.1: 1: 1.1 1: 1: 1 1.2: 1: 1.2 1: 1: 1	NaOH NaOH NaOH NaOH NaOH	9 9 9 9 8.5	30 15 30 3 -	- - - - - -	95 95 92 99 72	~100 ~99 - 86 -	Batch	[35] [36] [37] [38] [39]
Pretreated Landfill Leachate	Seawater Bittern	Bone Meal	1.3 : 1 : 1.3	NaOH	9	30	-	90	99	Batch	[40]
Synthetic Waste Water	$MgCl_2.6H_2O$ $MgSO_4$ $MgCl_2.6H_2O$ $MgCl_2.6H_2O$ $MgCl_2.6H_2O$ $MgCl_2.6H_2O$ $MgCl_2.6H_2O$	$Na_4H_2PO_4$ Na_2HPO_4 KH_2PO_4 $(NH_4)_2HPO_4$ KH_2PO_4 $(NH_4)_2HPO_4$ KH_2PO_4	0.9: 2.4: 1 1: 9.4: 1 1.3: 4: 1 1.6: NA: 1 1.75: NA: 1 1.6: NA: 1 1: NA: 1	NaOH NaOH NaOH NaOH NaOH NaOH Aeration	8.7 8.5 9.5 7.9 9.5 8.5 8	83 - - - - -	630 125 100 - 600 250 120	- - - - - - -	81.40 88 95.8 >90 >90 83 >91	Batch Cont. Stirred Tank Fluidized Bed Fluidized Bed Fluidized Bed Fluidized Bed Fluidized Bed	[41] [29] [42] [43] [44] [45] [46]
Sewage Sludge	$MgCl_2.6H_2O$	H_3PO_4	1 : 1.5 : 1	NaOH	9	30	-	89.35	95	Batch	[47]
Activated Sludge	$MgCl_2.6H_2O$	-	1.4:1.1:1	NaOH	9.5	-	-	53	75	-	[48]
Synthetic Urine	$MgCl_2.6H_2O$ MgO	KH_2PO_4	1: 4.3: 1 1.4: 1: 1	NaOH NaOH	8.8 9	- -	261 -	- -	90 96.70	Cont. Stirred Tank Batch	[49] [50]
Urine	Seawater	-	1.3: NA: 1	Aeration			120	-	>98.70	Fluidized Bed	[51]
Raw Swine Slurry	MgO	H_3PO_4	1 : 1 : 1	NaOH	10	240	-	77	80	Batch	[52]
Swine Lagoon Liquid	MgO	-	-	$Mg(OH)_2$	8	-	-	-	70	Fluidized Bed	[53]

Table 14.2 Literature data on the ionic concentration of MAP required reaching solubility product values.

pK_{sp}	K_{sp}	Ammonium (mg/L)	Phosphorous (mg/L)	Magnesium (mg/L)	References
12.60	2.50×10^{-13}	1.13	5.98	1.51	[54]
13.15	7.10×10^{-14}	0.74	3.93	0.99	[55]
9.40	3.90×10^{-10}	13.15	69.40	17.53	[56]
13.12	7.50×10^{-14}	0.76	4.01	1.01	[57]
12.60	2.50×10^{-13}	1.13	5.98	1.51	[58]
12.36	4.32×10^{-13}	1.36	7.18	1.81	[59]
12.94	1.5×10^{-13}	0.87	4.62	1.17	[60]
12.26	5.49×10^{-14}	0.68	3.61	0.91	[61]

Table 14.3 Literature data on adsorption of nutrients using different adsorbents.

Adsorbents	Nutrients recovered	Operating conditions		Nutrients adsorption capacity (mg/g)	References
		pH	Temp. (°C)		
Mg-Fe-Zr	Phosphate	7-7.5	50	24.5	[70]
La(III)-Modified Zeolite	Phosphate	6	20	24.6	[71]
Amphoteric Straw Cellulose (ASC)	Ammonium Phosphate	7 5	25 25	68.4 38.6	[72]
Synthetic Hydrotalcite	Phosphate	8.6	25	47.3	[73]
Amine-grafted Magnetic Absorbent	Nitrate Phosphate	-- --	-- --	3.87 0.5	[74]
Clinoptilolite	Ammonium	7-7.5	25	2.7	[75]
Waste Alumina	Nitrate Ammonium Phosphate	8.04	28	0.69 0.04 0.11	[76]
$Mg(OH)_2$ supported by biochar/ (PMRB) bentonite composite	Phosphate Ammonium Humic acid	8	25	125.36 58.20 34.57	[77]
Corncob-based modified Biochar	Ammonium	-	30	12.83	[78]
MgO-modified Palygorskite	Ammonium Phosphate	9.0	-	42.6 69.8	[79]

and 21.5 g/Kg of N for clinoptilolite [64]. The significant advantages of this process are reducing the concentration of P less than 0.1 mgPO_4-P/L from high concentration waste streams and achieving the highest P accumulation [65]. There are no byproducts produced, and the pH remains unchanged throughout the process. The drawbacks of implementing strategy at a large-scale are a requirement of chemicals for the sorbents recovery, bio-fouling, massive quantity of resins for the total removal, and limited life of the resin. Bio-regeneration could be a better strategy to lower down the cost of regeneration [66, 67]. The exchange media or sorbent can be used as nutrient based product finding its use in agricultural lands. Hence, adsorption and ion-exchange are classified as integrated nutrient recovery-accumulation processes [68, 69].

Tables 14.2 and 14.3 describes literature data on the ionic concentration of MAP required reaching solubility product values and adsorption of nutrients using different adsorbents, respectively.

14.3 Biological Processes for Resources Recovery

The nutrient recovery from biological processes involves nutrient accumulation by wetland plants and microorganisms to produce P and N rich biomass and bioenergy. Microphytes and duckweeds are wetland plants having a natural ability to absorb phosphorous from the wastewater. The photoautotrophic microorganisms (i.e., Microalgae) use solar energy to convert inorganic nutrients to organic matter and produce biomass. The biomass production from microalgae culture depends on the reactor technology, microalgae strain, and environmental conditions (Figure 14.2). The reports

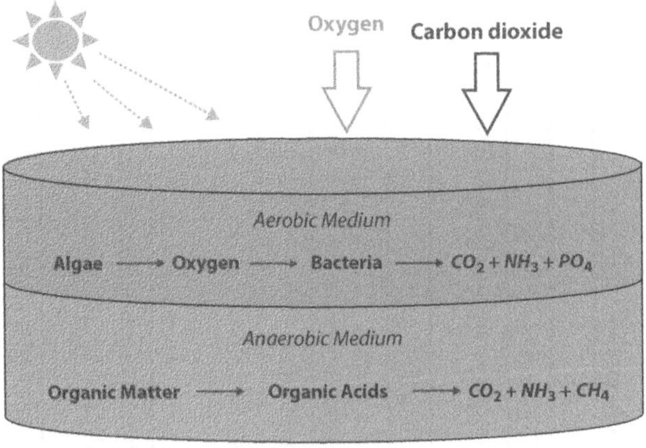

Figure 14.2 Schematic of a biological process for nutrients recovery.

[80, 81] have mentioned the value ranging from 40 to 150 tons/hectare.yr (dry matter). To achieve high production of biomass, a massive amount of nutrients is required. i.e., 100 tons of microalgae biomass production needs 200 tons of CO_2, 10 tons of N, and 1 ton of P. Generally, CO_2 is supplied as pure food-grade, N as nitrate salts, and P as phosphate salts. Chemicals fertilizers are readily used to fulfill the nutrients requirement. These have a production cost of 0.5 €/kg. This cost can be reduced using flue gases and wastewater as a source of nutrients [82]. The nutrient recovery from wastewaters is quite economical, which makes the microalgal process profitable.

The nutrients recovery through biological processes mainly depends on the nutrient composition in wastewater and suitable microorganisms for their removal and recovery. The phosphates can be recovered by the natural method through their accumulation in activated sludge. The polyphoshorous accumulating organisms perform well in the process. The anaerobic conditions are favorable for the cells to release phosphate into the solution by these organisms. Hence, the water gets rich in phosphate content. The Mg^{2+} and K^+ also release during the process and get enriched in the wastewater. The energy produced by this process is used to intake sources of carbon (some volatile fatty acids (VFAs)) and holds up in the form of poly-β-hydroxy alkanoates. The organisms hold the biomass in aerobic conditions, PHAs supply the energy required for the activity. The biomass carries the metal ions adsorbed on it and exists in the form of leftover sludge. On a final note, biological phosphate recovery can be categorized as (1) the release of phosphate and accumulation in anaerobic conditions in wastewater; (2) the storage of phosphate under the aerobic environment in the activated sludge. Several European countries have banned biological phosphate recovery as the sludge generated contains high phosphate. Apart from that, it includes pathogens, and heavy metals may cause the infertility of agricultural land. Hence, it is prohibited from direct use on land.

14.3.1 Anammox Process for Nutrients Recovery

One of the effective nitrogen removal processes is anammox (anaerobic ammonia oxidation). In 1999, the Delft Technical University had studied the efficacy of the process in an aerated reactor [83]. The anammox bacteria belongs to the category of chemolithoautotrophic bacteria (a phylum of Planctomycetes). NO_2^- is used as an electron acceptor prior to AAOB oxidizing NH_4^+ [84]. The AAOB converts NO_2^- into NH_2OH, besides when NH_4^+ gets oxidized to N_2H_4 with an electron acceptor produces N_2 [85]. The high-strength organic loading rate and ammonium, lesser carbon addition, minimal energy input, and low sludge production make the Anammox process more efficient than the other biological processes [86].

It offers good prospects and feasibility for the treatment of liquid manure. The researchers have treated different types of wastewater containing animal manure using this process and obtained excellent NO_2^--N and NH_4^+-N removal [87]. The report by Yagin and his team [88] mentioned 80% of NH_4^+-N and 60%-80% of COD removal from animal wastewater using the Anammox process. The microbial community, temperature, pH, nitrate concentration, ammonium, inorganic and organic carbon are the critical parameters for this process [89]. The temperature range of 35-40 °C and pH 6.5-8.8 are the optimum conditions for AAOB [84].

14.3.2 Algal Methods for Sewage Treatment and Nutrient Recovery

The algal methods to treat sewage waste have advantages over other bacterial processes like sludge accumulation and denitrification-nitrification processes. The stoichiometric C: N: P ratios are the main reason for sewage as it is deficient in carbon compared to the heterotrophic biomass. Hence, the heterotrophic metabolism of carbon observed is limited in N and P removal from the activated sludge [90, 91].

14.3.2.1 Nutrients Recovery from Micro-Algae Growth

Figure 14.3 explains the micro-algae culture reaction and resources recovery. Microalgae is one of the proven environmentally-friendly and traditional biofuel feedstocks. Sterile conditions favor their growth. It retains

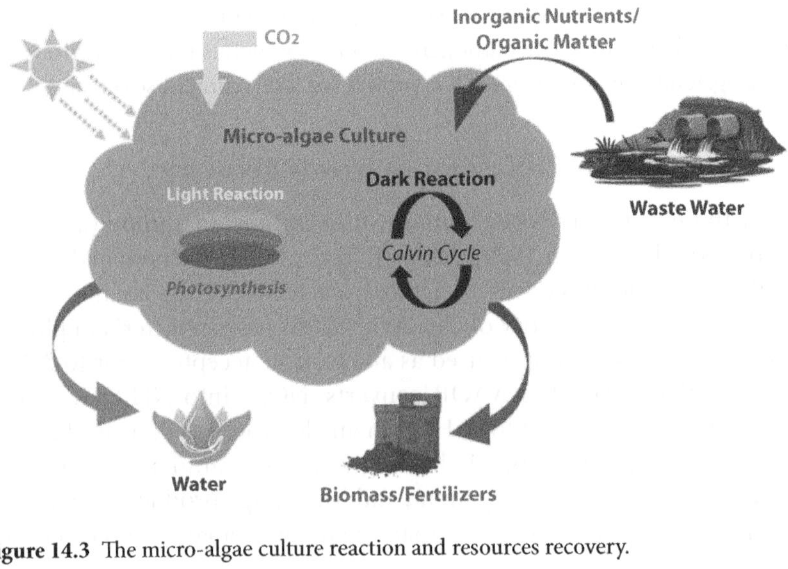

Figure 14.3 The micro-algae culture reaction and resources recovery.

Table 14.4 Literature data on nutrients recovery from microalgae.

Microalgae	Source of wastewater	Operating conditions			Performance			References
		Reactor configuration	Cultivation period (days)	pH	N-Recovery (%)	P-Recovery (%)	Specific growth rate (d^{-1})	
C. Vulgaris	Synthetic municipal wastewater Undiluted cattle wastewater	PBR PBR	2 4	8.8 8.1	92 73.38	87 81.39	0.09 0.617	[107] [108]
Chlorella sp.	Synthetic wastewater Digested manure	PBR Batch	17 21	8.7	52.6-55.6 76-83	20.4-67 63-75	0.07-0.09	[109] [110]
Desmodesmus sp.	Piggery wastewater	PBR	7	8.6	78.46	91.66	0.26-0.56	[111]
Acutodesmusdimorphus	Raw dairy wastewater	PBR	8	7.8	100	100	-	[112]
S. obliquus	Undiluted cattle wastewater Municipal wastewater	PBR Batch	4 0.2-8	8.1 -	71.19 79-100	62.37 49-78	0.551 -	[108] [113]

(Continued)

Table 14.4 Literature data on nutrients recovery from microalgae. (*Continued*)

Microalgae	Source of wastewater	Operating conditions			Performance			References
		Reactor configuration	Cultivation period (days)	pH	N-Recovery (%)	P-Recovery (%)	Specific growth rate (d^{-1})	
Chlorococcum sp.	Undiluted cattle wastewater	PBR	4	8.1	81.16	85.29	0.624	[108]
S. quadricauda	Undiluted cattle wastewater	PBR	4	8.1	77.23	81.51	0.572	[108]
Parachlorellakessleri	Undiluted cattle wastewater	PBR	4	8.1	73.23	80.69	0.618	[108]
Oscillatoria sp.	Municipal wastewater	Continuous	14	9.8	100	100	-	[114]
C. Sorokiniana	Winery wastewater	Batch	4	7.5	100	100	-	[115]

20-50% oil on a dry weight basis. The efficiency of microalgae strain is high for the removal of N and P from wastewater. The reports have identified the linear relation between nitrogen concentration (0.86 to 4.16%) and protein content (3.96 to 18.07%) [92, 93]. Ji *et al.* [94] recognized a species of microalgal family called Desmodesmus sp. EJ9-6 showed a massive self-growth rate along with a high potential for the reduction of nutrients from wastewater. The Chlorella sp showed phosphorous and nitrogen removal between 20%-100% and 23%-100%, respectively. Moreover, excellent tolerance has reported to NH_4^+-N [95, 96].

14.3.2.2 Nutrients Recovery from Wetland Plants Growth

The Duckweed has a natural ability to absorb nutrients from livestock wastewater. It has an excellent growth rate of 0.10–0.35 g/g·d in the presence of carbon dioxide. Carbon dioxide is a primary liquid feedstock for algal growth, which cultures the aquatic plants, and Duckweeds are one of them. It has low energy loss which directly converts ammonium into natural proteins and nucleic acid [97]. The swine wastewater is treated by Duckweed in countries such as the USA and Israel [98]. They can withstand a higher ammonia concentration environment. The Duckweed performs phosphorous conversion via the absorption mechanism. In average conditions, it absorbs around 0.9 kg/ha·d P from wastewater. The absorption rate of phosphorous and nitrogen reported 0.59 and 2.11 g/m^2 ·d, respectively, from sewage lagoon [99]. Moreover, the Duckweed has shown a maximum rate of recovery for TP and total Kjeldahl nitrogen (i.e., 98.8% and 98%, respectively) [100]. As Duckweeds easily digestible and are rich in vitamins and proteins, they can be used as a food supplement for animals after harvesting [101]. On drying, Duckweed is added to the anaerobic digester system for the production of methane [102].

The other wetland plants are Macrophytes that grow in repressed soil or water. The macrophytes are utilized in built-up wetlands as floating leaves, infant, repressed, or free-floating. These wetlands are aquatic green lives, unsaturated or saturated substrates, and microbial accumulation [103]. Due to the low cost and easy management, the built-up wetland application has been increasing rapidly over the past decade [104]. Typha, Phalaris arundinaceous, Iris species, and Phragmites are some of the common species of macrophytes [105]. These can be used directly into the plant biomass for the removal of nitrogen and phosphorous. The nitrogen and phosphorous can be recovered from the harvested macrophytes from wetlands prior to treatment in compost as anaerobic digestion or utilized as a fertilizer for soil [106]. Table 14.4 puts forth the data on nutrient recovery from microalgae.

14.4 Membrane-Based Hybrid Technologies for Nutrients, Energy, and Water Recovery

14.4.1 Membrane Based Nutrients Recovery

Three major membrane-based technologies are widely employed for the recovery of nutrients i.e., electrodialysis (ED), membrane distillation (MD), and forward osmosis (FO). The ED technology comprises the cation and anion exchange membranes whose sole function is to isolate phosphate and ammonium using energy in form of electricity from the inlet line and collect nutrients in separate chambers (Figure 14.4).

For recovery and concentration at cathodic chambers, phosphate and ammonium are placed at anode [116]. The MD technology comprises a membrane between moderately heated feed solution and the draw solution. Moderately heated feed solutions generate vapors which are transferred via a MD membrane to the draw solution. Hence, the ammonium ion from the feed solution gets converted into volatile ammonia through this process and collected to the draw solution side [117]. Acidic solutions are preferred for usage as draw solution because of their reaction with ammonium yields into volatile ammonium salts. Physical parameters such as pH, temperature etc. of feed affect transformation of ammonia

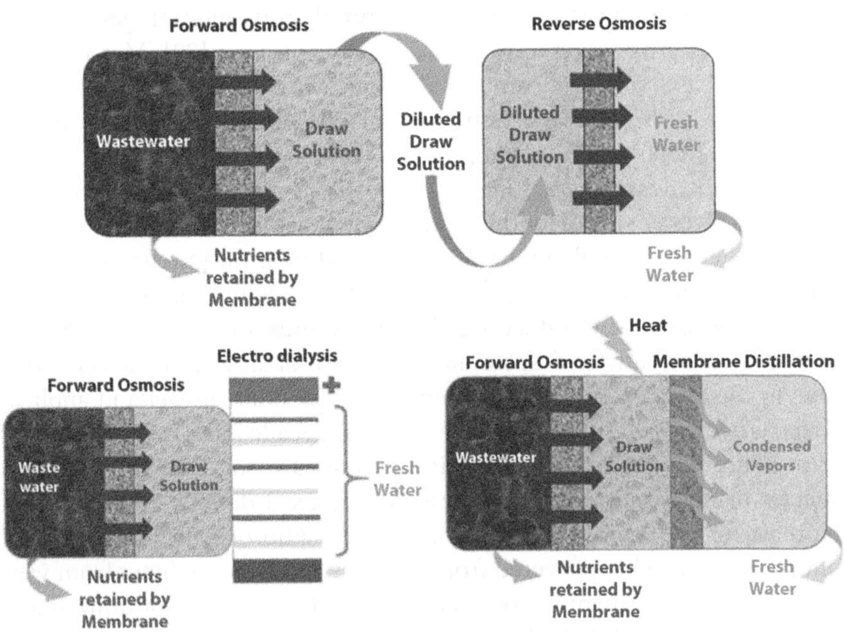

Figure 14.4 The membrane-based resources recovery processes.

and its transport. FO technology is similar to osmosis, in which the semipermeable membrane is kept between the draw and the feed solution. Driving force is identified as the osmotic pressure gradient which is responsible for driving the feed solution to the draw side. Nutrients on the other hand are rejected at FO membrane and are left behind in the feed side in this process. In the end, water gets enriched by the nutrients [118, 119]. Concentration of phosphates would be dense at the feed side due to its bigger hydrated radiuses as compared to the ammonium [119, 120]. In basic medium the surface of membrane gets negatively chargedas a result of which phosphate ions retain at the feed side pertaining to electrostatic repulsions between membranes and phosphate ions [121, 122]. However, in this scenario, the ammonium enrichment is repressed. For both the nutrients enrichment, neutral pH is quite beneficial, high pH converts ammonium to volatile ammonia. Table 14.5 sums up the resource recovery from membrane-based processes.

Table 14.5 The literature data on resources recovery from membrane based processes.

Wastewater source	Membrane process	Highlights and remarks	References
Urine	Reverse Osmosis	• 70% Ammonium and 73% Phosphate recovery • Pressure = 50 Bar	[123]
Synthetic urine	Nano-Filtration	• 55% Ammonium and 94 % Phosphate recovery • Pressure= 20 Bar	[124]
Digested Sludge	Nano-Filtration	• 50% Phosphate recovery • Pressure = 25 Bar	[125]
Raw Sewage	Forward Osmosis-Membrane Distillation	• More than 99% Phosphate recovery • Draw solution temperature maintained at 40 °C	[126]
Activated sludge	Forward Osmosis	• More than 96% and 98% Ammonium and Phosphate recovery, respectively	[127]

(Continued)

Table 14.5 The literature data on resources recovery from membrane based processes. (*Continued*)

Wastewater source	**Membrane process**	**Highlights and remarks**	**References**
Activated sludge	Forward Osmosis–Nano-Filtration	• More than 97% and 99% Ammonium and Phosphate recovery, respectively • Pressure = 80psi	[128]
Activated sludge	Forward Osmosis–Reverse Osmosis	• More than 92.1% and 99.8% Ammonium and Phosphate recovery using • Energy Consumed = 4 kWh/m^3	[129]
Anaerobic sludge	FO-MD	• More than 90% and 97% of Ammonium and Phosphate recovery, • Draw solution temperature maintained at 40 °C	[130]
Urine	Electrodialysis	• Concentration factors: Ammonia = 2.9; Potassium = 3.1; Phosphate = 2.7, • Current density: 22.5 mA/cm^2 • Current efficiency: 50%	[124]
Municipal Wastewater	Electrodialysis	• Concentration Factor of Phosphate = 6.5 • Current Density = 31.25 A/cm^2 • Current efficiency = 72% • Energy consumed = 16.7 kWh/m^3	[131]

14.4.2 Bio Electrochemical Systems (BES) for Resources Recovery

BES comprises microbial electrolysis cell (MEC) and microbial fuel cell (MFC); here, the microorganisms' function as catalysts for electrochemical reactions. MECs needs electrical connections and can give high-value resources at the cathode, whereas MFC utilizes chemical energy from organic substances and produces bio-energy. These systems work well for the recovery of high-end resources and the removal of inorganic and organic pollutants like sulfide, nitrogen, and phosphorous (Figure 14.5) [132, 133]. MFC devices can generate electric current by utilizing wastewater [133]. Microorganisms such as bacteria consume organics present in the wastewater for energy generation and release CO_2, proton, and electrons as byproducts. Through the conductive media/wire, the electrons pass from anode to cathode, and protons pass through the proton exchange membrane to the cathode chamber. The oxygen at the cathode acts as an oxidant; it reacts with hydrogen and electron from the cathode, which completes the circuit and produces water. Yokoyama *et al.* utilized MFC to remove phosphorous, nitrogen, and potassium from cow liquid

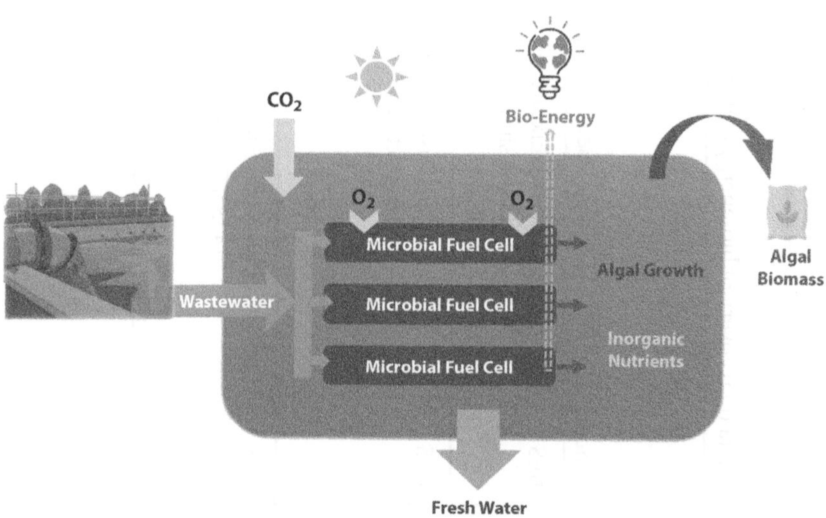

Figure 14.5 Schematic of bio-electrochemical system.

Table 14.6 Literature data on recovery of bio-energy and nutrients from waste water sources.

Wastewater type	Process	Substrate used for operation	Maximum power/ current density	Highlights and remarks	References
Urine-containing wastewater	BES	50:50 (v/v) mixture of anaerobic digestion sludge and growth medium	1300 mW/m^2	• Recovered 42% Nitrogen and 37% Phosphate	[142]
Urine	BES	Inoculated using effluent of another active MFC	34.22 mW/m^2	• Ammonium recovery rate was 3.29 g/m^2day at a current density of 0.50 A m^2	[143]
Synthetic wastewater	BES	with organic loading rate (OLR) (435–870 mg COD/L·d)	253.84 mW/m^2	• 91% Ammonium and 95% Phosphate recovery	

(*Continued*)

Table 14.6 Literature data on recovery of bio-energy and nutrients from waste water sources. (*Continued*)

Wastewater type	Process	Substrate used for operation	Maximum power/ current density	Highlights and remarks	References
Swine wastewater	BES	Inoculated only with paddy field soil (1:25, v/v)	1.3 W/m^2	• Recovered 70-82% phosphorous from influent • Achieved 37–47% of Columbic efficiency	[144]
Digested sewage sludge	BES	$FePO_4$ was reduced through electrons and protons obtained from metabolic activity of *Escherichia coli*	0.030 W/m^2	• Struvite obtained with higher phosphorous content than commercial superphosphate (18–20%).	[145]
Synthetic wastewater	BES	Inoculated with part of a biofilm formed from a previously operated MFC	--	• 27% Phosphorus recovery	[146]

(*Continued*)

Table 14.6 Literature data on recovery of bio-energy and nutrients from waste water sources. (*Continued*)

Wastewater type	Process	Substrate used for operation	Maximum power/ current density	Highlights and remarks	References
Synthetic wastewater	R^2 - BES	Inoculated with the anaerobic sludge from a municipal wastewater treatment plant	--	• Recovered 83.4% Ammonium and 52.4% Phosphorous • When applied 0.8-V produced a current of 68.7 ± 4.6 mA, without any external voltage produced a current of 22.4 ± 0.5 mA	[147]
Organic-rich sediment obtained from a natural lake	SMFC – BES	*Chlorella vulgaris* assisted cathode electrode	48.5 mA/m^2	• Recovered 420 mg/L of algae (dry cell weight) and Produced biomass	[148]

(*Continued*)

Table 14.6 Literature data on recovery of bio-energy and nutrients from waste water sources. (*Continued*)

Wastewater type	Process	Substrate used for operation	Maximum power/ current density	Highlights and remarks	References
Real domestic wastewater	AMNRC - BES	Aerobic bio cathode utilized microorganism as the cathodic catalyst	--	• Recovered 96.9% Ammonium Nitrogen, and 97.1% Phosphorous; • Current produced was the driving force for the movement of ions. Hence, decreased from 39.8 mA (0 h) to 13.9 mA (168 h) during operation.	[149]
Mixed anaerobic sludge	BES	*Chaetoceros* assisted anode to suppress methane production	21.43 W/m^3	• The maximum Current Efficiency of 45.18% was observed.	[150]

(*Continued*)

Table 14.6 Literature data on recovery of bio-energy and nutrients from waste water sources. (*Continued*)

Wastewater type	Process	Substrate used for operation	Maximum power/ current density	Highlights and remarks	References
Wastewater inoculum collected from the primary clarifier	BES	*Rhodopseudomonaspalustris* assisted anode for electricity generation	2.72 W/m^2	• It was reported first time that the purple bacterium had a potential of transferring electrons to produce huge current and power densities.	[151]
Municipal wastewater	Three-Chamber-BES	Hydrogenophilic chemoautothrophic microorganisms were used to inoculate cathodic chamber	--	• 32% of Total Nitrogen removal • Produced average current of 87 mA with Current efficiency of 86 %.	[152]

(*Continued*)

Table 14.6 Literature data on recovery of bio-energy and nutrients from waste water sources. (*Continued*)

Wastewater type	Process	Substrate used for operation	Maximum power/ current density	Highlights and remarks	References
Activated sludge	BES	Ammonium and Magnesium chloride solutions addition into Cathode Chamber	72 mW/m^2	• Recovery of 38% P as struvite	[153]
Anaerobic sludge	BES	The cathode compartment was initially inoculated with the green algae *Pseudokirchneriellasubcapitata*	2.2 W/m^3	Recovered 98% ammonium nitrogen and 82% phosphate and produced 128 mg/L of algal biomass	[154]

digestate and achieved 70%, 84%, and 91% removal, respectively [134]. Wu *et al.* integrated MFC with the wetland for ammonium removal of 92% from pig manure digestate [135]. Some researchers coupled MFC with anaerobic digestion, which results in a good strategy for removing nutrients from manure digestate [136]. Removal of phosphate using MFC-MEC was an excellent approach for anaerobic digestate; It could achieve 90% phosphorous [137].

The transfer of electrons starts at the anode in an oxidation reaction and finishes at the cathode in the reduction reaction. For anodic oxidation, microorganisms are utilized as catalysts [133]. A double chamber MEC reactor for digested pig slurry treatment has shown an excellent removal of nitrogen and other organics around 40% N recovery [138]. Sotres *et al.* reported that the recovery of nitrogen from pig slurry using MEC with nitrogen flux achieved 10.3 g $N/d{\cdot}m^2$ [139]. An integrated ammonia stripping can recover N up to 94.3% as the pH remains high at the cathode. A single-chamber MEC investigation proved the recovery of phosphorous by the formation of struvite at the cathode electrode. The production of hydrogen at the cathode electrode of MEC had increased the pH, favorable for the formation of struvite [140]. The crystallization of struvite studied for phosphorous recovery by integrating MEC and magnesium salt addition. The P recovery achieved was 18-60% at varying N/P ratios in the catholyte [141]. Table 14.6 mentions the data on recovery of bio-energy and nutrients from wastewater sources collected from literatures.

14.4.3 Nutrients Recovery via Osmotic Membrane Bioreactor

Qiu and Ting [155] had developed an osmotic membrane bioreactor for the recovery of nutrients. The forward osmosis process, along with the aerobic biological process, comprises OMBR. In this process, the nutrients recovery could be achieved more than 95% of P and N and enriched in the feed side then recovered in terms of struvite or calcium-based P precipitates. The most important aspect of this process is zero use of chemicals for nutrient recovery, but alkaline chemicals are necessary to adjust pH. The addition of $MgCl_2$ as draw solute for FO supplements the magnesium ions required for struvite recovery. Microfiltration [156] and ultrafiltration [157] membranes can be installed to the feed side for increasing the purity of the nutrients recovered in OMBR. These membranes extract the concentrated nutrients easily and ensure low permeation of unwanted substances to the nutrients side. This proves that in terms of economic and technical feasibility, the potential of recovering nutrients through OMBR is high. In the OMBR feed side [158], an addition of a fixed bed biofilm can reduce suspended growth and fouling of the membrane. The economic feasibility of the process can be increased by using RO or FO membranes for water extraction and recovering draw solution [159, 160]. There are specific gaps in the OMBR process, such as lower rejection rates to nutrients and the control over membrane fouling.

14.4.4 Economics and Feasibility of Processes

The summary of the technologies and their comparison for the recovery of nutrients present in the Table 14.7. The recovery of resources produces fertilizers, water, and energy for useful purposes and lowers down the cost of wastewater treatment [161, 162]. One such example is in Europe. The annual expenditure on wastewater treatment plants is around €66,000 to overcome the removal of struvite precipitation issues in machinery and pipes (≈ US $72,284.55) [163]. The recovery of nutrients could significantly reduce the same cost. As compared to the removal of phosphate, the recovery was less, i.e., (≈ US $2.22–3.33)/kg·P [164]. In this case, the excess sludge production could be reduced on a dry mass basis by 2-8%. Hence, the cost associated with the management of sludge reported AUD 1.13 (≈ US$0.78)/kg. struvite [165] and UK 100 pounds (≈ US$130.57)/ton. P [163].

The precipitation of struvite and calcium phosphates is the major product recovered from the nutrient recovery system either at lab-level or bulk. At present, these nutrients are not fully commercialized. Hence, their market values are not available. Some of the data can be found online. The market price of struvite mentioned as €310 (≈US$344.72)/ton [166] and €2.6 (≈ US$2.89)/kg·P [167]. Another report [168] mentioned that in Japan, the price of struvite estimated at around €220 (≈US$244.64)/ton, and in Australia, it was AUS$300–500 (≈ US $207.04–345.07)/ton. The price of phosphate rock (€0.6–1.2 (≈ US$0.67–1.33)/kg·P) is relatively lesser than the struvite [164, 167], and TSP (€ 1.2–2.2 (≈ US$1.33–2.45)/kg·P) [169]. An average market cost of fertilizers containing phosphate is €1.9–3.3 (≈ the US $2.11–3.67)/kg·P, which signifies that the fertilizers already available in the market are more economical for the farmers whereas, nutrients recovered are still costlier.

Table 14.7 Summary of the resource's recovery processes.

Processes	Technologies	Resources recovery	Major highlights	R&D need for the adoption of technology [14]		
				Existing knowledge	Application development	Product development
Chemical Process	Chemical Precipitation	Struvite, Hydroxyapatite	• Low capital cost simple and easy operation • High chemical requirement • High efficiency	***	*	*
	Adsorption	Phosphate, Ammonium	• High stability • The necessity of desorption and specific adsorbents requirement	*	**	**
	Wet-Chemical	Phosphates	• Downstream process requirement • Chemical requirement for extraction	**	*	*

(*Continued*)

Table 14.7 Summary of the resource's recovery processes. (*Continued*)

Processes	Technologies	Resources recovery	Major highlights	R&D need for the adoption of technology [14]		
				Existing knowledge	Application development	Product development
	Thermochemical	Phosphates	• High energy input • High chemical input • Low stability	**	**	***
Biological process	Algal or Wetland Plants Accumulation	Phosphates, Biomass	• Environmental footprints: low • Low applicability of recovered products containing foreign substances	**	***	**

(*Continued*)

Table 14.7 Summary of the resource's recovery processes. (*Continued*)

Processes	Technologies	Resources recovery	Major highlights	R&D need for the adoption of technology [14]		
				Existing knowledge	Application development	Product development
	Forward Osmosis	Phosphate, Biomass, Solutes	• Low Energy requirement • Low fouling and Easy cleaning • Low pressure requirement • Reconcentration of draw solute needed	**	**	**
Membrane-Based Process	Membrane Distillation	Nutrients water	• Renewable energy available for usage • Organic substrate accumulation • Membrane wetting issues	***	**	**

(*Continued*)

Table 14.7 Summary of the resource's recovery processes. (*Continued*)

Processes	Technologies	Resources recovery	Major highlights	R&D need for the adoption of technology [14]		
				Existing knowledge	Application development	Product development
	Electro dialysis	Nutrients water	• High nutrient enrichment • Low current efficiency • High energy input	*	***	***
Membrane-based Hybrid Process	Osmotic Membrane Bioreactor	Nutrients Water Energy Solutes	• Low membrane fouling • Organic removal • Low energy input • Low sanity level • Reconcentration of draw solute needed	***	**	**

(*Continued*)

Table 14.7 Summary of the resource's recovery processes. (*Continued*)

Processes	Technologies	Resources recovery	Major highlights	R&D need for the adoption of technology [14]		
				Existing knowledge	Application development	Product development
	Bio-Electrochemical System	Bio-energy Biomass Water	• Low chemical input • Possibility of neutral energy balance • Recovered products accumulation on cathode surface was an issue	***	**	**
	Membrane Photo bioreactor	Bio-energy Biomass Water	• Low chemical input • Environmental • Low stability • Long time requirement • Light needed	***	**	**

Note: *, **, *** indicates that for the adoption of the process, low, moderate, and high research and development is required.

14.5 Conclusion

Minerals such as phosphorous and nitrogen present in abundance in domestic, municipal and industrial effluents can be extracted economically, thereby delaying several natural processes such as eutrophication and reduce the demands on natural reserves. In this review the authors have aimed towards sustainable extraction and reutilization of resource methodologies to be used to extract resource from wastewater instead of directly discarding them for non-potable means. Methodologies discussed here are divided on the basis of process followed and quantized on the efficiency of the process. Citing the selectivity of the processes, for maximum situations of resource recovery, hybrid model based on membrane technology can be utilized. Also, the chemical and biological treatment reject laden with resources can be used in agricultural fields. Moreover, the detailed explanation as discussed in article can be utilized for further studies to improvise on existing methods to tailor fit the treatment of different effluents.

Overall, the review article has a positive aim for the resource recovery and successive reuse of water and other rejects into constructive usage thereby converting the wastewater into a useful resource. This conversion can be used for both (I) delayed eutrophication (II) reutilization of resources.

Acknowledgements

The authors would like to express sincere gratitude towards Director, CSIR-IICT [Ref. IICT/Pubs./2022/108] and Director, BITS Pilani Goa Campus for the continuous support and motivation.

Disclosure

Dr. Siddhartha Moulik would like to thank CSIR-MLP-0079 Grant for carrying out the work.

References

1. S. Chakraborty, A. Mukherjee, S. Das, N.R. Maddela, S. Iram, P. Das, Study on isotherm, kinetics, and thermodynamics of adsorption of crystal violet

dye by calcium oxide modified fly ash, Environ. Eng. Res. 26 (2021) 1–9. https://doi.org/10.4491/eer.2019.372.

2. A. Bittsánszky, K. Pilinszky, G. Gyulai, T. Komives, Overcoming ammonium toxicity, Plant Sci. 231 (2015) 184–190. https://doi.org/10.1016/j.plantsci.2014.12.005.
3. Y.H. Liu, J.H. Kwag, J.H. Kim, C.S. Ra, Recovery of nitrogen and phosphorus by struvite crystallization from swine wastewater, Desalination. 277 (2011) 364–369. https://doi.org/10.1016/j.desal.2011.04.056.
4. Y.H. Song, G.L. Qiu, P. Yuan, X.Y. Cui, J.F. Peng, P. Zeng, L. Duan, L.C. Xiang, F. Qian, Nutrients removal and recovery from anaerobically digested swine wastewater by struvite crystallization without chemical additions, J. Hazard. Mater. 190 (2011) 140–149. https://doi.org/10.1016/j.jhazmat.2011.03.015.
5. Y.D. Yilmazel, G.N. Demirer, Removal and recovery of nutrients as struvite from anaerobic digestion residues of poultry manure, Environ. Technol. 32 (2011) 783–794. https://doi.org/10.1080/09593330.2010.512925.
6. A. Mukherjee, A. Mullick, R. Teja, P. Vadthya, A. Roy, S. Moulik, Performance and energetic analysis of hydrodynamic cavitation and potential integration with existing advanced oxidation processes: A case study for real life greywater treatment, Ultrason. Sonochem. 66 (2020) 105116. https://doi.org/10.1016/j.ultsonch.2020.105116.
7. S. Lanjewar, A. Mukherjee, P. Khandewal, A.K. Ghosh, A. Mullick, S. Moulik, A. Roy, Thermodynamics of synthesis and separation performance of Interfacially polymerized "loose" reverse osmosis membrane: Benchmarking for greywater treatment, Chem. Eng. J. (2020). https://doi.org/10.1016/j.cej.2020.127929.
8. R. Otterpohl, Options for alternative types of sewerage and treatment systems directed to improvement of the overall performance, Water Sci. Technol. 45 (2002) 149–158. https://doi.org/10.2166/wst.2002.0074.
9. B.B. Lind, Z. Ban, S. Bydén, Nutrient recovery from human urine by struvite crystallization with ammonia adsorption on zeolite and wollastonite, Bioresour. Technol. 73 (2000) 169–174. https://doi.org/10.1016/S0960-8524(99)90157-8.
10. R. Mohammadi, W. Tang, M. Sillanpää, A systematic review and statistical analysis of nutrient recovery from municipal wastewater by electrodialysis, Desalination. 498 (2021) 114626. https://doi.org/10.1016/j.desal.2020.114626.
11. Y.L. Yang, Y. Wu, Y.X. Lu, Y. Cai, Z. He, X.L. Yang, H.L. Song, A comprehensive review of nutrient-energy-water-solute recovery by hybrid osmotic membrane bioreactors, Bioresour. Technol. 320 (2021) 124300. https://doi.org/10.1016/j.biortech.2020.124300.
12. R.L. Newcombe, R.A. Rule, B.K. Hart, G. Möller, Phosphorus Removal from Municipal Wastewater by Hydrous Ferric Oxide Reactive Filtration and Coupled Chemically Enhanced Secondary Treatment: Part I-Performance, Water Environ. Res. (2008). https://doi.org/10.2175/106143007x221003.

13. A.D. Gómez-Suárez, C. Nobile, M.P. Faucon, O. Pourret, D. Houben, Fertilizer potential of struvite as affected by nitrogen form in the rhizosphere, Sustain. 12 (2020) 1–11. https://doi.org/10.3390/su12062212.
14. C.M. Mehta, W.O. Khunjar, V. Nguyen, S. Tait, D.J. Batstone, Technologies to recover nutrients from waste streams: A critical review, Crit. Rev. Environ. Sci. Technol. 45 (2015) 385–427. https://doi.org/10.1080/10643389.2013.866621.
15. S. Antonini, M.A. Arias, T. Eichert, J. Clemens, Greenhouse evaluation and environmental impact assessment of different urine-derived struvite fertilizers as phosphorus sources for plants, Chemosphere. (2012). https://doi.org/10.1016/j.chemosphere.2012.07.026.
16. L.E. De-Bashan, Y. Bashan, Recent advances in removing phosphorus from wastewater and its future use as fertilizer (1997-2003), Water Res. 38 (2004) 4222–4246. https://doi.org/10.1016/j.watres.2004.07.014.
17. P. Simha, M. Ganesapillai, Ecological Sanitation and nutrient recovery from human urine: How far have we come? A review, Sustain. Environ. Res. 27 (2017) 107–116. https://doi.org/10.1016/j.serj.2016.12.001.
18. R. Kumar, P. Pal, Assessing the feasibility of N and P recovery by struvite precipitation from nutrient-rich wastewater: a review, Environ. Sci. Pollut. Res. (2015). https://doi.org/10.1007/s11356-015-5450-2.
19. K. Moriyama, T. Kojima, Y. Minawa, S. Matsumoto, K. Nakamachi, Development of artificial seed crystal for crystallization of calcium phosphate, Environ. Technol. (United Kingdom). (2001). https://doi.org/10.1080/09593330.2001.9619163.
20. M.J. Baker, D.W. Blowes, C.J. Ptacek, Laboratory development of permeable reactive mixtures for the removal of phosphorus from onsite wastewater disposal systems, Environ. Sci. Technol. (1998). https://doi.org/10.1021/es970934w.
21. S.H. Huang, B. Chiswell, Phosphate removal from wastewater using spent alum sludge, Water Sci. Technol. 42 (2000) 295–300. https://doi.org/10.2166/wst.2000.0394.
22. D. Donnert, M. Salecker, Elimination of phosphorus from waste water by crystallization, Environ. Technol. (United Kingdom). 20 (1999) 735–742. https://doi.org/10.1080/09593332008616868.
23. M. Özacar, I.A. Şengil, Enhancing phosphate removal from wastewater by using polyelectrolytes and clay injection, J. Hazard. Mater. (2003). https://doi.org/10.1016/S0304-3894(03)00070-0.
24. N.K. Lazaridis, Sorption removal of anions and cations in single batch systems by uncalcined and calcined Mg-Al-CO3 hydrotalcite, Water. Air. Soil Pollut. (2003). https://doi.org/10.1023/A:1023999303895.
25. K. Sakadevan, H.J. Bavor, Phosphate adsorption characteristics of soils, slags and zeolite to be used as substrates in constructed wetland systems, Water Res. (1998). https://doi.org/10.1016/S0043-1354(97)00271-6.

26. Y. Seida, Y. Nakano, Removal of phosphate by layered double hydroxides containing iron, Water Res. (2002). https://doi.org/10.1016/S0043-1354(01)00340-2.
27. E. Frossard, J.P. Bauer, F. Lothe, Evidence of vivianite in FeSO4 - Flocculated sludges, Water Res. 31 (1997) 2449–2454. https://doi.org/10.1016/S0043-1354(97)00101-2.
28. R. Yamamoto-Ikemoto, S. Matsui, T. Komori, E.K. Bosque-Hamilton, Control of filamentous bulking and interactions among sulfur oxidation-reduction and iron oxidation-reduction in activated sludge using an iron coagulant, Water Sci. Technol. (1998). https://doi.org/10.2166/wst.1998.0785.
29. I. Stratful, M.D. Scrimshaw, J.N. Lester, Removal of Struvite to Prevent Problems Associated with its Accumulation in Wastewater Treatment Works, Water Environ. Res. 76 (2004) 437–443. https://doi.org/10.2175/106143004x151491.
30. S. Regy, Phosphate Recovery By Stuvite Precipitation in a Stirred Reactor, Ceep. (2001).
31. Y. Jaffer, T.A. Clark, P. Pearce, S.A. Parsons, Potential phosphorus recovery by struvite formation, Water Res. (2002). https://doi.org/10.1016/S0043-1354(01)00391-8.
32. K. Suzuki, Y. Tanaka, T. Osada, M. Waki, Removal of phosphate, magnesium and calcium from swine wastewater through crystallization enhanced by aeration, Water Res. (2002). https://doi.org/10.1016/S0043-1354(01)00536-X.
33. Y. Ueno, M. Fujii, Three years experience of operating and selling recovered struvite from full-scale plant, Environ. Technol. (United Kingdom). 22 (2001) 1373–1381. https://doi.org/10.1080/09593332208618196.
34. P. Battistoni, R. Boccadoro, F. Fatone, P. Pavan, Auto-nucleation and crystal growth of struvite in a demonstrative fluidized bed reactor (FBR), Environ. Technol. 26 (2005) 975–982. https://doi.org/10.1080/09593332608618486.
35. C. Di Iaconi, M. Pagano, R. Ramadori, A. Lopez, Nitrogen recovery from a stabilized municipal landfill leachate, Bioresour. Technol. 101 (2010) 1732–1736. https://doi.org/10.1016/j.biortech.2009.10.013.
36. A. Siciliano, Assessment of fertilizer potential of the struvite produced from the treatment of methanogenic landfill leachate using low-cost reagents, Environ. Sci. Pollut. Res. (2016). https://doi.org/10.1007/s11356-015-5846-z.
37. X.Z. Li, Q.L. Zhao, Recovery of ammonium-nitrogen from landfill leachate as a multi-nutrient fertilizer, Ecol. Eng. (2003). https://doi.org/10.1016/S0925-8574(03)00012-0.
38. D. Kim, H.D. Ryu, M.S. Kim, J. Kim, S.I. Lee, Enhancing struvite precipitation potential for ammonia nitrogen removal in municipal landfill leachate, J. Hazard. Mater. (2007). https://doi.org/10.1016/j.jhazmat.2006.11.054.
39. S. Ramaswami, J. Behrendt, G. Wang, S. Eggers, R. Otterpohl, Combining magnesium ammonium phosphate precipitation with membrane processes for ammonia removal from methanogenic leachates, Water Environ. J. (2016). https://doi.org/10.1111/wej.12210.

40. A. Siciliano, C. Ruggiero, S. De Rosa, A new integrated treatment for the reduction of organic and nitrogen loads in methanogenic landfill leachates, Process Saf. Environ. Prot. (2013). https://doi.org/10.1016/j.psep.2012.06.008.
41. L. Pastor, D. Mangin, R. Barat, A. Seco, A pilot-scale study of struvite precipitation in a stirred tank reactor: Conditions influencing the process, Bioresour. Technol. (2008). https://doi.org/10.1016/j.biortech.2007.12.003.
42. Y.J. Shih, R.R.M. Abarca, M.D.G. de Luna, Y.H. Huang, M.C. Lu, Recovery of phosphorus from synthetic wastewaters by struvite crystallization in a fluidized-bed reactor: Effects of pH, phosphate concentration and coexisting ions, Chemosphere. (2017). https://doi.org/10.1016/j.chemosphere.2017.01.088.
43. A. Adnan, F.A. Koch, D.S. Mavinic, Pilot-scale study of phosphorus recovery through struvite crystallization – II: Applying in-reactor supersaturation ratio as a process control parameter, J. Environ. Eng. Sci. (2003). https://doi.org/10.1139/s03-048.
44. A. Guadie, S. Xia, W. Jiang, L. Zhou, Z. Zhang, S.W. Hermanowicz, X. Xu, S. Shen, Enhanced struvite recovery from wastewater using a novel cone-inserted fluidized bed reactor, J. Environ. Sci. (China). (2014). https://doi.org/10.1016/S1001-0742(13)60469-6.
45. M.S. Rahaman, N. Ellis, D.S. Mavinic, Effects of various process parameters on struvite precipitation kinetics and subsequent determination of rate constants, Water Sci. Technol. (2008). https://doi.org/10.2166/wst.2008.022.
46. E.M. Jordaan, B. Rezania, N. Ciçek. Investigation of chemical-free nutrient removal and recovery from CO 2-rich wastewater, Water Sci. Technol. 67 (2013) 2195–2201. https://doi.org/10.2166/wst.2013.116.
47. A. Uysal, Y.D. Yilmazel, G.N. Demirer, The determination of fertilizer quality of the formed struvite from effluent of a sewage sludge anaerobic digester, J. Hazard. Mater. (2010). https://doi.org/10.1016/j.jhazmat.2010.05.004.
48. A. Giesen, Crystallisation process enables environmental friendly phosphate removal at low costs, Environ. Technol. (United Kingdom). 20 (1999) 769–775. https://doi.org/10.1080/09593332008616873.
49. D. Aguado, R. Barat, A. Bouzas, A. Seco, J. Ferrer, P-recovery in a pilot-scale struvite crystallisation reactor for source separated urine systems using seawater and magnesium chloride as magnesium sources, Sci. Total Environ. (2019). https://doi.org/10.1016/j.scitotenv.2019.03.485.
50. S. Xu, L. Luo, H. He, H. Liu, L. Cui, Nitrogen and phosphate recovery from source-separated urine by dosing with magnesite and zeolite, Polish J. Environ. Stud. (2015). https://doi.org/10.15244/pjoes/43611.
51. X. Liu, Z. Hu, J. Mu, H. Zang, L. Liu, Phosphorus recovery from urine with different magnesium resources in an air-agitated reactor, Environ. Technol. (United Kingdom). (2014). https://doi.org/10.1080/09593330.2014.921732.
52. R. Taddeo, R. Lepistö, Struvite precipitation in raw and co-digested swine slurries for nutrients recovery in batch reactors, Water Sci. Technol. (2015). https://doi.org/10.2166/wst.2015.045.

53. K.E. Bowers, P.W. Westerman, Performance of cone-shaped fluidized bed struvite crystallizers in removing phosphorus from wastewater, Trans. Am. Soc. Agric. Eng. (2005). https://doi.org/10.13031/2013.18523.
54. K. Bube, Über Magnesiumammoniumphosphat, Zeitschrift Für Anal. Chemie. 49 (1910) 525–596. https://doi.org/10.1007/BF01306233.
55. A.W. Taylor, A.W. Frazier, E.L. Gurney, Solubility products of magnesium ammonium and magnesium potassium phosphates, Trans. Faraday Soc. (1963). https://doi.org/10.1039/tf9635901580.
56. J. Borgerding, Phosphate deposits in digestion systems., J. Water Pollut. Control Fed. (1972).
57. J.R. Burns, B. Finlayson, Solubility product of magnesium ammonium phosphate hexahydrate at various temperatures, J. Urol. (1982). https://doi.org/10.1016/S0022-5347(17)52952-3.
58. R.E. Loewenthal, U.R.C. Kornmuller, E.P. van Heerden, Modelling struvite precipitation in anaerobic treatment systems, in: Water Sci. Technol., 1994. https://doi.org/10.2166/wst.1994.0592.
59. J.R. Buchanan, C.R. Mote, R.B. Robinson, Thermodynamics of struvite formation, Trans. Am. Soc. Agric. Eng. (1994). https://doi.org/10.13031/2013.28121.
60. H.K. Aage, B.L. Andersen, A. Blom, I. Jensen, The solubility of struvite, J. Radioanal. Nucl. Chem. (1997). https://doi.org/10.1007/BF02223387.
61. K.N. Ohlinger, T.M. Young, E.D. Schroeder, Predicting struvite formation in digestion, Water Res. (1998). https://doi.org/10.1016/S0043-1354(98)00123-7.
62. S. Sengupta, A. Pandit, Selective removal of phosphorus from wastewater combined with its recovery as a solid-phase fertilizer, Water Res. (2011). https://doi.org/10.1016/j.watres.2011.03.044.
63. B.K. Biswas, K. Inoue, K.N. Ghimire, H. Harada, K. Ohto, H. Kawakita, Removal and recovery of phosphorus from water by means of adsorption onto orange waste gel loaded with zirconium, Bioresour. Technol. 99 (2008) 8685–8690. https://doi.org/10.1016/j.biortech.2008.04.015.
64. M. Sprynskyy, M. Lebedynets, R. Zbytniewski, J. Namieśnik, B. Buszewski, Ammonium removal from aqueous solution by natural zeolite, Transcarpathian mordenite, kinetics, equilibrium and column tests, Sep. Purif. Technol. 46 (2005) 155–160. https://doi.org/10.1016/j.seppur.2005.05.004.
65. D. Donnert, M. Salecker, Elimination of phosphorus from municipal and industrial waste water, in: Water Sci. Technol., 1999. https://doi.org/10.1016/S0273-1223(99)00501-6.
66. J.Y. Jung, Y.C. Chung, H.S. Shin, D.H. Son, Enhanced ammonia nitrogen removal using consistent biological regeneration and ammonium exchange of zeolite in modified SBR process, Water Res. (2004). https://doi.org/10.1016/j.watres.2003.09.025.
67. Y.X. Wei, Z.F. Ye, Y.L. Wang, M.G. Ma, Y.F. Li, Enhanced ammonia nitrogen removal using consistent ammonium exchange of modified zeolite and

biological regeneration in a sequencing batch reactor process, Environ. Technol. 32 (2011) 1337–1343. https://doi.org/10.1080/09593330.2010.536784.
68. G. Kocar, The use of anaerobically digested slurry combined with natural zeolite for rapeseed production, Energy Educ. Sci. Technol. Part A Energy Sci. Res. 30 (2012) 545–552.
69. R. Das, A. Mukherjee, I. Sinha, K. Roy, B.K. Dutta, Synthesis of potential bio-adsorbent from Indian Neem leaves (Azadirachta indica) and its optimization for malachite green dye removal from industrial wastes using response surface methodology: kinetics, isotherms and thermodynamic studies, Appl. Water Sci. 10 (2020) 1–18. https://doi.org/10.1007/s13201-020-01184-5.
70. A. Drenkova-Tuhtan, M. Schneider, M. Franzreb, C. Meyer, C. Gellermann, G. Sextl, K. Mandel, H. Steinmetz, Pilot-scale removal and recovery of dissolved phosphate from secondary wastewater effluents with reusable ZnFeZr adsorbent @ Fe3O4/SiO2 particles with magnetic harvesting, Water Res. 109 (2017) 77–87. https://doi.org/https://doi.org/10.1016/j.watres.2016.11.039.
71. P. Ning, H.J. Bart, B. Li, X. Lu, Y. Zhang, Phosphate removal from wastewater by model-La(III) zeolite adsorbents, J. Environ. Sci. (2008). https://doi.org/10.1016/S1001-0742(08)62111-7.
72. X. Wang, S. Lü, C. Gao, C. Feng, X. Xu, X. Bai, N. Gao, J. Yang, M. Liu, L. Wu, Recovery of Ammonium and Phosphate from Wastewater by Wheat Straw-based Amphoteric Adsorbent and Reusing as a Multifunctional Slow-Release Compound Fertilizer, ACS Sustain. Chem. Eng. (2016). https://doi.org/10.1021/acssuschemeng.5b01494.
73. K. Kuzawa, Y.J. Jung, Y. Kiso, T. Yamada, M. Nagai, T.G. Lee, Phosphate removal and recovery with a synthetic hydrotalcite as an adsorbent, Chemosphere. (2006). https://doi.org/10.1016/j.chemosphere.2005.04.015.
74. J. Kim, M.J. Hwang, S.J. Lee, W. Noh, J.M. Kwon, J.S. Choi, C.M. Kang, Efficient recovery of nitrate and phosphate from wastewater by an amine-grafted adsorbent for cyanobacterial biomass production, Bioresour. Technol. (2016). https://doi.org/10.1016/j.biortech.2016.01.055.
75. A. Hedström, L.R. Amofah, Adsorption and desorption of ammonium by clinoptilolite adsorbent in municipal wastewater treatment systems, J. Environ. Eng. Sci. (2008). https://doi.org/10.1139/S07-029.
76. S.J. Lee, S. Park, W. Noh, D.H. Yeom, S. Kim, D.W. Kim, J. Kim, Regeneration of nitrate and phosphate from toilet wastewater using waste alumina adsorbent for cultivation of spirulina platensis, Environ. Eng. Res. 25 (2020) 393–399. https://doi.org/10.4491/eer.2018.406.
77. H.-P. Jing, Y. Li, X. Wang, J. Zhao, S. Xia, Simultaneous recovery of phosphate{,} ammonium and humic acid from wastewater using a biochar supported Mg(OH)2/bentonite composite, Environ. Sci. Water Res. Technol. 5 (2019) 931–943. https://doi.org/10.1039/C8EW00952J.
78. L.H. Nguyen, T.M. Vu, T.T. Le, V.T. Trinh, T.P. Tran, H.T. Van, Ammonium removal from aqueous solutions by fixed-bed column using corncob-based

modified biochar, Environ. Technol. (United Kingdom). (2019). https://doi.org/10.1080/09593330.2017.1404134.
79. H. Wang, X. Wang, J. Zhao, Application of MgO-modified palygorskite for nutrient recovery from swine wastewater: effect of pH, ions, and organic acids, Environ. Sci. Pollut. Res. (2019). https://doi.org/10.1007/s11356-019-05254-3.
80. J. Benemann, Microalgae for biofuels and animal feeds, Energies. (2013). https://doi.org/10.3390/en6115869.
81. Y. Chisti, Raceways-based production of algal crude oil, Green. 3 (2013) 195–216. https://doi.org/10.1515/green-2013-0018.
82. F.G. Acién, J.M. Fernández, J.J. Magán, E. Molina, Production cost of a real microalgae production plant and strategies to reduce it, Biotechnol. Adv. (2012). https://doi.org/10.1016/j.biotechadv.2012.02.005.
83. H. Yoo, K.H. Ahn, H.J. Lee, K.H. Lee, Y.J. Kwak, K.G. Song, Nitrogen removal from synthetic wastewater by simultaneous nitrification and denitrification (SND) via nitrite in an intermittently-aerated reactor, Water Res. (1999). https://doi.org/10.1016/S0043-1354(98)00159-6.
84. A. Magrí, F. Béline, P. Dabert, Feasibility and interest of the anammox process as treatment alternative for anaerobic digester supernatants in manure processing - An overview, J. Environ. Manage. (2013). https://doi.org/10.1016/j.jenvman.2013.09.021.
85. A.A. Van De Graaf, P. De Bruijn, L.A. Robertson, M.S.M. Jetten, J.G. Kuenen, Metabolic pathway of anaerobic ammonium oxidation on the basis of 15N studies in a fluidized bed reactor, Microbiology. (1997). https://doi.org/10.1099/00221287-143-7-2415.
86. H. Meng, Y.C. Yang, J.G. Lin, M. Denecke, J.D. Gu, Occurrence of anammox bacteria in a traditional full-scale wastewater treatment plant and successful inoculation for new establishment, Int. Biodeterior. Biodegrad. 120 (2017) 224–231. https://doi.org/10.1016/j.ibiod.2017.01.022.
87. M. Waki, T. Tokutomi, H. Yokoyama, Y. Tanaka, Nitrogen removal from animal waste treatment water by anammox enrichment, Bioresour. Technol. (2007). https://doi.org/10.1016/j.biortech.2006.09.031.
88. C. Yangin-Gomec, G. Pekyavas, T. Sapmaz, S. Aydin, B. Ince, Ç. Akyol, O. Ince, Microbial monitoring of ammonia removal in a UASB reactor treating pre-digested chicken manure with anaerobic granular inoculum, Bioresour. Technol. (2017). https://doi.org/10.1016/j.biortech.2017.05.070.
89. B. Ma, S. Wang, S. Cao, Y. Miao, F. Jia, R. Du, Y. Peng, Biological nitrogen removal from sewage via anammox: Recent advances, Bioresour. Technol. (2016). https://doi.org/10.1016/j.biortech.2015.10.074.
90. T. Selvaratnam, A.K. Pegallapati, F. Montelya, G. Rodriguez, N. Nirmalakhandan, W. Van Voorhies, P.J. Lammers, Evaluation of a thermotolerant acidophilic alga, Galdieria sulphuraria, for nutrient removal from urban wastewaters, Bioresour. Technol. (2014). https://doi.org/10.1016/j.biortech.2014.01.075.

91. S.M. Henkanatte-Gedera, T. Selvaratnam, N. Caskan, N. Nirmalakhandan, W. Van Voorhies, P.J. Lammers, Algal-based, single-step treatment of urban wastewaters, Bioresour. Technol. 189 (2015) 273–278. https://doi.org/10.1016/j.biortech.2015.03.120.
92. T. Cai, S.Y. Park, Y. Li, Nutrient recovery from wastewater streams by microalgae: Status and prospects, Renew. Sustain. Energy Rev. (2013). https://doi.org/10.1016/j.rser.2012.11.030.
93. A.J. Cole, R. de Nys, N.A. Paul, Biorecovery of nutrient waste as protein in freshwater macroalgae, Algal Res. (2015). https://doi.org/10.1016/j.algal.2014.12.005.
94. F. Ji, Y. Liu, R. Hao, G. Li, Y. Zhou, R. Dong, Biomass production and nutrients removal by a new microalgae strain Desmodesmus sp. in anaerobic digestion wastewater, Bioresour. Technol. (2014). https://doi.org/10.1016/j.biortech.2014.03.034.
95. M. Khan, N. Yoshida, Effect of l-glutamic acid on the growth and ammonium removal from ammonium solution and natural wastewater by Chlorella vulgaris NTM06, Bioresour. Technol. (2008). https://doi.org/10.1016/j.biortech.2006.12.031.
96. R.B. Levine, M.S. Costanza-Robinson, G.A. Spatafora, Neochloris oleoabundans grown on anaerobically digested dairy manure for concomitant nutrient removal and biodiesel feedstock production, Biomass and Bioenergy. (2011). https://doi.org/10.1016/j.biombioe.2010.08.035.
97. J. Cheng, L. Landesman, B.A. Bergmann, J.J. Classen, J.W. Howard, Y.T. Yamamoto, Nutrient removal from swine lagoon liquid by Lemna minor 8627, Trans. Am. Soc. Agric. Eng. (2002). https://doi.org/10.13031/2013.9953.
98. Z. Zhao, H. Shi, Y. Liu, H. Zhao, H. Su, M. Wang, Y. Zhao, The influence of duckweed species diversity on biomass productivity and nutrient removal efficiency in swine wastewater, Bioresour. Technol. (2014). https://doi.org/10.1016/j.biortech.2014.06.031.
99. G.J. Alaerts, M.D. Rahman Mahbubar, P. Kelderman, Performance analysis of a full-scale duckweed-covered sewage lagoon, Water Res. (1996). https://doi.org/10.1016/0043-1354(95)00234-0.
100. R.A. Mohedano, R.H.R. Costa, F.A. Tavares, P. Belli Filho, High nutrient removal rate from swine wastes and protein biomass production by full-scale duckweed ponds, Bioresour. Technol. (2012). https://doi.org/10.1016/j.biortech.2012.02.083.
101. D.D. Culley, E. Rejmánková, J. Květ, J.B. Frye, Production, chemical quality and use of ducweeds (LEMNACEAE) in aquaculture, waste management, and animal feeds, J. World Maric. Soc. (2009). https://doi.org/10.1111/j.1749-7345.1981.tb00273.x.
102. J. Xu, G. Shen, Growing duckweed in swine wastewater for nutrient recovery and biomass production, Bioresour. Technol. (2011). https://doi.org/10.1016/j.biortech.2010.09.003.

103. T. Saeed, G. Sun, A review on nitrogen and organics removal mechanisms in subsurface flow constructed wetlands: Dependency on environmental parameters, operating conditions and supporting media, J. Environ. Manage. (2012). https://doi.org/10.1016/j.jenvman.2012.08.011.
104. T.D.H. Vo, X.T. Bui, C. Lin, V.T. Nguyen, T.K.D. Hoang, H.H. Nguyen, P.D. Nguyen, H.H. Ngo, W. Guo, A mini-review on shallow-bed constructed wetlands: a promising innovative green roof, Curr. Opin. Environ. Sci. Heal. (2019). https://doi.org/10.1016/j.coesh.2019.09.004.
105. R.D. Sooknah, A.C. Wilkie, Nutrient removal by floating aquatic macrophytes cultured in anaerobically digested flushed dairy manure wastewater, Ecol. Eng. (2004). https://doi.org/10.1016/j.ecoleng.2004.01.004.
106. A.N. Shilton, L.M. Blank, Phosphorus biotechnology, Curr. Opin. Biotechnol. (2012). https://doi.org/10.1016/j.copbio.2012.09.007.
107. G. Mujtaba, M. Rizwan, G. Kim, K. Lee, Removal of nutrients and COD through co-culturing activated sludge and immobilized Chlorella vulgaris, Chem. Eng. J. 343 (2018) 155–162. https://doi.org/10.1016/j.cej.2018.03.007.
108. J. Lv, Y. Liu, J. Feng, Q. Liu, F. Nan, S. Xie, Nutrients removal from undiluted cattle farm wastewater by the two-stage process of microalgae-based wastewater treatment, Bioresour. Technol. 264 (2018) 311–318. https://doi.org/10.1016/j.biortech.2018.05.085.
109. M.B. Benítez, P. Champagne, A. Ramos, A.F. Torres, V. Ochoa-Herrera, Wastewater treatment for nutrient removal with Ecuadorian native microalgae, Environ. Technol. (United Kingdom). 40 (2019) 2977–2985. https://doi.org/10.1080/09593330.2018.1459874.
110. L. Wang, Y. Li, P. Chen, M. Min, Y. Chen, J. Zhu, R.R. Ruan, Anaerobic digested dairy manure as a nutrient supplement for cultivation of oil-rich green microalgae Chlorella sp., Bioresour. Technol. (2010). https://doi.org/10.1016/j.biortech.2009.10.062.
111. L. zao Luo, Y. Shao, S. Luo, F. jian Zeng, G. ming Tian, Nutrient removal from piggery wastewater by Desmodesmus sp.CHX1 and its cultivation conditions optimization, Environ. Technol. (United Kingdom). 40 (2019) 2739–2746. https://doi.org/10.1080/09593330.2018.1449903.
112. K. Chokshi, I. Pancha, A. Ghosh, S. Mishra, Microalgal biomass generation by phycoremediation of dairy industry wastewater: An integrated approach towards sustainable biofuel production, Bioresour. Technol. 221 (2016) 455–460. https://doi.org/10.1016/j.biortech.2016.09.070.
113. A. Ruiz-Marin, L.G. Mendoza-Espinosa, T. Stephenson, Growth and nutrient removal in free and immobilized green algae in batch and semi-continuous cultures treating real wastewater, Bioresour. Technol. 101 (2010) 58–64. https://doi.org/10.1016/j.biortech.2009.02.076.
114. R.J. Craggs, P.J. McAuley, V.J. Smith, Wastewater nutrient removal by marine microalgae grown on a corrugated raceway, Water Res. 31 (1997) 1701–1707. https://doi.org/10.1016/S0043-1354(96)00093-0.

115. B.T. Higgins, I. Gennity, P.S. Fitzgerald, S.J. Ceballos, O. Fiehn, J.S. VanderGheynst, Algal–bacterial synergy in treatment of winery wastewater, Npj Clean Water. 1 (2018). https://doi.org/10.1038/s41545-018-0005-y.
116. A.T.K. Tran, Y. Zhang, D. De Corte, J.-B. Hannes, W. Ye, P. Mondal, N. Jullok, B. Meesschaert, L. Pinoy, B. Van der Bruggen, P-recovery as calcium phosphate from wastewater using an integrated selectrodialysis/crystallization process, J. Clean. Prod. 77 (2014) 140–151. https://doi.org/https://doi.org/10.1016/j.jclepro.2014.01.069.
117. Y.T. Ahn, Y.H. Hwang, H.S. Shin, Application of PTFE membrane for ammonia removal in a membrane contactor, Water Sci. Technol. (2011). https://doi.org/10.2166/wst.2011.141.
118. W. Xue, T. Tobino, F. Nakajima, K. Yamamoto, Seawater-driven forward osmosis for enriching nitrogen and phosphorous in treated municipal wastewater: Effect of membrane properties and feed solution chemistry, Water Res. (2015). https://doi.org/10.1016/j.watres.2014.11.007.
119. J. Zhang, Q. She, V.W.C. Chang, C.Y. Tang, R.D. Webster, Mining nutrients (N, K, P) from urban source-separated urine by forward osmosis dewatering, Environ. Sci. Technol. (2014). https://doi.org/10.1021/es405266d.
120. M.Y. Kiriukhin, K.D. Collins, Dynamic hydration numbers for biologically important ions, Biophys. Chem. (2002). https://doi.org/10.1016/S0301-4622(02)00153-9.
121. J.L. Cartinella, T.Y. Cath, M.T. Flynn, G.C. Miller, K.W. Hunter, A.E. Childress, Removal of natural steroid hormones from wastewater using membrane contactor processes, Environ. Sci. Technol. (2006). https://doi.org/10.1021/es060550i.
122. A. Mukherjee, A. Satish, A. Mullick, J. Rapolu, S. Moulik, A. Roy, A.K. Ghosh, Paradigm Shift toward Developing a Zero Liquid Discharge Strategy for Dye-Contaminated Water Streams: A Green and Sustainable Approach Using Hydrodynamic Cavitation and Vacuum Membrane Distillation, ACS Sustain. Chem. Eng. (2021). https://doi.org/10.1021/acssuschemeng.1c00619.
123. M. Maurer, W. Pronk, T.A. Larsen, Treatment processes for source-separated urine, Water Res. (2006). https://doi.org/10.1016/j.watres.2006.07.012.
124. W. Pronk, M. Biebow, M. Boller, Electrodialysis for recovering salts from a urine solution containing micropollutants, Environ. Sci. Technol. (2006). https://doi.org/10.1021/es051921i.
125. C. Blöcher, C. Niewersch, T. Melin, Phosphorus recovery from sewage sludge with a hybrid process of low pressure wet oxidation and nanofiltration, Water Res. (2012). https://doi.org/10.1016/j.watres.2012.01.022.
126. M. Xie, L.D. Nghiem, W.E. Price, M. Elimelech, A forward osmosis-membrane distillation hybrid process for direct sewer mining: System performance and limitations, Environ. Sci. Technol. (2013). https://doi.org/10.1021/es404056e.

127. N.C. Nguyen, S.S. Chen, H.Y. Yang, N.T. Hau, Application of forward osmosis on dewatering of high nutrient sludge, Bioresour. Technol. (2013). https://doi.org/10.1016/j.biortech.2013.01.028.
128. N.T. Hau, S.S. Chen, N.C. Nguyen, K.Z. Huang, H.H. Ngo, W. Guo, Exploration of EDTA sodium salt as novel draw solution in forward osmosis process for dewatering of high nutrient sludge, J. Memb. Sci. (2014). https://doi.org/10.1016/j.memsci.2013.12.068.
129. R.W. Holloway, A.E. Childress, K.E. Dennett, T.Y. Cath, Forward osmosis for concentration of anaerobic digester centrate, Water Res. (2007). https://doi.org/10.1016/j.watres.2007.05.054.
130. M. Xie, L.D. Nghiem, W.E. Price, M. Elimelech, Toward Resource Recovery from Wastewater: Extraction of Phosphorus from Digested Sludge Using a Hybrid Forward Osmosis-Membrane Distillation Process, Environ. Sci. Technol. Lett. (2014). https://doi.org/10.1021/ez400189z.
131. Y. Zhang, E. Desmidt, A. Van Looveren, L. Pinoy, B. Meesschaert, B. Van Der Bruggen, Phosphate separation and recovery from wastewater by novel electrodialysis, Environ. Sci. Technol. (2013). https://doi.org/10.1021/es4004476.
133. A. Bayrakdar, E. Tilahun, B. Çalli, Simultaneous nitrate and sulfide removal using a bio-electrochemical system, Bioelectrochemistry. (2019). https://doi.org/10.1016/j.bioelechem.2019.06.003.
133. P. Kuntke, T.H.J.A. Sleutels, M. Rodríguez Arredondo, S. Georg, S.G. Barbosa, A. ter Heijne, H.V.M. Hamelers, C.J.N. Buisman, (Bio)electrochemical ammonia recovery: progress and perspectives, Appl. Microbiol. Biotechnol. (2018). https://doi.org/10.1007/s00253-018-8888-6.
134. H. Yokoyama, H. Ohmori, M. Ishida, M. Waki, Y. Tanaka, Treatment of cow-waste slurry by a microbial fuel cell and the properties of the treated slurry as a liquid manure, Anim. Sci. J. (2006). https://doi.org/10.1111/j.1740-0929.2006.00395.x.
135. Wu, T. Lv, Q. Lu, Z. Ajmal, R. Dong, Treatment of anaerobic digestate supernatant in microbial fuel cell coupled constructed wetlands: Evaluation of nitrogen removal, electricity generation, and bacterial community response, Sci. Total Environ. (2017). https://doi.org/10.1016/j.scitotenv.2016.11.138.
136. T. Kim, J. An, J.K. Jang, I.S. Chang, Coupling of anaerobic digester and microbial fuel cell for COD removal and ammonia recovery, Bioresour. Technol. (2015). https://doi.org/10.1016/j.biortech.2015.06.009.
137. T. Pepè Sciarria, G. Vacca, F. Tambone, L. Trombino, F. Adani, Nutrient recovery and energy production from digestate using microbial electrochemical technologies (METs), J. Clean. Prod. (2019). https://doi.org/10.1016/j.jclepro.2018.10.152.
138. M. Cerrillo, M. Viñas, A. Bonmatí, Overcoming organic and nitrogen overload in thermophilic anaerobic digestion of pig slurry by coupling a microbial electrolysis cell, Bioresour. Technol. (2016). https://doi.org/10.1016/j.biortech.2016.05.085.

139. A. Sotres, M. Cerrillo, M. Viñas, A. Bonmatí, Nitrogen recovery from pig slurry in a two-chambered bioelectrochemical system, Bioresour. Technol. (2015). https://doi.org/10.1016/j.biortech.2015.07.036.
140. R.D. Cusick, B.E. Logan, Phosphate recovery as struvite within a single chamber microbial electrolysis cell, Bioresour. Technol. (2012). https://doi.org/10.1016/j.biortech.2011.12.038.
141. X. Wang, R. Yang, Z. Zhang, J. Wu, S. Chen, Mass balance and bacterial characteristics in an in-situ full-scale swine wastewater treatment system occurring anammox process, Bioresour. Technol. (2019). https://doi.org/10.1016/j.biortech.2019.122005.
142. S. Lu, H. Li, G. Tan, F. Wen, M.T. Flynn, X. Zhu, Resource recovery microbial fuel cells for urine-containing wastewater treatment without external energy consumption, Chem. Eng. J. (2019). https://doi.org/10.1016/j.cej.2019.05.130.
143. P. Kuntke, K.M. Śmiech, H. Bruning, G. Zeeman, M. Saakes, T.H.J.A. Sleutels, H.V.M. Hamelers, C.J.N. Buisman, Ammonium recovery and energy production from urine by a microbial fuel cell, Water Res. (2012). https://doi.org/10.1016/j.watres.2012.02.025.
144. O. Ichihashi, K. Hirooka, Removal and recovery of phosphorus as struvite from swine wastewater using microbial fuel cell, Bioresour. Technol. (2012). https://doi.org/10.1016/j.biortech.2012.02.124.
[145] F. Fischer, C. Bastian, M. Happe, E. Mabillard, N. Schmidt, Microbial fuel cell enables phosphate recovery from digested sewage sludge as struvite, Bioresour. Technol. (2011). https://doi.org/10.1016/j.biortech.2011.02.089.
146. K. Hirooka, O. Ichihashi, Phosphorus recovery from artificial wastewater by microbial fuel cell and its effect on power generation, Bioresour. Technol. (2013). https://doi.org/10.1016/j.biortech.2013.03.067.
147. F. Zhang, J. Li, Z. He, A new method for nutrients removal and recovery from wastewater using a bioelectrochemical system, Bioresour. Technol. (2014). https://doi.org/10.1016/j.biortech.2014.05.105.
148. H.J. Jeon, K.W. Seo, S.H. Lee, Y.H. Yang, R.S. Kumaran, S. Kim, S.W. Hong, Y.S. Choi, H.J. Kim, Production of algal biomass (Chlorella vulgaris) using sediment microbial fuel cells, Bioresour. Technol. (2012). https://doi.org/10.1016/j.biortech.2011.06.039.
149. X. Chen, H. Zhou, K. Zuo, Y. Zhou, Q. Wang, D. Sun, Y. Gao, P. Liang, X. Zhang, Z.J. Ren, X. Huang, Self-sustaining advanced wastewater purification and simultaneous *in situ* nutrient recovery in a novel bioelectrochemical system, Chem. Eng. J. (2017). https://doi.org/10.1016/j.cej.2017.07.130.
150. P.P. Rajesh, D.A. Jadhav, M.M. Ghangrekar, Improving performance of microbial fuel cell while controlling methanogenesis by Chaetoceros pretreatment of anodic inoculum, Bioresour. Technol. (2015). https://doi.org/10.1016/j.biortech.2014.12.095.

151. D. Xing, Y. Zuo, S. Cheng, J.M. Regan, B.E. Logan, Electricity generation by Rhodopseudomonas palustris DX-1, Environ. Sci. Technol. (2008). https://doi.org/10.1021/es800312v.
152. M. Zeppilli, A. Mattia, M. Villano, M. Majone, Three-chamber Bioelectrochemical System for Biogas Upgrading and Nutrient Recovery, Fuel Cells. (2017). https://doi.org/10.1002/fuce.201700048.
153. A. Almatouq, A.O. Babatunde, Concurrent phosphorus recovery and energy generation in mediator-less dual chamber microbial fuel cells: Mechanisms and influencing factors, Int. J. Environ. Res. Public Health. (2016). https://doi.org/10.3390/ijerph13040375.
154. L. Xiao, E.B. Young, J.A. Berges, Z. He, Integrated photo-bioelectrochemical system for contaminants removal and bioenergy production, Environ. Sci. Technol. (2012). https://doi.org/10.1021/es303144n.
155. G. Qiu, Y.P. Ting, Direct phosphorus recovery from municipal wastewater via osmotic membrane bioreactor (OMBR) for wastewater treatment, Bioresour. Technol. (2014). https://doi.org/10.1016/j.biortech.2014.07.103.
156. G. Qiu, Y.M. Law, S. Das, Y.P. Ting, Direct and complete phosphorus recovery from municipal wastewater using a hybrid microfiltration-forward osmosis membrane bioreactor process with seawater brine as draw solution, Environ. Sci. Technol. (2015). https://doi.org/10.1021/es504554f.
157. R.W. Holloway, A.S. Wait, A. Fernandes da Silva, J. Herron, M.D. Schutter, K. Lampi, T.Y. Cath, Long-term pilot scale investigation of novel hybrid ultra-filtration-osmotic membrane bioreactors, Desalination. (2015). https://doi.org/10.1016/j.desal.2014.05.040.
158. G. Qiu, S. Zhang, D.S. Srinivasa Raghavan, S. Das, Y.P. Ting, The potential of hybrid forward osmosis membrane bioreactor (FOMBR) processes in achieving high throughput treatment of municipal wastewater with enhanced phosphorus recovery, Water Res. (2016). https://doi.org/10.1016/j.watres.2016.09.017.
159. H.M. Chang, S.S. Chen, N.C. Nguyen, W.S. Chang, S. Sinha Ray, Osmosis membrane bioreactor–microfiltration with magnesium-based draw solute for salinity reduction and phosphorus recovery, Int. Biodeterior. Biodegrad. (2017). https://doi.org/10.1016/j.ibiod.2017.05.022.
160. W. Luo, F.I. Hai, W.E. Price, W. Guo, H.H. Ngo, K. Yamamoto, L.D. Nghiem, Phosphorus and water recovery by a novel osmotic membrane bioreactor-reverse osmosis system, Bioresour. Technol. (2016). https://doi.org/10.1016/j.biortech.2015.10.029.
161. A. Mukherjee, A. Mullick, P. Vadthya, S. Moulik, A. Roy, Surfactant degradation using hydrodynamic cavitation based hybrid advanced oxidation technology: A techno economic feasibility study, Chem. Eng. J. 398 (2020) 125599. https://doi.org/10.1016/j.cej.2020.125599.
162. A. Mukherjee, A. Mullick, R. Teja, P. Vadthya, A. Roy, S. Moulik, Performance and energetic analysis of hydrodynamic cavitation and potential integration with existing advanced oxidation processes: A case study for real life

greywater treatment, Ultrason. Sonochem. 66 (2020) 105116. https://doi.org/10.1016/j.ultsonch.2020.105116.
163. N. Jeanmaire, T. Evans, Technico-economic feasibility of p-recovery from municipal wastewaters, Environ. Technol. (United Kingdom). (2001). https://doi.org/10.1080/09593332208618194.
164. T. Dockhorn, About the economy of phosphorus recovery.pdf, in: Int. Conf. Nutr. Recover., 2009.
165. L. Shu, P. Schneider, V. Jegatheesan, J. Johnson, An economic evaluation of phosphorus recovery as struvite from digester supernatant, Bioresour. Technol. (2006). https://doi.org/10.1016/j.biortech.2005.11.005.
166. B. Etter, E. Tilley, R. Khadka, K.M. Udert, Low-cost struvite production using source-separated urine in Nepal, Water Res. (2011). https://doi.org/10.1016/j.watres.2010.10.007.
167. E. Desmidt, K. Ghyselbrecht, Y. Zhang, L. Pinoy, B. Van Der Bruggen, W. Verstraete, K. Rabaey, B. Meesschaert, Global phosphorus scarcity and full-scale P-recovery techniques: A review, Crit. Rev. Environ. Sci. Technol. (2015). https://doi.org/10.1080/10643389.2013.866531.
168. E. V. Munch, K. Barr, Controlled struvite crystallisation for removing phosphorus from anaerobic digester sidestreams, Water Res. (2001). https://doi.org/10.1016/S0043-1354(00)00236-0.
169. B.K. Mayer, L.A. Baker, T.H. Boyer, P. Drechsel, M. Gifford, M.A. Hanjra, P. Parameswaran, J. Stoltzfus, P. Westerhoff, B.E. Rittmann, Total Value of Phosphorus Recovery, Environ. Sci. Technol. (2016). https://doi.org/10.1021/acs.est.6b01239.

15

Sustainable Desalination: Future Scope in Indian Subcontinent

Rudra Rath[1], Asim K. Ghosh[2] and Anirban Roy[1]*

[1]Water-Energy Nexus Lab, Department of Chemical Engineering Goa, BITS Pilani, Goa, India
[2]Desalination and Membrane Technology Division, Bhabha Atomic Research Centre, Trombay, Mumbai, India

Abstract
The book chapter will be presenting a scope and perspective on the sustainable desalination processes for the Indian subcontinent. As water woes plague the world and as India is poised to be the next destination as manufacturing hub, water generation, conservation and recycling is the greatest challenge which is being posed. In this regard, desalination is the only possible route to generate water to meet demands. As different desalination technologies mature around the world, scalability and translation to large scale water production will be governed by technological maturity and acceptability. In this chapter, the author(s) will discuss the few technologies which have the potential to make inroads in the Indian market with respect to desalination, where, centralized desalination plants can be based on membrane distillation and decentralized treatments can be based on solar stills. The technological and economic challenges will be the drivers for sustaining these in India.

Keywords: Desalination, sustainability, solar, decentralization

15.1 Introduction

Traditionally, India has always been blessed with water. The perennial rivers like the Ganga and the annual monsoon have always helped the

**Corresponding author*: anirbanr@goa.bits-pilani.ac.in

Siddhartha Moulik, Aditi Mullick and Anirban Roy (eds.) Sustainable Water Treatment: Advances and Technological Interventions, (567–590) © 2022 Scrivener Publishing LLC

subcontinent in sustaining its water demand. However, due to an explosion in population as well as overexploitation, water availability stands challenged [1, 2]. Apart from these issues, changes in the socio-economic pattern, as well as climate, are posing an insurmountable challenge to water availability [3]. Several climatic prediction modeling efforts [4] have revealed that freshwater availability will be challenged due to uncertainties in precipitation patterns around the globe. This poses an intimidating challenge, since generating water independent of climatic changes for an entire nation is a nearly impossible task. However, since Loeb-Sourirajan [5] designed the first reverse osmosis membrane phase inversion and Israel started desalination to quench the thirst, the impossible was in fact possible. Desalinating seawater is one of the most sustainable ways to provide a steady and consistent supply of water. However, as desalination involves separation, it is thermodynamically a non-spontaneous process [6] and requires energy input in form of either work (membrane-based) or heat (thermodynamics) [7]. India has historically been a power-deficient nation [8] and therefore, desalination can be challenged. Thus, for desalination to be a sustainable solution, energy penalties need to be understood and minimized as well as the cost and price of water production [9–11]. This book chapter explores the background of desalination technology implementation in India and the technologies which have the potential to successfully make inroads. The technologies with their viability, challenges and implementation strategies for sustainable use of available desalination technologies have been discussed in depth.

15.2 Water Supply and Demand in India

According to the Aqueduct Water Risk Atlas released by the World Resources Institute (WRI) [1], India ranks 13th among the seventeen extremely water-stressed nations while the groundwater levels have declined at a rate of more than 8 cm/year from 1990-2014, Chandigarh followed by Haryana, Rajasthan and Uttar Pradesh being the most water-stressed states [1]. In the Annual report on Ground Water Year Book 2019-2020, water from 37% of the groundwater sources in India is unsafe for drinking [12]. 70% of water sources are polluted, with arsenic and fluoride being the major cause of concern [13]. In major parts of northwestern India, a groundwater level of more than 40 m bgl (below ground level) was recorded (as shown in Figure 15.1) and a maximum depth of 121.60 m bgl was observed in Rajasthan [12]. The stage of extraction of groundwater is more than 100 % clearly showing an offset between the supply and demand for potable water

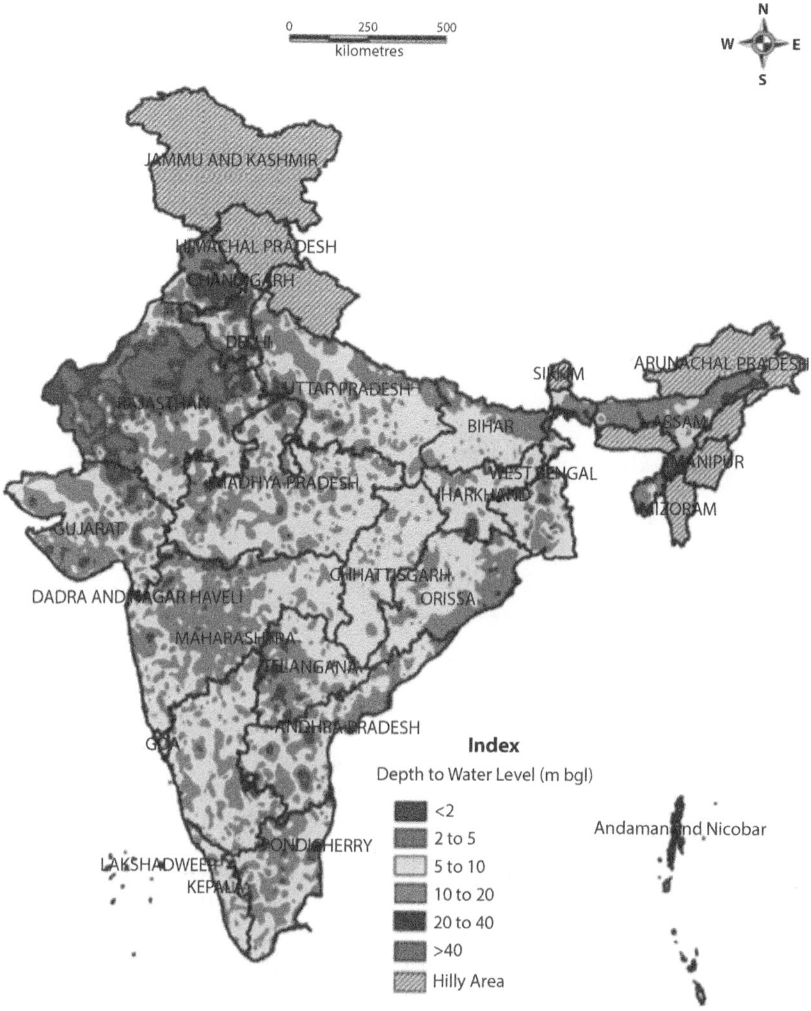

Figure 15.1 Map of ground water level in India 2019-20 (adapted from [12]).

as (shown in Figure 15.2) [12]. Post-independence, the Indian government has been introducing schemes like the Accelerated Rural Water Supply Programme of 1972 to the recent Jal Jeevan Mission (JJM) in 2019 under the Department of Drinking Water & Sanitation to provide functional household tap connections by 2024, yet in several states more than half of the rural population doesn't have access to the Government's benchmark of 55 LPCD (liters per capita per day) [14]. To cope with this demand-supply offset of drinking water, the Ministry of Drinking Water and

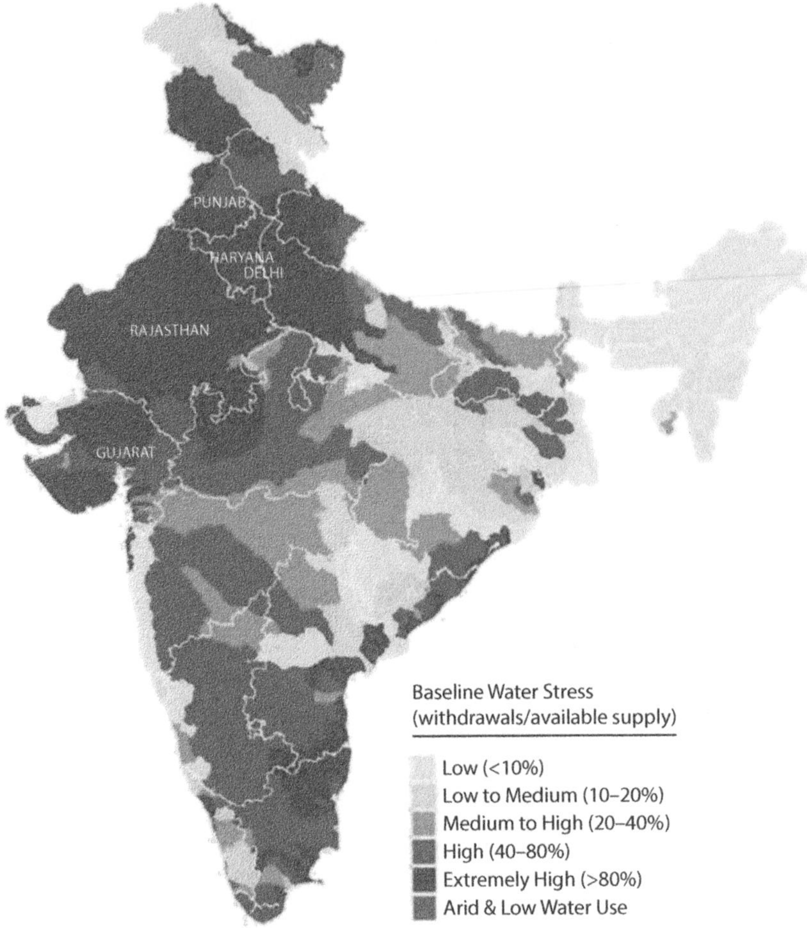

Figure 15.2 Water stress in India 2015 (adapted from [17]).

Sanitation (MoDWS) has proposed installing desalination plants developed by CSIR-Central Salt and Marine Chemicals Research Institute (CSMCRI) in coastal areas in the states of Gujarat, Tamil Nadu, and Andhra Pradesh [15]. The Ministry of Earth Sciences (MoES) has also proposed installation of six desalination plants of 1.5 LPCD in Lakshadweep islands [16], a Low-Temperature Thermal Desalination (LTTD) plant of 10 MLD (million liters per day) capacity, a floating offshore plant, and developing solar desalination plants for commercial use [15].

15.3 Current Status of Desalination in India

Desalination is a great alternative water source as it is drought-proof, independent of river flows, reservoir levels, or climate change, and 75% of the country's desalinated water is used in industries [19]. Figure 15.3 shows a classification of major desalination technologies. India's coastline is 7517 km and 25% of India's population lives in coastal areas [20] which makes desalination a preferable option. Poor water availability, low groundwater levels, and high demand have made states of Gujarat and Tamil Nadu the pioneers in desalination technology in India, which are followed by the coastal regions of Maharashtra, Andhra Pradesh, Karnataka, and West Bengal. Tamil Nadu was the first state in India where rainwater harvesting was made compulsory. Gujarat, Andhra Pradesh, Madhya Pradesh, Kerala, New Delhi, Rajasthan, and Haryana are making some legislative reforms for rainwater harvesting [21]. Gujarat has planned five new hybrid desalination plants along with solar and wind power generation at Mandvi (100 MLD), Dahej (100 MLD), Dwarka (70 MLD), Ghogha Bhavnagar (70 MLD), and Gir Somnath (30 MLD) [22, 23]. Currently, Reverse Osmosis

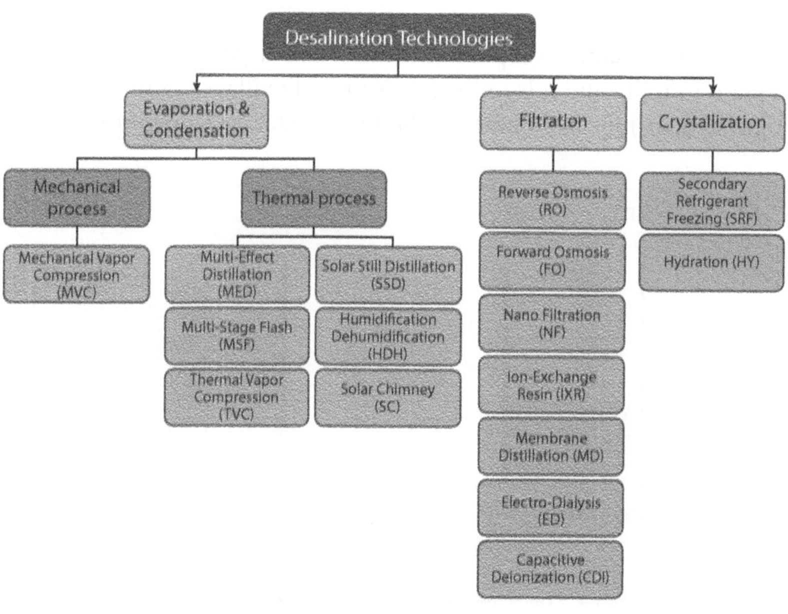

Figure 15.3 Classification of water desalination technologies (adapted from [18]).

(RO) is the most popular technology in India followed by Multiple Effect Distillation (MED) mostly being used for big capacity plant.

15.4 Commercially Available Technologies

15.4.1 Reverse Osmosis (RO)

Reverse osmosis (RO) is a pressure-driven process in which a higher concentration solution is separated from a lower concentration solution by a semi-permeable membrane as shown in Figure 15.4 [24, 25]. A pressure higher than the osmotic pressure of the high concentration solution is applied and it permeates through the membrane to give freshwater. Most desalination plants in India are based on this technology and it is the only commercially available technology for in-home desalination which can recover 25-50% of water supplied as product water [10, 26]. According to the Department of Drinking Water and Sanitation, there are 23457 RO community water purification plants in India, 75% of which are in Karnataka [27]. RO, with its superior chemical removal capacity, is also a fast-growing market for household water purification system in India [28] with the average prices of RO filters being ₹6895 to ₹11758, and the annual servicing costs range from ₹2075 to ₹3124 [29].

The Minjur desalination plant is the first and largest RO-based desalination plant in India and it is located in Kattupalli village, Tamil Nadu. It was inaugurated in July 2010 and has 8600 seawater RO membranes having

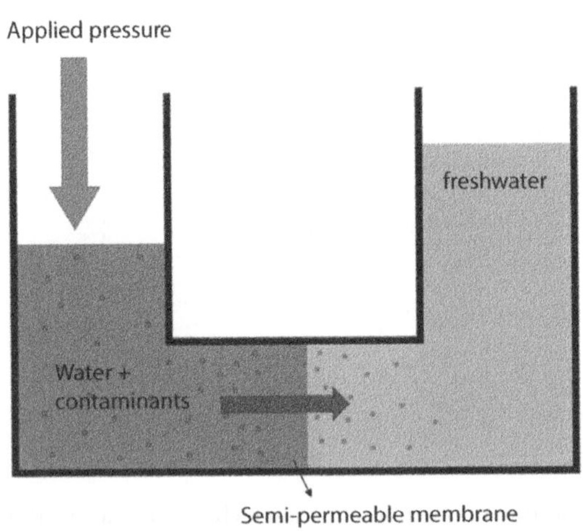

Figure 15.4 Schematic of reverse osmosis.

a capacity of 100 MLD [30]. The Nemmeli desalination plant was India's second plant located in Chennai to provide water to the southern parts of Chennai with 100 MLD capacity [31–33]. Bhabha Atomic Research Centre (BARC) has developed domestic water purifiers, membranes, as well as desalination plants in Andhra Pradesh, Gujarat, Rajasthan, Tamil Nadu, Port Blair, and Lakshadweep [31, 34, 35]. Bharat Heavy Electricals Limited (BHEL) is operating 12 desalination plants in Tamil Nadu using RO technology [34]. Gujarat Power Limited (CGPL) has a 25,200 m^3/day seawater RO unit located in Mundra, coastal Gujarat [36].

Major disadvantages of RO are its high maintenance cost, disposal of highly concentrated reject brine, and low water recovery rates [37]. Removal of microorganisms is not recommended and since it removes minerals and reduces pH, it makes water taste sour [28]. Some concerns have been raised regarding the adverse effects of drinking RO-treated water in the long run with claims of vitamin B12, calcium, magnesium deficiency which has made RO with TDS controller a possible alternative [29, 38].

15.4.2 Electrodialysis (ED)

ED consists of an anode and a cathode along with a series of cation or anion-selective membranes as shown in Figure 15.5. Ions are transported through a semipermeable membrane under the influence of electric potential and this method can be used to remove ions from water [26]. India's primary chemical water contaminants are charged ions like arsenic, fluoride, iron, nitrates, and dissolved salts, all of which are removable by ED [39]. In water-scarce regions, maximizing the recovery ratio is important and ED has a recovery ratio of nearly double that of RO systems [39]. The average life of ED membranes is 10 years, which is 2–3 times longer than

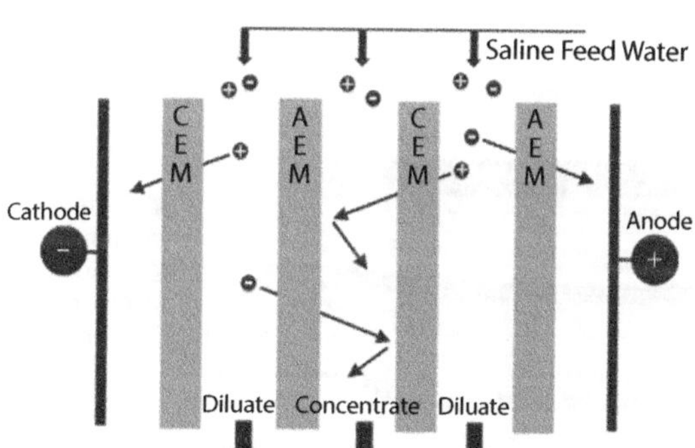

Figure 15.5 Schematic of electrodialysis (adapted from [39] with permission).

that of RO membranes (5-7 years) [39]. This gives ED plants an advantage in a community-scale water purification system. The first ED plant was constructed in Kavarrati island, Lakshadweep having a capacity of 5.5 m^3/day [40]. For in-home applications, the ED system is estimated to cost ₹ 19,869 producing desalinated water at a rate of 12 L/hr [26]. However, ED is only suitable for the treatment of water having TDS up to 12,000 mg/L and since it does not remove bacterial contaminants, post-treatment is required [20].

15.4.3 Membrane Capacitive Deionization (MCDI)

Membrane capacitive deionization (MCDI) is an ion-removal process as shown in Figure 15.6. Ion-exchange membranes are placed in front of two oppositely placed porous electrodes and saline water flows in between. Electrical potential difference is applied due to which the ions are adsorbed in the electrodes and the resultant product stream has a reduced salt concentration [41]. Idropan Dell'Orto of Italy is partnering with AquaSphere and has bought Plimmer MCDI units to India. There are currently 150 such MCDI units in India [37].

15.4.4 Thermal Desalination

Thermal desalination is the process of applying heat to the saline water to induce phase changes so that it gets evaporated and then condenses to produce freshwater [37].

Mechanical vapor compression (MVC) is a distillation process as shown in Figure 15.7 (a) where the water vapor is compressed which in turn delivers heat for the vaporization of saline water [34]. A 50 m^3/day two-stage MED-MVC unit was set up by BARC in November 2006 [43].

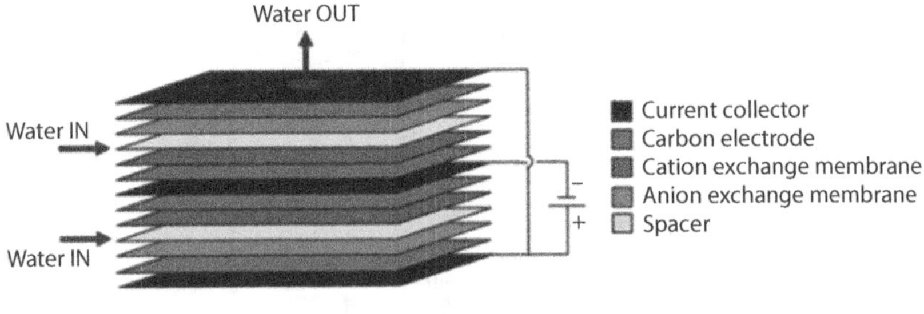

Figure 15.6 Schematic of membrane capacitive deionization (MCDI) (adapted from [41] with permission).

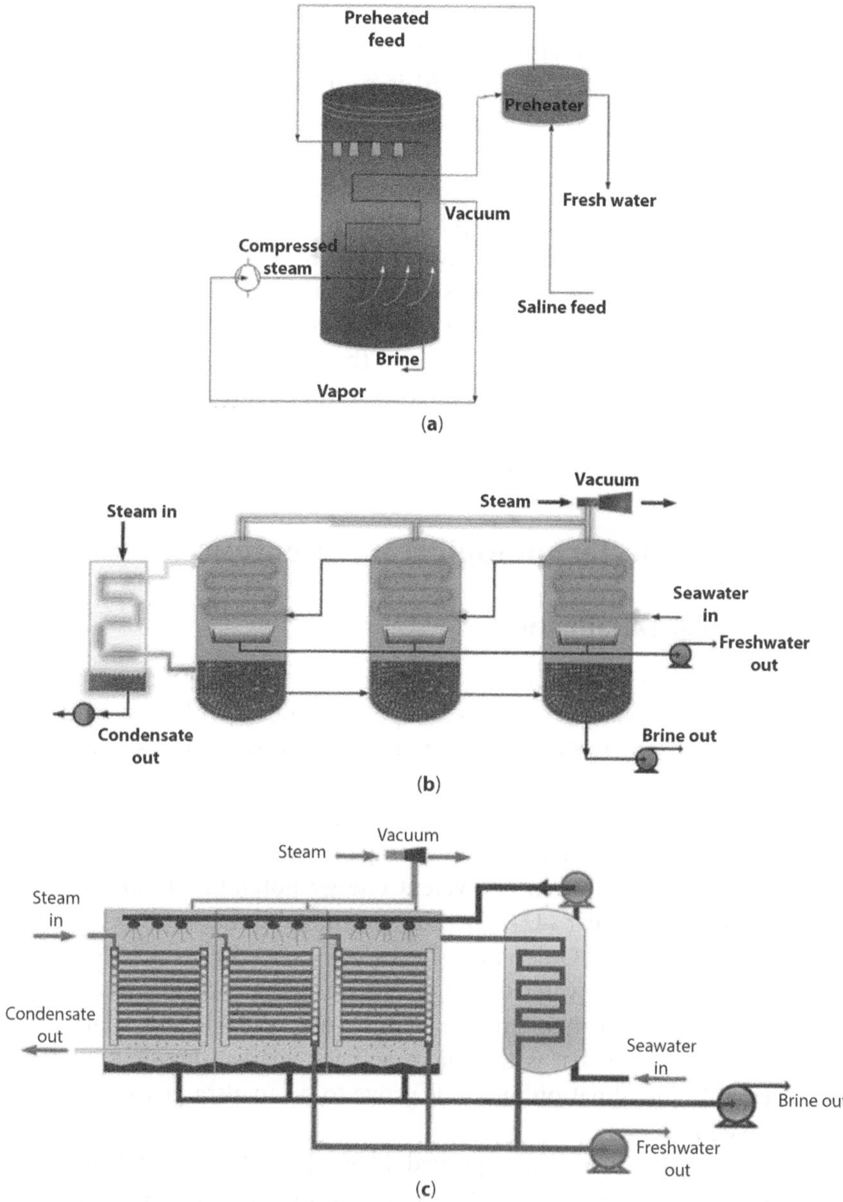

Figure 15.7 Schematic of (a) Mechanical vapor compression (MVC) (b) Multi-stage flash distillation (MSF) (c) Multi-effect distillation (MED) (adapted from [42] with permission).

The government of Andhra Pradesh is considering the installation of an MVC unit at Krishnapatnam port [44].

Multi-stage flash distillation (MSF) is a form of distillation as shown in Figure 15.7 (b) where heated saline water is allowed to flow through a series of chambers having progressively decreasing pressures. The water evaporated in each section is then condensed to be used as freshwater [37]. BARC has built a 425 KLD MSF pilot plant in Trombay, and a hybrid 4.5 MLD MSF unit as a part of the 6.3 MLD MSF-RO plant [45].

Multi-effect distillation (MED) as shown in Figure 15.7 (c) consists of a series of chambers with decreasing pressure and a heat source is used to increase the temperature of the feed and the vapor thus produced in the first stage is passed on to the next stage [42]. A 0.16 MLD MED plant is in operation in the Jamnagar oil refinery [46].

15.5 Possible Technological Intervention

15.5.1 Solar Desalination

Solar desalination is the method of producing potable water from brine with the help of solar energy. It is an age-old process with records dating back to 287 BC when Archimedes had used concave mirrors to concentrate solar radiation to boil seawater, Persian alchemists used it to produce medicinal distillates, and even in World War II as an emergency source of water [48, 49]. India receives an average intensity of solar radiation of 200 MW/Km^2 [50] with an equivalent energy potential of 6,000 million GWh of energy per year [47]. The highest annual radiation is received by states of Rajasthan and Gujarat (as shown in Figure 15.8) which being arid regions are plagued by water shortage, especially in the rural areas. In such locations where fewer maintenance facilities are available and in underdeveloped areas where cost is the major setback against desalination technologies, solar desalination methods prove to be an attractive economical alternative [49].

The Indian Government approved Phase-II of the "Grid Connected Rooftop and Small Solar Power Plants Programme", on 19th February 2019 for achieving a cumulative capacity of 40 GW rooftop solar plants by 2022 [51] and the Union Budget 2021 announced the allocation of ₹ 1.97 trillion from the financial year 2021-22 to help the country to move towards solar power [52].

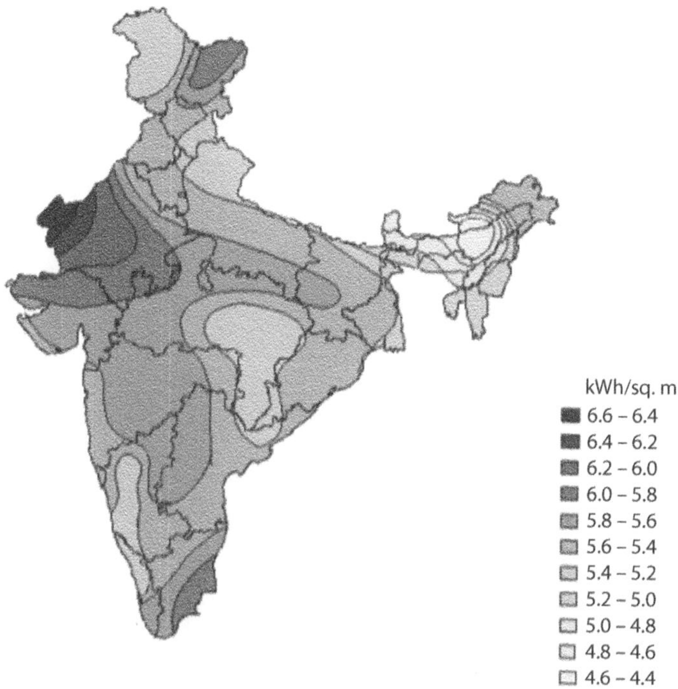

Figure 15.8 Map of solar radiation in India 2012 (adapted from [47]).

15.5.1.1 Solar Stills

A solar still (SS) (Figure 15.9) mimics the natural phenomena of rain formation. SS consists of an air-tight container covered with a sloped glass roof or plastic film through which sunlight enters and gets absorbed by the saline water, raising its temperature above the temperature of the glass cover while still being below the boiling point of water. The water gets evaporated and it condenses on the film, drips down, and is collected to be used as potable water [49]. SS can be single-effect or multi-effect stills and can be further classified as active or passive [54]. SS can produce good quality drinking water falling within the guidelines of the World Health Organization (WHO), US Environmental Protection Agency (EPA), and Bureau of Indian Standards (BIS) for safe drinking water from feed water from any source as long as it is free from volatile contaminants [55].

India's first-ever solar distillation plant was constructed in 1978 by Central Salt and Marine Chemical Research Institute (CSMCRI) in Awania

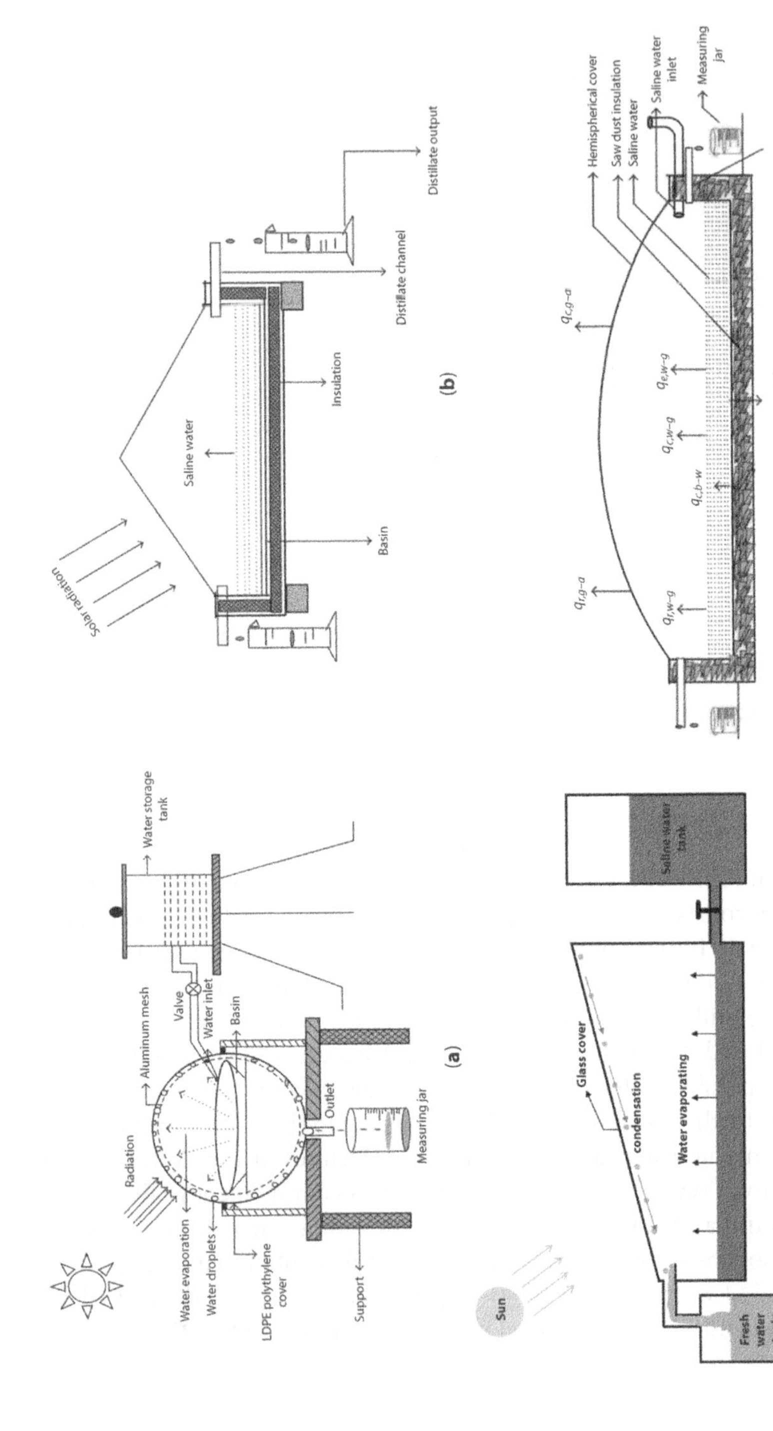

Figure 15.9 Schematic of single effect solar still (a) spherical solar still (b) pyramid solar still (c) single slope solar still (d) hemispherical solar still (adapted from [53]).

village, Bhavnagar Gujarat. It had 90 solar stills with a capacity of producing drinking water at the rate of 5000 L/day [56]. The cost of SS system installation for a whole village is reported to be ₹ 1528132.12 [57] with zero operating cost and the water can be as cheap as ₹ 2 /L - ₹ 5 /L [58, 59].

The major drawback of SS is its efficiency which varies between 25-50% depending on the design and material of construction and has productivity ranging from 4–5 l/m²/day [59]. It is greatly dependent on the weather conditions, time of the year, and geographic location. Also, standing water may promote algal growth, glass cover may break allowing dust to enter, distilled water will lack salts and minerals, and is not effective for community-scale water purification [39]. The efficiency of SS can be improved by waste hot water from plants as feed, using it with solar pond, and using multi-effect SS [54].

15.5.1.2 Photovoltaic (PV) Powered Desalination in India

Photovoltaics are made up of semiconductor materials which on exposure to photons form electron-hole pairs. Thus, they generate electricity by converting solar energy into a flow of electrons by photovoltaic effect [60] as shown in Figure 15.10.

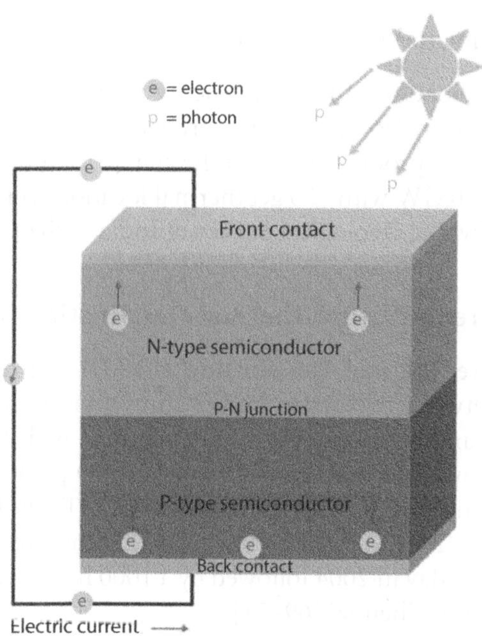

Figure 15.10 Schematic of a photovoltaic (PV) cell (adapted from [60] with permission).

Tata projects have partnered with MIT to facilitate solar-powered community-scale ED units in India and a photo voltaic-powered electrodialysis reversal (PV-EDR) having 500 LPH capacity is being investigated in Chelluru [37, 61, 62].

Community-scale PV-RO units have been installed in the Sambhar Salt Lake region of Rajasthan by an NGO called the Barefoot College where a family has to pay ₹ 50 per month and can use up to 50 L water per day [37].

15.5.2 Wave Power Desalination

Wave energy research in India was initiated in 1983 under the sponsorship of the Department of Ocean Development and the first wave energy plant was constructed at Vizhinjam in Kerala in 1990 [63]. It was based on the Oscillating Water Column (OWC) principle where a column of air gets compressed and decompressed with the rise and fall of waves [63, 64]. However, it was decommissioned later as the power generation varied a lot based on weather conditions with the maximum power being generated in the monsoon seasons. A new pilot plant is under construction in the same location by the Agency for New and Renewable Energy Research and Technology (ANERT) [65, 66].

15.5.3 Geothermal Desalination

Geothermal desalination utilizes the energy from high-temperature geothermal fluids like hydrothermal, geo pressurized brines, hot dry rocks, and magma for the production of freshwater [67]. India has a geothermal potential of 10 GW with 350 geothermal locations (Figure 15.11) have been identified by The Geological Survey of India (GSI) [31].

15.5.4 Low-Temperature Thermal Desalination (LTTD)

Low-Temperature Thermal Desalination (LTTD) utilizes the temperature gradient between two water bodies to evaporate the warmer water at low pressure and condense the resultant vapor with the colder water from which freshwater is obtained. The first LTTD plant was installed by the National Institute of Ocean Technology (NIOT) under the Ministry of Earth Sciences (MoES) at Kavaratti in the Lakshadweep islands with a capacity of 100 m^3/day in 2004 followed by a 1000 m^3/day capacity unity in 2007 on the coast of Chennai [69–71].

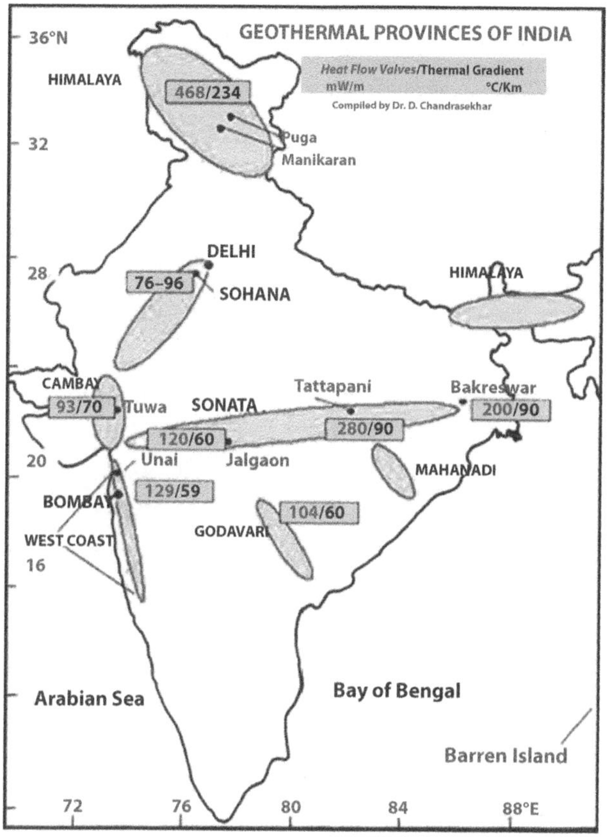

Figure 15.11 Geothermal sources in India (adapted from [68]).

15.5.5 Membrane Distillation (MD)

Membrane distillation is a thermally driven process in which a hot brine is passed on one side of a hydrophobic membrane which acts as a barrier to the liquid phase allowing only the vapors phase to pass through the membrane [73]. Figure 15.12 depicts direct contact membrane distillation (DCMD), air gap membrane distillation (AGMD), sweeping gas membrane distillation (SGMD), and vacuum membrane distillation (VMD) [72]. It can treat water having salinity up to 200,000 ppm and the water recovery can reach up to 90% [74]. Fouling is significantly lower and recovery can be increased to crystallization as a part of zero liquid discharge (ZLD) scheme [74].

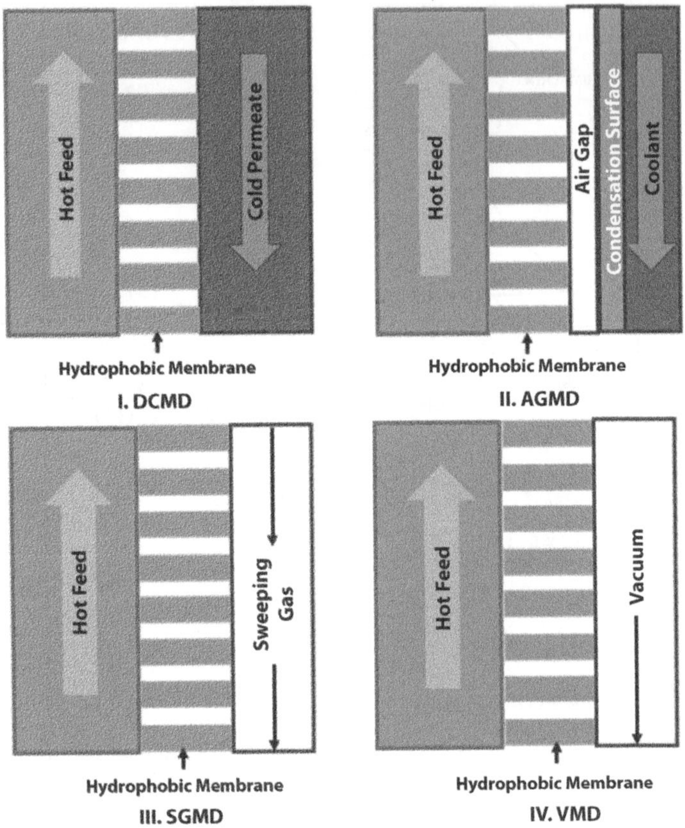

Figure 15.12 Schematic of membrane distillation (adapted from [72] with permission).

15.5.6 Forward Osmosis (FO)

Forward osmosis (FO) is the natural process of osmosis as shown in Figure 15.13. It relies on the osmotic pressure difference between the saline water feed solution and highly concentrated draw solution separated by a semi-permeable membrane causing water to flow from the feed solution into the draw solution [75]. A plant operating on FO technology can be constructed at 90% of the construction cost and operated at 80% of the operation cost of an SWRO plant [16].

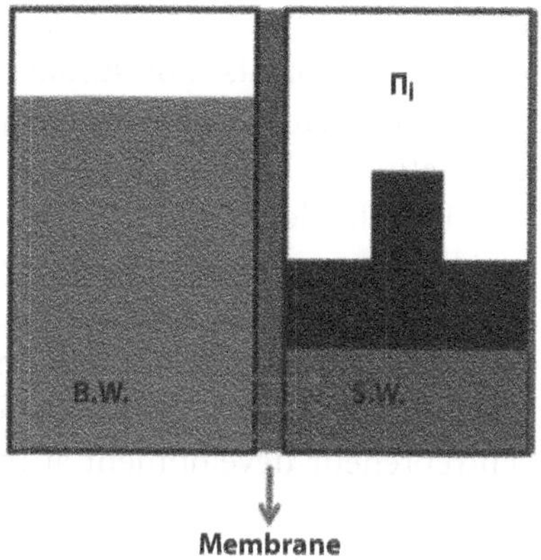

Figure 15.13 Schematic of forward osmosis (FO).

15.6 Challenges and Implementation Strategies for Sustainable Use of Desalination Technologies

In spite of technological advances, several constraints are currently holding back the wider use of desalination as an alternate technology for water purification. The cost of desalinated water is twice that of conventional water supply sourced from groundwater or rainwater while energy cost is 5 to 25 times higher in the case of desalination [76] thus it is usually implemented as a last resort [19]. Desalination indirectly uses fossil fuels for electricity source leading to greenhouse gas emissions. discharge from thermal driven processes is slightly higher in temperature and a significant number of chemicals are used for pretreatment and post-treatment especially in pressure-driven processes and discharge of these pretreatment chemicals like chlorine, antiscalants, antifoaming agents, dissolved organic carbon, and disinfectant chemicals like chloroform, bromoform, halophenols, phenols, the highly concentrated reject streams into water bodies disrupts ecosystems [76]. Brine disposal methods like brine lines, surface water discharge, disposal to sewer, sub-surface injection, evaporation ponds, landfills, land applications, and zero liquid discharge (ZLD) can be considered to overcome this [37]. Renewable sources of power for desalination such as solar, wave, wind, and geothermal power can be taken into consideration. Another challenge is revolving around the operation and maintenance of desalination plants. For sustainable use of desalination, implementation strategies across India are far more important than the latest available desalination technologies, particularly in rural India. The consumers need to adopt the developed technology for specific needs and this can be achieved by various marketing strategies with keeping the local tastes and preferences known as technology adaptation. In order to facilitate the adaptability of technologies in rural India, (a) desalinated water may be offered on a nominal chargeable basis for sustainability and its judicious use, (b) entrepreneur development at the respective sites to ensure better sustainability & (c) making of mobile desalination & water purification plants in force for specific zones.

References

1. R.W. Hofste, P. Reig, L. Schleifer, 17 Countries, Home to One-Quarter of the World's Population, Face Extremely High Water Stress, (2019). https://www.wri.org/insights/17-countries-home-one-quarter-worlds-population-face-extremely-high-water-stress (accessed May 5, 2021).
2. U.A. Amarasinghe, T. Shah, B.K. Anand, India' s Water Supply and Demand from 2025-2050 : Business- as- Usual Scenario and Issues, Water. (2007) 23–61.
3. M. Rodell, J.S. Famiglietti, D.N. Wiese, J.T. Reager, H.K. Beaudoing, F.W. Landerer, M.H. Lo, Emerging trends in global freshwater availability, Nature. 557 (2018) 651–659. https://doi.org/10.1038/s41586-018-0123-1.
4. J. Schewe, J. Heinke, D. Gerten, I. Haddeland, N.W. Arnell, D.B. Clark, R. Dankers, S. Eisner, B.M. Fekete, F.J. Colón-González, S.N. Gosling, H. Kim, X. Liu, Y. Masaki, F.T. Portmann, Y. Satoh, T. Stacke, Q. Tang, Y. Wada, D. Wisser, T. Albrecht, K. Frieler, F. Piontek, L. Warszawski, P. Kabat, Multimodel assessment of water scarcity under climate change, Proc. Natl. Acad. Sci. U. S. A. 111 (2014) 3245–3250. https://doi.org/10.1073/pnas.1222460110.
5. S. Loeb, The Leob-Sourirajan membrane: How it came about, Am. Chem. Soc. 135 (1981) 1198–1200. https://doi.org/10.1021/bk-1981-0153.ch001.
6. S. Chakraborty, A. Mukherjee, S. Das, N.R. Maddela, S. Iram, P. Das, Study on isotherm, kinetics, and thermodynamics of adsorption of crystal violet dye by calcium oxide modified fly ash, Environ. Eng. Res. 26 (2021) 1–9. https://doi.org/10.4491/eer.2019.372.
7. A. Mukherjee, S. Lanjewar, R. Kumar, A. Chakraborty, A. Abdelrasoul, A. Roy, Role of Thermodynamics and Membrane Separations in Water-Energy Nexus, Model. Membr. Membr. Process. (2020) 145–199. https://doi.org/10.1002/9781119536260.CH4.

8. A. Kemmler, Factors influencing household access to electricity in India, Energy Sustain. Dev. 11 (2007) 13–20. https://doi.org/10.1016/S0973-0826(08)60405-6.
9. A. Mukherjee, A. Satish, A. Mullick, J. Rapolu, S. Moulik, A. Roy, A.K. Ghosh, Paradigm Shift toward Developing a Zero Liquid Discharge Strategy for Dye-Contaminated Water Streams: A Green and Sustainable Approach Using Hydrodynamic Cavitation and Vacuum Membrane Distillation, ACS Sustain. Chem. Eng. 9 (2021) 6707–6719. https://doi.org/10.1021/acssuschemeng.1c00619.
10. R. Das, A. Mukherjee, I. Sinha, K. Roy, B.K. Dutta, Synthesis of potential bio-adsorbent from Indian Neem leaves (Azadirachta indica) and its optimization for malachite green dye removal from industrial wastes using response surface methodology: kinetics, isotherms and thermodynamic studies, Appl. Water Sci. 10 (2020) 1–18. https://doi.org/10.1007/s13201-020-01184-5.
11. S. Lanjewar, A. Mukherjee, P. Khandewal, A.K. Ghosh, A. Mullick, S. Moulik, A. Roy, Thermodynamics of synthesis and separation performance of Interfacially polymerized "loose" reverse osmosis membrane: Benchmarking for greywater treatment, Chem. Eng. J. 417 (2021) 127929. https://doi.org/10.1016/j.cej.2020.127929.
12. GW YEAR BOOK 2019-20 ALL INDIA FINAL 752021, n.d. http://cgwb.gov.in/Ground-Water/GW YEAR BOOK 2019-20 ALL INDIA FINAL 752021 (1).pdf (accessed April 27, 2021).
13. Asian Development Research Institure (ADRI), (n.d.). https://www.adriindia.org/adri/india_water_facts.
14. S. Chaudhuri, M. Roy, L.M. McDonald, Y. Emendack, Water for All (Har Ghar Jal): Rural Water Supply Services (RWSS) in India (2013–2018), Challenges and Opportunities, Int. J. Rural Manag. 16 (2020) 254–284. https://doi.org/10.1177/0973005220946661.
15. Strategy 3.7 | National Water Mission, Ministry of Jal Shakti, Department of Water Resources, RD & GR, Government of India, (n.d.). http://www.nationalwatermission.gov.in/?q=strategy-37 (accessed May 6, 2021).
16. Indian Scientists Develop World's First Low Temperature Thermal Desalination Plant, (n.d.). https://www.indiatimes.com/news/india/indian-scientists-develop-world-s-first-low-temperature-thermal-desalination-plant-357286.html (accessed May 14, 2021).
17. T. Shiao, A. Maddocks, C. Carson, E. Loizeaux, 3 Maps Explain India's Growing Water Risks, (2015). https://www.wri.org/insights/3-maps-explain-indias-growing-water-risks (accessed May 28, 2021).
18. D. Curto, V. Franzitta, A. Guercio, A review of the water desalination technologies, Appl. Sci. 11 (2021) 1–36. https://doi.org/10.3390/app11020670.
19. N. Dhakal, S.G. Salinas Rodriguez, J.C. Schippers, M.D. Kennedy, Perspectives and challenges for desalination in developing countries, IDA J. Desalin. Water Reuse. 6 (2014) 10–14. https://doi.org/10.1179/2051645214y.0000000015.

20. T. Abraham, A. Luthra, Socio-economic & technical assessment of photovoltaic powered membrane desalination processes for India, Desalination. 268 (2011) 238–248. https://doi.org/10.1016/j.desal.2010.10.035.
21. Legislation on Rainwater Harvesting, (n.d.). https://www.cseindia.org/legislation-on-rainwater-harvesting-1111 (accessed April 18, 2021).
22. "World's Largest Renewable Energy Park" Opens in Kutch, (n.d.). https://thewire.in/politics/worlds-largest-renewable-energy-park-opens-in-kutch (accessed May 19, 2021).
23. PM Modi to lay foundation stone of many projects including desalination plant of 100 MLD capacity in Gujarat - The Financial Express, (n.d.). https://www.financialexpress.com/infrastructure/pm-modi-to-lay-foundation-stone-of-many-projects-including-desalination-plant-of-100-mld-capacity-in-gujarat/2149594/ (accessed May 19, 2021).
24. C. Fritzmann, J. Löwenberg, T. Wintgens, T. Melin, State-of-the-art of reverse osmosis desalination, Desalination. 216 (2007) 1–76. https://doi.org/10.1016/j.desal.2006.12.009.
25. S. Lanjewar, A. Mukherjee, L. Muzamil Rehman, A. Roy, Blue Energy and Its Potential: The Membrane Based Energy Harvesting, in: A. Abdelrasoul (Ed.), Adv. Membr. Technol., IntechOpen, 2020. https://doi.org/10.5772/intechopen.86953.
26. K.G. Nayar, P. Sundararaman, C.L. O'Connor, J.D. Schacherl, M.L. Heath, M.O. Gabriel, S.R. Shah, N.C. Wright, A.G. Winter, V, Feasibility study of an electrodialysis system for in-home water desalination in urban India, Dev. Eng. 2 (2016) 38–46. https://doi.org/10.1016/j.deveng.2016.12.001.
27. Ministry of Drinking Water and Sanitation, Community Water Purification Plants, n.d. https://ejalshakti.gov.in/IMISReports/Reports/BasicInformation/rpt_RWS_CommunityWaterPurificationPlant_S.aspx?Rep=0&RP=Y.
28. N. Talat, S. Bhaduri, Incremental innovations, information contagion, and path dependence: The case of drinking water purification technologies in urban India, Ind. Corp. Chang. 26 (2017) 1089–1108. https://doi.org/10.1093/icc/dtx010.
29. L. Annala, A. Sarin, J.L. Green, Co-production of frugal innovation: Case of low cost reverse osmosis water filters in India, J. Clean. Prod. 171 (2018) S110–S118. https://doi.org/10.1016/j.jclepro.2016.07.065.
30. Minjur Desalination Plant, Tamil Nadu, India - Water Technology, (n.d.). https://www.water-technology.net/projects/minjurdesalination/(accessed May 5, 2021).
31. S. Manju, N. Sagar, Renewable energy integrated desalination: A sustainable solution to overcome future fresh-water scarcity in India, Renew. Sustain. Energy Rev. 73 (2017) 594–609. https://doi.org/10.1016/j.rser.2017.01.164.
32. S. Roumeau, A. Seifelislam, S. Jameson, L. Kennedy, S. Roumeau, A. Seifelislam, S. Jameson, L. Kennedy, W. Governance, L. Kennedy, Chennai To cite this version:, (2015).

33. N. Paul, L. Elango, Predicting future water supply-demand gap with a new reservoir, desalination plant and waste water reuse by water evaluation and planning model for Chennai megacity, India, Groundw. Sustain. Dev. 7 (2018) 8–19. https://doi.org/10.1016/j.gsd.2018.02.005.
34. D. Mehta, L. Gupta, R. Dhingra, Forward Osmosis in India: Status and Comparison with Other Desalination Technologies, Int. Sch. Res. Not. 2014 (2014) 1–9. https://doi.org/10.1155/2014/175464.
35. Bhabha Atomic Research Centre (BARC): Development & Deployment of Desalination & Water Purification, (n.d.). http://www.barc.gov.in/pubaware/water.html (accessed May 11, 2021).
36. Seawater Reverse Osmosis Plant for India's First 4,000 MW Ultra Mega Power Project | Aquatech, (n.d.). https://www.aquatech.com/project/sea-water-reverse-osmosis-plant-for-indias-first-4000-mw-ultra-mega-power-project/ (accessed May 19, 2021).
37. K.S. Boden, C. V Subban, A Road Map for Small Scale Desalination, 2018. https://www.ctc-n.org/sites/d8uat.ctc-n.org/files/resources/roadmap_for_small-scale_desalination_southeast_asia.pdf.
38. R.D. Saini, Health Risks from Long Term Consumption of Reverse Osmosis Water, Int. J. Appl. Chem. 13 (2017) 293–301. http://www.ripublication.com.
39. N.C. Wright, A.G. Winter, Justification for community-scale photovoltaic-powered electrodialysis desalination systems for inland rural villages in India, Desalination. 352 (2014) 82–91. https://doi.org/10.1016/j.desal.2014.07.035.
40. P.K. Narayanan, S.K. Thampy, N.J. Dave, D.K. Chauhan, B.S. Makwana, S.K. Adhikary, V.K. Indusekhar, Performance of the first sea water electrodialysis desalination plant in India Central Salt 8 Marine Chemicals Research Institute, SUMMARY The problem of potable water shortage in some of the islands and coastal areas of India has led to consider sea w, 84 (1991) 201–211.
41. P.M. Biesheuvel, A. van der Wal, Membrane capacitive deionization, J. Memb. Sci. 346 (2010) 256–262. https://doi.org/10.1016/j.memsci.2009.09.043.
42. H. Rabiee, K.R. Khalilpour, J.M. Betts, N. Tapper, Energy-water nexus: Renewable-integrated hybridized desalination systems, Elsevier Inc., 2018. https://doi.org/10.1016/B978-0-12-813306-4.00013-6.
43. A.K. Adak, G. Kishore, V.K. Srivastava, P.K. Tewari, Mechanical vapour compression desalination plant at Trombay, Int. J. Nucl. Desalin. 2 (2007) 353–362. https://doi.org/10.1504/IJND.2007.015801.
44. Committee to study feasibility for renewable energy based desalination plant at Krishnapatnam port - The Hindu BusinessLine, (n.d.). https://www.thehindubusinessline.com/news/committee-to-study-feasibility-for-renewable-energy-based-desalination-plant-at-krishnapatnam-port/article33859783.ece (accessed May 13, 2021).
45. Bhabha Atomic Research Center, Desalination & Water Purification Technologies: Technical Information Document, (2010) 40. http://www.barc.gov.in/publications/desalination.pdf.

46. Gujarat Reliance Project | IDE Technologies, (n.d.). https://www.ide-tech.com/en/our-projects/gujarat-reliance-project/?data=item_1 (accessed May 19, 2021).
47. S. Garud, I. Purohit, Making solar thermal power generation in India a reality – Overview of technologies , opportunities and challenges, Energy. (2012) 1–9.
48. M.T. Chaibi, An overview of solar desalination for domestic and agriculture water needs in remote arid areas, Desalination. 127 (2000) 119–133. https://doi.org/10.1016/S0011-9164(99)00197-6.
49. A.A. Delyannis, E. Delyannis, Solar desalination, Desalination. 50 (1984) 71–81. https://doi.org/10.1016/0011-9164(84)85018-3.
50. India Energy Portal, (n.d.). http://www.indiaenergyportal.org/subthemes.php?text=solar (accessed April 24, 2021).
51. MNRE, Guidelines on implementation of Phase-II of Grid Connected Rooftop Solar Programme for achieving 40 GW capacity from Rooftop Solar by the year 2022, Minist. New Renew. Energy, Gov. India. (2019). https://mnre.gov.in/img/documents/uploads/7ccd3b4b3bb94a51af516e2ee4fdede3.pdf (accessed April 27, 2021).
52. Technology in solar power transforms Indian energy sector - People - Construction Week Online India, (n.d.). https://www.constructionweekonline.in/people/17444-technology-in-solar-power-transforms-indian-energy-sector (accessed April 27, 2021).
53. T. Arunkumar, K. Vinothkumar, A. Ahsan, R. Jayaprakash, S. Kumar, Experimental Study on Various Solar Still Designs, ISRN Renew. Energy. 2012 (2012) 1–10. https://doi.org/10.5402/2012/569381.
54. P. Vishwanath Kumar, A. Kumar, O. Prakash, A.K. Kaviti, Solar stills system design: A review, Renew. Sustain. Energy Rev. 51 (2015) 153–181. https://doi.org/10.1016/j.rser.2015.04.103.
55. T. Arunkumar, K. Raj, M. Chaturvedi, A. Thenmozhi, D. Denkenberger, G. Tingting, A review on distillate water quality parameter analysis in solar still, Int. J. Ambient Energy. 0 (2019) 1–8. https://doi.org/10.1080/01430750.2019.1587722.
56. G.L. Natu, H.O. Goghari, S.D. Gomkale, Solar distillation plant at awania, Gujarat, India, Desalination. 31 (1979) 435–441. https://doi.org/10.1016/S0011-9164(00)88548-3.
57. S. Jasrotia, A. Kansal, V.V.N. Kishore, Application of solar energy for water supply and sanitation in Arsenic affected rural areas: A study for Kaudikasa village, India, J. Clean. Prod. 37 (2012) 389–393. https://doi.org/10.1016/j.jclepro.2012.07.030.
58. R.K. Khanna, R.S. Rathore, C. Sharma, Solar still an appropriate technology for potable water need of remote villages of desert state of India - Rajasthan, Desalination. 220 (2008) 645–653. https://doi.org/10.1016/j.desal.2007.01.060.

59. M. Sanserwal, A. Kumar Singh, P. Singh, Impact of materials and economic analysis of single slope single basin passive solar still: A review, Mater. Today Proc. 21 (2020) 1643–1652. https://doi.org/10.1016/j.matpr.2019.11.289.
60. O.O. Ogbomo, E.H. Amalu, N.N. Ekere, P.O. Olagbegi, A review of photovoltaic module technologies for increased performance in tropical climate, Renew. Sustain. Energy Rev. 75 (2017) 1225–1238. https://doi.org/10.1016/j.rser.2016.11.109.
61. W. He, S. Amrose, N.C. Wright, T. Buonassisi, I.M. Peters, A.G. Winter, Field demonstration of a cost-optimized solar powered electrodialysis reversal desalination system in rural India, Desalination. 476 (2020) 114217. https://doi.org/10.1016/j.desal.2019.114217.
62. D.W. Bian, S.M. Watson, N.C. Wright, S.R. Shah, T. Buonassisi, D. Ramanujan, I.M. Peters, A.G. Winter, Optimization and design of a low-cost, village-scale, photovoltaic-powered, electrodialysis reversal desalination system for rural India, Desalination. 452 (2019) 265–278. https://doi.org/10.1016/j.desal.2018.09.004.
63. N. Sharmila, P. Jalihal, A.K. Swamy, M. Ravindran, Wave powered desalination system, Energy. 29 (2004) 1659–1672. https://doi.org/10.1016/j.energy.2004.03.099.
64. P.A. Davies, Wave-powered desalination: Resource assessment and review of technology, Desalination. 186 (2005) 97–109. https://doi.org/10.1016/j.desal.2005.03.093.
65. Kerala to gauge wave power potential off Vizhinjam coast | Thiruvananthapuram News - Times of India, (n.d.). https://timesofindia.indiatimes.com/city/thiruvananthapuram/kerala-to-gauge-wave-power-potential-off-vizhinjam-coast/articleshow/76363299.cms (accessed May 28, 2021).
66. Wave power plant at Vizhinjam soon - The Hindu, (n.d.). https://www.thehindu.com/news/national/kerala/wave-power-plant-at-vizhinjam-soon/article31674205.ece (accessed May 28, 2021).
67. V.G. Gude, Geothermal source potential for water desalination - Current status and future perspective, Renew. Sustain. Energy Rev. 57 (2016) 1038–1065. https://doi.org/10.1016/j.rser.2015.12.186.
68. India Energy Portal, (n.d.). http://www.indiaenergyportal.org/subthemes_link.php?text=geothermal&themeid=13 (accessed May 28, 2021).
69. S. Kathiroli, P. Jalihal, P.V.S. Sistla, Barge Mounted Low Temperature Thermal Desalination Plant, (2008).
70. P.V.S. Sistla, G. Venkatesan, P. Jalihal, S. Kathiroli, Low Temperature Thermal Desalination Plants, (2009).
71. Low Temperature Thermal Desalination (LTTD): new sustainable desalination process: International Journal of Nuclear Desalination: Vol 3, No 1, (n.d.). https://www.inderscienceonline.com/doi/abs/10.1504/IJND.2008.01893 (accessed May 14, 2021).

72. F.E. Ahmed, B.S. Lalia, R. Hashaikeh, N. Hilal, Alternative heating techniques in membrane distillation: A review, Desalination. 496 (2020) 114713. https://doi.org/10.1016/j.desal.2020.114713.
73. J. Ravi, M.H.D. Othman, T. Matsuura, M. Ro'il Bilad, T.H. El-badawy, F. Aziz, A.F. Ismail, M.A. Rahman, J. Jaafar, Polymeric membranes for desalination using membrane distillation: A review, Desalination. 490 (2020) 114530. https://doi.org/10.1016/j.desal.2020.114530.
74. G. Amy, N. Ghaffour, Z. Li, L. Francis, R.V. Linares, T. Missimer, S. Lattemann, Membrane-based seawater desalination: Present and future prospects, Desalination. 401 (2017) 16–21. https://doi.org/10.1016/j.desal.2016.10.002.
75. A. Achilli, T.Y. Cath, A.E. Childress, Power generation with pressure retarded osmosis: An experimental and theoretical investigation, J. Memb. Sci. 343 (2009) 42–52. https://doi.org/10.1016/j.memsci.2009.07.006.
76. V.G. Gude, Desalination and sustainability - An appraisal and current perspective, Water Res. 89 (2016) 87–106. https://doi.org/10.1016/j.watres.2015.11.012.

16

Desalination: Thermodynamic Modeling and Energetics

Shubham Lanjewar[1], Ridhish Kumar[1], Kunal Roy[2], Rudra Rath[1], Anupam Mukherjee[1,3*] and Anirban Roy[1†]

[1]*Water-Energy Nexus Lab, Department of Chemical Engineering, BITS-Pilani K K Birla Goa Campus, Goa, India*
[2]*Department of Chemical Engineering, Haldia Institute of Technology, Purba Midnapur, West Bengal, Haldia, India*
[3]*Cavitation & Dynamics Lab, CSIR-Indian Institute of Chemical Technology, Hyderabad, India*

Abstract

Increasing water supply demand is the major issue nowadays all over the world. It is difficult to fulfill this demand on existing water resources. Hence, the desalination of seawater has been polished consistently for more than 60 years and is an entrenched method for water supply. Two primary bearings endure the significant development of desalination innovation, Thermal and membrane strategies. Membrane desalination is operating successfully for commercial production of fresh water such as Reverse osmosis. The energetics of desalination and how Reverse osmosis is a better choice over thermal desalination is the major focus of this chapter. The Thermodynamic modeling of the processes is presented in the first two sections to get an insight into related thermodynamic parameters and their significance in desalination. After that, the advantages of RO over other desalination techniques were explained on the basis of energy and cost analysis. lastly, an overview of exergy analysis of reverse osmosis is explained to prove that reverse osmosis can reduce the energy demand of desalination effectively as compared to other desalination techniques.

***Keywords*:** Desalination, membrane, thermal, energy, exergy

**Corresponding author*: p20170408@goa.bits-pilani.ac.in
†*Corresponding author*: anirbanr@goa.bits-pilani.ac.in

Siddhartha Moulik, Aditi Mullick and Anirban Roy (eds.) Sustainable Water Treatment: Advances and Technological Interventions, (591–642) © 2022 Scrivener Publishing LLC

16.1 Introduction

Water scarcity and its demand are increasing day by day worldwide which becomes a most severe challenge presently and over 1/3rd of the population lives in such countries worldly suffering from this problem. Moreover, population growth, industrialization, climate change and contamination of available freshwater resources are increasing the demand in safe drinking water also and therefore the challenge in getting potable water is very tough now [1, 2]. 99% of the water, available on the earth's surface is salty in nature [3]. Since clean water is mainly required to minimize pollution, therefore water reuse is the best option and it cannot be used only for irrigation or drinking purposes, every clean industry like semiconductor manufacturing plants can also use the treated wastewater to face the challenge of high water demand [4].

Today reverse osmosis becomes one of the most widely used technologies and it is gradually replacing the highly available conventional thermal desalination technologies like multi-stage flash distillation (MSF), multi-effect distillation (MED), mechanical vapor compression (MVC) etc [5] due to their higher energy consumption with installation cost and maintenance. Though there are some other new membrane-based new technologies available like membrane distillation [6] and forward osmosis [7], it is expected to maintain the leadership of RO in future also. The continuous advancement in reverse osmosis technology leads it to significant cost reductions due to the growing improvement in module design, process design, feed pre-treatment, and energy recovery, or reduction in energy consumption etc. The enhancement in the strength of RO module mechanically, biologically and chemically with the development in purification capacity by increasing its permeability have minimized the membrane and system cost per unit volume of output more than ten times since 1978 [8].

From the last few decades, reverse osmosis has reduced its specific energy consumption significantly is gradually moving towards the minimum theoretical thermodynamic limit [9]. Recent development in the large pump with an efficiency of 90% has been achieved using various energy recovery devices.

At high-pressure pumps, the energy consumption is generally high to overcome the osmotic pressure barrier of the salty water, which is a function of feed solute concentration and temperature [10]. The RO module is the most energy efficient pressure driven membrane process for the desalination of sea and brackish water among the other available membrane technologies which can operate in the range of 50-80 bar and 20-40 bar respectively.

As per the literature, a single stage RO module requires the minimum theoretical energy for desalination should be equal to the energy needed to bring the seawater to its osmotic pressure across the membrane. But in practice, to ensure that water permeates across the membrane at all points along the membrane channel, the pressure applied to the feed volume must be at least equal to the osmotic pressure leaving the membrane channel [1].

16.2 Thermodynamics Modeling of Desalination

Before going for the desalination through the various processes as discussed in this chapter, there is a need of understanding thermodynamic properties of seawater or brackish water first to develop a bridge between the actual and thermodynamic minimum energy of desalination. Seawater is mainly a mixture of various salts but the composition of sodium chloride is found more in the analysis. At very high concentration, the sodium chloride solution can also behave as seawater [11]. Along with the pressure, volume, temperature enthalpy, and entropy etc. salinity is also a thermodynamic property which plays an important role in desalination.

The solutions would behave as ideal, only if the enthalpy and the volume will remain the same before and after mixing (i.e. Change in Enthalpy and change in volume = constant or zero). The Ideal solutions are different from Ideal gases in only one important aspect that is, for Ideal gases intermolecular interactions can be neglected but in case of liquids intermolecular bonding is very strong and cannot be simply neglected and therefore we assume that the mean strength of the intermolecular interactions is same between all the molecules of the solution. If the change in enthalpy of solutions is closer to zero, more the solution tends towards the ideality and thereby the Gibbs free energy of mixing will solely depend on entropy change of mixing in an ideal solution.

$$G = \sum_i n_i \mu_i \tag{16.1}$$

Where, n_i and μ_i are no. of moles and the chemical potential of species i respectively.

Chemical potential can be related with activity as

$$\mu_i = \mu_i^0 + RT\ln a_i \tag{16.2}$$

where, a_i is the activity of species i

The scenario is different in case of Non-Ideal Solution, the intermolecular interactions can never be the same and hence deviation from ideality can be accounted by activity coefficient as per the modified Raoult's law and can be given as

$$a_i = \gamma_i x_i \tag{16.3}$$

where γ_i and x_i are the activity, activity coefficient and mole fraction of species i respectively. The value of γ_i is zero in case of ideal solutions. On comparing Eq. (16.1), (16.2) and (16.3) it can be concluded that the Gibbs free energy is a function of the activity coefficient of particular species.

16.2.1 Electrolyte Solutions

The electrolyte solutions are the one in which along with intermolecular interactions, interionic interactions were also present and due to that the deviation from ideality has to be addressed. The feed that has to desalinate are mostly electrolyte solutions which contain $MgCl_2$, NaCl. etc. and while dealing with this type of solutions it is more advantageous and practical to consider the free energy of the ion pairs that generated from electrolyte salt. In order to do that, the mean concentration and mean activity coefficients has been discussed. An electrolyte salt of neutral nature when dissociates fully in the solvent can be written as [12, 13].

$$A_{\nu+}B_{\nu-} = \nu_+ A^{z+} + \nu_- B^{z-} \tag{16.4}$$

The mean activity of dissociated salt molecule can be derived from the activities of the individual ions from Eq. 16.1 and 16.2.

$$RT\ln a_{A_{\nu+}B_{\nu-}} = \nu_+ RT\ln a_+ + \nu_- RT\ln a_- \tag{16.5}$$

$$a_{A_{\nu+}B_{\nu-}} = a_+^{\nu+} a_-^{\nu-} \tag{16.6}$$

$$\text{where } \nu = \nu_+ + \nu_- \tag{16.7}$$

$$\gamma_{m,\pm}^{\nu} = \gamma_{m,+}^{\nu+} \gamma_{m,-}^{\nu-} \tag{16.8}$$

$$m_{\pm}^{\nu} = m_{+}^{\nu+} m_{-}^{\nu-} \tag{16.9}$$

Where, $\gamma_{m,\pm}$ is the mean molal activity coefficient and $m_{\pm}$ is the mean molal concentration. Mole fraction can also be used to define the mean mole fraction $x_{\pm}$, and the mean rational activity coefficient, $\gamma_{x,\pm}$.

For neutral electrolytes such as for eg. NaCl (1:1), $MgCl_2$ (2:2)) in which $\nu_{+} = \nu_{-} = 1$, the mean molality of the salt is equivalent to the molality of the individual ions.

$$m_{A_{\nu+}B_{\nu-}} = m_{\pm} = m_{+} = m_{-} \text{ (Only if } \nu_{+} = \nu_{-} = 1)$$

The mean concentration has been derived, now the standard state is defined as a hypothetical solution at a mean concentration of unity referenced to infinite dilution such that the mean activity coefficient is unity when mean concentration is zero, regardless of temperature and pressure [12, 14, 15] as a direct significance of the definition of standard state it is clear that $\mu_s = \mu_s^0$ at the standard state and that $\mu_{m,s}^0 \neq \mu_{x,s}^0$.

For distinct concentration scales, activity coefficients corresponding to that can never be the same. The relation between molal and rational activity coefficient can be written as [12, 14].

$$\gamma_{x,\pm} = \gamma_{m,\pm}(1 + \nu m M_0) \tag{16.10}$$

Where m is molality of the electrolyte, M_0 is molecular weight of solvent, ν is the sum of stoichiometric coefficients.

The pure solvent coexists with its vapor when at equilibrium, the activity of the solvent is referenced to the pure vapor at the system temperature and 1 bar. Using modified Raoult's law in which nonidealities are accounted within the liquid mixture phase [13, 15], the ratio of the partial pressure of the vapor over the solution and the partial pressure of the vapor over pure solvent can be written in terms of the mole fractions.

$$\mu_0 = \mu_0^0 + RT\ln\gamma_{f,0}x_0 \tag{16.11}$$

The activity of the solvent is

$$a_0 = \gamma_{f,0}x_0 \tag{16.12}$$

$\gamma_{f,0}$ is the fugacity coefficient of the solvent.

16.2.2 Generalized Minimum Work of Separation

The minimum work of separation is a general metric used in desalination that defines the minimum amount of work required to separate a feed stream into two streams of different composition in the thermodynamic limit of reversible operation [13, 17]. As for power plants, Carnot efficiency is an ideal benchmark similarly minimum work of separation is the benchmark for the desalination system. The minimum work of separation is derived for a control volume containing a general separation unit and then explained from both a mass and mole basis [16].

Consider a General Separation Unit for desalination system as shown in Figure 16.1 the work of separation is denoted by $\dot{W}_{sep}$ and heat transfer within the system is denoted by $\dot{Q}$.

The feed, product and concentrated brine streams denoted by *f*, *p* and *c* are entering and leaving the system at different salinities *y*, at ambient temperature *T* and Pressure *p*. The first and second law of thermodynamics for the control volume can be written as

$$\dot{W}_{sep} + \dot{Q} + \dot{H}_f = \dot{H}_p + \dot{H}_c \tag{16.13}$$

$$\frac{\dot{Q}}{T} + \dot{S}_f + \dot{S}_{gen} = \dot{S}_p + \dot{S}_c \tag{16.14}$$

Here the convention that work and heat input to the system is positive is used. The first and second laws are combined by multiplying Eq. 16.13 by *T* and subtracting it from Eq. 16.14. Gibbs free energy rate is $\dot{G} = \dot{H} - T\dot{S}$ at $T = T_a$

$$\dot{W}_{sep} + \dot{G}_p + \dot{G}_c - \dot{G}_f + T\dot{S}_{gen} \tag{16.15}$$

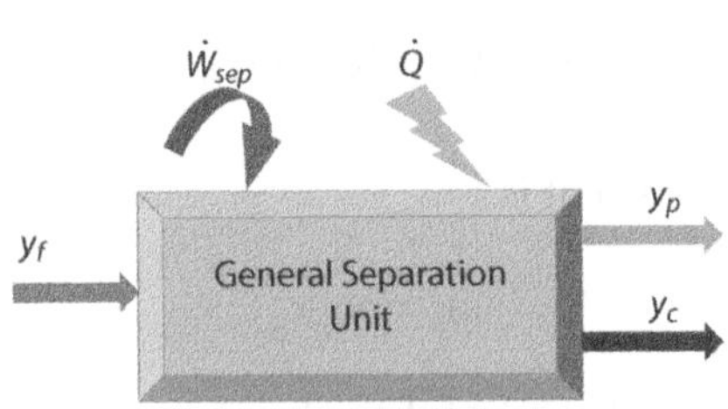

Figure 16.1 A control volume representation of the desalination system.

Considering the process is reversible, entropy generation will be zero and the work of separation becomes the reversible work of separation also called as minimum work of separation.

$$\dot{W}_{sep} = \dot{W}_{sep}^{rev} = \dot{G}_p + \dot{G}_c - \dot{G}_f \tag{16.16}$$

In the real world desalination systems, the major sources of entropy generation are viscous losses especially for membrane systems and heat transfer across finite temperature differences for thermal systems [13, 17]. While the composition is important in mixing processes, entropy generation is not strongly related to compositional effects in separation systems. Hence the least work is a relevant parameter for evaluating the impact of nonidealities on the performance of the system.

16.2.2.1 Mass Basis

To derive properties per unit mass of solution, Eq. 16.26 can be best written on mass flow rate basis as

$$\dot{W}_{min} = \dot{m}_p g_p - \dot{m}_c g_c - \dot{m}_f g_f \tag{16.17}$$

where g_j is specific Gibbs free energy per kilogram of solution.

The recovery ratio is defined as the ratio of the mass flow rate of product water to the mass flow rate of feed seawater and it can be written as

$$r = \frac{\dot{m}_p}{\dot{m}_f} = \frac{\text{the mass flow rate of product}}{\text{the mass flow rate of feed}} \tag{16.18}$$

The conservation of mass for the mixture and the salts can be written as

$$\dot{m}_f = \dot{m}_p + \dot{m}_c \tag{16.19}$$

$$\dot{m}_f y_f = \dot{m}_p y_p + \dot{m}_c y_c \tag{16.20}$$

Where, $\dot{m}_j$ is the mass flow rate in kg/s.

The minimum work of separation per unit mass of product can be obtained by using Eqs. 16.17, 16.18 and 16.19

$$\frac{\dot{W}_{min}}{\dot{m}_p} = (g_p - g_c) - \frac{1}{r}(g_f - g_c) \qquad (16.21)$$

The Gibbs free energy of each stream in Eq. 16.31 can be evaluated using seawater properties as a function of temperature and salinity [13]. The feed y_f and product y_p salinities are known properties, the concentrated brine salinity y_c can be evaluated as

$$y_c = \frac{y_f}{1-r} - \frac{r y_p}{1-r} \qquad (16.22)$$

16.2.2.2 *Mole Basis*

Physical properties can also be defined on a mole basis, hence Eq. 16.27 on substituting Eq. 16.1 can be written as

$$\dot{W}_{min} = [\dot{n}_{wt}\mu_{wt} + \dot{n}_{st}\mu_{st}]_p + [\dot{n}_{wt}\mu_{wt} + \dot{n}_{st}\mu_{st}]_c - [\dot{n}_{wt}\mu_{wt} + \dot{n}_{st}\mu_{st}]_f \qquad (16.23)$$

Conservation of moles for water and solute can be written as

$$\dot{n}_{wt,f} = \dot{n}_{wt,p} + \dot{n}_{wt,c} \qquad (16.24)$$

$$\dot{n}_{st,f} = \dot{n}_{st,p} + \dot{n}_{st,c} \qquad (16.25)$$

Where, $\dot{n}_j$ is the mole flow rate.

On substituting Eqs. 16.2, 16.24 and 16.25 into Eq. 16.23

$$\dot{W}_{min} = [\dot{n}_{wt}RT\ln a_{wt} + \dot{n}_{st}RT\ln a_{st}]_p + [\dot{n}_{wt}RT\ln a_{wt} + \dot{n}_{st}RT\ln a_{st}]_c - [\dot{n}_{wt}RT\ln a_{wt} + \dot{n}_{st}RT\ln a_{st}]_f \qquad (16.26)$$

The molar recovery ratio can be defined as

$$\bar{r} = \frac{\dot{n}_{wt,p}}{\dot{n}_{wt,f}} = \frac{\text{mole flow rate of water in the product}}{\text{mole flow rate of water in the feed}} \tag{16.27}$$

Using Eq. 16.25, 16.26 and 16.28 following result can be obtained

$$\frac{\dot{n}_{st,J}}{\dot{n}_{wt,J}} = m_{st,J} M_{wt} \tag{16.28}$$

And normalizing the least work by $\dot{n}_{wt,p}RT$ Eq. 16.27 becomes

$$\frac{\dot{W}_{min}}{\dot{n}_{wt,p}RT} = \left(\ln\frac{a_{wt,p}}{a_{wt,c}} + m_{st,p} M_{wt} \ln\frac{a_{st,p}}{a_{st,c}} \right) - \frac{1}{r}\left(\ln\frac{a_{wt,f}}{a_{wt,c}} + m_{c,f} M_{wt} \ln\frac{a_{st,f}}{a_{st,c}} \right) \tag{16.29}$$

Hence the generalized minimum work of separation for various salts or solute present in the feed can be written as

$$\frac{\dot{W}_{min}}{\dot{n}_{wt,p}RT} = \left(\ln\frac{a_{wt,p}}{a_{wt,c}} + \Sigma_{st}\, m_{st,p} M_{wt} \ln\frac{a_{st,p}}{a_{st,c}} \right) - \frac{1}{r}\left(\ln\frac{a_{wt,f}}{a_{wt,c}} + \Sigma_{st}\, m_{st,f} M_{wt} \ln\frac{a_{st,f}}{a_{st,c}} \right) \tag{16.30}$$

16.3 Modeling of Major Thermal Desalination Techniques

Presently, the fresh water demand is increasing day by day due to high scarcity in water sources mainly in developing countries and therefore it has become a bird's eye to the researchers for brine and brackish water desalination by more and more energy efficient ways [18]. In the last four decades, the water desalination industry has developed at a very high rate. The number of desalination plants in 1960 was 5 units with a total production rate of 5000 m^3/d, while in 1995 the number of units was 11000 with a capacity of 18.6×10^6 m^3/d [19]. The Fujairah 2 desalination plant in the United Arab Emirates is one of the biggest desalination facilities in the world, with a capacity of 591,000 m^3/day [20]. In spite of the huge development in reverse osmosis based desalination system in the last few years, thermal desalination techniques are still necessary because of their higher purification capacity at high temperatures, salinity or contamination [21]. While multistage flash is the dominant type of large-scale thermal desalination currently in use, multiple effect distillation is thermodynamically superior and is currently receiving considerable attention as a strong competitor to MSF, especially in Middle-East Asia. The MED process is characterized by lower energy consumption (2 kWh/m^3) compared to the MSF process (4 kWh/m^3) since recirculating large quantities of brine is not required. Additionally, MED provides higher overall heat transfer coefficients by utilizing primarily latent heat transfer and avoiding the lower specific heat transfer surface areas associated with sensible heat transfer found in MSF [22]. The ability to operate at low temperature and use low-grade heat from power station turbines as the primary heat source for MED yield very low specific energy costs for seawater desalination [23].

The MSF desalination configuration was first made as a patent by Silver in 1957 [24]. The main feature of this patent is the optimization of the number of stages versus the heat transfer area. Accordingly, it was found that the use of a large number of stages, i.e., above 20, resulted in the optimum cost for the MSF process. The MSF desalination plants are divided into two models, Once-through MSF and Brine recirculation MSF designs. The Once-through MSF plants consist of an evaporation section (heat recovery section) and a brine heater and the condenser tubes arrangement. The Brine recirculation MSF consists of brine heater, heat recovery section, and heat rejection section. The role of the rejection section is to remove the surplus thermal energy from the plant, thus cooling the distillate product and the concentrated brine to the lowest possible temperature [25]. In last few decades, modeling of these thermal desalination techniques (MSF, MED, MVC etc) has played a very significant role in their simulation, optimization and control. Many models have been developed to find a functional relationship between the design and operating variables [26] because optimized operating variables of a given design always help to increase in distillate production rates by lowering operating costs. While other parameters like top brine temperatures, brine circulation rate, intake flow etc help to manipulate performance parameters for its enhancement [27].

16.3.1 A General Multi-Effect Distillation (MED) Process Configuration for Desalination

A general multi-effect distillation based desalination plant consists of a variable sequence of stages. A conventional MED configuration scheme is illustrated in Figure 16.2, where cold brine is introduced from the top of the effect, above the bundle, and falls on the surface of the tube.

The vapor produced by the film evaporation is separated by means of a demister and exits the effect. The remaining brine, together with the droplets separated in the demister, continuously falls down to the bottom column, where it is collected and exits the effect. Flash evaporation will take place only when the pressure will be lower than the brine pressure and at that time an additional amount of vapor will be produced. For effects different from the first, a condensate flash box, where the pressure is fixed equal to the one in the effect, is installed at bundle tubes exit. Here, part of condensate turns into vapor, and mixed with the previously produced stream and acts as a heating medium in the next effect [28].

16.3.1.1 *Steady State Process Model of a MED System*

A thermal model of a MED system relies on the following considerations and simplifications:

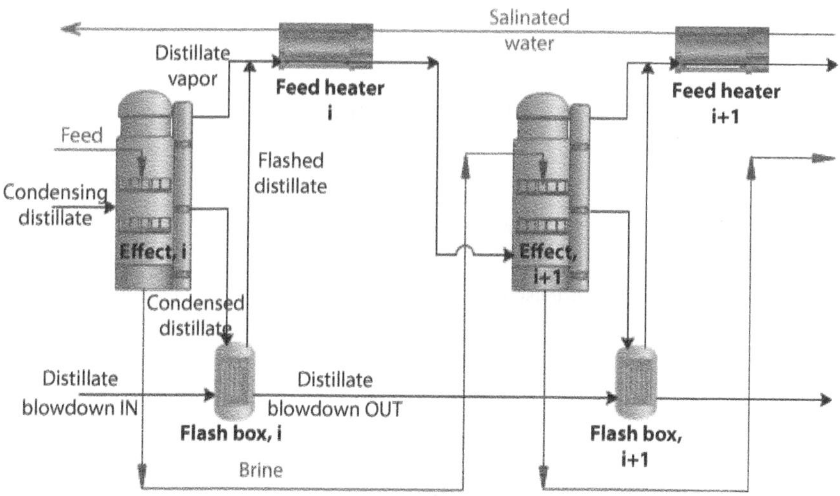

Figure 16.2 A general multi-effect distillation (MED) desalination process scheme [21].

- No energy losses with the surroundings.
- Steady state operation.
- Exchanger area in the effects is just large enough to condense vapor to saturated liquid at the previous effect's pressure.
- The overall heat transfer coefficient is averaged over the length of an exchanger.
- The final distillate production is the sum of the distillate produced in each effect.

16.3.1.1.1 Modeling of the Effects

One of the major components in any multi-effect distillation system is its effect in which feed water is entered into the column effect on the series of tubes and the distillate vapor coming from the previous effect gets condensed within these tubes. Generally, a pressure just below the saturation pressure of the feed water is maintained in the column which is used to flash evaporate a small fraction of the feed. During the condensation of distillate, it releases the heat of vaporization which is transferred to the feed to generate a high quantity of vapor. Vapor generated from the flashing, boiling and the brine solution is extracted from the effect. The effect's inlet and outlet streams are represented by the control volume shown in Figure 16.3.

Water balance:

The feed stream is distributed into a distillate (vapor) stream and a brine stream. The total distillate produced is the sum of the stream comes after flashing and boiling.

$$\dot{m}_{SW} = \dot{m}_B + \dot{m}_D \tag{16.31}$$

$$\dot{m}_{SW} = \dot{m}_{B_e} + \dot{m}_{D_f} \tag{16.32}$$

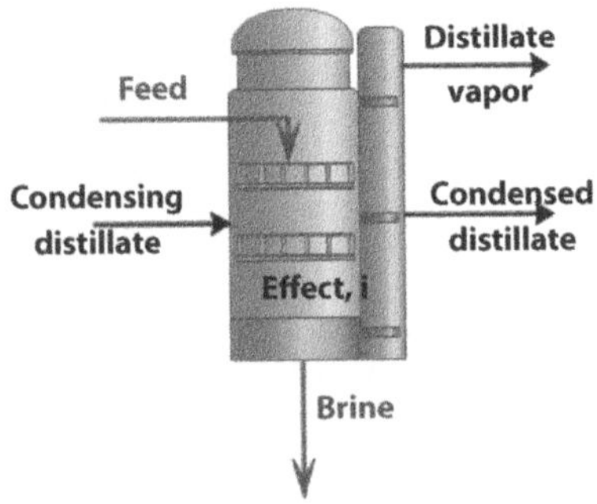

Figure 16.3 Detailed components stream in i-th effect.

$$\dot{m}_D = \dot{m}_{D_b} + \dot{m}_{D_f} \tag{16.33}$$

Salt balance:
The salinity of the brine stream within the effect and the brine stream leaving the effect is found through a salt balance assuming, $\lambda_{D_f} = \lambda_{D_b} = 0$;

$$\dot{m}_{SW}\lambda_{SW} = \dot{m}_B\lambda_B \tag{16.34}$$

$$\dot{m}_{SW}\lambda_{SW} = \dot{m}_{B_e}\lambda_{B_e} \tag{16.35}$$

Energy balance:
The enthalpy change due to condensation of the distillate from the previous effect helps to split the brine stream and distillate streams.

$$\dot{m}_{DC}H_{DC} = \dot{m}_D H_D + \dot{m}_B H_B - \dot{m}_{SW}H_{SW} \tag{16.36}$$

Temperatures:
Salinity causes the boiling point to be elevated. Distillate formed in the effect is superheated by an amount of temperature equal to the BPE (Boiling point elevation). The distillate will condense at the saturation temperature in the following feed heater and effect.

$$T_{D_{sat}} = T_D - BPE \tag{16.37}$$

Terminal temperature difference and the temperature difference between the two effects are as follows:

$$T_{TTD_e} = T_c - T_e \tag{16.38}$$

$$\Delta T_e = T_e^{Prev} - T_e \tag{16.39}$$

Heat transfer area:
The condensate tube surface area must be large enough to ensure that the distillate vapor from the previous effect condenses completely while flashing and evaporating the feed.

$$\dot{m}_{DC}\Delta H_{DC} = A_e h_e \left(T_{D_{sat}}^{Prev} - T_e\right) \tag{16.40}$$

The temperature at which the distillate from the previous effect condenses is equal to the saturation temperature of the previous effect, $T_c = T_{D_{sat}}^{Prev}$.

First effect:
In fact, there are just two minor differences between the first column effect and the others (up to n-th effect) i.e. feed enters the first effect just below the saturation temperature and the feed enters the others, up to n-th effects just above the saturation temperature. Another one is, steam is used to heat up the feed in the first effect only and the vapor produced in the previous the effect is used to heat the feed in all the following effects. Flashing does not occur in the first effect due to the sub-cooled condition of the feed.

$$\dot{m}_{D_f} = 0 \tag{16.41}$$

Steam input to the first effect can be expressed by modifying the equation (16.36) as

$$\dot{m}_{DC} H_{DC} \dot{m}_s L_{fg,s} \tag{16.42}$$

Second to the nth effect:
Here, an additional energy balance is necessary since a portion of the feed steam flashes.

$$\dot{m}_{SW} H_{SW} \dot{m}_{B_e} H_{B_e} + \dot{m}_{D_f} H_{D_f} \tag{16.43}$$

Assuming the complete condensation occurs, the change in enthalpy in equation (16.36) can be expressed as

$$\Delta H_{D_C} = H_{D_C} - H_{D_{c,sat}} \tag{16.44}$$

16.3.1.1.2 Modeling of Flash Box

Due to a collection of distillate in each stage in flash box, part of the distillate blow-down from the previous effect and the distillate used for condensing in the current effect is flashed during the depressurization. The flash box's inlet-outlet streams are represented in Figure 16.4.

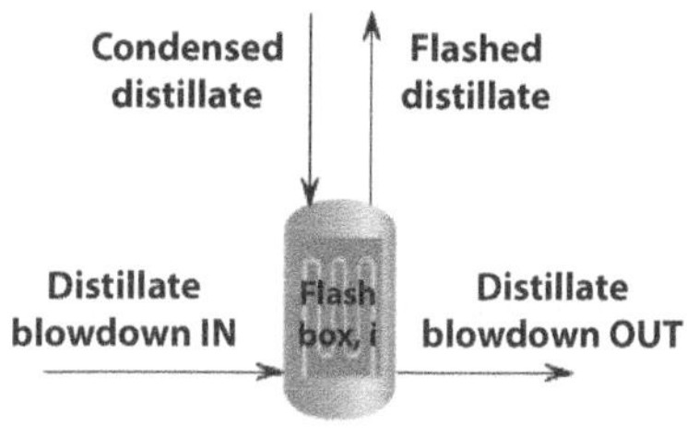

Figure 16.4 Detailed components stream in the i-th flash box.

$$\dot{m}_{D_{bd}} + \dot{m}_{D_{fb}} = \dot{m}_{D_{bd}^{in}} + \dot{m}_{DC} \tag{16.45}$$

$$\dot{m}_{D_{bd}} H_{D_{bd}} + \dot{m}_{D_{fb}} H_{D_{fb}} = \dot{m}_{D_{bd}^{in}} H_{D_{bd}^{in}} + \dot{m}_{DC} H_{DC} \tag{16.46}$$

16.3.1.1.3 Modeling of Mixing Box

These flashing and mixing processes are governed by the mass conservation law and 1st law of thermodynamics. No flashing occurs in the flash box when all inlet and outlet streams are at same pressure while flash box works as a mixing system.

$$\dot{m}_{D_{fb}} = 0 \tag{16.47}$$

$$H_{D_{fb}} = \text{Undefined} \tag{16.48}$$

16.3.1.1.4 Modeling of the Feed Heater

Feed heaters are used to recover energy and reduce the amount of steam required for heating the feed in the first effect. Inlet-outlet streams of the feed heater are characterized by a control volume presented in Figure 16.5.

Calculation of heat transfer area using energy balance and LMTD

$$\dot{m}_{DC}\left(H_{DC}^{in} - H_{DC}^{out}\right) = \dot{m}_{SW}\left(H_{\dot{m}SW}^{in} - H_{\dot{m}SW}^{out}\right) \tag{16.49}$$

$$\dot{m}_{DC}\left(H_{DC}^{in} - H_{DC}^{out}\right) = A_{fh} h_{fh} \frac{\Delta T_{\dot{m}SW}^{in} - \Delta T_{\dot{m}SW}^{out}}{\ln\left(\dfrac{T_{D_{c,sat}} - T_{\dot{m}SW}^{out}}{T_{D_{c,sat}} - T_{\dot{m}SW}^{in}}\right)} \tag{16.50}$$

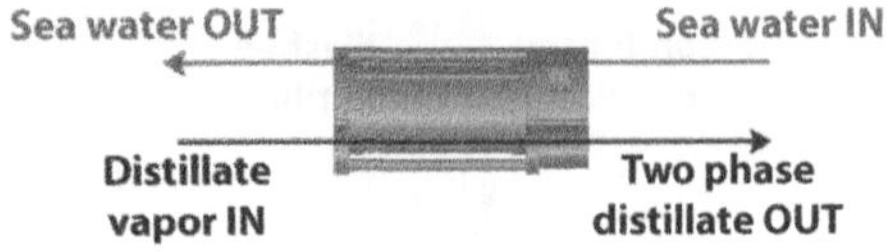

Figure 16.5 Detailed components stream in i-th feed heater.

The minimum temperature difference in the feed heater occurs at the outlet of the seawater.

$$T_{DC} - T_{\dot{m}SW}^{out} = \Delta T_{TTDfh} \tag{16.51}$$

16.3.1.1.5 Modeling of Condenser

Energy balance and heat transfer area calculations for the condenser are similar to those for the feed heaters.

$$\dot{m}_{DC} \Delta H_{DC} = \dot{m}_C \left(H_{sw}^{out} - H_{sw}^{in} \right) \tag{16.52}$$

$$\dot{m}_C \left(H_{sw}^{out} - H_{sw}^{in} \right) = A_C h_C \frac{\Delta T_{sw}^{out} - \Delta T_{sw}^{in}}{\ln \left(\dfrac{T_D - T_{sw}^{in}}{T_D - T_{sw}^{out}} \right)} \tag{16.53}$$

16.3.1.2 Performance Parameters Analysis

After performing all the above correlations and solving them, the following analysis are necessary:

$$PR = \frac{\dot{m}_D}{\dot{m}_s} \tag{16.54}$$

$$RR = \frac{\dot{m}_D}{\dot{m}_{sw}} \tag{16.55}$$

$$SA = \frac{\sum A_e + \sum A_{fh} + A_C}{\dot{m}_D} \tag{16.56}$$

A typical multi-effect distillation desalination system utilizes flash boxes and feeds heaters with the aim of accumulating the distillate and preheat the seawater prior to injection into the first effect [29–32]. The principal advantage of this particular configuration is its high energy recovery ability.

16.3.2 A General Process Configuration of Multi-Stage Flash (MSF) Desalination

Mainly, three sections are there in a typical multi-stage flash configuration system i.e. feed heater or brine heater, recovery stage with NR-stages, and rejection stage with NJ-stages represented in Figure 16.6. Now seawater comes from the last stage of the rejection section and passes through the series of coolant systems to remove heat from the stages. Before transferring to the recovery section, a fraction of seawater is discharged to the sea to adjust the heat. The remaining part of the heated seawater is mixed with recycled brine coming from the last stage rejection part and subsequently fed to the last stage of the recovery section. Then seawater flows through the tubes of different stages to recover heat from the stages and brine heater raises the temperature to the maximum attainable temperature i.e. top brine temperature (TBT).

Then it helps produce flashing vapor in flashing stage. This process goes on until the last stage of the rejection section. The concentrated brine from the last stage is partly discharged to the sea as blow down part and the left is recycled. The vapor from each stage is collected in a distillate tray to finally produce the fresh water [34].

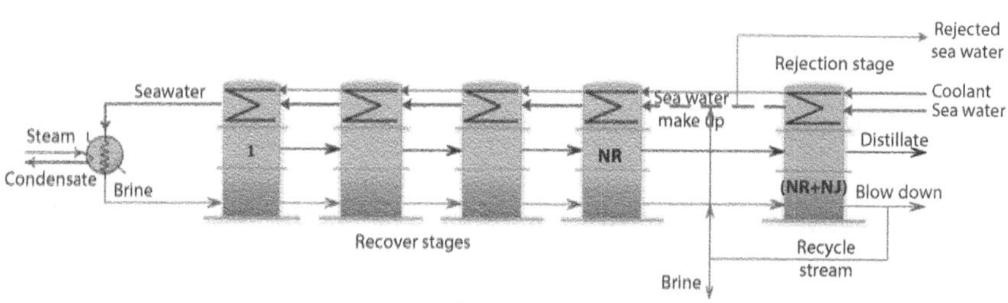

Figure 16.6 A typical multi-stage flash desalination process scheme [33].

16.3.2.1 Steady State Process Model of an MSF System

Based on the following assumptions and considerations, modeling of the MSF system is depicted.

- The output of every stage is pure distillate i.e. salt-free.
- The heat of mixing is not considered i.e. negligible.
- No sub-cooled condensate leaving the brine heater.
- No energy losses with the surroundings.
- Entrainment of mist by the flashed vapor is taken zero.

16.3.2.1.1 Modeling of MSF Stages

Mass balance in the flash chamber

$$\dot{m}_{B_{j-1}} = \dot{m}_{B_j} + \dot{m}_{V_j} \tag{16.57}$$

Mass balance on the distillate tray

$$\dot{m}_{D_{j-1}} = \dot{m}_{D_j} + \dot{m}_{V_j} \tag{16.58}$$

Enthalpy balance on flashed brine

$$\dot{m}_{B_j} = \frac{\left(H_{B_{j-1}} - H_{V_j}\right)}{\left(H_{B_j} - H_{V_j}\right)\dot{m}_{B_{j-1}}} \tag{16.59}$$

$$H_{B_j} = f\left(T_{V_j}\right) \tag{16.60}$$

$$H_{B_j} = f\left(\lambda_{B_j}, T_{B_j}\right) \tag{16.61}$$

Overall energy balance

$$\dot{m}_R c_{P R_j}\left(T_{m/SW_j} - T_{m/SW_{j+1}}\right) = \dot{m}_{D_{j-1}} c_{P D_{j-1}}\left(T_{D_{j-1}} - T^*\right) + \dot{m}_{B_{j-1}} c_{P B_{j-1}}\left(T_{B_{j-1}} - T^*\right)$$
$$\dot{m}_{D_j} c_{P D_j}\left(T_{D_j} - T^*\right) - \dot{m}_{B_j} c_{P B_j}\left(T_{B_j} - T^*\right) \tag{16.62}$$

Heat transfer equation

$$\dot{m}_R c_{P_{R_j}} \left(T_{m/SW_j} - T_{m/SW_{j+1}}\right) = h_j A_j (LMTD)_j \qquad (16.63)$$

In equation (16.76), replace $\dot{m}_R$ by $\dot{m}_{SW}$ for rejection stage calculation.

$$h_j = \frac{4.8857}{\left(x_j + Z_j + 4.8857 \times R_{f_j}\right)} \qquad (16.64)$$

$$\text{Where, } R_{f_j} = 3 \times 10^{11} e^{-\left(\frac{42000}{R_g T_{m/SW_j}}\right)} \qquad (16.65)$$

$$Z_j = 0.102 \times 10^{-2} - 0.747 \times 10^{-5} \times T_{D_j} + 0.997 \times 10^{-7} \times T_{D_j}^2 - 0.430 \times 10^{-9} \times T_{D_j}^3 + 0.620 \times 10^{-12} \times T_{D_j}^4 \qquad (16.66)$$

$$x_j = \frac{[u \times d]^2}{\left[\left(160 + 1.92 \times T_{m/SW_j}\right) \times u\right]} \qquad (16.67)$$

LMTD between recovery stage and rejection stages

$$LMTD_j = \left(T_{D_j} - T_{m/SW_{j+1}}\right) - \left(T_{D_j} - T_{m/SW_j}\right) / \ln\left\{\frac{\left(T_{D_j} - T_{m/SW_{j+1}}\right)}{\left(T_{D_j} - T_{m/SW_j}\right)}\right\} \qquad (16.68)$$

The heat capacity of cooling brine leaving stage j

$$c_{P_{R_j}} = f\left(T_{SW_j}, T_{m/SW_{j+1}}, \lambda_R\right) \qquad (16.69)$$

The heat capacity of distillate leaving stage j

$$c_{P_{D_j}} = f\left(T_{D_j}\right) \qquad (16.70)$$

The heat capacity of flashing brine leaving stage j

$$c_{P_{B_j}} = f\left(T_{B_j}, \lambda_{B_j}\right) \tag{16.71}$$

Correlation between distillate and flashing brine temperature

$$T_{B_j} = T_{D_j} + BPE_j + NEA_j + \mu_j \tag{16.72}$$

Distillate and flashing steam correlation

$$T_{V_j} = T_{D_j} + \mu_j \tag{16.73}$$

Temperature loss due to demister

$$\theta_j = f\left(T_{D_j}\right) \tag{16.74}$$

Boiling point elevation (BPE) at stage j

$$BPE_j = f\left(T_{D_j}, \lambda_{B_j}\right) \tag{16.75}$$

16.3.2.1.2 Modeling of Brine Heater

Mass and salt balance of brine heater

$$\dot{m}_{B_0} = \dot{m}_R \tag{16.76}$$

$$\lambda_{B_0} = \lambda_R \tag{16.77}$$

Overall enthalpy balance

$$\dot{m}_{B_0} c_{P_{Htr}} \left(T_{B_0} - T_{m/SW_1}\right) = \dot{m}_{steam} L_S \tag{16.78}$$

Heat transfer equation

$$\dot{m}_R c_{P_{Htr}} \left(T_{B_0} - T_{SW_1}\right) = h_{Htr} A_{Htr} (LMTD)_{Htr} \tag{16.79}$$

LMTD in brine heater

$$\mathrm{LMTD_{Htr}} = \frac{(T_{steam} - T_{m/SW_1}) - (T_{sTeam} - T_{B_0})}{\ln\dfrac{(T_{steam} - T_{m/SW_1})}{(T_{steam} - T_{B_0})}} \tag{16.80}$$

Overall heat transfer coefficient in brine heater

$$h_H = f\left(T_{steam}, T_{B_0}, T_{m/SW_1}, T_{steam}, d^{i}_{Htr}, d^{o}_{Htr}, \Pi^{i}_{Htr}\right) \tag{16.81}$$

The heat capacity of brine in brine water

$$c_{P_H} = f\left(T_{B_0}, T_{m/SW_1}\right) \tag{16.82}$$

16.3.2.1.3 Modeling of Splitter

Mass balance on the sea-water splitter

$$\dot{m}_{B_D} = \dot{m}_{B_N} - \dot{m}_R \tag{16.83}$$

$$\dot{m}_W = \dot{m}_{SW} - \dot{m}_{m/SW} \tag{16.84}$$

16.3.2.1.4 Modeling of Mixer

Mass balance on the mixer

$$m_R = \dot{m}_R - \dot{m}_{m/SW} \tag{16.85}$$

$$\lambda_R + \dot{m}_{m/SW}.\lambda_{m/SW} = m_R \lambda_{SW} \tag{16.86}$$

Enthalpy balance on the mixer

$$m_R H_{T_{m/SW}} = \dot{m}_R H_{flash} + \dot{m}_{m/SW} H_{\dot{m}_{m/SW}} H_{t_{m/SW}} = f\left(T_{m/SW_m}, \lambda_{SW}\right) \tag{16.87}$$

$$H_{m/SW} = f\left(T_{m/SW_{NR+1}}, \lambda_{m/SW}\right) \tag{16.88}$$

$$H_{m/SW} = f\left(T_{BR}, \lambda_R\right) \tag{16.89}$$

16.3.3 A General Process Configuration of Mechanical Vapor Compression (MVC) Desalination

A single effect mechanical vapor compression desalination system has been illustrated in Figure 16.7. The primary equipment which are mainly used for this desalination process, are evaporator, compressor, pumps and heat exchanger/pre-heaters. As depicted in Figure 16.6, the incoming feed water i.e. saline water is passed through heat exchangers where it is pre-heated by the outgoing brine stream and distillate steam (Product stream). Then the sea water is sprayed through the nozzle on the outer part of a bundle of evaporator tubes and it partially evaporates. After that, a centrifugal compressor draws the produced steam through the demister where the steam temperature and pressure are increased by the compression. Then the compressed steam is transferred to the heat exchanger where it condenses into distillate and a fraction of brine stream is recycled back to the evaporator channel. The energy performance of the system depends on the pressure increment in the mechanical compressor, on the thermodynamic efficiency of the polytrophic process and on the efficiency of the electric motor [35].

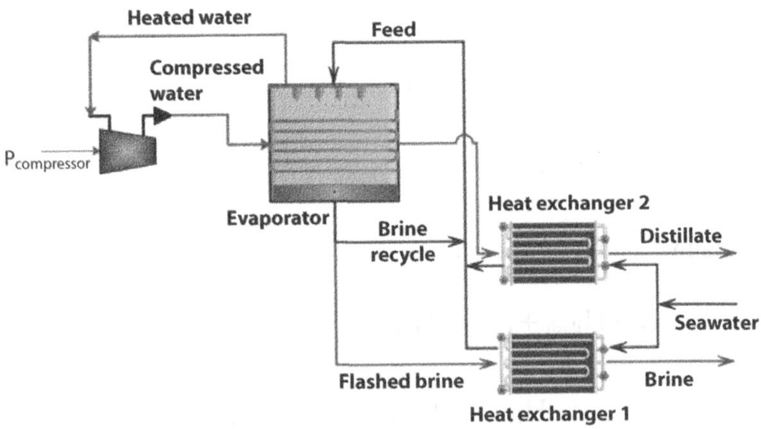

Figure 16.7 A typical single stage mechanical vapor compression desalination process scheme [36].

16.3.3.1 Steady State Process Model of an MVC System

The resulting mathematical modeling to describe the MVC desalination process is based on the following the basic assumptions:

- The system operates under steady state.
- No energy losses with the surroundings i.e. negligible.
- The product is considered as pure water.
- Heat losses from evaporator surface are negligible.
- No recycle is considered.
- Overall heat transfer coefficients are taken as equal for the heat exchangers.

16.3.3.1.1 Mathematical Model

Total mass and energy balance:

$$\dot{m}_{SW} = \dot{m}_B + \dot{m}_D \tag{16.90}$$

$$\dot{m}_{SW}\lambda_{SW} = \dot{m}_B\lambda_B + \dot{m}_D\lambda_D \tag{16.91}$$

The outside temperature at the compressor is computed as

$$\frac{T_{comp}}{T_{evap}} = Z_R^{\frac{\gamma-1}{\gamma}} \tag{16.92}$$

where
$$Z_R \cdot p_{evap} = p_{comp} \tag{16.93}$$

The mechanical power consumed by the compressor is

$$P_{comp}\ \eta\ MW\ 3600 = \dot{m}_D \frac{\gamma-1}{\gamma} RT_{evap} \frac{T_{comp}}{T_{evap}-1} \tag{16.94}$$

The energy balance on the evaporator is

$$\dot{m}_D\left(H_{comp} - H_{vap_{sat}}\right) + \dot{m}_D L_{T_D} = \dot{m}_{SW}\left(H_{T_{boil}} - H_{T_{SW}}\right) + \dot{m}_D L_{T_{evap}} \tag{16.95}$$

The energy balances on the two pre-heaters are

$$t_{HX_1} = \dot{m}_B(H_{boil} - H_{B_{disc}}) \tag{16.96}$$

$$t_{HX_1} = \dot{m}_{s_1}(H_{s_1} - H) \tag{16.97}$$

The mass balance on the sea water splitter is

$$\dot{m}_{SW} = \dot{m}_{S_1} + \dot{m}_{S_2} \tag{16.98}$$

$$t_{HX_2} = \dot{m}_D\left(H_{t_D} - H_{t_{Ddisc}}\right) \tag{16.99}$$

$$t_{HX_2} = \dot{m}_{S_2}\left(H_{S_2} - t_{SW}\right) \tag{16.100}$$

$$\dot{m}_{SW}H_{tSW} = \dot{m}_{S_1}H_{S_1} + \dot{m}_{S_2}H_{S_2} \tag{16.101}$$

The temperature differences in pre-heaters from both sides are

$$\Delta T_{HX_1}^{ho} = T_{boil} - T_{HX_1} \tag{16.102}$$

$$\Delta T_{HX_1}^{co} = T_{B_{disc}} - T_{SW} \tag{16.103}$$

$$\Delta T_{HX_2}^{ho} = T_D - T_{HX_2} \tag{16.104}$$

$$\Delta T_{HX_2}^{ho} = T_{D_{disc}} - T_{SW} \tag{16.105}$$

The LMTDs (Logarithmic mean temperature difference) to compute pre-heaters heat transfer areas are

$$\mathrm{LMTD}_j = \frac{\Delta T_j^{ho} - \Delta T_j^{co}}{\ln \dfrac{\Delta T_j^{ho}}{\Delta T_j^{co}}} \tag{16.106}$$

Now, the heat transfer areas are

$$t_j = h_j A_j (LMTD)_j \qquad (16.107)$$

Where,

$$j = HX_1 \text{ or } HX_2 \qquad (HX= \text{Heat exchanger}) \qquad (16.108)$$

Consequently, the evaporator area is

$$A_{evap} = \frac{\dot{m}_D L_{T_D}}{h_{evap}(T_D - T_{boil})} + \frac{\dot{m}_D c_{p_{evap}}(T_{evap} - T_{boil})}{h.LMTD} \qquad (16.109)$$

16.4 Advantage of RO Above Other Mentioned Technologies

The first introduction of the synthetic membrane in separation processes occurred in the 1960s. Later on, they started playing important role in water desalination in the 1980s. Membrane technologies include Nano-filtration, ultra-filtration, micro-filtration, membrane distillation, Reverse Osmosis (RO) and electrodialysis. RO has become quite an economical alternative all over the world for traditional water softening processes. RO is replacing phase change desalting technologies for supplying water to coastal and inland communities. The first commercial successful RO process was introduced in the early 1970s. A membrane separation process in which the water from a pressurized saline solution is separated from the solutes by flowing through a membrane is what we called as RO process. Most of the energy requirement in the RO process is for pressurizing the feed water. Alternatively, this process can also be described as a process of forcing a solvent from a region of higher solute concentration through a membrane to a region of low solute concentration under the influence of pressure greater than osmotic pressure. This leads to a reverse flow of water in the direction opposite to natural flow across the membranes leaving the dissolved salts behind with an increase in salt concentration [37].

Typically, large saline water RO plants consist of 5 major components which include a saline water supply system, a feed water treatment system, a feed water pretreatment system, high-pressure pumping, RO modules, and post-treatment system. There are huge installations of RO plant in recent years in the Middle East and particularly in Saudi Arabia. RO plant installed in Jeddah has the capability of producing a million gallons per day (MGD), while Al Jubail and Yanbu RO plants have capacities of 24 and 33.8 MGD [37]. Figure 16.8 portrays the distribution of installed desalination process.

16.4.1 Advantages of RO Process

1. There are significantly fewer chances of material corrosion as compared to MSF and MED processes due to the ambient temperature conditions.
2. Mostly polymeric materials are utilized as much as possible rather than the use of metal alloys.
3. Few developments have been made in the past decade to reduce the operating cost of RO plant

- Cheaper and highly durable membranes have been developed
- Use of energy recovery devices that are connected to the concentrated stream while it leaves the pressure vessel

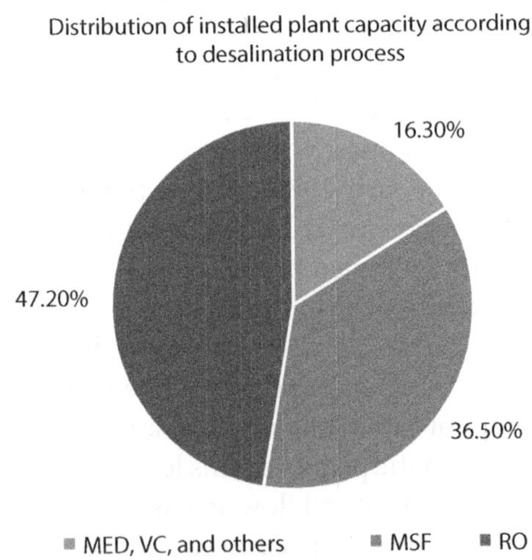

Figure 16.8 Distribution of installed plant capacity according to the desalination process [38].

16.4.2 Energy Requirement in Desalination by an Evaporation Technique

The energy requirement by single stage evaporation of seawater is high as 650 kWh/m^3. Although, the main evaporation techniques like MSF and MED have overcome the problem of high energy consumption by reusing energy through multiple stages [39]. The number of times heat evaporation is reused is represented by the "gained output ratio (GOR)". GOR is defined as the ratio of the number of tons of water produced per ton of steam as energy utilized.

Mostly, MSF manufacturers provide a GOR design in the range of 8-12 which depends on the steam feed temperature [40]. Although, MED manufactures put claims of GOR of 10-16 for their working units and GOR can go up to 30 in their designed prototypes. Thermal energy requirement based on fuel consumption in typical "single-purpose" plants comes around in the range of 55-80 kWh/m^3 of product for MSF process and 40-65 kWh/m^3 of product for MED. Based on the efficiency of 45% in a modern power station, the work equivalent of this energy consumption is 24-37 kWh/m^3 for MSF and 18-30 kWh/m^3 for MED. There is pumping energy requirement for both techniques which claimed to be around 1.2 - 4.5 kWh/m^3.

16.4.3 Energy Requirements for Desalination by Reversible RO Process

The energy efficiency of RO desalination process is expressed in terms of the specific energy that is defined as the energy requirement of a process for generating the one-unit volume of permeate as illustrated in Figure 16.9. The most convenient energy unit used in desalination application is (KWh), which is equal to 3.6×10^6 Joules [41]. The specific energy of the reversible RO process is given by:

$$W_1 = 2.78 \times 10^{-7} \pi \frac{1}{r} ln \frac{1}{1-r} \tag{16.110}$$

where, W_1 is the specific energy for desalination in the reversible RO processes, π is the osmotic pressure of the feed water, r is the water recovery that is defined as the ratio of the permeate flow rate to the feed flow rate in the RO process.

RO membranes have a finite resistance. There is a requirement of non-zero permeate flux for practical RO processes which can be demonstrated

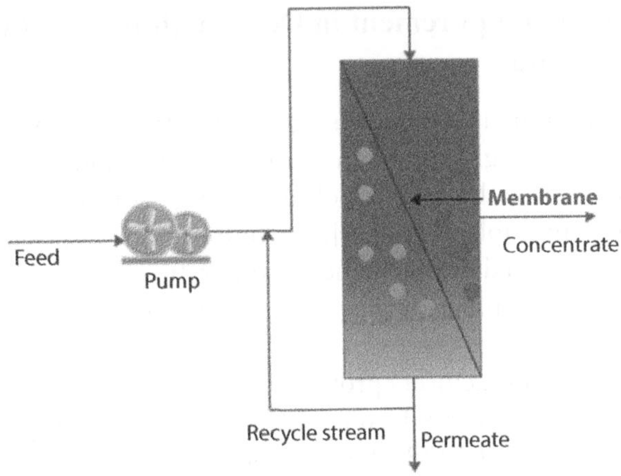

Figure 16.9 Flow sheet for energy use in single stage RO desalination operation.

by the fact that permeate flux of a finite value actually minimizes the combined cost of membrane and energy. The specific energy for the ideal RO process is given by:

$$W_2 = 2.78 \times 10^{-7} \left(\pi_0 \frac{1}{r} ln \frac{1}{1-r} + \Delta P_{net} \right) \tag{16.111}$$

Where, W_2 is the specific energy for desalination in ideal RO processes, ΔP_{net} is the net driving pressure for the required permeate flux. The ideal RO is considered to be more appropriate than compared to reversible RO in the desalination process.

The osmotic pressure of the feed water can be linearly related to its salt concentration that is;

$$\pi_o = \varphi C_o \tag{16.112}$$

where, C_o is the concentration of feed water in mg/lit. and φ is the osmotic pressure coefficient. The osmotic pressure coefficient has a value of 73.9 Pa. The value can be used for both sea water and brackish water.

It is quite evident from Figure 16.10, that the Energy consumption in the RO process has shown a decreasing trend over years. With the advancement in science and technology and the development of efficient membranes

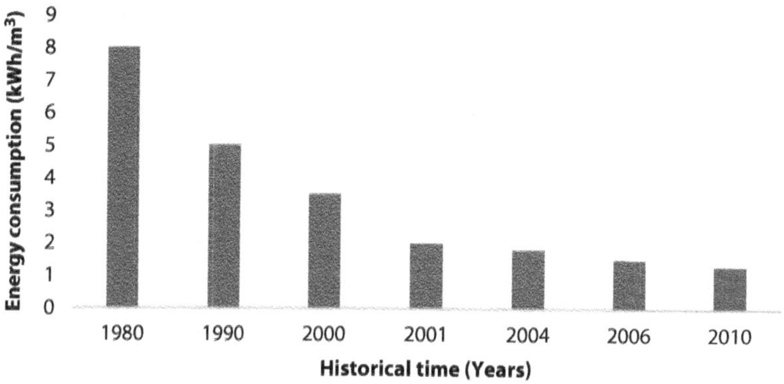

Figure 16.10 Development of achievable energy consumption in RO desalination process [37].

made it possible to reduce energy consumption by RO processes. Because of this trend, RO is now gaining wide applications for desalination purpose.

16.4.4 Energy Analysis of Different Desalination Techniques

Currently, we are facing huge challenges in mitigating and optimizing our energy consumption. The earth's resources have limitation and we need to come up with an alternative solution for different processes to operate in the safe zone. Thus, energy analysis plays a very critical role in seawater desalination because water is the crucial element for human survival. There are countries which are mostly dependent on seawater to fulfill their need for water. They burn a huge chunk of their energy reservoir to generate drinking water via MSF, MED or MEB. In such a scenario, RO acts as a boon for them as it drastically reduces energy consumption. It can be seen from Table 16.1, that total electrical equivalent required for MD, MSF, MED and MEB process is in between 15-60 KWh/m^3 but BWRO & SWRO requires only 0.5-6 KWh/m^3 of electrical equivalent. RO process requires 10 times less energy as compared to MSF & MED. This is a huge achievement for the current generation. Because of the RO process, our natural reservoir can stay a little longer and less environmental emission can also be observed.

Table 16.1 Comparison of the desalination process by thermal and membrane technology.

Comparison thermal desalination vs. membrane desalination		
	MSF	**RO**
Energy consumption	~13 kWh/ m^3	4-5 kWh/m^3
Recovery	10% - 20%	30% - 50%
Investment [$/(m^3/day)]	~1,000 – 1,500	~7,00 – 1,500
Chemicals	~ 0.03 to 0.05	~0.06 to 0.1
Brine quantity	Distillate x 4 to 9	Permeate x 1 to 4
Brine quality	Chemicals, Heat	Chemicals
Robustness	High	Fouling Sensitivity, Feedwater monitoring
Improvement potential	Low	High

16.4.5 Economic Analysis of Different Desalination Techniques

With the evolution of science and technology, and the growing demand for cheaper water led the concentration of researchers in minimizing the price of SWRO process. In the last 20 years, the desalination process by SWRO saw a decreasing trend. Researchers are working still on it to reduce the further cost of desalination techniques. Cost evaluation of SWRO and cost composition of MSF and RO is shown in Figures 16.11–16.13.

Here, the comparison is made in Table 16.2 and Table 16.3 which shows the cost of desalination on basis of the capacity of plants. Detailed cost analysis of RO plant is presented in Tables 16.4 and 16.5. It is extremely economical to use RO process for brackish water for handling the larger capacity of the desalination plant as it reduces the cost to half of that cost of MED process. Similarly, the RO process can be used for desalination of seawater when capacity is large such as (>15000 m^3/day), as it will be an economically sound decision. Vapor compression is among costly method for desalination which is not economically sound suggestions for installing desalination plant.

This table analyses the overall investment and operation cost of the different desalination process. It shows that the RO process is the cheapest among all from a capital investment point of view to energy-related costs. Thus, RO is a winner among all desalination processes available because of its performance and infrastructural investment.

Table 16.2 Different energy requirements in desalination techniques in plant [42].

Technology	Electrical energy (KWh/m^3)	Thermal energy (KWh/m^3)	Total equivalent electrical
BWRO	0.5-3	-	0.5-3
SWRO	3-6	0	3-6
ED	1-3.5	-	1-3.5
EDR	1-2	-	1-2
MVC	7-15	-	7-15
FO	0.2-0.5	20-150	10-68
MD	1.5-4	4-40	3-22
MSF	2.5-5	40-120	21-59
MED	2-2.5	30-120	15-57
MEB	2	60	30

Notes: BWRO = Brackish water reverse Osmosis; SWRO = Sea water reverse Osmosis; ED = Electrodialysis; EDR = Electrodialysis reversal; MVC = Mechanical vapor compression; FO = Forward Osmosis; MD = Membrane distillation; MSF = Multi-stage flash; MED = Multiple effect distillation; MEB = Multi-effect boiling.

Table 16.3 Cost of desalinated water in thermal processes [37].

Desalination process	The capacity of desalination plant (m^3/day)	Desalination cost per m^3 (US $)
Multi-effect distillation, MED	<100	2.5-10
	12000-55000	0.95-1.95
	>91000	0.52-1.01
Multi-stage flash, MSF	23000-528000	0.52-1.75
Vapor compression	1000-1200	2.01-2.66

Table 16.4 Cost of desalinated water in the membrane (RO) Plants [37].

Type of feed water	The capacity of desalination plant (m³/day)	Desalination cost per m³ (US $)
Brackish water	<20	5.63-12.9
	20-1200	0.78-1.33
	40000-46000	0.26-0.54
Seawater	<100	1.5-18.75
	250-1000	1.25-3.93
	15000-60000	0.48-1.62
	100000-320000	0.45-0.66

Table 16.5 Cost with relative to RO [38].

Cost component	MSF	MED	MVC	RO
Capital Investment	120	114	118	100
Energy Related costs	215	175	140	100
Membrane replacement	-	-	-	100
Other remaining costs	103	89	100	100
Overall product costs	14	109	107	100

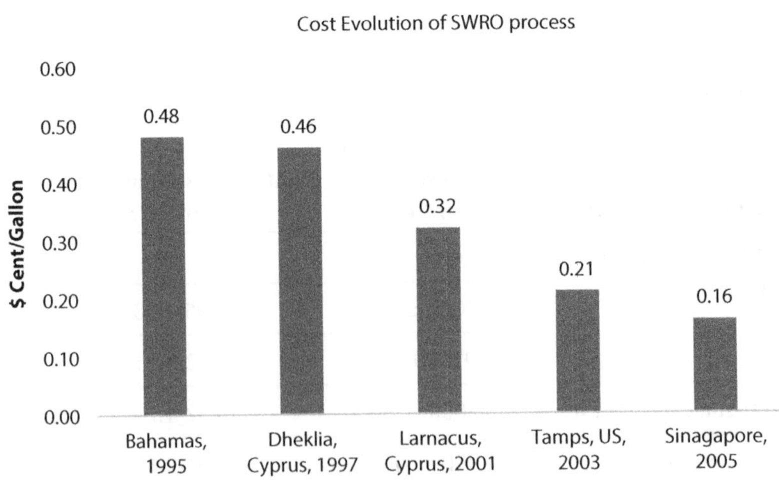

Figure 16.11 Cost evolution of SWRO process [38].

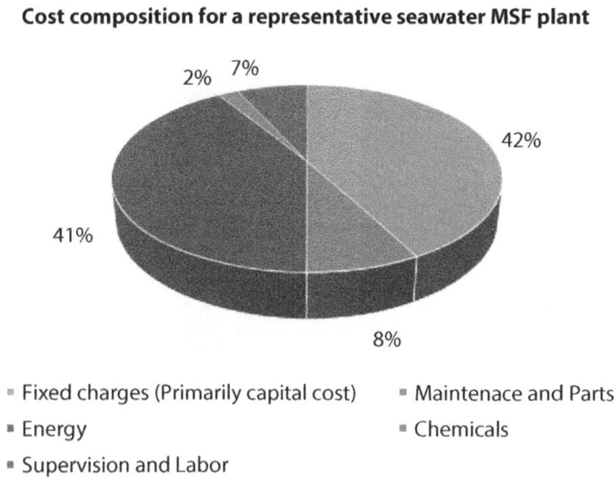

Figure 16.12 Cost composition for a representative seawater MSF plant. (Ebensperger and Phyllis Isley [75].

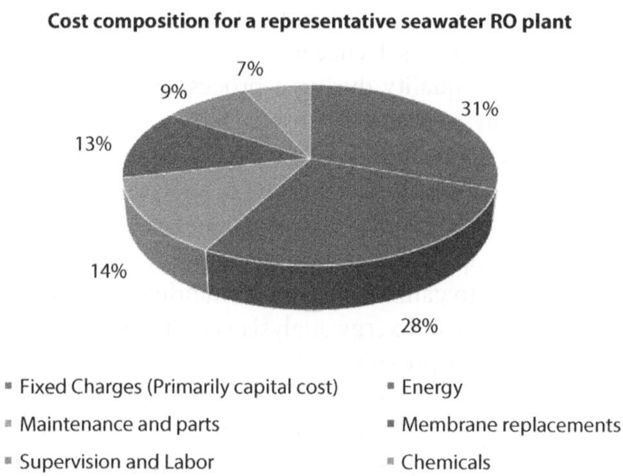

Figure 16.13 Cost composition for a representative seawater RO plant. (Ebensperger and Phyllis Isley [75].

16.5 Exergy Analysis of Reverse Osmosis

The first law of thermodynamics mainly explains the traditional process analysis methods like use of energy balance on the system which generally determines the heat transfer between system and environment whereas the second law is only brought into the formulation of criteria of performance

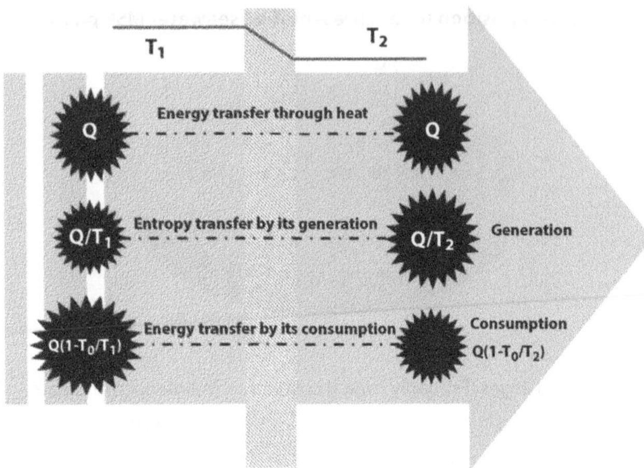

Figure 16.14 The relation between various forms of energy transfers (heat energy, entropy, and exergy transfer) [44].

as isentropic efficiencies, even though it governs the limits of convertibility between different forms of energy and determines the relative grades or quality of these energy forms. Hence it can be taken as a failure to consider the changes in energy quality during a process which makes traditional thermodynamic analysis methods unsatisfactory [43]. Relation between various forms of energy transfer is illustrated in Figure 16.14.

According to the conventional methods of analysis, the first law of thermodynamics has been considered to evaluate the energy conversion or utilization processes up to the certain period. But in recent periods, the use of exergy analysis to gain a better understanding of such processes has become a point of interest. Exergy analysis is derived from the second law of thermodynamics and provides better perceptions into the identifying and quantifying sources of inefficiencies, selecting optimal process parameters, and in assessing resource utilization efficiency with its environmental impacts [45]. Exergy of a system at a certain thermodynamic state can be expressed as the maximum amount of work that can be obtained when the system changes from that particular state to an equilibrium state with the surroundings [44]. According to the 1st thermodynamic law, energy neither can be generated nor destroyed, it just can be transferred from one form to another. But with increasing disorders of a system or irreversibility of a process, exergy is always destroyed and therefore only, exergy can be explained in terms of entropy and it obeys the 2nd law of thermodynamics.

In simulation thermodynamic analysis of energy systems, an exergy analysis (Second law analysis or irreversibility analysis) has been considered as a powerful tool. In other words, it has been widely used in the design, simulation and performance evaluation of energy systems. Exergy analysis method is employed to detect and to evaluate quantitatively the causes of the thermodynamic imperfection of the process under consideration [46, 47].

16.5.1 General Exergy Analysis in Desalination and Its Necessity

Exergy of a system is always zero at an equilibrium state and hence, it is measured as the maximum work output for a system to reach equilibrium with the surroundings. The exergy destruction of a cycle is the sum of the exergy destruction of the processes that compose the cycle. The exergy destruction of a cycle can also be determined without tracing the individuals by considering the entire cycle as a single process. From the second law thermodynamic, we can say that the decrease in exergy (Ex) is equivalent to an increase in entropy (S) w.r.t. a time when the reference environment is considered to be a large system:

$$0 \geq \frac{\partial Ex}{\partial t} \cong \frac{\partial S}{\partial t} \geq 0 \tag{16.113}$$

According to the first law of thermodynamics, the total energy is always constant while exergy output will never be the balanced by its input because some part of the input is destroyed during the process according to the second law of thermodynamics. Now, in any engineering applications, the selectivity of any process is carried out based on the efficiency of that process. In this case, the first law will determine the most efficient process w.r.t. the loss of minimum energy relative to its inputs and consequently, the second law exergy analysis will find out the most energy efficient way w.r.t. the destruction of minimum available work from given input.

Among various technologies (MED, MVC, MSF etc) employed for the desalination of seawater till now, reverse osmosis is one of the most prominent and popular technologies that uses a membrane to separate salt (Salinity removal) from seawater [48, 49]. Because as much we discussed the various premium thermal desalination technologies (MED, MSF and MVC), reverse osmosis is quite different from them due to its membrane driven system but compared to them, reverse osmosis can be taken as better technology for its low cost driven and maintenance system, higher purification capacity and highly energy efficient. The running of these desalination plants incurs a cost due to the energy they consume. There are many ways to reduce this, such as i) integrating the plant with other systems [50–56], ii) development of superior membranes [57–59], iii) usage of higher efficiency pumps, and iv) employing new or improved energy recovery technologies [60–62]. Performance of seawater reverse osmosis (SWRO) plants can be assessed using the first and second laws of thermodynamics. A complete steady-state analysis of energy conversion or utilization processes can be made based on mass, energy (first law), entropy and exergy (second law) balances for a control volume to find out the work and heat interactions, the rate of exergy decrease, the rate of irreversibility, the energy and exergy efficiencies [63–65].

Representation of general mass balance equation:

$$\sum \dot{m}_{in} = \sum \dot{m}_{out} \tag{16.114}$$

Representation of the general energy balance equation i.e. total energy inputs equal to total energy outputs:

$$\sum \dot{E}_{in} = \sum \dot{E}_{out} \tag{16.115}$$

In the absence of electricity, magnetism, surface tension, and nuclear reaction, the total exergy of a system $\dot{E}_x$ can be divided into four components, namely (i) physical exergy, $\dot{E}x^{PH}$ (ii) kinetic exergy, $\dot{E}x^{KN}$, (iii) potential exergy, $\dot{E}x^{PT}$, and (iv) chemical exergy, $\dot{E}x^{CH}$ [66].

$$\dot{E}x = \dot{E}x^{PH} + \dot{E}x^{KN} + \dot{E}x^{PT} + \dot{E}x^{CH} \tag{16.116}$$

Although exergy is extensive property, it is often convenient to work with it on a unit of mass or molar basis. The total specific exergy on a mass basis may be written as follows:

$$ex = ex^{PH} + ex^{KN} + ex^{PT} + ex^{CH} \tag{16.117}$$

Representation of general exergy balance equation:

$$\sum \dot{E}x_{in} - \sum \dot{E}x_{out} = \sum \dot{E}x_{dest} \tag{16.118}$$

$$\Rightarrow \dot{E}x_{heat} - \dot{E}x_{work} + \dot{E}x_{mass,in} - \dot{E}x_{mass,out} = \dot{E}x_{dest} \tag{16.119}$$

$$\text{where,}\quad \dot{E}x_{heat} = \sum\left(1 - \frac{T^*}{T_k}\right)\dot{Q}_k \tag{16.120}$$

$$\dot{E}x_{work} = P \tag{16.121}$$

$$\dot{E}x_{mass,in} = \sum \dot{m}_{in}\,\xi_{in} \tag{16.122}$$

$$\dot{E}x_{mass,out} = \sum \dot{m}_{out}\,\xi_{out} \tag{16.123}$$

where $\dot{Q}_k$ is the heat transfer rate through the boundary at temperature T_k at location k. Representation of flow exergy calculation:

$$\xi = (H - H^*) - T^*(S - S^*) \tag{16.124}$$

where H^*, S^* are the properties at the restricted dead state of P^* and T^*.

The rating form of the entropy balance can be represented as

$$\dot{S}_{in} - \dot{S}_{out} + \dot{S}_{gen} = 0 \tag{16.125}$$

where the rates of entropy transfer by heat transferred at a rate of $\dot{Q}_k$ and mass flowing at a the rate of $\dot{m}$ are $\dot{S}_{heat} = \dot{Q}_k/T_k$ and $\dot{S}_{mass} = \dot{m}S$, respectively.

Taking the positive direction of heat transfer to be to the system, the rate form of the general entropy relation given in Eq. (16.124) can be rearranged to give

$$\dot{S}_{gen} = \sum \dot{m}_{out}S_{out} - \sum \dot{m}_{in}S_{in} + \sum \frac{\dot{Q}_k}{T_k} \tag{16.126}$$

Anyhow, it is usually more convenient to find $\dot{S}_{gen}$ first and then to evaluate the exergy destroyed or the irreversibility rate $\dot{I}$ directly from the following equation, which is called Gouy-Stodola relation [67]:

$$\dot{I} = \dot{E}x_{dest} = T^*\dot{S}_{gen} \tag{16.127}$$

The specific exergy (flow exergy) of an incompressible substance (i.e., water) is given by [67]

$$\xi_{wt} = c_p\left(T - T^* - T^* \ln\frac{T}{T^*}\right) \tag{16.128}$$

Assuming air to be a perfect gas, the specific physical exergy of air is calculated by the following relation [68]

$$\xi_{wt,per} = c_{p,a}\left(T - T^* - T^* \ln\frac{T}{T^*}\right) + R_a T^* \ln\frac{p}{p^*} \tag{16.129}$$

Ideal and actual desalination system performances vary significantly. The actual energy cost of desalination is about 5-10 times the cost under ideal operation. This corresponds to a second law efficiency of under 20% and points out that there are tremendous opportunities in RO plant for improvement. The first step in any improvement or enhancement project is diagnostics, and the most powerful diagnostics tool in thermodynamics is second law analysis. Exergy destruction in a desalination process depends on the process and operational parameters related to feeding water, heat source, process type, and withdrawal and discharge conditions and system configuration such as hybrid or combine power and water production [44].

16.5.1.1 *Exergy Efficiency and Its Improvement Potential Analysis*

Thermodynamic analysis of energy resources may be performed using the following parameters [69]:

Relative irreversibility:

$$\chi_j = \frac{\dot{I}_j}{\dot{I}_T} \tag{16.130}$$

Fuel depletion rate:

$$\delta_j = \frac{\dot{I}_j}{\dot{F}_T} \tag{16.131}$$

Exergetic factor:

$$\varepsilon_j = \frac{\dot{F}_j}{\dot{F}_T} \tag{16.132}$$

The following measures can now be defined to assess the thermodynamic performance of the components of a system and the entire system [70]:

Exergy destruction ratio for a component (let, q-component) of the system:

$$y_{D,q} = \frac{\text{Exergy destruction in the component}, \dot{E}_{D,q}}{\text{Exergy destruction in the system}, \dot{E}_D} \tag{16.133}$$

Exergy destruction ratio for a complete system:

$$y_D = \frac{\text{Exergy destruction in system}, \dot{E}_D}{\text{Exergy of fuel supplied}, \dot{E}_F} \tag{16.134}$$

Exergy loss ratio for the complete system:

$$y_L = \frac{\text{Exergy loss in the system}, \dot{E}_L}{\text{Exergy of fuel supplied}, \dot{E}_F} \tag{16.135}$$

Exergetic efficiency (η_{exergy}):

$$\eta_{exergy} = 1 - \left(\frac{\dot{E}_D + \dot{E}_L}{\dot{E}_F} \right) \tag{16.136}$$

It has been also proposed by Van Gool [71] that maximum improvement in the exergy efficiency for a process or system is obviously achieved when the exergy loss or irreversibility $\left(\dot{E}x_{in} - \dot{E}x_{out}\right)$ is minimized.

Representation of improvement potential rate:

$$I\dot{P} = \left(1 - \eta_{exergy}\right)\left(\dot{E}x_{in} - \dot{E}x_{out}\right) \tag{16.137}$$

The relation between SEC (Specific Exergy Consumption) and Exergetic efficiency [72]:

$$\eta_{exergy_{II,R0}} = \frac{\rho_{perm}\xi_{perm}}{36(SEC)} \tag{16.138}$$

Where, $\eta_{II,RO}$ = Exergetic efficiency according to second law (II) and R0 indicates, based on infinitesimal recovery ratio.

Specific entropy production (SEP) is defined as the amount of entropy production per unit mass of product water in a membrane desalination plant which is equal to the amount of lost work in the desalination plant per unit mass of water product [73]:

$$SEP = \frac{R_S T_o}{Q_{\bar{P}}} \tag{16.139}$$

Also, specific energy consumption (SEC) is defined and calculated as [73]:

$$SEC = \frac{\Delta p}{\eta_{pump}\phi} \tag{16.140}$$

Where, $\phi = Q_{\bar{P}}/Q_{\bar{F}}$ (recovery).

The ratio of the product to inlet feed water mass flow rates is taken as the recovery ratio, RR:

$$RR = \frac{\dot{m}_{\bar{P}}}{\dot{m}_{\bar{F}}} \tag{16.141}$$

16.5.2 A Case Study on Reverse Osmosis Based Desalination Unit Reporting Exergy Performance

A study reported the exergy analysis of a reverse osmosis desalination plant in Jordan which received a brackish water as feed water with a salinity of 2500 ppm [74]. There were two reverse osmosis units in this plant which could produce 90 ppm and 70 ppm permeates respectively.

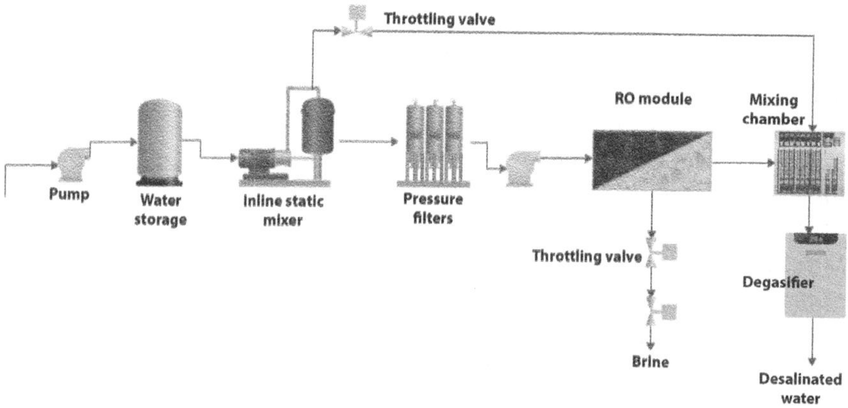

Figure 16.15 Schematic flow of RO-module based desalination plant for saline water purification [44].

As shown in the flow scheme in Figure 16.15, the unit had pressure filters and cartridge filters followed by RO modules, decarbonator, and a hydrostatic water tank. From the process, it was observed that the major dissipation of exergy occurred in the throttling valves followed by membrane units which accounted up to 56.8% with 21% of total exergy destruction. Whereas, the pumps and the motors caused only 19.6% of exergy consumption. The second law efficiency of this plant was 4.1%. Therefore, from this analysis, it was shown that the filters had higher exergy efficiencies and the analysis recommended to replace the throttling valves with energy recovery devices and pumps equipped with variable frequency motors which might reduce the exergy consumption in the whole unit.

16.6 Conclusion

Here are a few conclusions which can be drawn from the analysis done above like energy analysis and economic analyses. (1) Energy consumption for the desalination process using RO technology is almost close to common household utilities. (2) RO process utilizes very minimal energy close to thermodynamic benchmarks; going below this limit is close to impossible. (3) Integration of energy recovery devices with the RO desalination has led to the achievement of lowest energy consumption. (4) Energy consumption of distillation technique for desalination is quite high. There should be the implementation of new technology in desalination process which could enhance the efficiency of the plant and lead to the minimization of energy consumption.

Nomenclature

Abbreviations

PR = Productivity ratio
SA = Specific area
RR = Recovery ratio
MED = Multi-effect distillation
MVC = Mechanical vapor compression
MSF = Multi-stage flash
SEC = Specific energy consumption
SEP = Specific entropy production
BPE = Boiling point elevation
NEA = Non-equilibrium allowance
TBT = Top brine temperature
LMTD = Log mean temperature difference

2. Roman symbols

$\dot{m}$ = Mass flow rate, kg/h
$\dot{W}$ = work rate (power), J/s
$\dot{n}$ = Mole flow rate, mol/hr
$\dot{E}$ = Energy rate, kW
$\dot{E}x$ = Energy rate, kW
ex = Specific exergy, kJ/kg
T = Temperature, K
$\dot{Q}$ = Total heat transfer rate, kW
P = Work rate/Power consumption
H = Specific enthalpy, kJ/kg
S = Specific entropy, kJ/kg-K
$\dot{S}$ = Entropy rate, kW/K
$\dot{I}$ = Irreversibility rate, kW
c_p = Specific heat capacity, kJ/kg.K
$\bar{R}$ = Universal gas constant J/mol.K
p = Pressure, kPa
δ_i = Fuel depletion ratio
$\dot{F}$ = Exergy rate of fuel, kW
$y_{D,q}$ = Exergy destruction ratio for component – q
$\dot{E}_{D,q}$ = Exergy destruction in component – q
$\dot{E}_D$ = Exergy destruction in the system
y_D = Exergy destruction ration for complete system
$\dot{E}_F$ = Exergy of fuel supplied

y = Salinity
y_D = Exergy loss ratio in system
$\dot{E}_L$ = Exergy loss in system
$\dot{I}_P$ = Improvement potential rate, kW
R_sT_0 = Entropy production rate
Δp = Pressure difference
Q = Flow rate, m^3/sec
W_1 = specific energy for desalination in the reversible RO processes
W_2 = specific energy for desalination in ideal RO processes
r = mass recovery ratio
$\bar{r}$ = molar recovery ratio
C_o = concentration of feed water (mg/lit)
ΔP_{net} = net driving pressure for the required permeate flux
A = Area, m^2
L = Latent heat, kJ/kg
t = Thermal load of heat exchanger, kJ/kg-K
u = Velocity, m/sec
d = Diameter
h = Overall heat transfer coefficient, W/m^2-K
c_p = Specific heat capacity, kJ/kg-K
$\dot{M}$ = Molecular weight
m = Mass, kg
m = molality kg/mol
$\dot{m}_{s_1}$ = Seawater mass flow rate to heat exchanger – 1, kg/hr
$\dot{m}_{s_2}$ = Seawater mass flow rate to heat exchanger – 2, kg/hr
H_{s_1} = Enthalpy of seawater at heat exchanger – 1, kJ/kg
H_{s_2} = Enthalpy of seawater at heat exchanger – 2, kJ/kg
Z_R = Compression ratio
Z_f, x_f, R_f = Constants used in Heat transfer calculation
x = mole fraction

3. Subscript
in = Input
out = Output
dest = Destroyed
heat = Heat energy
work = work energy
mass,in = Incoming mass
mass,out = Outgoing mass
pump = for pump information
k = Location

a = Ambient
wt = Water
sep = Separation
per = Permeate
gen = Generation
min = minimum
exergy = exergetic
$\overline{P}$ = *Water/permeate production*
p = product
f = feed
c = concentrate
$\overline{F}$ = *Feed entering*
j = Successive nos/stages
T = Thermal system
B = Brine
D = Distillate
Htr = Heater
R = Recycle stream
W = Rejected seawater
C = Condensation
e = Effect/column
s = Steam
st = solute
fg,s = Saturated vapor/steam
fh = Feed heater
boil = Boiling
comp = Compressor
evap = Evaporator
SW = Seawater/feed
B_D = Blow down
B_N = Concentrated brine from the last stage
B_0 = Flashing brine leaving brine heater
B_e = Brine in the effect
B_j = Brine at j^{the} stage
B_{j-1} = Brine at $(j-1)^{th}$ stage
B_{disc} = Discharge brine
D_b = Distillate from boiling
D_c = Condensed distillate
D_f = Distillate from flashing
D_{fb} = Distillate from the flash box
D_{sat} = Distillate at saturation temperature

$D_{c,sat}$ = Condensing distillate at saturation condition
D_{j-1} = Distillate at (j – 1)th stage
D_{disc} = Discharged distillate
HX_1 = Heat exchanger – 1
HX_2 = Heat exchanger – 2
$\dot{m}_{SW}$ = Seawater mass flow rate
m/SW_j = Make up seawater at stage j
$m/SW_{(j+1)}$ = Make up water at stage (j+1)
$m/SW_{(NR+1)}$ = Make up water at stage (NR+1) of rejection section
SW_1 = Input seawater
T_D = Distillate temperature
T_{boil} = Boiling temperature
T_{evap} = Evaporator temperature
T_{SW} = Seawater temperature
TTD_e = Terminal temperature difference in the effect
TTD_{fh} = Terminal temperature difference in the feed heater
v_j = Vaport at jth stage
vap_{sat} = Staurated vapor

4. Superscript
PH = Physical energy
KN = Kinetic energy
CH = Chemical energy
PT = Potential energy
* = Dead state/reference value
co = Cold side
ho = Hot side
in = Input
Out = Output
i = Inside
o = Outside
Prev = Previous stage

5. Greek symbols
γ = Polynomial gas coefficient
λ = Salt concentration, kg/kg
θ = Temperature loss due to demister, K
Π = Fouling factor
η = Efficiency
ϕ = Recovery
ξ = Specific exergy, kJ/kg

χ = Relative irreversibility
ε = Exergetic factor
Δ = difference operator
γ_f = fugacity coefficient
γ_m = molal activity coefficient
γ_x = rational activity coefficient
μ = chemical potential, J/mol
φ = osmotic coefficient
υ = sum of stoichiometric coefficients
π = osmotic pressure, Pa
ρ = density, kg/m^3

References

1. M. Elimelech and W. A. Philip, "The Future of Seawater Desalination: Energy, Technology and the Environment," *Science*, vol. 333, pp. 712-717, 2011.
2. R. F. Service, "Desalination Freshens Up," *Science*, vol. 313, pp. 1088-1090, 2006.
3. D. Halder, S. Mondal, A. Mukherjee and A. Ghosh, "An Overview on the Treatment of Wastewater using Renewable Energy," *International Journal of Research in Engineering and Technology*, vol. 1, pp. 78-83, 2016.
4. R. Semiat, "Energy Isuues in Desalination Processes," *Environmental Science & Technology*, vol. 22, pp. 8193-8201, 2008.
5. S. Chakraborty, A. Mukherjee, S. Das, N.R. Maddela, S. Iram, P. Das, Study on isotherm, kinetics, and thermodynamics of adsorption of crystal violet dye by calcium oxide modified fly ash, Environ. Eng. Res. 26 (2021) 1–9. https://doi.org/10.4491/eer.2019.372
6. S. T. Hsu, K. T. Cheng and J. S. Chiou, "Seawater desalination by direct contact membrane distillation," *Desalination*, vol. 143, pp. 279-287, 2002.
7. R. L. McGinnis and M. Elimelech, "Energy requirements of ammonia-carbon dioxide forward osmosis desalination," *Desalination*, vol. 207, pp. 10-29, 2007.
8. A. Mukherjee, S. Lanjewar, R. Kumar, A. Chakraborty, A. Abdelrasoul, A. Roy, Role of Thermodynamics and Membrane Separations in Water-Energy Nexus, Model. Membr. Membr. Process. (2020) 145–199. https://doi.org/10.1002/9781119536260.CH4
9. S. Raphael, "Energy Issues in Desalination Processes," *Environmental Science & Technology*, vol. 42, pp. 8193-8201, 2008.
10. A. Mukherjee, A. Satish, A. Mullick, J. Rapolu, S. Moulik, A. Roy, A.K. Ghosh, Paradigm Shift toward Developing a Zero Liquid Discharge Strategy for Dye-Contaminated Water Streams: A Green and Sustainable Approach Using Hydrodynamic Cavitation and Vacuum Membrane Distillation,

ACS Sustain. Chem. Eng. 9 (2021) 6707–6719. https://doi.org/10.1021/acssuschemeng.1c00619.
11. M. H. Sharqawy, J. H. Lienhard and S. M. Zubair, “Thermophysical properties of sea water:A Review of Existing Correlations and Data,” *Desalination and water treatment,* pp. 354-380, 2010.
12. R. A. Robinson and R. H. Stokes, Electrolyte solutions, vol. 2nd Revised, Mineola,New York: Dover Publications, 2002.
13. K. H. Mistry and J. H. Lienhard, “Effect of Non Ideal Solution Behaviour on Desalination of a Sodium Chloride Solution and Comparison to seawater,” *Journal of Energy Resources Technology,* 2013.
14. R. Das, A. Mukherjee, I. Sinha, K. Roy, B.K. Dutta, Synthesis of potential bio-adsorbent from Indian Neem leaves (Azadirachta indica) and its optimization for malachite green dye removal from industrial wastes using response surface methodology: kinetics, isotherms and thermodynamic studies, Appl. Water Sci. 10 (2020) 1–18. https://doi.org/10.1007/s13201-020-01184-5.
15. S. Lanjewar, A. Mukherjee, P. Khandewal, A.K. Ghosh, A. Mullick, S. Moulik, A. Roy, Thermodynamics of synthesis and separation performance of Interfacially polymerized “loose” reverse osmosis membrane: Benchmarking for greywater treatment, Chem. Eng. J. 417 (2021) 127929. https://doi.org/10.1016/j.cej.2020.127929
16. K. S. Pitzer, J. C. Peiper and R. H. Bussey, "Thermodynamic Properties of Aqueous Sodium Chloride Solutions," Journal of Physical Chemistry, vol. 13, no. 1, pp. 1-102, 1984.
17. K. H. Mistry, R. K. McGovern, G. P. Thiel, E. K. Summers, S. M. Zubair and J. H. Lienhard, “Entropy Generation Analysis of Desalination Technologies,” *Entropy,* vol. 13, no. 10, pp. 1829-1864, 2011.
18. G. Filippini, M. A. Al-Obaidi, F. Manenti and I. M. Mujtaba, “Performance analysis of hybrid system of multi effect distillation and reverse osmosis for seawater desalination via modelling and simulation,” *Desalination,* vol. 448, pp. 21-35, 2018.
19. K. Wangnick, “IDA, Int. Desalination Association, Worldwide Desalting Plants Inventory,” Desalination Association, Gnarrenburg, 1995.
20. S. Lanjewar, A. Mukherjee, L. Muzamil Rehman, A. Roy, Blue Energy and Its Potential: The Membrane Based Energy Harvesting, in: A. Abdelrasoul (Ed.), Adv. Membr. Technol., IntechOpen, 2020. https://doi.org/10.5772/intechopen.86953.
21. K. H. Mistry, M. A. Antar and J. H. V. Lienhard, “An improved model for multiple effect distillation,” *Desalination and Water Treatment,* vol. 51, pp. 807-821, 2013.
22. F. Alasfour, M. Darwish and A. B. Amer, “Thermal analysis of ME-TVC+MEE desalination systems,” *Desalination,* vol. 174, no. 1, pp. 39-61, 2005.
23. A. Ophir and F. Lokeic, “Advanced MED process for most economical sea water desalination,” *Desalination,* vol. 182, pp. 187-198, 2005.

24. R. S. Silver, "Multi stage flash distillation- The first 10 years," in *3rd Int. Sym. on Fresh Water from the Sea*, Athens, Greece, 1970.
25. H. F. Al-Fulaij, "Dynamic Modelling of Multi Stage Flash Desalination Plant," 2011.
26. F. N. Alasfour and H. K. Abdulrahim, "Rigoroussteady state modelling of MSF-BR desalination system," *Desalination and Water Treatment*, vol. 1, pp. 259-276, 2009.
27. S. A. Said, M. Emtir and I. M. Mujtaba, "Flexible Design and Operation of Multi Stage Flash (MSF) Desalination Process Subject to Variable Fouling and Variable Freshwater Demand," *Processes*, vol. 1, pp. 279-295, 2013.
28. A. Nannarone, C. Toro and E. Scriubba, "Multi-Effect Distillation Process: Modelling and Simulation," in *Proceeedings: Efficicency, cost, optimization, simulation and environmental impact of energy systems*, San Diego, California, 2017.
29. H. El-Dessouky, I. Alatiqi, S. Bingulac and H. Ettouney, "Steady state analysis of the multiple effect evaporation desalination process," *Chemical Engineering & Technology*, vol. 21, no. 5, pp. 437-451, 1998.
30. H. T. El-Dessouky and H. M. Ettouney, Fundamentals of salt water desalination, Amsterdam: Elseveir, 2002.
31. M. Darwish, F. Al-Juwayhel and H. K. Abdulraheim, "Multi-effect boiliing systems from an enegy viewpoint," *Desalination*, vol. 194, no. 1-3, pp. 22-39, 2006.
32. Y. M. El-sayed and R. S. Silver, Fundamentals ofdistillation, 2nd ed., vol. Principles of Desalination, A. D. K. Laird, Ed., New York: Academic Press, 1980, pp. 55-109.
33. "Improved Process to Convert Saline Water: Multi-effect, multi stage, flash distillation technique uses brine recirculation loops to increase operating efficiency," C&EN (Chemical & Engineering News, 1964.
34. S. A. Said, M. Emtir and I. M. Mujtaba, "Flexible Design and Operation of Multi Stage Flash (MSF) Desalination Process to Variable Fouling and Variable Freshwater Demand," *Processes*, vol. 1, pp. 279-295, 2013.
35. M. Marcovecchio, P. Aguirre, N. Scenna and S. Mussati, "Global Optimal Design of Mechanical Vapor Compression (MVC) Desalination Process," in *20th European Symposium on Computer Aided Process Engineering*, 2010.
36. S. Mussatia, N. Scennaa, N. Tarifac, S. Francoc and J. A. Hernandez, "Optimization of the mechanical vapor compression (MVC) desalination process using mathematical programming," *Desalination and Water Treatment*, vol. 5, pp. 124-131, 2009.
37. M. Shatat and S. B. Riffat, "Water desalination technologies utilizing conventional and renewable energy sources," *International Journal of Low-carbon Technologies*, pp. 1-19, 2014.
38. F. Banat, "Economic and technical assessmet of desalination technologies," Geneva, 2007.

39. A.-K. A. and K. L.L., "Energy consumption and water production cost of conventional and renewable-energy-powered desalination processes," *Renewable and sustainable energy reviews,* pp. 343-356, 2013.
40. S. R., "Energy Issues in Desalination Processes," *Environmental Science & Technologies,* pp. 8193-8201, 2008.
41. K. R. S. Cui Liu, "Calculation of energy consumption for cross flow RO desalination Processes," *Desalination and Water Treatment,* pp. 295-303, 2012.
42. M. E. Ashlynn S.Stillwell, "Predicting the Specific Energy Consumption of Reverse Osmosis Desalination," *Water,* pp. 1-18, 2016.
43. T. J. Kotas, The Exergy Method of Thermal Plant Analysis., Essex: Anchor Brendon Ltd, Tiptree, 1985.
44. V. G. Gude, "Use of Exergy Tools in Renewable Energy Driven Desalination," *Thermal Science and Engineering Progress,* 2018.
45. V. G. Gude, N. Nirmalakhandan, S. Deng and A. Maganti, "Desalination at low temperatures: an exergy analysis," *Desalination and Water Treatment,* vol. 40, pp. 272-281, 2012.
46. J. Szargut, D. R. Morris and F. R. Stewart, Exergy analysis of thermal, chemical, and metallurgical processes., USA: Edwards Brothers Inc. work, 1998.
47. M. A. Rosen and I. Dincer , "Exergy methods for assessing and comparing thermal storage systems.," *International Journal of Energy Resources,* vol. 27, no. 4, pp. 415-430, 2003.
48. "Water: Our Thirsty World," *National Geographic Magazine,* 2010.
49. R. W. Baker, Membrane Technology and Applications, Second Ed ed., Chichester: John Wiley & Sons Ltd, 2004.
50. A. S. Nafey, M. A. Sharaf and L. Garcia-Rodriguez, "Thermo-economic analysis of a combined solar organic Rankine-cycle reverse-osmosis desalination process with different energy recovery configurations," *Desalination,* vol. 261, pp. 138-147, 2010.
51. G. P. Narayan, R. K. McGovern, S. M. Zubair and J. H. Lienhard, "High-temperature steam driven, varied pressure, humidification-dehumidification system coupled with reverse osmosis for energy-efficient seawater desalination," *Energy,* vol. 37, pp. 482-493, 2012.
52. H. Cherif and J. Belhadj, "Large-scale time evaluation for energy estimation of standalone hybrid photovoltaic-wind system feeding: a reverse osmosis desalination unit," *Energy,* vol. 36, pp. 6058-6067, 2011.
53. E. S. Hrayshat, "Brackish water desalination by a stand-alone reverse osmosis desalination unit powered by photovoltaic solar energy," *Renewable Energy,* vol. 33, pp. 1784-1790, 2008.
54. A. M. Delgado-Torres and L. Garcia-Rodriguez, "Design recommendations for solar organic Rankine cycle powered reverse osmosis desalination," *Renewable and Sustainable Energy Reviews,* vol. 16, pp. 44-53, 2012.

55. M. Ibarra, A. Rovira, D. C. Alarcon-Padilla, G. Zaragoza and J. Blanco, "Performance of a 5kW solar-only organic Rankine unit coupled to a reverse osmosis plant," *Energy Procedia,* vol. 49, pp. 2251-2260, 2014.
56. G. Xia, Q. Sun, X. Cao, j. Wang, Y. Yu and L. Wang, "Thermodynaic analysis and optimization of a solar-powered transcritical-CO2 power cycle for reverse osmosis desalination based on the recovery of cryogenic energy of LNG (Liquefied natural gas).," *Energy,* vol. 66, pp. 643-653, 2014.
57. L. F. Liu, Z. B. Cai, J. N. Shen, L. X. Wu, E. M. V. Hoek and C. J. Gao, "Fabrication and characterization of a novel poly(amide-urethane-amide) TFC reverse osmosis membranc with chlorine-tolerant property," *Journal of Membrane Science,* vol. 469, pp. 397-409, 2014.
58. M. Safarpour, A. Khatee and V. Vatanpour, "Thin film nanocomposite reverse osmosis membrane modified by reduced graphene oxide/TiO2 with improved desalination performance," *Journal of Membrane Science,* vol. 489, pp. 43-54, 2015.
59. J. Duan, Y. Pan, F. Pacheco, E. Litwiller, Z. Lai and I. Pinnau, "High-performance polyamide thin-film-nanocomposite reverse osmosis membranes containing hydrophobic zeolitic imidazolate framework-8," *Journal of Membrane Science,* vol. 476, pp. 303-310, 2015.
60. M. H. Sharqawy, S. M. Zubair and J. H. Lienhard, "Second analysis of reverse osmosis desalination plants: an alternative design using pressure retarded osmosis," *Energy,* vol. 36, pp. 6617-6626, 2011.
61. C. Knutson, "Discussion of "Second law analysis of reverse osmosis desalination plants: An alternative design using pressure retarded osmosis"," *Energy,* vol. 46, pp. 688-690, 2012.
62. M. H. Sharqawy, S. M. Zubair and J. H. Lienhard, "Rebuttal to "Discussion of 'Second law analysis of reverse osmosis desalination plants: An alternative design using pressure retarded osmosis,"" *Energy,* vol. 46, pp. 691-693, 2012.
63. F. Balkan, N. Colak and A. Hepbasli, "Performance evaluation of a triple effect evaporator with forward feed using exergy analysis," *International Journal of Eergy Resources,* vol. 29, pp. 455-470, 2005.
64. R. L. Cornelissen, "Thermodynamics and sustainable development: the use of exergy analysis and the reduction of irreversibility.," The Netherlands, 1997.
65. G. Wall, "Exergy tools," *Proc Inst Mech Eng,* pp. 125-136, 2003.
66. I. Dincer, M. M. Hussain and I. Al-Zaharnah, "Energy and exergy use in public and private sector of Saudi Arabia," *Energy Policy,* vol. 32, no. 141, pp. 1651-1624, 2994.
67. J. Szargut, Exergy method: technical and ecological applications, Southampton: WIT Press, 2005.
68. T. J. Kotas, The exergy method of thermal power plants, Malabal: FL: Krieger Publishing Company, 1995.

69. A. Hepbasli, "A key review on exergetic analysis and assessment of renewable energy resources for a sustainable future," *Renewable and Sustainable Energy Reviews,* vol. 12, pp. 593-661, 2008.
70. A. Bejan, G. Tsatsaronis and M. J. Moran, Thermal design and optimization, John Wiley & Sons, 1996.
71. G. W. Van, Energy policy: fairly tales and factualities. Innovation and technology strategies and policies, Dordrecht:Kluwer, 1997, pp. 93-105.
72. B. A. Qureshi and Q. S. Zubair, "Exergetic efficiency of NF, RO, and EDR desalination plants," *Desalination,* vol. 378, pp. 92-99, 2016.
73. H. Mehdizadeh, "Membrane desalination plants from an energy-exergy viewpoint," *Desalination,* vol. 191, pp. 200-209, 2006.
74. I. H. Aljundi, "Second-law analysis of a reverse osmosis plant in Jordan," *Desalination,* vol. 239, no. 1, pp. 207-215, 2009.
75. Ebensperger, U., & Isley, P. (2005). Review of the current state of desalination. Water policy working paper, 8, 2005.

Index

Also of Interest

Check out these other related titles from Scrivener Publishing

From the Same Editors

MODELING IN MEMBRANES AND MEMBRANE-BASED PROCESSES, Edited by Anirban Roy, Siddhartha Moulik, Reddi Kamesh, and Aditi Mullick, ISBN: 9781119536062. This latest volume in the new "Advances in Membrane Processes" series from Wiley-Scrivener contains the most comprehensive and up-to-date coverage of modeling in membranes and membrane-based processes, a "one stop shop" on the subject. *NOW AVAILABLE!*

MEMBRANE PROCESSES: Pervaporation, Vapor Permeation and Membrane Distillation for Industrial Scale Separations and Water/Wastewater Treatment, Edited by Sundergopal Sridhar and Siddhartha Moulik, ISBN: 9781119418221. A reference for engineers, scientists, and academics who want to be abreast of the latest industrial separation/treatment technique, this new volume aims at providing a holistic vision on the potential of advanced membrane processes for solving challenging separation problems in industrial applications. *NOW AVAILABLE!*

Other books in the same subject area

SUSTAINABLE WATER PURIFICATION, By M. Safiur Rahman and M. R. Islam, ISBN: 9781119650997. This is the only book that takes a zero-waste approach to propose 100% sustainable water purification techniques. *NOW AVAILABLE!*

APPLIED WATER SCIENCE VOLUME 1: Fundamentals and Applications, Edited by Inamuddin, Mohd Imran Ahamed, Rajender Boddula and Tauseef Ahmad Rangreez, ISBN: 9781119724766. Edited by one of the most well-respected and prolific engineers in the world and his team, this is the first volume in a two-volume set that is the most thorough, up-to-date, and comprehensive volume on applied water science available today. *NOW AVAILABLE!*

APPLIED WATER SCIENCE VOLUME 2: Remediation Technologies, Edited by Inamuddin, Mohd Imran Ahamed, Rajender Boddula and Tauseef Ahmad Rangreez, ISBN: 9781119724735. The second volume in a new two-volume set on applied water science, this book provides understanding, occurrence, identification, toxic effects and control of water pollutants in aquatic environment using green chemistry protocols. *NOW AVAILABLE!*

Nano and Bio-Based Technologies for Wastewater Treatment: Prediction and Control Tools for the Dispersion of Pollutants in the Environment, edited by Elvis Fosso-Kankeu, ISBN 9781119577096. Presents recent challenges related to new forms of pollution from industries and discusses adequate state-of-the-art technologies capable to remediate such forms of pollution.

SUSTAINABLE SOLUTIONS FOR ENVIRONMENTAL POLLUTION VOLUME 1: Waste Management and Value-Added Products, Edited by Nour Shafik El-Gendy, ISBN: 9781119785354. This first volume in a broad, comprehensive two-volume set, Sustainable Solutions for Environmental Pollution, concentrates on the role of waste management in solving pollution problems and the value-added products that can be created out of waste, turning a negative into an environmental and economic positive. *NOW AVAILABLE!*

Handbook of Environmental Management Practices, by Nicholas P. Cheremisinoff and Anton Davletshin ISBN 9781118946350. The most up-to-date and thorough coverage of the environmental management practices used in hydraulic fracturing operations or "fracking." *NOW AVAILABLE!*

Printed and bound by CPI Group (UK) Ltd, Croydon, CR0 4YY

17/09/2026

14964245-0007

Printed and bound by CPI Group (UK) Ltd, Croydon, CR0 4YY
17/09/2026
14964261-0004